Fundamentals of Stability Theory

Since their inception, the Perspectives in Logic and Lecture Notes in Logic series have published seminal works by leading logicians. Many of the original books in the series have been unavailable for years, but they are now in print once again.

In this volume, the 12th publication in the Perspectives in Logic series, John Baldwin presents an introduction to first order stability theory, organized around the spectrum problem: calculate the number of models a first order theory T has in each uncountable cardinal. The author lays the groundwork and then moves on to three sections: independence, dependence and prime models, and local dimension theory. The final section returns to the spectrum problem, presenting complete proofs of the Vaught conjecture for ω-stable theories for the first time in book form. The book provides much-needed examples, and emphasizes the connections between abstract stability theory and module theory.

JOHN T. BALDWIN works in the Department of Mathematics, Statistics and Computer Science at the University of Illinois, Chicago.

PERSPECTIVES IN LOGIC

The *Perspectives in Logic* series publishes substantial, high-quality books whose central theme lies in any area or aspect of logic. Books that present new material not now available in book form are particularly welcome. The series ranges from introductory texts suitable for beginning graduate courses to specialized monographs at the frontiers of research. Each book offers an illuminating perspective for its intended audience.

The series has its origins in the old *Perspectives in Mathematical Logic* series edited by the Ω-Group for "Mathematische Logik" of the Heidelberger Akademie der Wissenschaften, whose beginnings date back to the 1960s. The Association for Symbolic Logic has assumed editorial responsibility for the series and changed its name to reflect its interest in books that span the full range of disciplines in which logic plays an important role.

For more information, see www.aslonline.org/books_perspectives.html

PERSPECTIVES IN LOGIC

Fundamentals of Stability Theory

JOHN T. BALDWIN
University of Illinois, Chicago

ASSOCIATION FOR SYMBOLIC LOGIC

CAMBRIDGE
UNIVERSITY PRESS

CAMBRIDGE
UNIVERSITY PRESS

University Printing House, Cambridge CB2 8BS, United Kingdom

One Liberty Plaza, 20th Floor, New York, NY 10006, USA

477 Williamstown Road, Port Melbourne, VIC 3207, Australia

4843/24, 2nd Floor, Ansari Road, Daryaganj, Delhi – 110002, India

79 Anson Road, #06–04/06, Singapore 079906

Cambridge University Press is part of the University of Cambridge.

It furthers the University's mission by disseminating knowledge in the pursuit of education, learning, and research at the highest international levels of excellence.

www.cambridge.org
Information on this title: www.cambridge.org/9781107168091
10.1017/9781316717035

Association for Symbolic Logic
Richard A. Shore, Publisher
Department of Mathematics, Cornell University, Ithaca, NY 14853
http://www.aslonline.org

A catalogue record for this publication is available from the British Library.

ISBN 978-1-107-16809-1 Hardback

To Sharon, Katie,
and my parents

Preface to the Series

Perspectives in Mathematical Logic

(Edited by the Ω-group for "Mathematische Logik" of the Heidelberger Akademie der Wissenschaften)

On Perspectives. *Mathematical logic arose from a concern with the nature and the limits of rational or mathematical thought, and from a desire to systematise the modes of its expression. The pioneering investigations were diverse and largely autonomous. As time passed, and more particularly in the last two decades, interconnections between different lines of research and links with other branches of mathematics proliferated. The subject is now both rich and varied. It is the aim of the series to provide, as it were, maps of guides to this complex terrain. We shall not aim at encyclopaedic coverage; nor do we wish to prescribe, like Euclid, a definitive version of the elements of the subject. We are not committed to any particular philosophical programme. Nevertheless we have tried by critical discussion to ensure that each book represents a coherent line of thought; and that, by developing certain* themes, *it will be of greater interest than a mere assemblage of results and techniques.*

The books in the series differ in level: some are introductory, some highly specialised. They also differ in scope: some offer a wide view of an area, others present a single line of thought. Each book is, at its own level, reasonably self-contained. Although no book depends on another as prerequisite, we have encouraged authors to fit their book in with other planned volumes, sometimes deliberately seeking coverage of the same material from different points of view. We have tried to attain a reasonable degree of uniformity of notation and arrangement. However, the books in the series are written by individual authors, not by the group. Plans for books are discussed and argued about at length. Later, encouragement is given and revisions suggested. But it is the authors who do the work; if, as we hope, the series proves of value, the credit will be theirs.

History of the Ω-Group. *During 1968 the idea of an integrated series of monographs on mathematical logic was first mooted. Various discussions led to a meeting at Oberwolfach in the spring of 1969. Here the founding members of the group (R.O. Gandy, A. Levy, G.H. Müller, G. Sacks, D.S. Scott) discussed the project in earnest and decided to go ahead with it. Professor F.K. Schmidt and Professor Hans Hermes gave us encouragement and support. Later Hans Hermes joined the group. To begin with all was fluid. How ambitious should we be? Should we write the books ourselves? How long would it take? Plans for authorless books were promoted, savaged and scrapped. Gradually there emerged a form and a method. At the end of an infinite discussion we found our name, and that of the series. We established our centre in Heidelberg. We agreed to meet twice a year together with authors, consultants and assistants, generally in Oberwolfach. We soon found the value of collaboration: on the one hand the permanence of the founding group gave coherence to the overall plans; on the other hand the stimulus of new contributors kept, the project alive and flexible. Above all, we found how intensive discussion could modify the authors' ideas and our own. Often the battle ended with a detailed plan for a better book which the author was keen to write and which would indeed contribute a perspective.*

Oberwolfach, September 1975

Acknowledgements. *In starting our enterprise we essentially were relying on the personal confidence and understanding of Professor Martin Barner of the Mathematisches Forschungsinstitut Oberwolfach, Dr. Klaus Peters of Springer-Verlag and Dipl.-Ing. Penschuck of the Stiftung Volkswagenwerk. Through the Stiftung Volkswagenwerk we received a generous grant (1970–1973) as an initial help which made our existence as a working group possible.*

Since 1974 the Heidelberger Akademie der Wissenschaften (Mathematisch-Naturwissenschaftliche Klasse) has incorporated our enterprise into its general scientific program. The initiative for this step was taken by the late Professor F.K. Schmidt, and the former President of the Academy, Professor W. Doerr.

Through all the years, the Academy has supported our research project, especially our meetings and the continuous work on the Logic Bibliography, in an outstandingly generous way. We could always rely on their readiness to provide help wherever it was needed.

Assistance in many various respects was provided by Drs. U. Felgner and K. Gloede (till 1975) and Drs. D. Schmidt and H. Zeitler (till 1979). Last but not least, our indefatigable secretary Elfriede Ihrig was and is essential in running our enterprise.

We thank all those concerned.

Heidelberg, September 1982

R. O. Gandy *H. Hermes*
A. Levy *G.H. Müller*
G. Sacks *D.S. Scott*

Acknowledgements

The subject of this book was created by Michael Morley and Saharon Shelah. As anyone familiar with the area could easily guess, results which are not explicitly credited elsewhere are due to Shelah. I thank, Alistair Lachlan not only for introducing me to the area but for his own seminal contributions. The development of this book has been mightily influenced by the French school, notably the work of Lascar and Poizat. The discussion of the spectrum problem depended not only on the original papers of Shelah but from the thorough reworking already accomplished by Harrington and Makkai.

During the preparation of this book the support and encouragement of G.H. Müller, the Ω group and the Heidelberg Academy have been invaluable. The ability to make legible early versions of the text and to cope with the ensuing multiple revisions arose only from the generous computer facilities of the University of Illinois at Chicago. I must particularly thank George Yanos and Richard Larson for their efforts in making TEX available. Work proceeded in Jerusalem for one year with the support of Institute for Advanced Studies of the Hebrew University. Many of the final preparations were made on a computer at Carnegie Mellon University in Pittsburgh. My thanks to Dana Scott for making these arrangements. The Minio and Manders-Boelema families are thanked for their hospitality. Roberto Minio translated my unformatted TEX into a file which produces acceptable output. I greatly appreciate his generosity and his patience with the interminable revisions.

For four summers I lectured one week to the ten to twenty participants in the 'Forking Festival' on the material written the previous year. The details of this work have benefitted immensely from the supervision of these model theorists. I do not have a list of all the participants. But I must thank first Bruce Rose for his help in beginning the process and some of the most persistent critics. These include Gisela Ahlbrandt, Steve Buechler, Andrew Glass, Bradd Hart, Ward Henson, Matt Kaufmann, Julia Knight, David Kueker, Doug Miller, Anand Pillay, Mike Prest, Charles Steinhorn, Bill Tait, and Carol Wood.

Finally, I want to thank those who at one time or another read over large sections of the text and provided me with detailed comments. The introduction benefitted from the criticism of Joel Berman, Wim Blok, Greg Cherlin, and Sol Feferman. Major commentators on the rest of the book include Steve Buechler, Greg Cherlin, Victor Harnik, Wilfrid Hodges, Matt Kaufmann, Jürgen Saffe, Charles Steinhorn, John Vaughn, and Martin Ziegler. They have not caught all the errors. But if I had a penny for each one they caught I could refuse the royalties.

My wife and daughter have borne up well in my absence. It will be nice to see them again.

Table of Contents

Part D. The Number of Models

Chapter I
Groundwork

The first section of this chapter purports to be a nontechnical introduction to stability theory. It attempts to provide an overview where a neophyte can find his bearings when the subsequent discussions threaten to lose the forest for the trees. The second section contains the necessary list of prerequisites and fixing of notation. We attempt to explain a few of the conventions which have become second nature to specialists but are unfamiliar to many model theorists and unknown to recursion theorists. The third section is an attempt to both summarize the results that a reader should at least be aware of and to show how some of the particular constructions of this book fit into the proof of a more familiar result, Morley's categoricity theorem. One aim of this book is to show the relationship between stability theory and classical algebra. Although various illustrations of this relation will appear in the book, the simplest continuing example is the theory of modules. Thus, in Section 4 of the first chapter we relate the basic properties of the model theory of modules. As we explain in Section 1 the subject of stability theory contains both a structure and a nonstructure side. Most of this book concerns the structure theory. Section 5 contains a brief survey of the methods and results on the nonstructure side.

1. *An Overview of Stability Theory*

An *algebraic structure theorem* (Ulm's theorem) assigns to each member of a class K of algebras (countable Abelian groups) a system of invariants (the Ulm invariants) which determine the structure up to isomorphism. But many familiar classes, for example, the class of linear orderings, have no such structure theorem. A *classification theorem for universal algebra* determines for each member K of a family $\mathcal{K}$ of classes of algebras whether or not such an algebraic structure theorem holds for K, and, if there is one, exhibits it. This book expounds the fundamentals of what is, perhaps, the first such classification theorem for universal algebra. This theory applies to several families of classes, the most important of which is the family of first order definable classes of algebras or more generally, relational structures.

To aid this exposition and as a step towards the eventual extension of the theory to more general families of classes, we have adopted an axiomatic framework for this book. These axioms describe notions of independence and generation which when satisfied yield the structure theory. Parts A and B of this book describe these axioms and the partition of first order theories which they determine. In Part C we develop the richer part of the structure theory and in Part D use the invariants assigned by the theory to count the number of models of a theory. We outline briefly at the end of Chapter I how to construct the maximum number of nonisomorphic models for each class K which does not have a structure theorem in our sense. However, since a very different and rather set theoretic group of methods is used to prove these results we do not develop them in this book.

The use of the phrase 'system of invariants' in the previous paragraph begs a fundamental question. The usual mathematical meaning of this phrase is a single number (the dimension of a vector space) or a finite sequence of cardinals (the orders of the generators of a finitely generated Abelian group) or an infinite sequence of cardinals (Ulm invariants). We must make a major generalization of this notion which is suggested by the following simple example. Let T be the theory of two equivalence relations E_1, E_2 such that E_1 has infinitely many classes and each E_1 class is refined into infinitely many infinite E_2 classes. A model of T is determined by a set of sets of cardinals: for each E_1 class the set of cardinalities of E_2 classes contained in it. Since it is easy to iterate this example, we are led to the following inductive definition of a system of invariants. A *system of invariants* of rank 0 for a structure A is a set of at most $2^{|T|}$ cardinal numbers each less than or equal to $|A|$. A system of invariants of rank β is a set of at most $2^{|T|}$systems of invariants each of rank less than β. We discuss in the text some natural variants on this idea. The methodological significance of this notion is argued at greater length in [Shelah 1985] and [Baldwin 1986].

There are a number of other uses of the word invariant in mathematics. For example, one refers to the fundamental group of a topological space as an invariant of that space. We demand that a system of invariants for a structure determines that structure up to isomorphism. Still, it would be possible to provide more general systems than suggested here. For example, [Pillay & Steinhorn 1986], one could replace the cardinal numbers by linear orderings. It seems the definition given here produces as simple a set of invariants as one could envision if structures involving refining equivalence relations are to be covered by the theory.

The computation of the number of isomorphism types of members of a class K with cardinality at most $\aleph_\alpha$ provides a challenging test problem for a structure theory. We call the resulting function of α the spectrum function for the class K. We write $I^*(\aleph_\alpha, K) = |\{M \in K : |M| \le \aleph_\alpha\}|$. Our definition of a system of invariants has the following immediate consequence. If all members of K have a system of invariants of rank less than β, then $I^*(\aleph_\alpha, K) \le \beth_\beta(\sup(2^{\aleph_0}, |\alpha \dotplus \omega|))$. (The $|\alpha + \omega|$ arises here as the number of cardinals less than $\aleph_\alpha$.) A weak version of the contrapositive of this result is even more striking. If, for all sufficiently large κ, K has 2^κ models of power κ then K has no system of invariants of rank β for any β. This raises the question of whether there is a middle ground, a class K which has fewer than the maximal number of models but does not admit a system of invariants. The major conclusion about the spectrum function is that no such middle ground exists. To state this result precisely but simply, let $\mathcal{K}$ be the family of all elementary classes in a countable language. For every $K \in \mathcal{K}$, for all sufficiently large α, $I^*(\aleph_\alpha, K) = 2^{\aleph_\alpha}$ or for all α, $I^*(\aleph_\alpha, K) \leq \beth_\beta(|\alpha + \omega|)$ for some β less than ω_1. This dichotomy is known as 'the main gap'.

The ultimate regularity which arises from the classification of the finite simple groups is the existence of a finite number of families of simple groups. This regularity is obscured by the 26, relatively small, sporadic groups. Similarly, the classification of the spectrum function is obscured by some 'static' among the small models. The correct picture is obtained by first examining the eventual behavior of the spectrum function and then fine-tuning to find its values for small cardinals.

Thus, in this book we will first develop a classification of first order theories. Then for each class of theories we will prove that no structure theory exists for that class or develop one. This classification consists first of a partition of all theories into four stability classes. There are three other, mutually independent, relevant properties: the dimensional order property, the omitting types order property, and depth. A priori, these properties divide all theories into 32 classes. In fact, once a property can be shown to imply that there is no structure theorem we do not have to consider the partition of the theories with that property by the other properties.

In the remainder of this introduction we provide a geography of stable theories. That is, giving few formal definitions and no proofs, we list the types of theories, state the relations between the different types and provide a few examples of each kind. We indicate how invariants are attached to models of theories in the various classes.

In the following survey we deal only with first order theories with *countably* many relation and function symbols. Our ambivalent attitude toward uncountable languages in this book has two sources. First, most of the theories that arise in mathematical practice are naturally formalized in countable languages. However, some are not, notably the theory of vector spaces over the real numbers. Second, a number of important advances in the theory, and I mean important for the study of countable theories, arose from the consideration of uncountable theories. Thus, in the main text, we state most results in a form applicable to uncountable languages and we remark when this case involves added difficulties.

The first dichotomy in our classification of theories is between those which are *stable* and those which are *unstable*. Essentially, a theory T is unstable if it is possible to find in some model of T a linear order which is definable in T on some infinite set of n-tuples. Otherwise, T is stable. Thus, any theory of linear order, any extension of the theory of Boolean algebras, and the theory of arithmetic are all examples of unstable theories. Any unstable theory has 2^λ models of power λ for any uncountable λ ([Shelah 1978]). Thus, no structure theory of the sort described above is possible for the class of models of T.

We will discuss the class of stable theories from an historical viewpoint. The notion of stability arose in the study of categorical theories. A theory T is said to be categorical in power λ if all its models of power λ are isomorphic. Łos [Łos 1954] and Vaught [Vaught 1954] proved in the middle 1950's that if a countable theory with no finite models is categorical in some infinite power λ then it is complete. Łos conjectured in [Łos 1954] that if a countable theory T is categorical in one uncountable power then T is categorical in every uncountable power. (This conjecture is suggested by the Steinitz theorem that every algebraically closed field is determined by the cardinality of its transcendence basis over the prime field.) Morley [Morley 1965] proved the Łos conjecture; in the process, he introduced a wider class of theories than the $\aleph_1$-categorical theories, those that he called totally transcendental and which we now call ω-stable. Examples of ω-stable theories include, in addition to all $\aleph_1$-categorical theories, the theory of differentially closed fields of characteristic zero, the theory of an equivalence relation with infinitely many infinite classes, and the theory of any Abelian group which is the direct sum of a divisible Abelian group and one of bounded exponent. A more specific version of the last example is the theory of the Abelian group $Z_3^\omega \oplus Z_2^\omega$.

While extending Morley's theorem to theories in uncountable languages, Shelah discovered that the crucial properties of ω-stable theories could be generalized by localizing some of Morley's notions from types involving all formulas to ϕ-types, types which contain only instances of a single formula, ϕ. He was then able to isolate a larger class than the ω-stable theories which shared a number of the properties which make ω-stable theories tractable. He called these theories stable. At the same time he discovered a notion which lies between ω-stability and stability which he called the class of superstable theories. In [Shelah 1978], Shelah extends the nonstructure result by proving that any theory which is not superstable has 2^λ models of power λ for every uncountable λ. A natural example of a superstable but not ω-stable theory is the theory of $(Z, +)$. A slightly less natural example of a stable but not superstable theory is the theory of $(Z^\omega, +)$.

The effort to determine exactly which theories have the maximal number of models in each uncountable power required the introduction of two further concepts. A theory is said to have the *dimensional order property* if, speaking roughly and somewhat inaccurately, it is possible to define an ordering within the theory with the use of cardinality quantifiers (There exist κx such that ...). Any such theory has 2^λ models of cardinality λ for any uncountable λ. The simplest example of such a theory is the theory of two crosscutting equivalence relations E_1 and E_2 such that each equivalence class of the relation $E_1 \cap E_2$ is infinite.

Another complication appears exactly at the separation between superstable and ω-stable theories. It is much more difficult to work out the structure theory for superstable theories than it is for ω-stable theories. An interim solution is to restrict the class of models considered when T is only superstable. Of course, such a restriction actually strengthens the result when we prove the restricted class has the maximal number of models. There are a number of families of classes of models which can be treated in a nearly uniform way. The two families $\mathcal{K}$ which receive the most attention in this book are $\mathcal{K}_1 = \{K : K$ is the class of models of an ω-stable theory$\}$ and $\mathcal{K}_2 = \{K : K$ is the class of **S**-models of a superstable theory$\}$. The models in the second family, which we call the strongly ω-saturated models, are also known in the literature as F^a_ω-saturated models, $\aleph_\epsilon$-saturated models and a-models.

The discussion so far shows that if a theory is to have less than the maximal number of models in some uncountable power λ, then T must be superstable and must not have the dimensional order property. In this book we give a complete solution to the spectrum problem for countable ω-stable theories and for S-models of countable superstable theories. The full solution of the original problem, that is the assigning of invariants to the class $\mathcal{K}_3 = \{K :\ K$ is the class of models of a superstable theory$\}$ was obtained recently and will appear in the second edition of [Shelah 1978].

The solution of the spectrum problem for theories in the classes K_1 and K_2 depends upon a key construction which assigns to each model of power λ a skeleton of submodels. Each submodel has cardinality at most 2^ω and the skeleton is partially ordered by the natural tree order on a subset of $\lambda^{<\omega}$. The isomorphism type of the model is determined by the small submodels and this partial ordering. One more property of a theory, which distinguishes between structure and nonstructure, depends on the isomorphism types of trees associated with models of the theory. If one of these trees is not well-founded, the theory is said to be *deep* and has 2^λ models in every uncountable power λ. If not, the theory is *shallow* and the type of structure theory we have described exists. We are able to assign to each such shallow theory a depth β corresponding to the rank of a system of invariants, as discussed above, and to compute the spectrum function of K_1 or K_2 in terms of that depth. Both the dimensional order property and depth are properties which are independent of the stability hierarchy. A good example of a deep ω-stable theory which does not have the dimensional order property is the theory of a single unary function such that each element has infinitely many preimages.

There is a further subdivision of the shallow theories which do not have the dimensional order property. This division is determined by the number of 'bases' to which one can independently assign a dimension. Thus the theory of an equivalence relation with two infinite classes is 2-dimensional since to determine a model we must prescribe the number of elements in each of two classes. Two important classes arise in this context. A theory T is called *bounded* if there is a cardinal $\delta(T)$ such that every model of

T can be determined by specifying less than $\delta(T)$ dimensions. T is called *unbounded* or *multidimensional.* if no such $\delta(T)$ exists. If the theory T is bounded then T has fewer than $|\alpha|^{2^\omega}$ strongly ω-saturated models in power $\aleph_\alpha$. But if T is unbounded, T has at least 2^α strongly ω-saturated models of power $\aleph_\alpha$, where, for simplicity, $\alpha \geq 2^\omega$.

Shelah's solution of the problem for the class K_3 requires the introduction of a number of new ideas which will be treated only peripherally here. The one property necessary to state the final classification is the *omitting types order property.* Again, a non-first order property is used to define a linear ordering. The main gap becomes: If a superstable theory T does not have the dimensional order property, is shallow, and does not have the omitting types order property then $I^*(\aleph_\alpha, T) \leq \beth_\beta(|\alpha|)$ for some β less than ω_1.

The final solution of the spectrum problem can be summarised as follows. Shelah defines an extended logic $L_{\infty,c^+}(Q_D)$ where $c = 2^{\aleph_0}$ and Q_D quantifies over dimensions of independent sets. In [Shelah 198?], Shelah shows that if T is a countable superstable theory without either the dimensional order property or the omitting types order property then each model of T is characterized up to isomorphism by a sentence of $L_{\infty,c^+}(Q_D)$. Since validity in this logic is invariant under changes of the set theoretic universe which do not collapse cardinals and which fix the continuum, the isomorphism classes of such a theory are fairly absolute. If T does not satisfy these conditions, Shelah shows that one can extend the universe of sets by a notion of forcing which preserves cardinals and the continuum yet changes the isomorphism classes of T.

This completes our survey of the relation between the stability hierarchy and the number of models in an uncountable power. We have said nothing as yet about the impact of this theory on the problem of determining the number of countable models of a theory.

Categoricity in power $\aleph_0$ interacts with the stability hierarchy in an interesting way. There are $\aleph_0$-categorical theories which are unstable, for example, dense linear order without endpoints. At the other extreme are the theories which are totally categorical, that is, categorical in every infinite power. Theories in the latter class have a very rich structure which has prompted some of the most sophisticated of the stability theoretic investigations. Every such theory is closely tied to either the theory of an infinite set, the theory of infinite dimensional projective space over a finite field or the theory of infinite dimensional affine space over a finite field. We will summarize this connection and provide more detailed references in Chapter VIII. Lachlan [Lachlan 1974], and independently Shelah [Shelah 1978], proved that a superstable $\aleph_0$-categorical theory is ω-stable. Lachlan's investigation of the stronger conjecture that every $\aleph_0$-categorical stable theory is ω-stable resulted in the definition of a *pseudoplane* and the first introduction of geometry into the subject.

A second major consideration regarding countable models arose from Vaught's conjecture that an $\aleph_1$-categorical theory has either 1 or $\aleph_0$ countable models. Morley [Morley 1970] proved that such a theory has at most $\aleph_0$ countable models. Baldwin and Lachlan [Baldwin & Lachlan 1971] proved this conjecture of Vaught and established that the dimension theory of $\aleph_1$-categorical theories is extremely well behaved. Later Lachlan [Lachlan 1973] extended this result to superstable theories by proving that a countable superstable theory is either $\aleph_0$-categorical or has infinitely many countable models. Lascar [Lascar 1976] and Shelah [Shelah 1978] provided more conceptual proofs of this result. Finally, Pillay [Pillay 1983] gave the simplest and most general version of the theorem. As before, the attempt to extend the result from the superstable case to the stable, that is, to prove that a countable stable theory can not have a finite number of countable models, remains a major open question.

The third major question about countable models, Vaught's conjecture that a first order theory has either $2^{\aleph_0}$ or at most $\aleph_0$ countable models has also been attacked by stability-theoretic means. The only result proved for an arbitrary theory is Morley's theorem that any sentence of $L_{\omega_1,\omega}$ has either $\aleph_1$ or less countable models or exactly $2^{\aleph_0}$ countable models. This proof turns out to be a result in descriptive set theory. This has led to a number of attempts to prove the conjecture itself by the methods of descriptive set theory. So far, there has been no further success on this front. There have been a number of solutions to the problem for special theories such as linear orderings [Rubin 1974] and trees [Steel 1978]. The attempt to solve the problem by breaking it down according to the stability hierarchy has led to the following results. Vaught's conjecture is true for ω-stable theories [Shelah, Harrington, & Makkai 1984]. It is also true for stable theories with Skolem functions [Lascar 1981]. Finally, Vaught's conjecture is true for linearly ordered models with Skolem functions [Shelah 1978a]. The extension of the result in [Shelah, Harrington, & Makkai 1984] to superstable theories seems difficult. Even if it is managed, the usual problem of stable but not superstable theories will remain.

This introduction, and the book as a whole, emphasizes the rôle of the spectrum problem in the development of stability theory. This makes the organization of the discussion simpler while obscuring some of the other important contributions of the theory. In particular, the structure theory itself may have been lost in this discussion of its application. Many of the structure results in algebra are not actually stated in terms of assigning invariants but rather as the the decomposition of an arbitrary member of some class into a 'product' of known members of the class. The Wedderburn theory is a nice example of this sort. Of course, such a result leads to an assignment of invariants if the 'product' operation is sufficiently tight and the factors are well enough known. We have such a decomposition here, but the 'product' operation is extremely complicated because the factors are partially ordered by a tree. We have chosen to treat this decomposition as an 'internal direct sum'. In [Baldwin & Shelah 1985] it is treated as an 'external direct sum'. That approach emphasizes the significance of the structures $\lambda^{<\omega}$ and λ^{ω}. These are the partially ordered sets of all finite sequences (respectively all countable sequences) from a cardinal λ. The assertion that every sufficiently saturated model with power λ of a stable (superstable) theory can be decomposed over a skeleton of the form λ^ω ($\lambda^{<\omega}$) means that the relation between any two elements in M depends on only a countable amount of information even when λ is very large.

There have been major advances in applying these techniques to the study of infinitary languages and languages with generalized quantifiers. Although almost no mention of these matters is made here, the reason for our general approach is the hope of eventually incorporating all these seemingly diverse enterprises into a single framework.

The following table summarises the contents of this section and provides an overview of the stability classification and its relation to the number of countable models. The rows describe a place in the stability classification. A theory belongs in a particular row if it satisfies the condition at the left of the row and does not satisfy the condition in the previous row. The columns indicate the number of countable models. An entry of 'x' means there is no theory of this kind; an entry of '?' means the existence of such a theory is unknown; a numeric entry refers to a theory witnessing that property from the list of theories which follows the table. Many of these examples are discussed in more detail in the text.

An Array of Stable Theories

CLASS	1	$n<\omega$	$\aleph_0$	$\aleph_1$	$\aleph_1<\mu<2^{\aleph_0}$	$2^{\aleph_0}$
$\aleph_1$-categorical	1	x	2	x	x	x
ω-stable	3	x	4	x	x	5
superstable	x	x	6	?	x	7
stable	?	?	8	?	x	9
unstable	10	11	12	?	x	13

1. The theory of an infinite set. The theory of $Z^\omega_{p^k}$.
2. The theory of the integers with the successor function. The theory of an algebraically closed field. The theory of $(Q,+)$.
3. The theory of an equivalence relation with two infinite classes.
4. The theory of an equivalence relation with one class of cardinality n for each positive integer n. The theory of an equivalence relation with two classes and a copy of a theory of the second kind in each class.
5. The theory of an equivalence relation with infinitely many classes and a model of a theory of the second kind in each class.
6. The theory of the structure $(2^\omega, E_i, +)$ where $+$ denotes coordinatewise addition and $E_i(\sigma,\tau)$ holds if $\sigma|i=\tau|i$.
7. The theory of $(2^\omega, E_i)$ with E_i defined as in type 6. The theory of $(Z,+)$.
8. The theory of $(V, B_1, B_2, \ldots)$ where V is a vector space over a finite field and the B_i are a decreasing sequence of subspaces such that

$[B_i : B_{i+1}] = \aleph_0$. (The countable models are determined by the dimension of $\bigcap B_i$.)

9. The theory of (ω^ω, E_i) where E_i is defined as in type 6. The theory of $(Z^\omega, +)$.
10. The theory of dense linear order without endpoints. The theory of atomless Boolean algebras.
11. The theory of the structure $(Q, <, 0, 1, \ldots)$. All known examples are variants of this one.
12. Any disjoint union of an $\aleph_0$-categorical unstable theory with a theory with $\aleph_0$ countable models.
13. Any complete theory of linear orderings which is not $\aleph_0$-categorical [Rubin 1974]. The theory of $(Z, +, <)$.

2. *Basic Notions And Fundamental Conventions*

In this section we fix our notation. We assume the reader is familiar with such results as the compactness and Löwenheim-Skolem theorems and such notions as model completeness and quantifier elimination.

We use α, β, γ for ordinals and κ, μ, λ for cardinals. The notation λ^α generally means the set of functions from α into λ and $\lambda^{<\alpha}$ the set of functions from initial segments of α into λ. This notation should cause little confusion as we rarely use ordinal exponentiation. κ^λ denotes ambiguously the set of functions from λ into κ or the cardinality of that set of functions.

Throughout this book we will be considering models of a complete first order theory T which has no finite models. We denote by $|T|$ the *cardinality* of T, that is, the number of relation symbols, constant symbols, and function symbols in the *language* of T which we denote $L(T)$ or just L. We regard equality as a logical symbol and will always interpret it as identity.

We will write $\phi(\overline{x})$ for the first order formula $\phi(x_1, \ldots, x_n)$ but will not make the n explicit unless its value is crucial. We write M for the structure $(M; \langle R_i^M : R_i \text{ an } L(T)-\text{symbol}\rangle)$. We denote by $\overline{a} \in M$ the assertion that $\overline{a}$ is a finite sequence of elements from M, again suppressing the length of the sequence unless it is particuarly relevant. If it is relevant we denote this length by $\lg(\overline{a})$. Technically, such a sequence is a function and we should write $\text{rng}(\overline{a})$; however, we identify such a finite sequence with its range except on those few occasions when confusion is rife. In particular, we frequently write $\overline{a} \cup B$ to denote $\{a_0, \ldots, a_{n-1}\} \cup B$. If α is a function from A to B, we write $\alpha(\overline{a})$ to abbreviate $\langle \alpha(a_1), \ldots, \alpha(a_n)\rangle$. For $\overline{a} \in M$, we write $M \models \phi(\overline{a})$ to indicate that the sequence $\langle a_1, \ldots, a_n\rangle$ satisfies the formula $\phi(\overline{x})$ in the structure M. Let $\phi(\overline{x}; \overline{y})$ be an L-formula and let $\overline{a}$ denote a sequence of elements from M. We denote by $\phi(M; \overline{a})$ the set $\{\overline{m} \in M : \models \phi(\overline{m}; \overline{a})\}$. If $X = \phi(M; \overline{a})$ we say X is *definable over* any set which contains $\overline{a}$. For any structure M, we write $\text{Th}(M)$ for the set of L-sentences true in M. We denote by (M, B), the structure obtained by adding names for each element $b \in B \subsetneq M$. We write $M \models (\exists!^k x)\phi(x)$ if $|\phi(M)| = k$ and $M \models (\exists! x)\phi(x)$ if $|\phi(M)| = 1$.

We will be studying the relation between models $M, N, \ldots$ of the theory T. It is inconvenient if there is an embedding α of M into N such that $M \models \phi(\overline{m})$ but $N \models \neg\phi(\alpha(\overline{m}))$. Of course, this situation does not arise when M is an *elementary submodel* of N, written $M \prec N$. Specifically, this problem cannot arise if the theory T is *model complete.* We also deal with substructures, $A, B, C, \ldots$ of models of T. Again, it is inconvenient if the truth of $\phi(\overline{a})$ depends upon the particular embedding of $\overline{a}$ into a model of T. These problems disappear if we assume that T is *quantifier eliminable.* Fortunately, this is a harmless assumption in our context. For, given any theory T in a language L, we first extend L to a language $L^{\#}$ by adding a relation symbol $R_{\phi(\overline{x})}(\overline{x})$ for each formula $\phi(\overline{x})$. Now we extend T to a theory $T^{\#}$ in $L^{\#}$ by adding the axioms $(\forall\overline{x})[\phi(\overline{x}) \leftrightarrow R_{\phi(\overline{x})}(\overline{x})]$. The resulting theory $T^{\#}$ is equivalent to T for the properties in which we are interested but is quantifier eliminable. For example, for any κ, T and $T^{\#}$ have the same number of models of power κ. Thus we make the following fundamental convention.

2.1 Convention. The theory T is a complete quantifier eliminable theory with no finite models.

While we hold to this convention when proving the general theorems in the text, we will on occasion deal with specific examples of naturally occurring theories that do not admit elimination of quantifiers. Thus it will be essential, although never mentioned, to observe that the definitions we make are preserved in the extension from T to $T^{\#}$.

We may apply any definition made for a theory (e.g. quantifier eliminable) to a structure M by saying M has the property if $\mathrm{Th}(M)$ does.

By means of the translation discussed before Convention 2.1 we are at liberty to assume that all languages are relational. For simplicity of expression we will formulate many examples in languages with function symbols.

For the following exercises we write $M \cap N$ for the intersection of two structures only when the intersection is a common submodel of M and N. This detail will automatically be satisfied on the basis of Convention 2.1. These exercises either are direct translations of terms such as quantifier eliminable or can be deduced from these concepts by the 'method of diagrams'.

2.2 Exercise. If T admits elimination of quantifiers and $\overline{a} \in M \cap N$, where M and N are models of T, show that for any formula $\phi(\overline{x})$, $M \models \phi(\overline{a})$ if and only if $N \models \phi(\overline{a})$.

2.3 Exercise. Show that if T is complete then T has the *joint embedding property.* That is, if M_1 and M_2 are models of T then there is a model N and elementary embeddings α_1 and α_2 of M_1 and M_2, respectively, into N.

2.4 Exercise. Show that if T is model complete then T has the *amalgamation property.* That is, if M_1 and M_2 are models of T and $M_1 \cap M_2 \models T$ then there is a model N and elementary embeddings α_1 and α_2 of M_1 and M_2, respectively, into N which agree on $M_1 \cap M_2$.

2.5 Exercise. Show that if T admits elimination of quantifiers and if M_1, M_2 are models of T with $M_1 \cap M_2 = A$, there is a model N and elementary embeddings α_1 and α_2 of M_1 and M_2, respectively, into N so that α_1 and α_2 agree on A.

Many results in this book are assertions that a certain property holds for all cardinals κ. The proofs given will seem to depend upon the cardinal κ or even upon the existence of a 'well-behaved' model of power κ_1 which is larger than κ. In fact, this dependence is illusory; the arguments are uniform in κ. Rather than continually remark on this uniformity, we assume that there is a well-behaved 'monster' model $\mathcal{M}$ whose cardinality is greater than that of any object in our discussion. For the moment, we take well-behaved to simply mean '$\mathcal{M}$ is a model'. Thus our only use at the moment of this artifice is to be able to write $\mathcal{M} \models \phi$ or $\models \phi$ interchangeably rather than have truth depend on particular models. We have already explained that the assumption that T is quantifier eliminable justifies this simplification. In order to further specify what is meant by well-behaved we recall a few notions from basic model theory.

The Stone representation theorem (e.g. [Sikorski 1964]) associates with each Boolean algebra B a totally disconnected Hausdorff space $S(B)$, the Stone space of B. When B is the Boolean algebra of formulas with parameters from a subset of a model of T, this space has a particularly important interpretation. The analysis of this space is one of the key tools used in this book. We do not prove the Stone representation theorem but we review the principal ingredients of the proof in our context. We begin by defining the Boolean algebra of definable subsets of $\mathcal{M}$ (or $\mathcal{M}^n$). Then we will discuss the natural interpretation of the Stone space of this Boolean algebra. A *formula over* $B \subseteq \mathcal{M}$ is the result of substituting a sequence of parameters $\overline{b}$ from B into an L-formula $\phi(\overline{x};\overline{y})$ to obtain $\phi(\overline{x};\overline{b})$. For any $A \subseteq \mathcal{M}$ we write Diag(A) for $\{\phi(\overline{a}) : \models \phi(\overline{a})\}$.

2.6 Definition. $F^n(B)$ denotes the Boolean algebra of formulas with n free variables having parameters from B. The Boolean operations in $F^n(B)$ are just the Boolean operations on formulas. Note that we identify formulas up to equivalence in $\mathrm{Th}(\mathcal{M}, B)$.

We write $F(B)$ ambiguously for $F^n(B)$ for any finite n and $F^\omega(B)$ for $\bigcup\{F^n(B) : n < \omega\}$. When we want to emphasize the dependence on T, we write $F(T)$ or $F^n(T)$ for $F(\emptyset)$ or $F^n(\emptyset)$.

The type of $\overline{a}$ over B is a complete description of how $\overline{a}$ relates to the set B. Formally,

2.7 Definition. Let $\lg(\overline{a}) = n$. The (*complete*) *type of* $\overline{a}$ *over* B, denoted $t(\overline{a}; B)$, is

$$\{\phi(\overline{x}) : \models \phi(\overline{a}) \text{ for } \phi \in F^n(B)\}.$$

More generally an α-type p is a consistent set of formulas with α free variables usually denoted $\langle x_i : i < y\alpha \rangle$. We denote by $\operatorname{dom} p$ the *domain of* p, that is, the smallest subset of $\mathcal{M}$ which contains all parameters which occur in a formula in p. We say a type p is *realized* by $\overline{a}$ if $\models \phi(\overline{a})$ for each $\phi(\overline{x}) \in p$. For p a type, we let $p(M)$ equal

$$\{\overline{a} \in M : \models \phi(\overline{a}), \phi \in p\}.$$

If A is a set, by $t(A; B)$ we mean $t(\overline{a}; B)$ where $\overline{a}$ is a fixed enumeration of A. More detail on this convention occurs in Section II.2. The complete type p is a *principal type* if there is a formula $\phi(\overline{x}) \in p$ such that for all $\psi(\overline{x}) \in p$, $\models \phi \to \psi$.

2.8 Definition. The (nth) *Stone Space of* B, denoted $S^n(B)$, is the collection of all complete n-types over B. We make $S^n(B)$ into a topological space by specifying the collection of sets $U_\phi = \{p \in S^n(B) : \phi \in p\}$ where $\phi \in F^n(B)$ as a basis of open sets.

2.9 Exercise. Show that each set U_ϕ is actually clopen. Conversely, show that each clopen subset of $S^n(B)$ is definable by a single formula with parameters from B.

We write $S(B)$ ambiguously for $S^n(B)$ for any finite n and $S^\omega(B)$ for $\bigcup\{S^n(B) : n < \omega\}$. We sometimes write $S(T)$ for types over the empty set.

An n-*type over* A is a subset of an element of $S^n(A)$. Thus types may be incomplete but are always consistent. If p is a type over A and $B \subseteq A$ then $p|B = \{\phi(\overline{x}; \overline{b}) : \phi(\overline{x}; \overline{b}) \in p \text{ and } \phi(\overline{x}; \overline{b}) \in F(B)\}$. We could regard two types p and q over possibly different domains as equal if they are satisfied by the same elements of the monster model; that is, if $p(\mathcal{M}) = q(\mathcal{M})$. However, this identification destroys the notion of $\operatorname{dom} p$ and makes talk of the cardinality of a type impossible. Thus, we formally view a type as a collection of formulas which has a definite domain and cardinality. Abusing notation we write the expression $t(\overline{a}; A \cup \overline{d}) = t(\overline{a}'; A \cup \overline{d}')$ to mean there is an automorphism of $\mathcal{M}$ which fixes A, maps $\overline{a}$ to $\overline{a}'$ and $\overline{d}$ to $\overline{d}'$.

For $p(\overline{x})$ a type over A, we write $p \vdash \phi(\overline{x})$ if every realization of p satisfies ϕ. Note that this is the same as saying that if L' is formed by adding new constants $\overline{c}$ to L then in L', $\operatorname{Th}(\mathcal{M}, A) \cup p(\overline{c})$ implies $\phi(\overline{c})$. It is tempting to assume that types are closed under logical consequence. Unless this concept is suitably restricted, however, it makes nonsense of the notion of the cardinality of the domain of p. That is, for every p and for every $a \in \mathcal{M}$, $a = a$ would be in p. We sometimes explicitly assume that if $\operatorname{dom} p$ is specified as A, $\phi(\overline{x}; \overline{a})$ is over A and $p \vdash \phi(\overline{x}; \overline{a})$ then $\phi(\overline{x}; \overline{a}) \in p$. Then, types are closed under conjunction.

There is a complete discussion of the impact of various kinds of closure conditions on the definition of type in Chapter II of [Shelah 1978].

2.10 Exercise. Let $A \subseteq B$. Show the restriction map is continuous from $S(B)$ onto $S(A)$.

2.11 Exercise. Show that a point $p \in S(B)$ is isolated just if p is a principal type.

2.12 Exercise. Let (Z, S) denote the integers under successor. Show that $|S^1(Z)| = \aleph_0$ and that it contains a unique non-isolated point. Note that if we replace (Z, S) by $(Z, <)$ then there are two non-isolated points.

2.13 Exercise. Let T be a countable theory. Show that if $|S(A)| = 2^{\aleph_0}$ for some finite A then T has $2^{\aleph_0}$ countable models.

2.14 Exercise. Find an example showing the necessity of the assumption that A is finite in the previous exercise.

Regarding $S(A)$ as a topological space provides a certain insight into the behavior of types. Moreover, it provides a very compact notation for some operations which are model-theoretically important. We next summarize some topological background which is useful in this context.

2.15 Topological Facts. Suppose $A \subseteq B$ and r denotes the restriction map from $S(B)$ onto $S(A)$. Since $S(B)$ is compact, r is continuous, and $S(A)$ is Hausdorff, r is a closed map. Moreover, if X is a closed subset of $S(B)$ then $r(X)$ is compact.

The following definition lists those properties we will need to specify the 'well-behaved' nature of the monster model.

2.16 Definition. i) The model M is λ-*saturated* if for every $A \subseteq M$ with $|A| < \lambda$, every $p \in S(A)$ is realized by some $b \in M$.

ii) The model M is λ-*compact* if for every p with $\operatorname{dom} p \subseteq M$ and $|p| < \lambda$, p is realized in M.

iii) The model M is λ-*homogeneous* if for any subsets A and B of M, such that $|A|, |B| < \lambda$ and $\mathrm{Th}(M, A) = \mathrm{Th}(M, B)$, and for any $m \in M$ there is an $m' \in M$ such that $\mathrm{Th}(M, A, m) = \mathrm{Th}(M, B, m')$.

iv) The structure M is λ-*universal* if for every $N \equiv M$ with $|N| < \lambda$, there is an elementary embedding of N into M.

v) The structure M is *compact, saturated, homogeneous* if it is respectively, $|M|$-compact, $|M|$-saturated, $|M|$-homogeneous.

The following exercises just list some of the basic properties of these notions. They all occur in, for example, [Chang & Keisler 1973].

2.17 Exercise. Show that if M is λ-saturated then M is λ-homogeneous.

2.18 Exercise. Show that if M is homogeneous, $\overline{a}$ and $\overline{b}$ are sequences from M with cardinality less than $|M|$ and $t(\overline{a}; \emptyset) = t(\overline{b}; \emptyset)$ then there is an automorphism of M which takes $\overline{a}$ to $\overline{b}$.

2.19 Exercise. Show that any two saturated elementarily equivalent models with the same cardinality are isomorphic.

2.20 Exercise. Show that any two elementarily equivalent, homogeneous models of the same power which realize the same types (of finite sequences) over the empty set are isomorphic. (This problem is more difficult than the last; it needs to be done by induction on the cardinality of the structures.)

2.21 Exercise. Show that if $\lambda > |T|$ then λ-saturated and λ-compact are equivalent.

2.22 Exercise. Show that a model M is saturated if and only if M is both homogeneous and universal.

2.23 Exercise. Show every saturated model M is $|M|^+$-universal.

2.24 Exercise. Show that any theory T has a κ^+-saturated model of cardinality $\leq 2^\kappa$. (Hint: Choose any model of T with cardinality $\leq 2^\kappa$ and realize every type over each subset of M with cardinality $\leq \kappa$. Repeat this procedure κ^+ times and take the union of the chain of models constructed in this way.)

Assuming the continuum hypothesis one can deduce from Exercise 2.24 that every theory has a saturated model of power $\aleph_1$. Our approach here is to avoid such general set-theoretic hypotheses and search for conditions on a theory which allow a specific model theoretic construction, e.g. a saturated model of power $\aleph_1$. This particular problem is discussed fully in Theorems II.2.20 and III.4.37.

Now we can finish defining the concept of the monster model.

2.25 Convention. We assume that all models and sets that we deal with are contained in a saturated model, called the monster model and denoted $\mathcal{M}$.

As we mentioned earlier, the existence of this model is only assumed as a technical convenience. For those who prefer a more concrete characterization and who believe in large cardinals, we might take $\mathcal{M}$ as a saturated model of power κ where κ is the first strongly inaccessible cardinal.

Items i) and ii) in the following definition are standard notation throughout this book; item iii) fixes our notation for the remainder of this section.

2.26 Notation. i) If α is a partial automorphism of $\mathcal{M}$ and p is a type with $\operatorname{dom} p \subseteq \operatorname{dom} \alpha$ then $\alpha(p)$, sometimes written αp, is the type $\{\phi(\overline{x}; \alpha(\overline{a})) : \phi(\overline{x}; \overline{a}) \in p\}$.

ii) For $A \subseteq M$, $\operatorname{Aut}_A(M)$ denotes the set of automorphisms of M which fix A pointwise. We call an element of $\operatorname{Aut}_A(M)$ an *A-automorphism* of M.

iii) Fix $A \subseteq M \subseteq N$ and suppose M is saturated and $|A| < |M|$. Let r denote the projection map from $S(M)$ onto $S(A)$. Fix $X \subseteq S(M)$; let $P = r(X)$. Let X_N denote $\{\overline{n} \in N : t(\overline{n}; M) \in X\}$ and P_N denote $\{\overline{n} \in N : t(\overline{n}; A) \in P\}$.

Note that $X_N \subseteq P_N$.

For any B, the elements of $S(B)$ can be viewed either syntactically as maximal consistent sets of formulas and thus topologically as elements of the Stone space or semantically as equivalence classes of elements in the monster model. The next series of results show the advantage of adopting the right view at the right time. The following theorem shows that if a subset of $S(M)$ is definable in M but invariant under A-automorphisms of M then, if M is sufficiently saturated, the subset is actually definable over A. Its statement depends on the notation established in 2.26.

2.27 The Basic Definability Theorem. *If X is clopen in $S(M)$ and X is invariant under* $\mathrm{Aut}_A(M)$*then* $X_N = P_N$.

Proof. Let $\psi(\overline{x};\overline{m})$ define X. By assumption $\psi(\overline{x};\overline{m})$ must be invariant under A-automorphisms of M. Suppose $P_N - X_N$ is not empty; then for some $p \in X$, $q = p|A \cup \{\neg\psi(\overline{x};\overline{m})\}$ is consistent. Let $\overline{a}$ in M realize q and let $\overline{b}$ in M realize $p|A \cup \{\psi(\overline{x};\overline{m})\}$. Then there is an automorphism of M which fixes A and takes $\overline{a}$ to $\overline{b}$. This contradicts the invariance of $\psi(\overline{x};\overline{m})$.

2.28 Corollary. *Under the notation of 2.15 and the conditions of Theorem 2.27:*

i) $r^{-1}r(X) = X$.
ii) $r(X) = P$ *is clopen in* $S(A)$.

Proof. i) simply restates the conclusion of Theorem 2.27. ii) follows by an immediate topological argument noting $r(X)^c = r(X^c)$.

2.29 Exercise. Prove Corollary 2.28 ii) by logical means (a 'double compactness argument').

By taking $M = N = \mathcal{M}$ in Theorem 2.27 we obtain the following corollary which is used extensively in Chapter IV.

2.30 Corollary. *If X is a definable subset of $\mathcal{M}$ and X is fixed by every A-automorphism of $\mathcal{M}$ then X is definable over A.*

2.31 Historical Notes. The basic notions discussed at the beginning of this section can be found in such texts as [Chang & Keisler 1973], [Bridge 1977], [Hodges 198?], [Poizat 1985], [Shoenfield 1967], or [Sacks 1972]. The more advanced material can also be found in [Chang & Keisler 1973] and [Sacks 1972]. Morley pointed out that for the problems discussed here, one could assume without loss of generality that the theory admits elimination of quantifiers. The notion of a monster model was introduced by Shelah.

The vital role of types in model theory was recognized in the middle 1950's. Morley [Morley 1965] introduced the systematic study of Stone spaces. The important role of the basic definability lemma first surfaced in Poizat's proof of the symmetry lemma for forking (cf. exercises III.2.26 and III.2.27.). Its general significance became evident during the 1980-81 model theory year in Jerusalem. Thus, it plays an important role in [Harnik & Harrington 1984]. The formulation here was suggested by Dale Myers.

3. Categoricity Of Countable Theories

In this section we quickly survey the prehistory of stability theory by describing the main results known by 1965 about theories which are categorical in some infinite power. Recall that a theory T is *categorical in power* κ if all models of T which have cardinality κ are isomorphic. To organize this survey we prove Morley's theorem that a countable theory is $\aleph_1$-categorical if and only if it is categorical in every uncountable power. Since this result will appear naturally when we develop the solution to the general spectrum problem, the treatment here is more of an overview. There are several excellent accounts of the theorem in print [Morley 1965], [Chang & Keisler 1973], [Sacks 1972], [Baldwin & Lachlan 1971]; the proof here differs somewhat from all of these. We collect in this section those ingredients of the proof which do not arise in the natural development of the general theory. For other results we will refer to the place in this book where a proof is given. Thus, this survey will provide a point of reference for the more abstract developments later. *We assume in this section that each theory has a countable language.*

Before giving this proof we state a full characterization of countable theories which are $\aleph_0$-categorical. A proof of this characterization can be found in any of the general references cited above.

3.1 Theorem (Ryll-Nardzewski's Theorem). *Let T be a countable complete theory with no finite models. The following are equivalent:*

i) *T is $\aleph_0$-categorical.*
ii) *For each n, $S^n(T)$ is finite.*
iii) *For each n, $F^n(T)$ is finite.*

This proof of Morley's theorem falls into three parts. We first define the concept of a strongly minimal theory. In a strongly minimal theory we are able to directly generalize the notion of algebraic dependence from the theory of vector spaces or fields. Thus, the proof that such a theory is categorical in all uncountable powers just follows the usual argument that the transcendence degree of an algebraically closed field determines the field up to isomorphism. We then show that this result extends to any theory T such that:

i) there is a strongly minimal formula, ϕ, in the language of T.
ii) if M is a model of T, then M is prime and minimal (Definition 3.16) over $\phi(M)$.

It is easily seen that the second condition holds if ϕ is not a two-cardinal formula (see Definition 3.19). Finally, we show that if T is categorical in some uncountable power then it is possible to make a harmless extension of T which satisfies conditions i) and ii). Thus, we can deduce Morley's theorem.

3.2 Definition. i) The formula $\phi(x;\overline{a})$ is *strongly minimal* in T if $\phi(\mathcal{M};\overline{a})$ is infinite and if for every formula $\psi(x;\overline{c})$ (with $\overline{c}$ from the monster model $\mathcal{M}$) either $\phi(\mathcal{M};\overline{a}) \cap \psi(\mathcal{M};\overline{c})$ or $\phi(\mathcal{M};\overline{a}) \cap \neg\psi(\mathcal{M};\overline{c})$ is finite.

ii) The theory T is *strongly minimal* if the formula $x = x$ is strongly minimal in T.

iii) The formula $\phi(x;\overline{a})$ is minimal in M if for any formula $\psi(x;\overline{c})$ with parameters from M either $\phi(M;\overline{a}) \cap \psi(M;\overline{c})$ or $\phi(M;\overline{a}) \cap \neg\psi(Mx;\overline{c})$ is finite.

We call a set A strongly minimal when $A = \phi(M)$ for some strongly minimal formula ϕ.

The first important property of a strongly minimal formula is that it is possible to assign a dimension to the solution set of the formula in any model. We do this by introducing a notion of dependence which is a direct extension of the notion of an element of a field depending on several others.

3.3 Definition. i) The element a is in the *algebraic closure* of the set B if there is a formula $\phi(x;\overline{b})$ with $\overline{b} \in B$ such that $\phi(\mathcal{M},\overline{b})$ is finite and $\models \phi(a,\overline{b})$.

ii) We denote the set of all elements in the algebraic closure of B by $\mathrm{cl}(B)$. If $a \in \mathrm{cl}(B)$, we say $t(a;B)$ is an *algebraic type*; all realizations of such a type are in $\mathrm{cl}(B)$.

iii) A set E is *algebraically independent* if for every $e \in E$, $e \notin \mathrm{cl}(E - e)$.

iv) A set E is a *basis* for a set F if $E \subseteq F$, E is independent, and $\mathrm{cl}(E) \supseteq F$.

We may also describe this situation by saying a depends (algebraically) on B. Note that this dependence notion has finite character. That is, if a depends on B then a depends on a finite subset of B. In Lemma 3.6 we show that within a strongly minimal set this notion of dependence satisfies the axioms for vector space dependence. Chapter II is devoted to a full discussion of the abstract notion of dependence.

3.4 Exercise. Show that if T is $\aleph_0$-categorical and A is finite then $\mathrm{cl}(A)$ is finite.

3.5 Exercise. Show that if $t(a;B)$ is algebraic then $t(a;B)$ is principal. Show the converse fails.

3.6 Lemma. *Suppose $\phi(\mathcal{M})$ is strongly minimal and $(a \cup B \cup C) \subseteq \phi(\mathcal{M})$.*

i) $a \in \mathrm{cl}(a)$.

ii) *If $a \in \mathrm{cl}(B \cup C)$ and $B \subseteq \mathrm{cl}(C)$ then $a \in \mathrm{cl}(C)$.*

iii) *If $a \in \mathrm{cl}(C \cup b)$ and $a \notin \mathrm{cl}(C)$ then $b \in \mathrm{cl}(C \cup a)$.*

iv) *If $a \in \mathrm{cl}(B)$ and $B \subseteq C$ then $a \in \mathrm{cl}(C)$.*

All of these properties except iii) are easily verified by just chasing the definitions. A direct proof of iii) is in [Baldwin & Lachlan 1971]. We will see in Chapter III how to deduce iii) from the general properties of forking. It is now routine to deduce as in linear algebra the invariance of dimension for a strongly minimal set.

3.7 Lemma. If $\phi(M)$ is strongly minimal then all bases X for $\phi(M)$ have the same cardinality.

The second important property of a strongly minimal set is that if $\phi(x;\overline{a})$ is strongly minimal, then for any set A with $\overline{a} \in A$ there is a unique non-algebraic type $p \in S(A)$ with $\phi(x;\overline{a}) \in p$. The next two exercises highlight the ideas needed to prove Theorem 3.10.

3.8 Exercise. If X and Y are independent subsets of a strongly minimal set which have the same cardinality, then $t(X;\emptyset) = t(Y;\emptyset)$. (Hint: Work by induction on $|X|$, showing first that within a strongly minimal set there is a unique non-algebraic type.)

3.9 Exercise. If X and Y realize the same type over the empty set then there is an isomorphism between $\mathrm{cl}(X)$ and $\mathrm{cl}(Y)$. (Hint: Apply Zorn's lemma.)

3.10 Theorem. *If T is a strongly minimal theory then T is categorical in all uncountable powers.*

Proof. Let M and N be models of T and X, Y bases of M and N, respectively. By the first of the above exercises, there is an isomorphism between X and Y; by the second, it extends to an isomorphism between M and N.

We have established the first part of our planned proof. The next step is to extend Theorem 3.10 from strongly minimal theories to those where the universe of each model is determined by the solution set of a strongly minimal formula. For example, the universe might be the algebraic closure of a strongly minimal set. In our outline of the proof we referred to harmless extensions of a theory. The next definition makes precise the class of extensions to which we refer. The exercises show that these extensions are indeed harmless in our context.

3.11 Definition. i) For any theory T, any $M \models T$ and any sequence $\overline{a} \in M$, $T' = \mathrm{Th}(M, \overline{a})$ is an *inessential extension* of T. If $\overline{a}$ is finite, T' is a *finite inessential extension* of T.

ii) If the finite inessential extension T' of T is axiomatized by adding a single sentence to the axioms of T, then T' is a *principal extension* of T.

iii) If some principal extension T' of T has a strongly minimal formula $\phi \in L(T')$ such that if $M \models T'$, $M = \mathrm{cl}(\phi(M))$, the theory T is *almost strongly minimal.*

3.12 Exercise. Show that for any κ with $\kappa^{<\mu} = \kappa$ and any countable theory T, T has a μ-homogeneous model of cardinality κ.

The previous exercise is needed for one implication in the following exercise.

3.13 Exercise. Let T' be a principal extension of the countable theory T. Show that for any uncountable κ, T is κ-categorical if and only if T' is κ-categorical.

3.14 Exercise. Show that if T is almost strongly minimal then T is categorical in every uncountable power.

3.15 Exercise. Show the theory of the Abelian group Z_4^ω is categorical in every uncountable power but not almost strongly minimal. In particular, if $M \models T$ then M is not the closure of the elements of order 2. (Hint: This requires the results on quantifier eliminability of Abelian groups discussed in Section 4.)

We need to extend the notion of a model being generated by a set. The first two notions in the following definition are plausible candidates for the 'model generated by A'. In general, however, if 'generated model' is defined by either of these conditions there is no guarantee that the model generated by A will, in fact, exist. The third notion provides a condition on T which implies that prime models in the sense of Definition 3.16 i) exist. We will see that if T is κ-categorical for some uncountable κ and if A is an infinite definable set, both kinds of model generated by A do exist. Chapter IX contains a detailed study of possible formalizations of the notion of generation.

3.16 Definition. i) The model M is *prime* over $A \subseteq M$ if every elementary map of A into a model N of T extends to an elementary embedding of M into N. If A is empty we simply say M is a prime model of T.

ii) The model M is *minimal* over A if $A \subseteq M$ and there is no proper elementary submodel N of M with $A \subseteq N$.

iii) The theory T is *κ-stable* if for every $A \subseteq \mathcal{M}$ with $|A| \leq \kappa$, $|S(A)| \leq \kappa$.

iv) The theory T is *stable* if it is κ-stable for some κ.

Morley's definition of an ω-stable theory began stability theory. We will see an important source of ω-stable theories in Theorem 3.18 which shows that a countable theory which is categorical in some uncountable power is ω-stable.

The following result demonstrates the powerful effect of combining the existence of strongly minimal formulas and minimal prime models. The lemma generalizes the fact that if $M = \mathrm{cl}(X)$ then M is both prime and minimal over X.

3.17 Lemma. *If some principal extension T' of T has a strongly minimal formula ψ such that each model M of T' is minimal and prime over $\psi(M)$, then T is categorical in all uncountable powers.*

Proof. Let M and N be models of T with cardinality $\kappa > \aleph_0$. Let $\psi(x;\overline{y})$ be an $L(T)$-formula such that $\psi(x;\overline{c})$ is strongly minimal in T' for any $\overline{c}$ which realizes the principal type p. Then we can choose $\overline{c}_1 \in M$ and $\overline{c}_2 \in N$ to expand M and N respectively to models of T'. Then, if X and Y are bases for $\psi(M;\overline{c}_1)$ and $\psi(N;\overline{c}_2)$ respectively, $|X| = |Y| = \kappa$. Since both $\overline{c}_1$ and $\overline{c}_2$ realize p, this implies there is an isomorphism between X and Y which extends, as in the proof of Theorem 3.10, to an isomorphism between

$\psi(M;\bar{c}_1)$ and $\psi(N;\bar{c}_2)$. But since M is minimal and prime over $\psi(M;\bar{c}_1)$, this isomorphism extends further to an isomorphism between M and N.

To complete our proof of Morley's theorem we need to show that if T is categorical in some uncountable power then every model of T is prime and minimal over the interpretation in it of a strongly minimal formula. For 'prime', we quote the following theorem which is proved in Chapter IX. The first part is fairly complicated; the second depends only on the Vaught [Vaught 1961]characterization of prime models.

3.18 Theorem. i) *If T is ω-stable then for every A there is a prime model over A.*
ii) *If M is prime over $\emptyset$ and $\overline{m} \in M$, then* $\mathrm{Th}(M, \overline{m})$ *is a principal extension of T.*

To apply Lemma 3.17 we will show, in fact, that any uncountable model of T, or of any inessential extension of T, is minimal over any infinite definable subset. Moreover, we will find a principal extension of T which has a strongly minimal formula. The next definition and following two theorems accomplish the first aim.

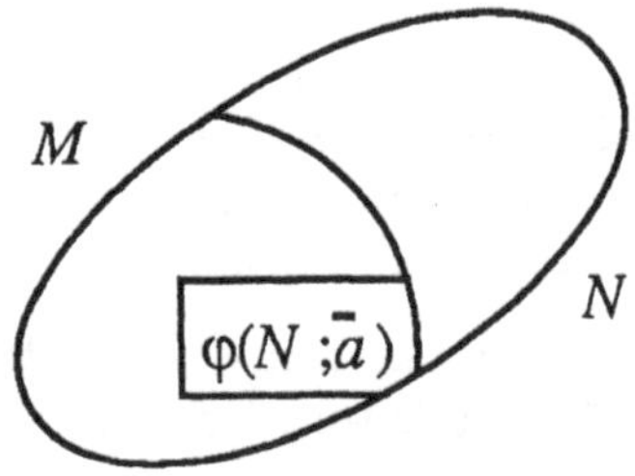

Fig. 1. A two cardinal model

3.19 Definition. (Fig. 1). The formula $\phi(x;\bar{a})$ is a *two-cardinal formula* for T if there exist distinct models $M \subset N$ with $\bar{a} \in M$, such that $\phi(M;\bar{a})$ is infinite and $\phi(M;\bar{a}) = \phi(N;\bar{a})$. Sometimes (M, N) is called a Vaughtian pair.

The next theorem yields that every model of a theory which is categorical in some power is prime and minimal over any definable subset.

3.20 Theorem. *Suppose the countable theory T is categorical in some power $\kappa > \aleph_0$.*

i) *No finite inessential extension of T has a two-cardinal formula.*
ii) *If $M \models T$ and M is prime over $\phi(M;\bar{a})$ then M is minimal over $\phi(M;\bar{a})$.*

Since we show below that a theory which is categorical in some uncountable power is ω-stable and since, in general, ω-stability implies stability Theorem 3.20 i) is easily deduced from the following important result (the two cardinal theorem)proved in Section IX.5.

3.21 Theorem. *If T is a countable stable theory and $\phi(x;\overline{a})$ is a two-cardinal formula in T, then for every $\lambda > \kappa \geq |T|$ there is a model, M, of T with cardinality λ and with $|\phi(M;\overline{a})| = \kappa$.*

Now, to derive Theorem 3.20 i) from Theorem 3.21, suppose that T is κ-categorical. If the finite inessential extension T' of T has a two-cardinal formula, Theorem 3.21 implies there is a model M of T' with cardinality κ but with $|\phi(M;\overline{a})| = \aleph_0$. Clearly there is another model N of T of power κ with $|\phi(N;\overline{b})| = \kappa$ for every choice of the parameters $\overline{b}$. Taking reducts of these two models to the original language contradicts the κ-categoricity of T.

Theorem 3.20 ii) follows immediately from Theorem 3.20 i) and the definition of a two-cardinal formula.

We have thus established condition ii) in our strategy for proving Morley's theorem proclaimed after Definition 3.1. In order to find the strongly minimal formula, required by condition i) of the strategy, we must first show that a theory which is categorical in some uncountable power is ω-stable. This requires the introduction of another powerful technique, the method of Ehrenfeucht-Mostowski models. Surprisingly, this method will play no further role in our development of the positive structure theory. As we indicate in Section 5, it is the principal tool in the construction of many non-isomorphic models. There are actually two ingredients to Ehrenfeucht-Mostowski models, Skolem hulls and indiscernibles. For any first order language L we define the Skolem language associated with L as follows.

3.22 Definition. i) Let $L^0 = L$ and for each $i < \omega$, let L^{i+1} denote the language obtained by adjoining to L^i the $\lg(\overline{y})$-ary function symbol $F_{\exists x\phi}$ for each L^i-formula $(\exists x)\phi(x,\overline{y})$ in which $\overline{y}$ does not occur bound. Let L^{sk}, the *Skolem language* associated with L, denote $\bigcup_{i<\omega} L^i$.

ii) For any L-theory T, the *Skolem theory*, T^{sk}, associated with T is obtained by adding to T the Skolem axioms:

$$(\forall\overline{y})(\exists x)\phi(x,\overline{y}) \rightarrow \phi(F_{\exists x\phi}(\overline{y}),\overline{y})$$

for each L^{sk}-formula $(\exists x)\phi$ in which $\overline{y}$ does not occur bound.

iii) If T is an L-theory and $M \models T$, then a *Skolem expansion*, M^{sk}, of M is any model obtained by interpreting each of the new functions so as to satisfy the Skolem axioms.

3.23 Exercise. For any T, T^{sk} admits elimination of quantifiers.

The next result follows immediately from the previous exercise. Both results are easily proved by induction on the complexity of formulas. See [Chang & Keisler 1973].

3.24 Lemma. *Let M be an L-structure, $X \subseteq M$ and let M_X^{sk} denotes the closure of X under the Skolem functions as interpreted in M^{sk}, then $M_X^{sk} \prec M^{sk}$. We call M_X^{sk} the Skolem Hull of X.*

3.25 Definition. Let $E \subseteq M$ be linearly ordered by $<$, which need not be a definable relation in M. The ordered sequence $(E, <)$ is a *sequence of order indiscernibles* if for any $\phi(\overline{x})$ and any pair $\overline{e}, \overline{e}'$ of finite sequences from E, which are both in increasing order, $\models \phi(\overline{e}) \leftrightarrow \phi(\overline{e}')$.

The following result is proved using the compactness theorem and Ramsey's theorem.

3.26 Theorem (The Ehrenfeucht-Mostowski Theorem). *If T is a theory with infinite models then for every infinite cardinal κ, T has a model which is the Skolem hull of a sequence of order indiscernibles with cardinality κ.*

3.27 Exercise. Prove Theorem 3.26. (Good accounts of this theorem appear in [Chang & Keisler 1973] and [Morley 1967].)

3.28 Exercise. Deduce from (the proof of) Theorem 3.26 that if T has infinite models then for every infinite cardinal κ, T has a model which contains a set of indiscernibles of order type κ.

We deduce from Theorem 3.26, using the last exercise and the properties of Skolem models, the following fundamental result.

3.29 Theorem. *For any theory T and any cardinal κ there is a model M of T with $|M| = \kappa$ such that for any subset A of M with $|A| \leq |T|$, at most $|T|$ types over A are realized in M.*

Proof. Choose by the proof of the last exercise a model, M, of T which is the Skolem hull of a well ordered set, Y, of length κ. Let A be a subset of M with $|A| \leq |T|$. Now each element m of M is determined by a Skolem term $t_m(\overline{x})$ and a finite sequence $\overline{y}_m$ from Y. For any $b \in M$, $t(b; A)$ is determined by the term t_b and the position of $\overline{y}_b$ with respect to the $|A|$ tuples $\overline{y}_a$ for $a \in A$. But there are only $|A|$ ways to interpolate a finite set in a well-ordered set of cardinality κ. Thus, there are at most $(|T| \times |A|) \leq |T|$ types over A realized in M.

Note that Theorem 3.29 does not depend on the countability of T.

3.30 Corollary. *If a countable theory is categorical in some uncountable power then it is ω-stable.*

Proof. Let T be categorical in $\kappa > \aleph_0$. Suppose T is not ω-stable; then there is a countable set A with $|S(A)| > \aleph_0$. By compactness there is a model N of T with $|N| = \kappa$, $A \subset N$ and with uncountably many types over A realized in N. But N must also be isomorphic to the model constructed in Theorem 3.29. This contradiction yields the result.

Many of the basic notions expounded in this book were developed by Shelah in his work on the generalized Łos conjecture: if a theory T is T^+-categorical then it is categorical in all powers. The following exercise shows that such a theory must be $|T|$-stable; in fact, it must be stable in all $\lambda \geq |T|$.

3.31 Exercise. Show that if the theory T (in a possibly uncountable language) is categorical in some $\kappa > |T|$ then T is $|T|$-stable.

We want now to combine 3.30 and 3.20 to show that if T is categorical in some uncountable power we can find a principal extension of T with a strongly minimal formula. We first draw a more precise syntactic consequence of the hypothesis that no finite inessential extension of T has a two-cardinal formula.

3.32 Lemma. *If no finite inessential extension of T has a two-cardinal formula then for every formula $\phi(x;\overline{y})$ there is an integer n such that for every $\overline{c}$, if $|\phi(\mathcal{M};\overline{c})| > n$ then $\phi(\mathcal{M};\overline{c})$ is infinite.*

Proof. Expand L by adding a new predicate symbol U and new constant symbols $\overline{a}$. Let Γ be a set of sentences in the new language which assert:

i) U defines a proper L-elementary submodel of the universe.
ii) Every solution of $\phi(x;\overline{a})$ is in U.
iii) There are infinitely many solutions of $\phi(x;\overline{a})$ in U.

Since no finite inessential extension of T has a two-cardinal model, Γ is inconsistent. But, if for arbitrarily large $m < \omega$, there is a $\overline{c}_m \in \mathcal{M}$ such that $|\phi(\mathcal{M};\overline{c}_m)| = m$, Γ is consistent. Thus, the required n must exist.

To guarantee condition i), denote the relativization of an L-sentence ψ to the predicate U by ψ^U. Place in Γ the sentence $\psi^U \to \psi$ for each L-sentence ψ.

The following observation is a straightforward consequence of compactness.

3.33 Lemma. *The formula $\phi(x;\overline{a})$ is strongly minimal if and only if for every formula $\psi(x;\overline{y})$ there is a positive integer n such that the two formulas*

$$(\exists^{\geq n}x)[\phi(x;\overline{a}) \wedge \psi(x;\overline{y})]$$

and

$$(\exists^{\geq n}x)[\phi(x;\overline{a}) \wedge \neg\psi(x;\overline{y})]$$

are not both consistent.

One of the basic applications of ω-stability is the construction of minimal sets. The proof is one of a number of variants on the proof that a closed uncountable set of reals contains a perfect subset. A somewhat more detailed explanation of this part of the next argument appears with Theorem III.1.7. We concentrate here on the fact that in the absence of a two cardinal model we can require the set to be strongly minimal.

3.34 Theorem. *If T is an ω-stable countable theory with no two-cardinal formulas then there is a strongly minimal formula with parameters in the prime model of T.*

Proof. Let M be a prime model of T. We can easily choose a formula $\phi(x;\overline{a})$ with parameters from M which is minimal in M (i.e which cannot be split into two infinite parts by a formula with parameters from M. (If not, we would construct an infinite binary tree and contradict ω-stability.) If this formula is not strongly minimal then, by Lemma 3.33 we can find a formula $\psi(x;\overline{y})$ and for each $n \in \omega$ a $\overline{d}_n$ such that both

$$(\exists^{\geq n}x)[\phi(x;\overline{a}) \wedge \psi(x;\overline{d}_n)]$$

and

$$(\exists^{\geq n}x)[\phi(x;\overline{a}) \wedge \neg\psi(x;\overline{d}_n)]$$

hold in $\mathcal{M}$. Now, for arbitrarily large n, either $\phi(\overline{x};\overline{a}) \wedge \psi(\overline{x};\overline{d}_n)$ is infinite and $\phi(\overline{x};\overline{a}) \wedge \neg\psi(\overline{x};\overline{d}_n)$ is finite or vice versa. In the first case $\chi(\overline{x};\overline{y}) = \phi(\overline{x};\overline{y}) \wedge \neg\psi(\overline{x};\overline{y})$ contradicts Lemma 3.32 and the second case is similar.

Now, given a theory T which is categorical in some uncountable power, we can choose by Theorem 3.34 and Theorem 3.20 ii) a principal extension T' which satisfies the hypothesis of Theorem 3.17 and thus is categorical in all uncountable powers. As observed in the Exercises 3.12 and 3.13, this implies T is categorical in all uncountable powers and we have established Morley's theorem.

3.35 Theorem (Morley's Categoricity Theorem). *Let T be a countable first order theory. Then T is categorical in one uncountable power if and only if T is categorical in all uncountable powers.*

The most subtle point of this argument is the choice of the additional parameters for the strongly minimal formula. Let T be an $\aleph_1$-categorical theory. As we noticed in Lemmas 3.6 and 3.7 it is rather easy to see that if $\phi(x)$ is strongly minimal in T then the dependence relation defined by algebraic closure allows us to assign a dimension to the strongly minimal set. If T is not $\aleph_0$-categorical, it is easy to conclude that infinitely many distinct finite dimensions are possible. Thus when ϕ contains no parameters one concludes that T has $\aleph_0$ countable models. Thus, by Theorem 3.34, some finite inessential extension of every $\aleph_1$ but not $\aleph_0$-categorical theory has infinitely many countable models. To prove that T itself has infinitely many models is much more difficult. It may be that different instances of $\phi(x;\overline{y})$ give rise to different dimensions so we cannot assign a dimension to each model. We will establish in Chapter XIII that not only every $\aleph_1$-categorical but every superstable countable theory which is not $\aleph_0$-categorical has $\aleph_0$ countable models.

3.36 Historical Notes. The characterization of $\aleph_0$-categorical theories is due independently to Ryll-Nardzewski [Ryll-Nardzewski 1959], Svenonius [Svenonius 1959], and Engeler [Engeler 1959].

There are at least three fairly distinct proofs in print of Morley's theorem; the one given here is a variant on one of them. The first proof by Morley [Morley 1965] makes substantial use of the rank of types, for example, in the construction of sequences of indiscernibles. The proof in [Chang & Keisler 1973] avoids the use of rank but depends, as does Morley's, on the isomorphism of elementarily equivalent saturated models with the same power. The proof here is closely based on that in [Baldwin & Lachlan 1971] but substitutes the two-cardinal theorem for stable theories (Theorem 3.21) for an *ad hoc* argument in that paper. The one feature which all the proofs have in common is Corollary 3.30, which was first proved by Morley, drawing on the fundamental paper [Ehrenfeucht & Mostowski 1956]. The notion of a strongly minimal set was introduced by Marsh [Marsh 1966]. The same concept was known to Morley as a rank 1 degree 1 set.

The fact that an ω-stable theory with no two-cardinal model has a strongly minimal formula is closely related to the *finite cover property.* This intriguing notion is explored in [Keisler 1967], Chapter II.4 of [Shelah 1978], [Poizat 1983], [Poizat 1984], [Hrushovski 1986], and [Baldwin & Kueker 1980].

4. *Introduction to the Model Theory of Modules*

One of the major examples used to illustrate the concepts introduced in this book is the theory of modules. In this section we explain the basic facts about the theory of modules which are needed below. Many of the proofs are sketchy; for details see [Ziegler 1984] or [Prest 198?]. By the 'theory of modules' we actually mean the theory of left R-modules for any specified ring R. The following definition makes this precise.

4.1 Definition. The language, L_R, for the theory of R-modules consists of a binary function symbol $+$, a constant symbol 0, and for each element of the fixed ring R a unary function symbol f_r.

The axioms assert that a structure $(M;+,0,\{f_r : r \in R\})$ is a unitary R-module. That is, $(M;+,0)$ is an Abelian group, each f_r is an endomorphism of this group, and f_1 is the identity map. We often abbreviate $f_r(y)$ as ry and $(\exists w)rw = y$ by $r|y$. We denote the theory of all R-modules by T_R.

4.2 Examples. The two most important examples are vector spaces over a division ring and modules over the ring of integers Z which are, of course, just Abelian groups. For many of our uses the reader will not suffer by restricting himself to these cases.

4.3 Positive Primitive Formulas. A formula ϕ is a *positive primitive* (p.p.) formula if it is equivalent in T_R to an existential quantification of a conjunction of atomic formulas. In L_R, ϕ is positive primitive just if ϕ has (or is equivalent to a formula with) the form:

$$(\exists \overline{y}) \bigwedge_{i<p} \Big(\sum_{j<m} a_{ij} y_j + \sum_{k<n} b_{ik} x_k = 0 \Big)$$

for some $m, n, p < \omega$. Thus, a positive primitive formula asserts the solvability of a finite system of linear equations. If we write A for the matrix of coefficients (a_{ij}) and B for the matrix of coefficients (b_{ik}) and think of $\overline{y}$ and $\overline{x}$ as column vectors, we can rewrite this p.p. formula as an existentially quantified matrix equation:

$$(\exists \overline{y}) A\overline{y} + B\overline{x} = 0.$$

This allows one to simplify the expression even further when R is a principal ideal domain and, in particular, when $R = Z$ and we are speaking of the integers. We require the following fact found in most graduate algebra texts.

4.4 Theorem. *Suppose H is a square matrix over the principal ideal domain R. Then there are invertible matrices X and Y such that XHY is a diagonal matrix.*

Using this fact it is straightforward to deduce the following normal form for p.p. formulas over a principal ideal domain. Remember that the usual definition of a prime number makes sense in any principal ideal domain.

4.5 Corollary. *If R is a principal ideal domain then every p.p. formula $\phi(\overline{v})$ is equivalent in T_R to a conjunction of formulas of the form $p^k | t(\overline{v})$ or of the form $t(\overline{v}) = 0$ where $t(\overline{v})$ denotes an R-linear combination of variables from $\overline{v}$, p is a prime and k a natural number.*

The following exercises outline a proof from Theorem 4.4 of Corollary 4.5.

4.6 Exercise. Show using Theorem 4.4 that the expression

$$(\exists \overline{y}) A\overline{y} + B\overline{x} = 0$$

can be transformed to

$$(\exists \overline{y}) A'\overline{y} + D\overline{x} = 0 \qquad (*)$$

where D is a diagonal matrix.

4.7 Exercise. Note that $(*)$ is equivalent to a conjunction of formulas of the form $r | t(\overline{v})$ or of the form $t(\overline{v}) = 0$. Factor the elements r to obtain the form in Corollary 4.5.

Note that the collection of p.p. formulas is closed under existential quantification and conjunction. Moreover, the standard preservation theorems imply that the set of solutions of a p.p. formula is preserved under extension, product and homomorphism of models.

Even more importantly any p.p. formula $\phi(\overline{x})$ with no parameters defines a subgroup of M^n (where $n = \lg(\overline{x})$). A p.p. formula $\phi(\overline{x};\overline{a})$ defines either $\emptyset$ or a coset of the subgroup defined by $\phi(\overline{x};\bar{0})$ where $\bar{0}$ defines a sequence of 0's of appropriate length. To see that $\phi(\overline{x};\bar{0})$ defines a subgroup, note that

$M \models \phi(\bar{0};\bar{0})$ and if $M \models \phi(\bar{a};\bar{0}) \wedge \phi(\bar{b};\bar{0})$ then $M \models \phi(\bar{a}+\bar{b};\bar{0})$. (We write $\bar{a}+\bar{b}$ to denote the coordinatewise sum of $\bar{a}$ and $\bar{b}$.) Summing up, we have:

4.8 Lemma. *If $\phi(\bar{x};\bar{y})$ is a p.p. formula then for any $\bar{a}$ such that $\phi(\bar{x};\bar{a})$ is consistent there is a $\bar{b}$ so that $\phi(M;\bar{a}) = \bar{b} + \phi(M;\bar{0})$.*

4.9 Exercise. Verify Lemma 4.8.

Note that these results imply that any p.p. formula $\phi(\bar{a};\bar{y})$ is either equivalent to a p.p. formula of the form $\phi(\bar{0};\bar{b}-\bar{y})$ or is inconsistent.

4.10 Exercise. Show that if R is commutative and $\phi(\bar{x})$ is a p.p. formula without parameters then $\phi(M)$ is a submodule of M^n.

The p.p. formulas determine the lattice of p.p. definable subgroups where the meet of two such subgroups is their intersection and the join is defined by $(\phi+\psi)(\bar{v}) = (\exists\bar{w})\phi(\bar{w}) \wedge \psi(\bar{v}-\bar{w})$.

The following exercise is crucial to our development.

4.11 Exercise. For any p.p. formula ϕ and any $\bar{a}$, $\bar{b}$, either $\phi(\bar{x};\bar{a})$ and $\phi(\bar{x};\bar{b})$ are equivalent or they are contradictory.

Here is the main theorem of this section.

4.12 Theorem. *Fix an R-module M. For every formula $\phi(\bar{x})$ there is a formula $\phi^*(\bar{x})$ which is a Boolean combination of p.p. formulas such that $M \models \phi(\bar{x}) \leftrightarrow \phi^*(\bar{x})$.*

The proof of this result depends on one algebraic result and one easy combinatorial lemma.

4.13 Lemma ([Neumann 1954]). *Suppose H_i for $i < l$ with $l < \omega$ are Abelian groups. Fix $k \leq l$ such that for each $i < k$, $[H_0 : H_0 \cap H_i]$ is finite while for $k \leq i < l$, $[H_0 : H_0 \cap H_i]$ is infinite. If $H_0 + a_0 \subseteq \bigcup_{i<l} H_i + a_i$ for some sequence $\langle a_i : i < l\rangle$, then $H_0 + a_0 \subseteq \bigcup_{i<k} H_i + a_i$.*

4.14 Exercise (A combinatorial lemma). Let A and $A_0, \ldots, A_{n-1}$ be finite sets. Show $A \subseteq \bigcup_{i<n} A_i$ iff

$$\sum_{\Delta \subseteq n} (-1)^{|\Delta|} |A \cap \bigcap_{i\in\Delta} A_i| = 0.$$

Proof of Theorem 4.12. We will proceed by induction on quantifiers. Thus, consider the formula $(\forall y)\psi(\bar{x};y)$ and suppose by induction that ψ is a Boolean combination of p.p. formulas. Since universal quantification distributes over conjunction and p.p. formulas are closed under conjunction we may assume ψ has the form:

$$\phi_0(\bar{x};y) \rightarrow \bigvee_{1\leq i<n} \phi_i(\bar{x};y)$$

where each ϕ_i is p.p. For each $i < n$, let H_i denote the subgroup $\phi_i(\bar{0};M)$. Then for any $\bar{b}$, $\psi(\bar{b};y)$ is equivalent (by the remark after Lemma 4.13) to

$$H_0 + a_0 \subseteq \bigcup_{i<n} H_i + a_i$$

for appropriate choice of the a_i. Note that some of the $H_i + a_i$ may be empty. If we reorder the a_i so that $[H_0 : H_0 \cap H_i] < \omega$ iff $i < k$ then apply Lemma 4.8, we see that $\psi(\overline{a}; y)$ is equivalent to:

$$H_0 + a_0 \subseteq \bigcup_{i<k} H_i + a_i. \tag{$*$}$$

Now the key fact is that the truth of $(*)$ can be expressed by a formula which does not mention the a_i. To see this let H denote $\bigcap\{H_i : i < k\}$ and let for $\Delta \subseteq k - \{0\}$

$$N_\Delta = [H_0 \cap \bigcap_{i\in\Delta} H_i : H].$$

Let $m = [H_0 : H]$. Then $H_0 + a_0$ is the union of m cosets of H. For each i, if $(H_i + a_i) \cap (H_0 + a_0) = B_i$ is not empty then B_i contains $n_i = [H_0 \cap H_i : H]$ cosets of H. Thus $H_0 + a_0 \subseteq \bigcup_{i<k} H_i + a_i$ if the number of distinct cosets of H which occur in $\bigcup\{B_i : i < k\}$ is greater than m. The number of these cosets is not $\Sigma\, n_i$ (since this counts various cosets more than once) but must be computed using the N_Δ and the combinatorial lemma. That is, if A_i is the collection of cosets of H in B_i, then $(*)$ is equivalent to $A_0 \subseteq A_1 \cup A_2 \cup \ldots \cup A_{k-1}$. By the combinatorial lemma, $(*)$ is equivalent to

$$\sum_{\Delta\subseteq\Gamma(\overline{a})} (-1)^{|\Delta|} N_\Delta = 0$$

where $\Gamma(\overline{a}) = \{i \in k - \{0\} : B_i \neq \emptyset\}$.

Now to remove the apparent dependence of $\Gamma(\overline{a})$ on $\overline{a}$, note that $\Gamma(\overline{a})$ is a subset of $k - \{0\}$ and the statement

$$\sum_{\Delta\subseteq\Gamma(\overline{a})} (-1)^{|\Delta|} N_\Delta = 0$$

depends not on $\overline{a}$ but only on this subset. Let Λ be the collection of subsets Γ of $2^{k-\{0\}}$ such that: $\sum_{\Delta\subseteq\Gamma} (-1)^{|\Delta|} N_\Delta = 0$. Let

$$\phi_\Gamma = (\exists y)\phi_0(\overline{x}; y) \wedge \bigwedge_{\substack{i\in\Delta \\ \Delta\in\Gamma}} \phi_i(\overline{x}; y).$$

Then $(\forall y)\psi(\overline{x};y)$ is equivalent to

$$\bigwedge_{\Gamma\in\Lambda} \phi_\Gamma \wedge \bigwedge_{\Gamma\notin\Lambda} \neg\phi_\Gamma$$

which is a Boolean combination of p.p. formulas.

Note that the definition of the Boolean combination of p.p. formulas which is equivalent to ψ depends only on the N_Δ and these numbers are the same for elementarily equivalent models. Conversely, if for each Δ, $N_\Delta(M) = N_\Delta(N)$ then M and N are elementarily equivalent. With some further notation we rephrase this result in a more elegant form.

4.15 Notation. If ϕ and ψ are p.p. formulas without parameters and N is a module let $n(\phi/\psi; N) = [\phi(N) : \phi(N) \cap \psi(N)]$. Note that $n(\phi/\psi; N)$ is a natural number or ∞.

For every ϕ, ψ and n, there is a sentence $\chi_{\phi,\psi,n}$ such that for any module N, $N \models \chi_{\phi,\psi,n}$ if and only if $n(\phi/\psi; N) \geq n$. Following [Prest 198?], we call any sentence which expresses these invariants of a module an *invariants sentence.*

4.16 Corollary. *Suppose N and M are R-modules and for each pair of p.p. formulas without parameters, ϕ and ψ, $n(\phi/\psi; M) = n(\phi/\psi; N)$, then $M \equiv N$.*

4.17 Exercise. Write out the general form of an invariants sentence. Notice that it is in AE form. Conclude that if M and N are modules and $M \equiv_{AE} N$ then $M \equiv N$.

A standard compactness argument solves the following exercise ([Prest 198?]).

4.18 Exercise. *Show that for any ring R and any formula $\chi(\overline{x})$ there is a p.p. formula $\theta(\overline{x})$ and an invariants sentence σ such that*

$$T_R \vdash (\forall \overline{x})[\chi(\overline{x}) \leftrightarrow \sigma \wedge \theta(\overline{x})].$$

4.19 Pure Submodules. The submodule M of N is *pure* in N if every p.p. formula with parameters from M which is satisfiable in N is satisfiable in M. Of course, this is a weakening of the notion of elementary submodel but in the light of Theorem 4.13 it is not a very great weakening. In fact, it is easy to see:

4.20 Lemma. *If $M \subseteq N$ and $M \equiv N$ then $M \prec N$ iff M is pure in N.*

4.21 Exercise. Show that if M is a direct factor of N then M is pure in N.

Derive the next exercise from the last one.

4.22 Exercise. Show that every subspace of a vector space is a pure submodule.

4.23 Purity in Abelian Groups. For Abelian groups or, more generally, modules over principal ideal domains, the notion of purity can be simplified: M is pure in N just if

$$N \models (\exists y) ry = m \text{ implies } M \models (\exists y) ry = m$$

for any integer r and any $m \in M$.

4.24 Exercise. Derive the remark in 4.23 from Corollary 4.5.

We denote by $M^{<\kappa}$ the direct sum of κ copies of M and by M^{κ} the direct product of κ copies of M.

4.25 Lemma. *For p.p. formulas without parameters, ϕ and ψ, and any module N, if $\kappa < \omega$ and $n(\phi/\psi; N) < \omega$ then $n(\phi/\psi; N^{\kappa}) = (n(\phi/\psi; N)^{\kappa})$. If $\kappa \geq \omega$, $n(\phi/\psi; N^{\kappa}) = \infty$ or $n(\phi/\psi; N^{\kappa}) = 0$. The analogous results hold for direct sums.*

Proof. For any p.p. formula ϕ, $\phi(N^\kappa) = \phi(N)^\kappa$.

4.26 Corollary. *For any module M and any κ, $M^{<\kappa} \prec M^\kappa$.*

4.27 Historical Notes. The logical analysis of the theory of modules begins with Szmielew's quantifier elimination theorem for Abelian groups ([Szmielew 1955]). The work of Eklof, Fisher, and Sabbagh ([Eklof & Fisher 1972], [Eklof & Sabbagh 1970/71], [Baur 1976]) took a more model theoretic turn. The p.p.-elimination theorem was proved at about the same time but independently by Baur, Garavaglia, and Monk; Garavaglia ([Garavaglia 1979], [Garavaglia 1980]) realized its significance for stability theory. The connections between stability theory and the theory of modules were first pointed out by Shelah ([Shelah 1975]). The exposition here generally follows [Ziegler 1984]. We also rely on [Prest 198?].

5. *Non-Structure Theory*

The bulk of this book is concerned with the development of structure theorems for stable theories. In this section we place that development in context by reciting some of the non-structure theorems and briefly outlining the method of proof. These proofs fall into two parts. The first shows if a theory T is not well behaved then some kind of ordering is definable within the monster model. Second there is a combinatorial argument showing that there are 2^λ of these 'orderings' with power λ for each uncountable λ. In fact, the combinatorial conclusion must be somewhat stronger. The structures constructed must be sufficiently unlike one another that even the Skolem hull of two of the nonisomorphic orderings cannot be isomorphic.

The first result of this kind was Shelah's proof for unstable theories.

5.1 Theorem. *An unstable countable first order theory has 2^λ nonisomorphic models in every uncountable power, λ.*

In this theorem, for example, the required orderings are indeed linear orderings and the first part of the proof depends on showing that every unstable theory contains a formula which linearly orders an infinite set of n-tuples.

A later and stronger result is:

5.2 Theorem. *An unsuperstable countable theory has 2^λ models in every uncountable power.*

This result is proved in Chapter VIII of [Shelah 1978]. The full proof requires a considerable combinatorial arsenal and breaks into cases depending on properties of the cardinal λ. We give a simpler proof for regular λ in Section IX.6 by using more of the forking machinery.

A similar difference occurs between the proof in [Shelah 1982] and the one in [Harrington & Makkai 1985] for showing a theory with the dimensional order property has 2^λ models in every uncountable power. Shelah's proof continues the theme mentioned above of showing an ordering is definable, this time using the language $L_{\omega_1,\omega}$, and then closing under Skolem functions. The more ad hoc argument given here has the advantage that the distinct models can be distinguished in the original language rather than being imposed from without.

5.3 Historical Notes. In addition to Shelah see [Hodges 1983], [Hodges 198?] and [Pillay 1981] for accounts of this technique. The techniques described in this section have been elaborated in several directions. There are two major lines in the study of first order theories. One is the refinement of the sufficient condition to apply the technique. Thus, unstability [Shelah 1971a] is replaced by unsuperstability [Shelah 1975b] and then by the omitting types order property [Shelah 198?]. Alternatively, the conclusion is strengthened to find models which are not only not isomorphic but are mutually non-embeddable. These arguments are of an even more set-theoretic nature [Shelah 1982b], [Shelah 198?d], [Shelah 1983]. In particular, independence results arise in this context [Shelah 1982b].

A further strengthening arises if the non-isomorphic models are required to be elementarily equivalent in various infinitary languages (see for example [Grossberg & Shelah]). Still another generalization considers infinitary conditions in the defining of the trees; here there is considerable work by both Shelah and Grossberg.

Part A

Independence

As we remarked in the introduction, to arrive at our structure theory we must develop the notions of independence and generation separately. Part A is devoted to the first of those tasks, Part B to the second. We begin by giving an axiomatic description of an independence or freeness relation. This description summarizes the properties of the nonforking relation just as the axioms of Whitney and van der Waerden summarize the properties of vector space independence. We adopt this axiomatic formulation for several reasons. First, it clarifies the principles applied in the various constructions and proofs later in the book. Second, it provides a general framework for the discussion of several of the main concepts of the book, notably nonforking and orthogonality. By allowing us to separate the arguments used to verify these axioms from the applications of the axioms, we take a step towards the generalization of this structure theory to other families of classes of structures. If $\mathcal{K}$ is the family of classes of models of first order theories, we show in Section III.4 and Chapter VII, that *all* the axioms are satisfied on a class $K \in \mathcal{K}$ only if the relation is nonforking and K is the class of models of a *stable* first order theory. However, some of the results proved here depend on proper subsets of the axioms listed and many of the axioms hold under less restrictive conditions ([Shelah 1980a], [Shelah 1986]). More importantly, many of the arguments from Shelah's extension of the theory to the nonelementary case [Shelah 1983a] can also be fit into this rubric. A unified account of the first order and infinitary case will undoubtedly require changes in the axioms proposed here; we regard this as simply a first step.

Chapter II

The Abstract Notion of Independence

In the 1930's van der Waerden [van der Waerden 1949] and Whitney [Whitney 1935] abstracted the following properties of linear independence in vector spaces and algebraic independence in fields and used them to define the general notion of an independence relation or matroid. One of the most important properties of the vector space notion is the immediate definition of dependent as 'not independent'. With this in mind, van der Waerden's notion is most easily described in terms of a point a depending on a set X. For all a,

i) (Reflexivity) a depends on $\{a\}$.
ii) (Monotonicity) If a depends on X and $X \subseteq Y$ then a depends on Y.
iii) (Transitivity of dependence) If a depends on X and each $x \in X$ depends on Y then a depends on Y.
iv) (Exchange axiom) If a depends on $X \cup \{b\}$ but a does not depend on X then b depends on $X \cup \{a\}$.
v) (Finite character) If a depends on X then a depends on a finite subset of X.

In this book we describe a similar notion of independence which also specializes to linear independence and algebraic independence. It is not, strictly speaking, a generalization of the usual notion since it is stronger in some respects, weaker in others. However, it retains the most important consequence of the usual theory, the ability to assign a dimension (or, more precisely in the general case, a family of dimensions) to each member of certain classes of models.

We deal with a ternary relation denoted $A \downarrow_C B$, which intuitively means, 'A is free from B relative to C' or 'A is no more restrained by B than by C'. This notion differs from the standard one in three fundamental respects. First, with C fixed the relation $A \downarrow_C B$ differs from normal independence relations only by obeying slightly different axioms. Second, we have replaced the element a in the usual definition by the set A. Finally, we deal with $t(A; B)$, that is, all B-automorphic images of A rather than just A itself.

The extension to families of independence relations depending on the parameter C is necessary to create the ranked systems of invariants described in the introduction. The theory of two equivalence relations, E_1 and E_2, with E_1 refining E_2, will illustrate the point. Under our definition, if $\langle a_\alpha : \alpha < \kappa \rangle$ is a sequence of E_1-inequivalent elements, they will be independent over $\emptyset$. For each α, a set $B_\alpha = \langle b_{\alpha,\beta} : \beta < \kappa_\alpha \rangle$ of E_2-inequivalent elements which are all equivalent to a_α will be independent over $\{a_\alpha\}$ but not over $\emptyset$. (With a_α we can impose the restraint 'xE_1a_α'.) To determine the isomorphism type of a model we must know the dimension of each set B_α, and, even more, we must know for each α, β, the dimension of the set $C_{\alpha,\beta}$ of distinct elements which are both E_1 and E_2 equivalent to $b_{\alpha,\beta}$. This set will be independent over $\{b_{\alpha,\beta}\}$ but not over $\{a_\alpha\}$.

The consideration of the type of an element rather than the element itself generalizes the passage in algebraic geometry from a point to the ideal of polynomials which are zero on that point. This observation is not merely an analogy. Let $\mathcal{M}$ denote the 'universal domain' of algebraic geometry (an algebraically closed field of immense transcendence degree), and k a subfield of $\mathcal{M}$ which need not be a model of $\mathrm{Th}(\mathcal{M})$. Then polynomials over k are special kinds of formulas over k, ideals in $k[\overline{x}]$ are (incomplete) types, and the spectrum of $k[\overline{x}]$, the set of maximal ideals in $k[\overline{x}]$, is in 1-1 correspondence with the Stone space of k. For $\overline{a}$, a sequence from $\mathcal{M}$, this correspondence sends the ideal of polynomials over k which vanish at $\overline{a}$ to $t(\overline{a}; k)$.

The extension from considering singletons to considering finite sequences parallels the extension from the study of solutions to equations in one variable to the study of higher dimensional surfaces. If one studies only single elements then the distinction between an element and its type is easily ignored, as the type is often algebraic. This is the case, for example, in both vector spaces and fields. As soon as the study of n-types is broached, the type is no longer algebraic (except in trivial cases) and this distinction becomes more important. In our study we posit certain symmetry and finite character properties which allow us to extend the basic notions from finite sequences to arbitrary sets.

We employ two notations for the notion of independence. One emphasizes the role of types; the other is more algebraic. Thus, we write $t(\overline{a}; B)\mathcal{F}C$ for the assertion that $t(\overline{a}; B)$ is abstractly free over C. From any freeness relation $\mathcal{F}$ one can derive a relation $A\downarrow^{\mathcal{F}}_{C}B$ among sets. We will never write the superscript $\mathcal{F}$, but in this chapter we use $A \downarrow_C B$ to refer to the relation on sets associated with an abstract freeness relation. In Chapter III we introduce the most useful example of an independence notion: $t(\overline{a}; B)$ does not fork over C. Thereafter, we write $A \downarrow_C B$ for the relation on sets associated with the nonforking relation. We introduce the axioms in terms of the type notation to emphasize that we consider all elements conjugate to $\overline{a}$ over B. This notation is convenient for establishing the basic properties of the nonforking relation; the set notation is more compact and flexible in applications. We now quickly summarize the axioms using the set notation. In Section II.1 we state the axioms carefully in the type notation and exhibit in Section II.2 the exact relation between the two notations.

The axioms fall naturally into three groups. The first contains those which arise from thinking of $A \downarrow_C B$ as an 'ordinary' independence relation with C fixed. Thus, we both require monotonicity and finite character of dependence in the second variable: if $A \downarrow_C B$ and $B' \subseteq B$ then $A \downarrow_C B'$; if $A \not\downarrow_C B$ then for some finite $B' \subseteq B$, $A \not\downarrow_C B'$. Similarly, we require a natural translation of the exchange axiom, the symmetry axiom: $A \downarrow_C B$ implies $B \downarrow_C A$. We also need a weak version of the reflexivity axiom. In slightly more generality than actually holds, we can state the reflexivity as: if $\overline{a} \not\in C$ then $\overline{a} \not\downarrow_C \overline{a}$. However, we can not require the obvious translation of the transitivity axiom for dependence. It is easy to see that the full strength of the transitivity axiom yields a contradiction if independence is defined for n-types. For this, suppose $\{a, b, c, d\}$ are four elements with $\{a, b\} \downarrow_C \{c, d\}$. Then monotonicity and reflexivity yield first $\{a, b\} \not\downarrow_C \{b, c\}$ and then $\{b, c\} \not\downarrow_C \{c, d\}$. But then, from transitivity, we derive a contradiction.

A second group of axioms arises from the consideration of the relationship between dependence over different third coordinates. Here, we have upward monotonicity in the third coordinate: if $A \downarrow_C B$ and $C \subseteq C' \subseteq C \cup B$, then $A \downarrow_{C'} B$. We also invoke the following 'transitivity of independence': if $D \subseteq C \subseteq B$, $A \downarrow_C B$, and $A \downarrow_D C$ then $A \downarrow_D B$. We demand a generalization of local character in the third variable by defining a cardinal $\overline{\kappa}(T)$ such that, for any finite A and any B, there is a $B' \subseteq B$ such that $A \downarrow_{B'} B$ and $|B'| < \overline{\kappa}(T)$. Our dimension theory is best worked out when this local character becomes finite character, i.e. $\overline{\kappa}(T) = \omega$ (equivalently, T is superstable).

Finally, our axioms make explicit a property of vector space independence which does not follow from the usual axioms. In a vector space all maximal independent sets can be mapped to one another by automorphisms of the vector space. This will not be true in our situation but it will be true of certain kinds of independent sets: those which are infinite and based on a stationary type (cf. Definition 1.24). This property can be expressed in another way. In a vector space there is, up to isomorphism, only one way to freely extend an infinite independent set. We will demand a similar but slightly weaker property here, namely, up to isomorphism there are only a bounded number of ways to freely extend an independent set. (We can remove the requirement that the set be infinite by requiring, in addition, that the type be regular. But this is a second problem which will be discussed in Chapter XII.)

We close Section 1 with an illustration of the power of these axioms. We show that if T is any theory admitting a freeness relation satisfying our axioms then certain unions of chains of models of T preserve saturation. This foreshadows our proof in Section III.4 that if T is stable in μ then T has a saturated model of power μ.

In addition to describing the second notation for independence, a number of basic lemmas for calculating with the notion of independence are discussed in Section II.2. Moreover, the important notions of an independent sequence and, more generally, a set which is independent with respect to a partial order are introduced. Although these constructions are not exploited in full until much later in the book, we expound them now to emphasize that they can be easily derived from the axioms and do not depend on the more subtle considerations of Chapter III.

1. *Axioms for Independence*

In this section we formulate precisely the axioms described above and illustrate them with various examples which satisfy certain combinations of the axioms. Only the rather complicated Example 1.4 provides an algebraic example which satisfies all the axioms. This quick survey, along with the proof in Theorem 1.30 that a union of saturated models is saturated, give an overview of the environment in which we work and some of the arguments available there. The major purpose of this section is to provide a framework for the exposition of the nonforking relation in Chapter III. A second purpose is to extend a relation originally defined between the type of a finite sequence $\overline{a}$ over a set B and a set C to one defined between the type of a possibly infinite set A over B and the set C.

In our discussion here we will formalize the notion '$\overline{a}$ does not depend on B over A' by defining: $t(\overline{a}; B)$ is free over A. Since we are working inside the monster model, which is saturated, there is actually little difference between dealing with a complete type, $t(\overline{c}; B)$, and the pair $(\overline{c}; B)$. For, all realizations of the type can be mapped to $\overline{c}$ by an automorphism which fixes B. However, dealing with types has two advantages.

i) It allows us to compare two pairs $(\overline{c}; B)$ and $(\overline{d}; E)$ by considering $t(\overline{c}; B)$, $t(\overline{d}; E)$.
ii) It makes it convenient to speak of approximations to the pair $(\overline{c}; B)$ by discussing subtypes of $t(\overline{c}; B)$.

In Section II.2 we explore the relation between this notation and the set notation used in the introduction.

1.1 Free Extensions. A *notion of freeness*, $\mathcal{F}$, is a collection of pairs of the form (p, A). Each p is a (not necessarily complete) type which determines its domain, B. A is a subset of $\mathcal{M}$. If $(p, A) \in \mathcal{F}$ we say *p is free over A* and write $p\mathcal{F}A$. Alternatively, we may deal with a sequence $\overline{c}$ realizing p and write $t(\overline{c}; B)\mathcal{F}A$. We freely switch between these notations depending on which is most convenient. Note that the first is more general since it permits p to be an incomplete type. If $t(\overline{c}; B)$ is not free over A, we say $\overline{c}$ *depends on B over A*.

For an initial intuition into the meaning of the axioms one can assume that $A \subseteq B$. Those familiar with ω-stable theories can think of $t(a; B)$ is free over A as meaning that the Morley ranks (cf. Chapter VII) of $t(a; A)$ and $t(a; A \cup B)$ are the same. We will see in Chapter VII how to use rank on n-types to define a relation satisfying these axioms.

Each exercise or proposition in this section assumes the axioms that have already been introduced unless the contrary is given as a hypothesis.

1.2 Isomorphism Invariance. Like any other reasonable model theoretic notion, we require that $\mathcal{F}$ be closed under isomorphism; i.e., if f is an isomorphism and $(p, A) \in \mathcal{F}$ then $(f(p), f(A)) \in \mathcal{F}$.

We provide in Example 1.3 the prototypic algebraic examples of a freeness relation. Naturally, they satisfy some but not all of our axioms. Accordingly, they can provide some intuition about both the similarities and distinctions of our approach from the traditional one. In particular, these examples are only defined for 1-types.

1.3 Examples. i) Let V be a vector space over a field F and let $p \in S(B)$. Define $(p, A) \in \mathcal{F}$ if and only if for any a realizing p, a is not in the span of $A \cup B$ unless a is in the span of A.

ii) Let M be a model of a first order theory, A and B subsets of M, and $c \in M$. Define $(t(c; B), A) \in \mathcal{F}$ if $t(c; A \cup B)$ is not algebraic unless $t(c; A)$ is algebraic.

The dependence relation in the first example could be extended to tuples by declaring $\overline{a}$ to be independent from B over C if and only if for each $\overline{a}' \subseteq \overline{a}$, if $\overline{a}'$ is linearly independent over C then $\overline{a}'$ is linearly independent over $B \cup C$.

The following example, worked out by Chantal Berline [Berline 1983], is the algebraic interpretation of the nonforking relation for the case of algebraically closed fields. Naturally, one can check that it satisfies all the axioms. (Note that there is still one restriction. We deal only with complete types over subfields, not types over arbitrary subsets.)

1.4 Example. Let T be the theory of algebraically closed fields of characteristic 0. If k is a field then each n-type $p \in S(k)$ corresponds naturally to an ideal $I(p) = \{p(\overline{x}) : p(\overline{a}) = 0\}$ where $p \in k[\overline{x}]$, the polynomial ring in n variables over k and $\overline{a}$ is an arbitrary realization of p. Now suppose $k \subseteq K$ and p extends to $p' \in S(K)$. Call p' a free extension of p if $t.d.(k[\overline{x}]/I(p)) = t.d.(K[\overline{x}]/I(p'))$, where $t.d.$ abbreviates transcendence degree.

Thinking of $t(\overline{a}; B)\mathcal{F}A$ as meaning '$\overline{a}$ is no more constrained by $A \cup B$ than it is by A' suggests the following properties.

1.5 Monotonicity Axioms. M_1. If $q \subseteq p$ and p is free over A then q is free over A.

M_2. If $A_1 \subseteq A_2$ and p is free over A_1 then p is free over A_2.

In this formulation, M_2, of monotonicity, we do not assume that A_2 is a subset of $\operatorname{dom} p$. A more restricted form of monotonicity holds when this assumption is built into the notation (cf. Section II.1).

There are two aspects to the first monotonicity axiom. On the one hand, it asserts that if you take away information about a sequence $\overline{a}$ by considering a subset of $t(\overline{a}; B)$, then the resulting type is at least as likely to be free as the first. On the other, it asserts that every subsequence of a 'free sequence' is free. That is, we can conclude from M_1 that if $t(\bar{a}\frown\bar{b}; B)\mathcal{F}A$ then $t(\bar{a}; B)\mathcal{F}A$.

There is no assumption that p and q are complete types; in fact, incomplete types play an important role. Intuitively, the incomplete type p is free over A if some extension of p to a complete type over $\operatorname{dom} p$ is free over A. Indeed, this intuition is embodied in Axiom M_1.

1.6 Exercise. Verify that the examples in Example 1.3 satisfy the monotonicity axioms if we restrict their application to 1-types. For M_1, assume $A \subseteq \operatorname{dom} p$ and for M_2 assume that $A_2 \subseteq \operatorname{dom} p$.

The following can be viewed as a converse to the monotonicity axioms.

1.7 Transitivity Axiom for Independence. Let $C \subseteq B \subseteq \operatorname{dom} p$. If p is free over B and $p|B$ is free over C then p is free over C.

We will discuss later the much stronger requirement that dependence be transitive. We will be unable to establish this stronger axiom in general but rather will show in Chapter XII that a variant of it (which suffices to define dimension) holds on the realizations of stationary regular types.

1.8 Exercise. Show that Axiom 1.7 holds for Example 1.3 i).

The following combination of Axioms 1.5 and 1.7 is frequently used.

1.9 Exercise. Suppose $A \subseteq B \subseteq C$, $p \in S(B)$, and $p \subseteq q \in S(C)$. If $q\mathcal{F}B$ then $q\mathcal{F}A$ iff $p\mathcal{F}A$.

The next axiom corresponds to the exchange axiom in the van der Waerden formulation.

1.10 Symmetry Axiom. If $t(\bar{a}; B \cup \{\bar{b}\})$ is free over B then $t(\bar{b}; B \cup \{\bar{a}\})$ is free over B.

1.11 Exercise. Show this axiom fails for Example 1.3 ii).

Using transitivity it is easy to extend the symmetry axiom as follows:

1.12 Generalized Symmetry Lemma. *If both $t(\bar{a}; B)$ is free over A and $t(\bar{b}; B \cup \{\bar{a}\})$ is free over B then $t(\bar{a}; B \cup \{\bar{b}\})$ is free over A.*

In set notation the generalized symmetry lemma reads: if $D \downarrow_A B \cup C$ and $B \downarrow_A C$ then $D \cup B \downarrow_A C$. We can translate this as a symmetry relation for dependency. Note that now the auxilliary hypothesis applies to elements on opposite sides of the dependency sign. If $D \not\downarrow_A B \cup C$ and $B \downarrow_A D$ then $D \cup B \not\downarrow_A C$.

The following exercise shows the connection between the symmetry property and the exchange principle.

1.13 Exercise. Show that the special case of Axiom 1.10 where $\bar{a}$ and $\bar{b}$ are singletons holds for Example 1.3 i). That is, show that the symmetry axiom generalizes the usual 'exchange axiom' for vector spaces.

The following axioms guarantee that our notion is nontrivial.

1.14 Existence and Extension Axioms. E_1. $t(\bar{b}; A)$ is free over A for any $\bar{b}$ and A.

E_2. If p is free over A and $\operatorname{dom} p \subseteq B$, there is a $p_1 \in S(B)$ which extends p such that p_1 is free over A.

We next consider the generalization of the finite character of dependence to our context. This also allows us to extend our relation to a relation among three arbitrary sets.

1.15 Axiom: The Finite Character of Nonfreeness. If $t(\bar{c}; B)$ is not free over A then there is a formula $\phi(\bar{x}; \bar{b}) \in t(\bar{c}; B)$ such that no type containing $\phi(\bar{x}; \bar{b})$ is free over A.

This axiom has an important logical, as opposed to algebraic, character because we insist that not only is there a finite subset B_0 of B such that every extension of $t(\bar{c}; B_0)$ is not free over A but that there is a specific formula which forces nonfreeness. The formulation is somewhat subtle because there is a hidden reference to $t(\bar{b}; A)$. The assertion does not imply that for any $\bar{b}'$, if $\phi(\bar{x}; \bar{b}')$ is in q then q depends on $\bar{b}'$ over A, but only for those $\bar{b}'$ realizing $t(\bar{b}; A)$.

Our definition of the freeness relation, $t(\bar{a}; B)\mathcal{F}A$, deals with the relation between a sequence $\bar{a}$ and sets A and B. The symmetry axiom moves the elements of the sequence $\bar{a}$ to the domain of a type where the order of the sequence does not matter. Thus, freeness becomes a property not of the sequence $\bar{a}$ but of the set $\{a_0, \ldots, a_{n-1}\}$. More formally, we have the following lemma whose proof would be obvious using the symmetry axiom if we assumed $B \subseteq C$. Without this assumption it requires a use of the extension axiom. In the proof of this lemma we emphasize the distinction between a sequence $\bar{b}$ and its range, $\operatorname{rng}(\bar{b})$. The lemma itself justifies our usual abuse of this distinction.

1.16 Lemma. *Let $\bar{a}$ be a finite sequence and $\bar{a}'$ a permutation of $\bar{a}$. Then for any B and C, $t(\bar{a}; B)\mathcal{F}C$ iff $t(\bar{a}'; B)\mathcal{F}C$.*

Proof. Choose $\bar{d}$ with $t(\bar{d}; B) = t(\bar{a}; B)$ and such that $t(\bar{d}; C \cup B)$ is free over C. If $\bar{a}'$ is the result of permuting the elements of $\bar{a}$, let $\bar{d}'$ be the result of applying the same permutation to the elements of $\bar{d}$. Since $t(\bar{d}; B \cup C)\mathcal{F}C$, for any $\bar{b} \in B$ monotonicity and symmetry yield $t(\bar{b}; \bar{d} \cup C)\mathcal{F}C$. Since in the second position the order of $\bar{d}$ makes no difference this implies, $t(\bar{b}; \bar{d}' \cup C)\mathcal{F}C$. Applying symmetry again we have $t(\bar{d}'; \bar{b} \cup C)\mathcal{F}C$. Since this holds for each $\bar{b} \in B$, $t(\bar{d}'; B \cup C)\mathcal{F}C$ as required.

We have justified the following definition.

1.17 Definition. For arbitrary sets C, B, and a finite set A, $t(A; B)\mathcal{F}C$ iff for some ordering $\bar{a}$ of A, $t(\bar{a}; B)\mathcal{F}C$.

Since formulas can contain only finitely many variables, finite character implies that if every finite subtype of p is free over A then p is free over A. There are two equivalent ways to extend the definition of freeness to allow the first coordinate to be infinite. We could extend the domain of the freeness relation $\mathcal{F}$ to pairs $(p; C)$ where p is an α-type over B for any ordinal α and then define A to be independent from B over C if for some ordering $\overline{a}$ of A, $t(\overline{a}; B)\mathcal{F}C$. It is easy to see, using the monotonicity and symmetry axioms and the axiom of finite character, that an equivalent notion is obtained by retaining the convention that $(p; A) \in \mathcal{F}$ implies p is a finite type and making the following definition.

1.18 Definition. For arbitrary sets A, B, C, $t(A; B)\mathcal{F}C$ iff for each finite sequence $\overline{a}$ from A, $t(\overline{a}; B)\mathcal{F}C$.

The finite character allows us to reformulate the symmetry axiom as follows.

1.19 Exercise. If $t(A; B \cup C)\mathcal{F}C$ then $t(B; A \cup C)\mathcal{F}C$.

Now, we can rephrase Definition 1.18 in somewhat more generality: $t(A; B)\mathcal{F}C$ iff for each finite $\overline{a} \in A$, $\overline{b} \in B$, $t(\overline{a}; C \cup \overline{b})\mathcal{F}C$.

Let X be a subset of a vector space V and $a \in V$. If a depends on X then, as in Axiom 1.15, a depends on a finite subset X_0 of X. Thus a depends no more on X than on X_0; in our notation, $t(a; X)\mathcal{F}X_0$. For vector spaces this is simply the assertion that a depends on X_0. But in the general case the assertion that a does not depend on X over X_0 has more content than this trivial assertion. We shall always be able to find such an X_0 and to bound its size uniformly over all types, thus providing a local character for freeness. More formally.

1.20 Definition. For any theory T, let $\overline{\kappa}(T)$ denote the least cardinal κ, if one exists, such that for any finite sequence $\overline{a}$ and any set A there is an $A_0 \subseteq A$ with $|A_0| < \kappa$ such that $t(\overline{a}; A)\mathcal{F}A_0$. If no such cardinal, κ, exists $\overline{\kappa}(T)$ is ∞.

In Chapter III, especially Definition III.4.18, we consider several variants on $\overline{\kappa}(T)$ and calculate bounds on its value for a stable theory T. For now, we simply require that it exist.

1.21 Axiom: The Local Character of Freeness. $\overline{\kappa}(T) < \infty$.

Note that if $\overline{\kappa}(T) = \omega$, we have full local, that is finite, character in the third coordinate. This additional condition corresponds to T being a superstable theory.

1.22 Exercise. For a sequence $\overline{a}$ with $\lg(\overline{a}) = \mu$ and for an arbitrary set A there is a subset A_0 of A with $|A_0| \leq \mu \times \overline{\kappa}(T)$ such that $t(\overline{a}; A)\mathcal{F}A_0$.

We have defined the monotonicity, transitivity (of independence), symmetry, existence, extension and finite character axioms. In fact, if we declare every type to be free over every set, all of these axioms are satisfied. The reflexivity axiom implies the existence of types which are not free.

1.23 Reflexivity Axiom. If $A \subseteq B$, $b \in B - A$ and $t(b; A)$ is not algebraic then $t(b; B)$ is not free over A.

The final axiom generalizes the fact that in vector spaces there is a unique extension of an independent set. We require one definition to state the axiom.

1.24 Definition. Let $\mathcal{F}$ be a notion of freeness. An n-type p is *stationary over A* if $p\mathcal{F}A$ and if for every B containing A, if $q_1, q_2 \in S^n(B)$ extend p and are free over A then $q_1 = q_2$. We say p is *stationary* if p is stationary over dom p.

This concept would be easier to formulate if we could insist that p is a type over A. However, in a very important case (strong types, Chapter IV) p is not a type over A.

In any complete theory of vector spaces and more generally in a strongly minimal theory every non-algebraic type is stationary. When every type is stationary an independent set can be freely extended in only one way (up to isomorphism). In our generalization we will weaken this requirement somewhat by allowing more than one such extension but bounding the number of contradictory free extensions of a type. This bound is a cardinal number which depends only on T, not on the particular type p. This aim can be realized by demanding that types over models be stationary.

The existence of stationary types guarantees that the freeness relation is nontrivial. For, any non-algebraic type $p \in S(A)$ has more than one extension to any B properly containing A so there are nonfree extensions. In the situation in this book we have the following strong existence axiom for stationary types.

1.25 Boundedness Axiom. Every type over a model is stationary.

The following simple fact plays an important role.

1.26 Proposition. *If $p \in S(B)$ extends $q \in S(A)$, q is stationary over A, and $p\mathcal{F}A$ then p is stationary over B.*

1.27 Exercise. Prove Proposition 1.26.

1.28 Exercise. Show that if $\mathcal{F}$ is a notion of freeness which satisfies all the axioms from this section and $\mathcal{F}' = \{(p, A) : p \text{ is } \mathcal{F}\text{-stationary over } A\}$ then $\mathcal{F}'$ satisfies all the axioms except E_1, M_1 and symmetry.

We now illustrate the strength of these axioms by considering the problem of constructing saturated models. The natural way to try to construct a saturated model of power λ is to start with an arbitrary model of power λ, realize all types over subsets with power less than λ and iterate this procedure, hoping that it will close off. If T is stable and λ is regular, this procedure yields a saturated model of power λ. When λ is not regular we can build a chain of models M_i where M_i is μ_i-saturated and the μ_i have limit λ. If $M = \bigcup_{i<\lambda} M_i$ is μ_i-saturated for each i, then we could conclude that M is λ-saturated. We show now that this kind of preservation of saturation under unions of chains is implied by the axioms we have discussed in this section.

1.29 Notation. We frequently construct sequences denoted $\langle \bar{e}_i : i < \alpha \rangle$. We let $E_i = \{\bar{e}_j : j < i\}$. Technically, in the definition of E_i we should replace $\bar{e}_j$ by $\mathrm{rng}(\bar{e}_j)$. As usual, we dispense with such pedantry.

1.30 Theorem. *Let T be a theory which admits a freeness relation satisfying the axioms listed in this section. Suppose λ and $\mathrm{cf}(\delta)$ are cardinals strictly greater than $|T|^+ + \bar{\kappa}(T)$. The union of an increasing chain of length δ of models of T, each of which is λ-saturated, is itself λ-saturated.*

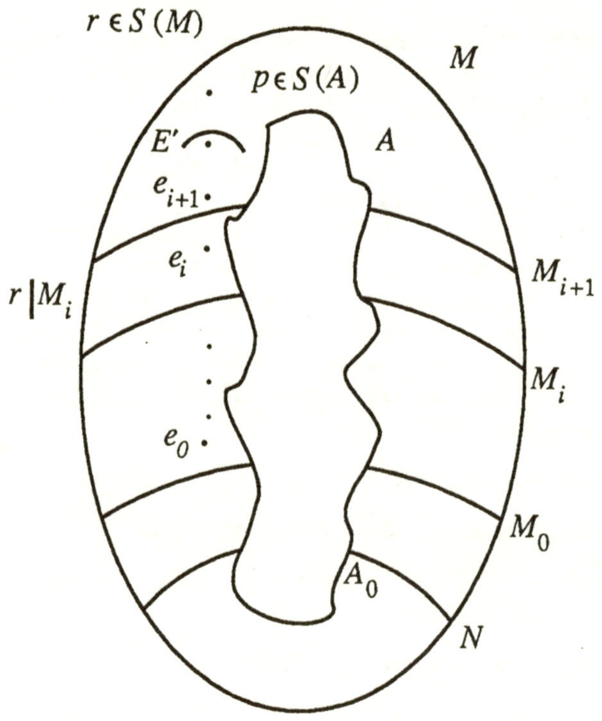

Fig. 1. Theorem 1.30

Proof. (Fig. 1). Let $M = \bigcup_{i<\delta} M_i$ and let $A \subseteq M$ with $|A| < \lambda$. Fix $p \in S(A)$; we must show p is realized in M. Extend p to $r \in S(M)$ and choose $A_0 \subseteq M$ with $|A_0| < \overline{\kappa}(T)$ such that $r \mathcal{F} A_0$. Choose N with $A_0 \subseteq N$ and $|N| \leq |T| + \overline{\kappa}(T)$. Without loss of generality, as $\mathrm{cf}(\delta) > |T| + \overline{\kappa}(T)$, N can be chosen inside M_0. Now choose by induction a sequence $E = \langle e_i : i < \lambda \rangle$ of elements of M_0 such that $t(e_i; E_i \cup N) = r|(E_i \cup N)$. The λ-saturation of M_0 guarantees the existence of this sequence. Now, by Exercise 1.22 we can choose an $E' \subseteq E$ such that $|E' \cup N| < (\overline{\kappa}(T) + |T|) \times |A| < \lambda$ and $t(A; E \cup N) \mathcal{F} (N \cup E')$. Thus $E - E' \neq \emptyset$. But, for $e \in E - E'$, the symmetry axiom implies that if q denotes $t(e; E' \cup N \cup A)$ then $q \mathcal{F} (E' \cup N)$. Now q and $r|(E' \cup N \cup A)$ are two free extensions of the stationary (since it is a free extension of a type over a model) type $r|(N \cup E')$. Thus they are equal and e realizes p.

This is a step towards proving that if T is stable in λ then T has a saturated model of power λ. A very similar proof suffices for this result after we compute in Section III.4 the relation between $\overline{\kappa}(T)$ and the set of cardinals in which T is stable.

1.31 Exercise. Suppose the countable theory T satisfies the axioms in this section with $\overline{\kappa}(T) = \omega$. Show that for every uncountable λ and any cardinal δ with $\mathrm{cf}(\delta) > \omega$, the union of a chain of length δ of λ-saturated models of T is λ-saturated.

1.32 Historical Notes. The abstract treatment of dependence relations began in the 1930's with the work of van der Waerden [van der Waerden 1949] and Whitney [Whitney 1935]. Our development here fuses these ideas with the axiomatic consideration of rank in [Baldwin & Blass 1974], [Lascar 1976] and Shelah's exposition of forking [Shelah 1978]. Theorem 1.30 is abstracted from a proof by Shelah. We show in Section III.4 that a proper subset of these axioms imply the rest. Our purpose here is to expound a set of principles which are often called on in the development of the theory.

Another approach to the study of independence was taken by universal algebraists in the 1960's. They attempted to axiomatize the properties of a dependence relation in a single algebra (which is free for some associated variety). This approach is summarized in Chapter 5 of [Grätzer 1979] and several survey articles [Marczewski 1966] and [Urbanik 1966]. This approach yielded a number of strong representation theorems ([Narkiewicz 1964], [Urbanik 1959/60], [Urbanik 1965]) which are very similar to those obtained later for strictly minimal sets by Zilber [Zilber 1981], [Zilber 1980b] and Cherlin [Cherlin, Harrington, & Lachlan 1985] (cf. Chapter VIII.3). These authors investigated the so-called v, v^* and v^{**} algebras. Urbanik proved that any v-algebra with a properly three-ary polynomial could be represented as an affine algebra. The connection between these results and work on categoricity is emphasized by the classification of categorical quasivarieties ([Givant 1973], [Givant 1976], [Palyutin 1973], [Palyutin 1976]) which relies on the representation theorems of Urbanik.

2. Further Properties of Independence

In this section we reformulate the freeness relation as a relation among three sets and then deduce some of the most important working tools of the theory. We wish to emphasize that these crucial tools depend only on the axiomatic properties of the dependence notions, and not on the intricacies of the proof in Chapter III that such a relation exists.

The notation is most compact if we adopt the following convention.

2.1 Definition. $C \downarrow_A B$ means $t(C; B \cup A) \mathcal{F} A$.

In light of the discussion after Definition 1.17 we can rephrase Definition 2.1 to read: for any sets A, B, and C

$$C \downarrow_A B \text{ iff for every finite } C_0 \subseteq C,\ C_0 \downarrow_A B$$

2.2 Exercise. Show $A \downarrow_C B$ iff for every D with $C \subseteq D \subseteq B \cup C, A \downarrow_C D$.

Note that in Definition 2.1, putting A into the domain of the type introduces an asymmetry between the treatment of the left and right side of the $\mathcal{F}$. This asymmetry is real; by burying it in the notation we can simplify our description of many situations. The price is a complicating of the monotonicity axiom (cf. Lemma III.3.7). The notation $A \downarrow_C B$ suggests two variants on the monotonicity axioms. We can restrict the extensions D of the base set C.

$$\text{If } A \downarrow_C B \text{ then for any } D \text{ with } C \subseteq D \subseteq B \cup C,\ A \downarrow_D B.$$

Alternatively, we can replace A by an A' which realizes the same type over B.

$$\text{If } A \downarrow_C B \text{ then for any } D \text{ with } D \supset C, \text{ there is an } A' \text{ with } t(A'; B) = t(A; B) \text{ and } A' \downarrow_D B.$$

The first of these assertions is an obvious consequence of the second monotonicity axiom. The second requires an application of the extension axiom (E_2) as well.

The following exercise spells out in a rather simple case the problems that can arise with monotonicity.

2.3 Exercise. Show that the 'freeness relation' of Example 1.3 i) does not satisfy the general form of the second monotonicity axiom. That is, the restriction in Exercise 1.6 that $A_2 \subseteq \operatorname{dom} p$ is necessary. (Hint: Let e_1, e_2, e_3 be independent elements and let $A_1 = \{e_1\}$, $A_2 = \{e_1, e_2\}$, $B = \{e_1, e_3\}$ and let $c = e_1 + e_2 + e_3$. Show that $t(e_2; B) = t(c; B)$ is free over A_1 but not over A_2. Thus, although e_2 and c realize the same type over A_1, one depends on B over A_2 and the other does not.)

It is natural to read $C \downarrow_A B$ as C is free from B over A. However, since we work primarily in contexts where the symmetry lemma holds, we will frequently write C and B are free over A or C and B are independent over A.

One value of this notation is indicated by the following simple statement of the symmetry axiom.

2.4 Exercise. Show that the symmetry axiom is equivalent to the following proposition: $B \downarrow_A C$ iff $C \downarrow_A B$.

The proofs of the next few propositions require the translation between the 'type' and the 'set' formulation of the freeness relation. This is an appropriate place for such a translation since these results are constantly used in the structure theory but in fact do follow from the properties listed in Section 1, rather than from special properties of the forking relation. Consider the relation between the assertion, '$t(\overline{a}; B)$ depends on A' and the assertion 'For each i, $t(a_i; B)$ depends on A.' where $\overline{a} = \langle a_0, \ldots, a_{k-1} \rangle$. It is perfectly possible to find c and d such that $t(c; B)\mathcal{F}A$ and $t(d; B)\mathcal{F}A$ but $t(c^\frown d; B)$ is not free over A. (e.g. Consider $e, e^{\sqrt{2}}$ as c, d, and $e + e^{\sqrt{2}}$ as B.) The situation is completely described by the following vital lemma.

2.5 Lemma. *The following are equivalent.*

i) $\overline{c}^\frown \overline{d} \downarrow_A B$.
ii) $\overline{c} \downarrow_A B$ *and* $\overline{d} \downarrow_{A \cup \overline{c}} B \cup \overline{c}$.
iii) $\overline{d} \downarrow_A B$ *and* $\overline{c} \downarrow_{A \cup \overline{d}} B \cup \overline{d}$.

Proof. We prove the equivalence of i) and ii); a permutation of the variables yields a proof of i) $\leftrightarrow$ iii). To facilitate the derivation from the axioms, we first restate the equivalence in the notation of Section 1.

i)′ $t(\overline{c}^\frown \overline{d}; B)$ is free over $A \subseteq B$.
ii)′ $t(\overline{c}; B)\mathcal{F}A$ and $t(\overline{d}; B \cup \{\overline{c}\})\mathcal{F}(A \cup \{\overline{c}\})$.

To simplify notation, we apply the finite character of dependence and assume $B - A$ is a finite sequence $\overline{b}$. By the symmetry axiom $t(\overline{c}^\frown \overline{d}; A \cup \overline{b})\mathcal{F}A$ iff $t(\overline{b}; A \cup \overline{c}^\frown \overline{d})\ \mathcal{F}A$ which, by monotonicity and transitivity, is equivalent to $t(\overline{b}; A \cup \overline{c}^\frown \overline{d})\mathcal{F}A \cup \overline{c}$ and $t(\overline{b}; A \cup \overline{c})\mathcal{F}A$. By the symmetry axiom again these are equivalent to $t(\overline{d}; A \cup \overline{b}^\frown \overline{c})\ \mathcal{F}(A \cup \overline{c})$ and $t(\overline{c}; A \cup \overline{b})\mathcal{F}A$ as required.

There are a number of important corollaries to this result. The first is best stated in the type notation.

2.6 Corollary. *If $p\mathcal{F}A$ and $q\mathcal{F}A$, where $p(\overline{x})$ and $q(\overline{y})$ are complete types over B, then there exists $r(\overline{x}; \overline{y}) \in S(B)$ such that $r(\overline{x}; \overline{y}) \vdash p(\overline{x}) \cup q(\overline{y})$ and $r\mathcal{F}A$.*

2.7 Exercise. Prove Corollary 2.6.

We can now derive a stronger version of the extension property.

2.8 Exercise. Prove the following Strong Extension Property: If $t(\overline{a}; B)$ is free over the subset A of B then for any $\overline{c}$ there is a $\overline{c}'$ such that $t(\overline{a}^\frown \overline{c}'; B)$ is free over A and $t(\overline{a}^\frown \overline{c}'; A) = t(\overline{a}^\frown \overline{c}; A)$.

It is easy using the extension property to find $\overline{a}'^\frown \overline{c}'$ free from B over A realizing $t(\overline{a}^\frown \overline{c}; A)$. The greater strength of this assertion arises from the demand that $\overline{a}' = \overline{a}$.

2.9 Exercise. Deduce from the extension axiom (E_2) that if $p \vdash q$ and $p\mathcal{F}A$ then $q\mathcal{F}A$.

The next corollary is stated using both notations in parallel. The type notation makes explicit the fact that the theorem deals not just with sequences $\overline{a}$ and $\overline{b}$ but with all realizations of the appropriate types. This corollary shows that free extensions preserve freeness.

2.10 Corollary. *Suppose $A \subseteq B$.*

i) *If $t(\overline{a}\frown\overline{b}; B)$ is free over A then $t(\overline{a}; B \cup \overline{b})$ is free over B iff $t(\overline{a}; A \cup \overline{b})$ is free over A.*
ii) *If $\overline{a}\frown\overline{b} \downarrow_A B$ then $\overline{a} \downarrow_B B \cup \overline{b}$ iff $\overline{a} \downarrow_A A \cup \overline{b}$.*

Proof. We prove the type version and leave the translation to the reader. By Lemma 2.5, $t(\overline{a}; B \cup \overline{b})$ is free over $A \cup \overline{b}$. Thus by transitivity and monotonicity $t(\overline{a}; B \cup \overline{b})$ is free over A iff $t(\overline{a}; A \cup \overline{b})$ is free over A. So if $t(\overline{a}; A \cup \overline{b})$ is free over A, we have $t(\overline{a}; B \cup \overline{b})$ is free over B by monotonicity. Conversely, again by Lemma 2.5 $t(\overline{a}; B)$ is free over A. So, if $t(\overline{a}; B \cup \overline{b})$ is free over B we conclude $t(\overline{a}; A \cup \overline{b})$ is free over A by transitivity.

Corollary 2.10 easily yields that if $A = \operatorname{rng} \overline{a}_1 \cup \operatorname{rng} \overline{a}_2 \cup \cdots \cup \operatorname{rng} \overline{a}_n$ then, for any B and C, $A \downarrow_C B$ iff $\{\overline{a}_1, \ldots, \overline{a}_n\} \downarrow_C B$.

The proofs of the last three results require only the monotonicity, transitivity, and symmetry axioms. Using only the first two we can reduce the third to a question of 'symmetry over models'. Thus when verifying in Chapter III that nonforking satisfies the symmetry axiom we will only have to check monotonicity, transitivity and 'symmetry over models'. The following technical lemma which we use to establish this reduction illustrates the power of the extension axiom combined with a judicious use of automorphisms.

2.11 Lemma. *Suppose $t(\overline{a}; A \cup \overline{b})$ is free over A. There exists a model M with $A \subseteq M$ such that $t(\overline{a}; M \cup \overline{b})$ is free over A and $t(\overline{b}; M)$ is free over A.*

Proof. We first guarantee that $t(\overline{b}; M)\mathcal{F}A$. We provide this guarantee in a devious way by working with models which are isomorphic over A to our eventual M. First, let M'' be an arbitrary model containing A. By the extension axiom choose $\overline{b}'$ with $t(\overline{b}'; A) = t(\overline{b}; A)$ and $t(\overline{b}'; M'')\mathcal{F}A$. Now let M' be the image of M'' under an automorphism which fixes A and maps $\overline{b}'$ to $\overline{b}$. Certainly, $t(\overline{b}; M')\mathcal{F}A$. Since $t(\overline{a}; A \cup \overline{b})$ is free over A, by the extension axiom there is an $\overline{a}'$ such that $t(\overline{a}'; A \cup \overline{b}) = t(\overline{a}; A \cup \overline{b})$ but $t(\overline{a}'; M' \cup \overline{b})$ is free over $A \cup \overline{b}$. Thus, by transitivity $t(\overline{a}'; M' \cup \overline{b})$ is free over A. Now choose an automorphism f of $\mathcal{M}$ which fixes $A \cup \overline{b}$ and maps $\overline{a}'$ to $\overline{a}$. Let M be $f(M')$.

There is an easier argument for the conclusion of Lemma 2.11 using the symmetry axiom. The point of this version is to simplify checking the symmetry axiom.

2.12 Lemma. *If a notion of freeness $\mathcal{F}$ satisfies the transitivity and monotonicity axioms and, in addition, for each model M*

$$t(\overline{a}; M \cup \overline{b})\mathcal{F}M \text{ iff } t(\overline{b}; M \cup \overline{a})\mathcal{F}M$$

then $\mathcal{F}$ satisfies the symmetry axiom.

Proof. Suppose $t(\overline{a}; A \cup \overline{b})$ is free over A. Applying Lemma 2.11 choose an M such that $t(\overline{b}; M)$ is free over A and $t(\overline{a}; M \cup \overline{b})$ is free over A. By monotonicity, $t(\overline{a}; M \cup \overline{b})$ is free over M. Now, by symmetry for types over models, $t(\overline{b}; M \cup \overline{a})$ is free over M. Since $t(\overline{b}; M)$ is free over A, we have by transitivity $t(\overline{b}; M \cup \overline{a})$ is free over A and by monotonicity $t(\overline{b}; A \cup \overline{a})$ is free over A.

We next describe a construction which will appear repeatedly in this book, the construction of an independent sequence. We emphasize here that it depends only the axioms we have enunciated. The straightforward exercises following the definition show that this construction yields more than is at first apparent. Recall from Notation 1.29 our notation for initial segments of a sequence of sequences. The most common use of this notation is in the construction of a sequence by repeatedly extending a type. We formally label the result of such a construction.

2.13 Definition. The sequence $\langle \overline{e}_i : i \in I \rangle$ is *coherent* over A if $k < j$ implies $t(\overline{e}_j; E_k \cup A) = t(\overline{e}_k; E_k \cup A)$.

2.14 Definition. The sequence E of sequences $\langle \overline{e}_i : i \in I \rangle$ is *independent* over A if for each i, $t(\overline{e}_i; E_i)$ is free over A.

Note that this definition only makes sense in the presence of the convention identifying the finite sequence $\overline{e}_i$ with its range. That is, the definition concerns the set of elements in the range of the function $\overline{e}_i$, not the finite function itself.

2.15 Exercise. Show the coherent sequence of sequences $E = \langle \overline{e}_i : i \in I \rangle$ is independent if and only if for each i, $t(\overline{e}_{i+1}; E_{i+1})$ is free over E_i.

The proof of the following result is given as an exercise because it is an entirely routine application of the symmetry and monotonicity axioms. Nevertheless, the result is vitally important. For example, it justifies our frequent reference to a *sequence* E satisfying Definition 2.14 as an independent *set* of sequences.

2.16 Lemma. *Let $E = \{\overline{e}_i : i < \omega\}$ be constructed so that $t(\overline{e}_i; E_i)\mathcal{F}A$ for each i. Then*

i) $t(\overline{e}_i \frown \overline{e}_{i+1}; E_i)\mathcal{F}A$ *for each* i.
ii) $t(\overline{e}_i; E - \{\overline{e}_i\})\mathcal{F}A$ *for each* i.

2.17 Exercise. Prove Lemma 2.16.

The following theorem plays a central role throughout this book. We introduced $\overline{\kappa}(T)$ in Definition 1.20. The theorem asserts that no finite sequence $\overline{b}$ can depend on each of $\overline{\kappa}(T)$ independent elements.

2.18 Theorem. *If $\langle \overline{e}_i : i < \overline{\kappa}(T)\rangle$ is an independent sequence over A then for any $\overline{b}$ there is an i with $t(\overline{b}; A \cup \overline{e}_i) \not\mathcal{F} A$.*

Proof. By the definition of $\overline{\kappa}(T)$ there exists a $B \subseteq E$ with $|B| < \overline{\kappa}(T)$ such that $t(\overline{b}; A \cup E) \not\mathcal{F} A \cup B$ and, a fortiori, $t(\overline{b}; A \cup B \cup \overline{e}) \not\mathcal{F} A \cup B$ for any $\overline{e} \in E - B$. Since E is an independent sequence, $t(\overline{e}; B \cup A) \not\mathcal{F} A$. By symmetry and transitivity this yields $t(\overline{e}; A \cup \overline{b}) \not\mathcal{F} A$ which by symmetry again yields the theorem.

(This argument presages the discussions of the relation between $\kappa(T)$ and $\overline{\kappa}(T)$ in Section III.4.)

Unlike the situation in vector spaces, there are different kinds of maximal independent sets. The most useful are those which are strongly independent in the sense described in Theorem 2.18.

2.19 Definition. The sequence $E = \langle \overline{e}_i : i \in I\rangle$ of sequences is *strongly independent* over A if E is independent over A and all the $\overline{e}_i$ realize the same stationary type over A.

2.20 Exercise. Show that if E is a strongly independent sequence then E is coherent.

The following exercises follow by formal manipulation from the principles we have described. They will prove to be extremely important in some of our later constructions.

2.21 Exercise. Suppose $\overline{a}_1 \downarrow_A \overline{a}_2$ and for each i, $t(a_i; A)$ is stationary. Show $t(a_1{}^\frown a_2; A)$ is stationary.

2.22 Exercise. If a, b and c are independent show $a^\frown b \downarrow_b b^\frown c$.

2.23 Exercise. Show $B \cup C \downarrow_A D$ implies $B \downarrow_{A \cup C} C \cup D$.

The next few definitions and results will illustrate the complicated constructions that can be made with just the machinery we have discussed in this chapter. We require the following special notions about partial orders. We will not make substantial use of many of these notions until Part D. We place them here to emphasize that this type of construction depends only on the abstract properties of independence and not on the detailed development of forking in Chapter III.

2.24 Definition. i) A subset B of a partial order $(A, <)$ is an *ideal* if $a \in B$ and $b < a$ implies $b \in B$.

ii) A partial order is *well-founded* if it contains no infinite descending chains.

iii) For an element a in a partially ordered set $(A, <)$, we denote by $a_{\#}$ the set of elements which are incomparable with a and write $a_{<}$ for $\{b : b < a\}$, and $a_{\geq}$ for $\{b : b \geq a\}$.

Note that $<$ is a free variable in this definition. Thus, in considering a partial order $\prec$, we may write, e.g., $a_{\prec}$ for the set of $\prec$-predecessors of a.

Actually, workers in partially ordered sets call what we have called an ideal a semi-ideal. But, since we never call on joins in this context, such pedantry seems inappropriate. The following notion is the natural extension of the concept of an independent set.

2.25 Definition. The set A is *independent with respect to the partial order* $<$ if for any pair of ideals B, C contained in A, $B \downarrow_{B \cap C} C$.

Note that we will often use this notation when A is a set of finite sequences.

As a device to simplify the construction of sets independent with respect to a given well-founded partial order, we extend it to a well-order and then construct free sequences relative to that well order. The next lemma shows that the resulting construction fulfills our intention.

2.26 Lemma. *Suppose the partial order $(A, <)$ is extended by the well-order $(A, \prec)$ and for each $a \in A$, $a \downarrow_{a_<} a_\prec$. Then A is independent relative to $<$.*

Proof. Let B, C be ideals of $(A, <)$. We must show $B \downarrow_{B \cap C} C$. We induct on the order type of $(A, \prec)$. By the finite character of dependence, we can assume that $(A, \prec)$ has a greatest element b. The result holds by induction unless $b \in B \cup C$. Suppose first that $b \in B - C$ and let $B' = B - \{b\}$. Since b is last with respect to $\prec$, $b \downarrow_{b_<} (B' \cup C)$. Two applications of the monotonicity axioms yield $((b \cup B') \downarrow C; b_< \cup B')$. Since $b \in B$ and B is an ideal with respect to $<$, it follows that $b_< \subseteq B'$. Thus, $B \downarrow_{B'} C$. By induction, $B' \downarrow_{B' \cap C} C$, so by transitivity of independence $B \downarrow_{B \cap C} C$ as required. The case $b \in C - B$ is completely symmetric so we are left with the case $b \in B \cap C$. Let $C' = C - \{b\}$. By the first case $B \downarrow_{B \cap C'} C'$. Applying monotonicity twice, we get $B \downarrow_{(B \cap C') \cup \{b\}} C' \cup \{b\}$. That is, $B \downarrow_{B \cap C} C$.

After solving the next exercise the second one is an easy application of Lemma 2.26.

2.27 Exercise. Show that any well-founded partial order $<$ on a set A can be extended to a well-ordering $\prec$ of A.

2.28 Exercise. Suppose $(A, <)$ is a well-founded partial order such that for each $a \in A$, $a \downarrow_{a_<} a_{\#}$. Show A is independent with respect to $<$.

The following theorem is the key to the construction of sets which are independent relative to some prescribed partial order.

2.29 Theorem. *Let $(A, <)$ be a well-founded partial order, $\mathcal{B}$ a collection of ideals of A, and for each $B \in \mathcal{B}$, p_B a nonalgebraic type over B. Then we can choose for each $B \in \mathcal{B}$ a c_B realizing p_B such that, letting C be $\{c_B : B \in \mathcal{B}\}$, if we extend $<$ to an ordering on $A \cup C$ so that $(c_B)_< = B$ then $A \cup C$ is independent relative to the extended order.*

Proof. Well order $\mathcal{B}$ by $\prec\prec$. By induction on this order choose c_B so that $c_B \downarrow_B A \cup \{c_{B'} : B' \prec\prec B\}$. Let $\prec$ be a well order of $A \cup C$ which satisfies the following conditions.

i) $\prec |A$ extends $<$.
ii) All elements of A precede all elements of C.
iii) $c_{B'} \prec c_B$ if and only if $B' \prec\!\!\prec B$.

Let $<$ be the partial order of $A \cup C$ obtained by extending the original order on A only by adding $b < c_B$ if $b \in B$. It is easy to check that, applying Lemma 2.26, we have the result.

The remaining exercise in this section describes a more concrete but also more complicated notation for the last several results. This notation appears in [Shelah 1982], [Makkai 1984], and [Harrington & Makkai 1985].

2.30 Exercise. Let S be the collection of finite subsets of a set X and $\mathcal{A} = \{A_s : s \in S\}$ a collection of subsets of a structure M such that if $t \subseteq s$ then $A_t \subseteq A_s$. Show that if $(A'_s \downarrow \bigcup\{A'_t : s \not\subseteq t\}; \bigcup\{A'_t : t \subseteq s \wedge t \neq s\})$ then under the partial ordering determined by the natural partial order of subsets of X, $\mathcal{A}$ is an independent partial order in the sense of Definition 2.25.

2.31 Historical Notes. An independent sequence is often called a Morley sequence. This notion, generalizes Morley's [Morley 1965] idea of finding indiscernibles by constructing sequences of elements such that the type of each element over its predecessors has a fixed rank. The notations $A \downarrow_C B$ and $(A \downarrow B; C)$ are variants of a notation introduced by Makkai. The development of 2.19 through 2.26 is buried in [Shelah 1978]. Except for simplifying the notation, we have followed the considerably clearer version of [Makkai 1984].

Chapter III

Forking

In this chapter we introduce the concept of a stable theory and expound Shelah's notion of forking. We will show that in a stable theory nonforking obeys the axioms described in Chapter II. One intuition behind this notion is that if $t(\overline{a}; B)$ is not free over A then $\overline{a}$ must satisfy more relations over B than it does over A. Each of $e, e^{\sqrt{2}}$,and $e^{1/2}$ realize the same type over Q, the field of rational numbers, namely the type p of a transcendental element. Both $p_1 = t(e^{\sqrt{2}}; Q \cup \{e\})$ and $p_2 = t(e^{1/2}; Q \cup \{e\})$ extend p, but p_1 is clearly a more generic or freer extension than p_2. We will give an account of the distinction between p_1 and p_2 which applies to any stable theory. A first approximation, '$e^{1/2}$ is in the algebraic closure of e while $e^{\sqrt{2}}$ is not,' works in a few cases. Thinking of the theory of algebraically closed fields of characteristic zero as a prototypical ω-stable theory, one can notice that the Morley rank of p_1 equals the Morley rank of p while the Morley rank of p_2 is less than that of p. This version will apply to any ω-stable theory. The extension to arbitrary stable theories requires some effort.

The most naive statement of the leitmotif of stability theory reads, 'many types implies many models; few types implies few models.' To make this notion precise we must specify what is meant by 'many types' and we must refine the phrase 'few models.' The appropriate rendering of 'few models' is 'admits a structure theory', and more specifically in this chapter, 'admits a freeness relation satisfying the axioms described in Chapter II.' We show in this chapter that a theory which has few types admits a freeness relation satisfying our axioms. We show this by introducing the notion of 'definability of types' and showing that if a theory has few types then every type is definable. From this we derive the existence of an appropriate freeness relation.

Thus in Section III.1 we show the equivalence of the two main characterizations of a stable theory.

i) There are few types, in the sense made precise in Section III.1.
ii) Every type is definable.

Thereafter, we develop the positive structure theory for stable theories solely from this definability property without further recourse to the number of types. We outlined in Section I.5 the proof that 'many types' implies there are 'many models.'

From one standpoint, a freeness relation can be seen as a means of picking out certain distinguished extensions of a type. That is, when $C \subseteq B$, the property, '$t(\overline{a}; B) \mathcal{F} C$' distinguishes $t(\overline{a}; B)$ among the extensions of $t(\overline{a}; C)$. We restrict ourselves in Section III.2 to searching for such a distinguished extension when C is the universe of a model. We note that the definable extension of a type provides such an extension and then investigate several characterizations of this extension. We establish the symmetry lemma for types over models, an important precursor of the exchange lemma. One important tool for these investigations is the fundamental order of Lascar-Poizat; another is the notion of a strongly saturated model. We conclude Section III.2 by combining these two ideas to provide a characterization of conjugate types.

In Section III.3 we show that if T is stable we can define a notion of independence satisfying the axioms of Section II.1. We relate this notion, nonforking, to the fundamental order and show that all nonforking extensions of a complete type over A to the monster model M are conjugate over A. We return to the 'distinguished extension' theme and show that the nonforking extensions (to a global type) of a type $p \in S(A)$ are exactly those which have few conjugates over A.

In Section III.4 we investigate the effect of those axioms which are true, but trivial, in the vector space case and so had not been singled out: the finite character of freeness and the existence of stationary types. We connect these axioms with the function which computes the number of types over a set of power κ. This leads to the division of stable theories into three classes, depending on the values of this function for the theory. We provide a number of examples of theories in the various classes and compute the spectrum of saturation of a stable theory. Section III.5 is devoted to a survey of algebraic examples illustrating the material in this chapter. We explain the meaning of forking in the context of modules and discuss various definable chain conditions in algebra.

1. *Stable Theories: ϕ-Types, Rank, and Definability*

In [Shelah 1971], Shelah took the crucial step for the development of stability theory. He specialized Morley's discussion of complete types to a study of types which consist of instances of a single formula or its negation. This localization distinguishes stable from ω-stable theories. For, the concept of a stable theory is defined by a condition on instances of single formulas. It is important to deal with instances of one formula in order to make some crucial compactness arguments.

In this section we lay the basis for our study. We show the notion of a stable theory can be defined by a definability criterion, by conditions on the cardinalities of Stone spaces, or by the existence of local rank functions.

The basic intuition is a familiar one. If there are few possibilities for a given phenomenon then each possibility is definable. Beth's Theorem [Chang & Keisler 1973] is a prototype for such results and, in fact, the key Theorem 1.22 of this section can be proved using Beth's theorem [Lascar & Poizat 1979]. However, we will follow here the original route of Shelah and introduce rank as the tool used to find the definition of a ϕ-type over a set A.

Recall the following convention.

1.1 Separation of Variables. When we write the formula ϕ in the form $\phi(\overline{x};\overline{y})$, this indicates that $\overline{x}$ should be regarded as a sequence of free variables in the usual way, but $\overline{y}$ is a placeholder for a sequence of parameters. Thus, in forming $\phi(\overline{x};\overline{y})$-types over A as in the following definition, we know from the expression $\phi(\overline{x};\overline{y})$ that we are defining a subset of $S^n(A)$ (where $n = \lg(\overline{x})$) and that sequences from A will be substituted for $\overline{y}$.

1.2 Definition. Let $\phi(\overline{x};\overline{y})$ be a formula. A *ϕ-type* over A is a consistent set of formulas, each formula having the form $\phi(\overline{x};\overline{a})$ or $\neg\phi(\overline{x};\overline{a})$, where $\overline{a}$ is a $\lg(\overline{y})$-sequence from A. The ϕ-type p is a *complete ϕ-type* if for each $\lg(\overline{y})$ sequence $\overline{a}$ from A, either $\phi(\overline{x};\overline{a})$ or $\neg\phi(\overline{x};\overline{a})$ is in p. We denote by $S_\phi(A)$ the set of all complete ϕ-types over A. If $p \in S(A)$ we write p_ϕ for the restriction of p to instances of ϕ and $\neg\phi$.

For any set of formulas $\Delta = \{\phi_i(\overline{x};\overline{y}) : i \in I\}$ we define a Δ-type over A to be a consistent collection of A-instances of formulas in Δ or their negations. In the natural sense we may say an arbitrary type p is complete for ϕ if p_ϕ is a complete ϕ-type.

The notion of a Δ-type is fundamental for stability theory. As will be seen through this book, much of the material about stable theories can be seen as localizing results about types in ω-stable theories to results about Δ-types for arbitrary finite sets of formulas Δ. Shelah in [Shelah 1978] and earlier [Shelah 1971] shows how to code Δ-types for a finite Δ as ψ-types for a single formula ψ. The idea is to consider the conjunction of formulas of the form $(z_i = b_0 \rightarrow \phi_i(\overline{x};\overline{y})) \wedge (z_i = b_1 \rightarrow \neg\phi_i(\overline{x};\overline{y}))$ where the z_i are a collection of new variables and b_0, b_1 are a pair of points used as indicators.

1.3 Exercise. Show that for any set A with $|A| \geq 2$ and any finite set Δ of formulas there is a single formula ψ such that each Δ-type over A is equivalent to a ψ-type.

1.4 Exercise. Show that if A is subset of a model of any theory with $2 \leq |A| < \omega$ then for each ϕ, $|S_\phi(A)| < \omega$.

The following definition provides a technical framework for discussing freeness of types.

1.5 Definition. (Fig. 1). Let p be a type and let κ be a cardinal. Then $\Gamma_p(\phi,\kappa)$ is the following set of sentences in the language obtained by adding

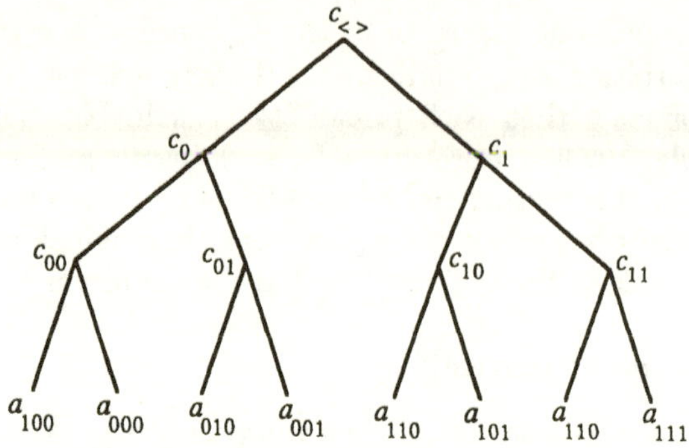

Fig. 1. $\Gamma_{\langle\,\rangle}(\phi, 2)$

to L additional constant symbols $\overline{a}_\tau, \tau \in 2^\kappa$ and $\overline{c}_s, s \in 2^{<\kappa}$.

$$\Gamma_p(\phi, \kappa) = \{p(\overline{a}_\tau) : \tau \in 2^\kappa\} \cup \{\phi(\overline{a}_\tau; \overline{c}_{\tau|i})^{\tau(i)} : i < \kappa\}.$$

Here $p(\overline{a}_\tau)$ denotes $\{\psi(\overline{a}_\tau) : \psi \in p\}$ and $\phi^{\tau(i)}$ denotes ϕ or $\neg\phi$ depending on whether $\tau(i)$ is 0 or 1.

The collection $\Gamma_p(\phi, \kappa)$ is consistent just if there is a complete binary tree of height κ of extensions of p by instances of ϕ such that each path is consistent but the paths are pairwise contradictory. Note we can easily extend our definition to $\Gamma_{\langle\,\rangle}(\phi, \kappa)$ by letting $\langle\,\rangle$ denote the empty type. The following diagram illustrates the definition of $\Gamma_p(\phi, \kappa)$. The $\overline{a}_\tau$ in the definition correspond to paths through the tree; the $\overline{c}_s$ correspond to nodes.

1.6 Theorem. *$\Gamma_p(\phi, n)$ is consistent for arbitrarily large finite n if and only if $\Gamma_p(\phi, \kappa)$ is consistent for all κ.*

Proof. The result is immediate by compactness noting that the consistency of an arbitrary n-element subset of $\Gamma_p(\phi, \kappa)$ is implied by the consistency of $\Gamma_p(\phi, n)$.

The key to this argument is the fact that all the nodes on the tree, except the base, are instances of ϕ or $\neg\phi$.

We now show that if for some infinite A, $|S_\phi(A)| > |A|$ then there is such an A of each infinite cardinality. The proof has two stages. The first is a variant on the proof that a closed uncountable set of reals has a perfect subset. While in that proof we choose a pair of disjoint open subsets which are each uncountable, here we show there is a single formula $\phi(\overline{x}; \overline{a})$ such that both ϕ and $\neg\phi$ are elements of more than $|A|$ distinct ϕ-types. Then, as completeness of the reals allows us to construct the perfect set, the compactness theorem allows us to find arbitrarily large A satisfying the condition.

1.7 Lemma. *Suppose that for some A and ϕ, $|S_\phi(A)| > |A| = \kappa \geq \omega$. Then for each infinite cardinal λ there exists a B with $|B| = \lambda$, $|S_\phi(B)| > \lambda$ and, a fortiori, $|S(B)| > \lambda$.*

Proof. Note first that for some $\overline{a} \in A$ both $X(\overline{a}) = \{p \in S_\phi(A) : \phi(\overline{x};\overline{a}) \in p\}$ and $Y(\overline{a}) = \{p \in S_\phi(A) : \neg\phi(\overline{x};\overline{a}) \in p\}$ have more than κ elements. If not, $S_\phi(A) = \bigcup_{\overline{a} \in A} Z(\overline{a}) \cup \{q\}$ where $Z(\overline{a})$ is whichever of $X(\overline{a})$, $Y(\overline{a})$ has $\leq \kappa$ elements, while q is the unique complete ϕ-type such that $\phi(\overline{x};\overline{a}) \in q$ if and only if $|X(\overline{a})| > \kappa$. But then $|S_\phi(A)| \leq \kappa$. Relativizing this argument to $X(\overline{a})$ and $Y(\overline{a})$ and proceeding by induction, we see $\Gamma_{\langle\,\rangle}(\phi, \aleph_0)$ is consistent. Fix λ and let μ be the least cardinal with $2^\mu > \lambda$. Then, by compactness, $\Gamma_{\langle\,\rangle}(\phi, \mu)$ is consistent. Let B be a set of interpretations for $\{\overline{c}_s : s \in 2^{<\mu}\}$; we have the lemma. For, $2^{<\mu}$ has at most λ nodes while 2^μ has more than λ.

The preceding argument of Shelah localized to a single formula the following seminal result of Morley.

1.8 Exercise. Let T be a countable theory. Show that if for some infinite λ there is a set B with $|B| = \lambda$ and $|S(B)| > \lambda$, then there is a countable set A with $|S(A)| > \aleph_0$. In fact, $|S(A)| = 2^{\aleph_0}$.

We can now relate the cardinalities of $S(A)$ and $S_\phi(A)$.

1.9 Theorem. *Let T be an arbitrary complete first order theory. The following are equivalent:*

i) *For every A, $|S(A)| \leq |A|^{|T|}$.*
ii) *For every A, if $|A|^{|T|} = |A|$ then $|S(A)| \leq |A|$.*
iii) *For every countable A and each ϕ, $|S_\phi(A)| \leq \aleph_0$.*
iv) *For every infinite A and every ϕ, $|S_\phi(A)| \leq |A|$.*

Proof. Clearly, i) implies ii). By Lemma 1.7, ii) implies iii) and iii) implies iv). As $S(A)$ can be embedded into the Cartesian product of the $S_\phi(A)$ for $\phi \in F(\emptyset)$ by mapping p to $\langle p_\phi : \phi \in F(\emptyset)\rangle$, iv) implies i).

We now introduce a notion of local rank. While the entire subject of stability theory can be erected on the basis of this notion, we use it primarily as a tool to obtain the definability of types in this section and use definability as our basic notion. Our axioms for independence guide the choice of properties derived from the definability of all types in a stable theory. The relation between the formulation here and the various rank notions is explained in Chapter VII.

1.10 Definition. Let p be a type over A. The ϕ-rank of p is defined by induction as follows.

- $R(p, \phi) \geq 0$ for all p.
- $R(p, \phi) \geq \delta$ if $R(p, \phi) \geq \alpha$ for all $\alpha < \delta$ when δ is a limit ordinal.
- $R(p, \phi) \geq \beta + 1$ just if for each finite $q \subseteq p$ there is a sequence $\overline{a} \in \mathcal{M}$ such that:

$$R(q \cup \{\phi(\overline{x};\overline{a})\}, \phi) \geq \beta \text{ and } R(q \cup \{\neg\phi(\overline{x};\overline{a})\}, \phi) \geq \beta.$$

- $R(p, \phi) = \alpha$ if $R(p, \phi) \geq \alpha$ but $R(p, \phi) \not\geq \alpha + 1$.
- $R(p, \phi) = \infty$ if $R(p, \phi) > \alpha$ for every ordinal α.

1.11 Exercise. Show that for $n < \omega$, $R(p, \phi) \geq n$ iff $\Gamma_p(\phi, n)$ is consistent.

1.12 Exercise. For every p and ϕ, $R(p, \phi) < \omega$ or $R(p, \phi) = \infty$.

1.13 Exercise. For every formula ϕ

i) If $p \subseteq q$, $R(p, \phi) \geq R(q, \phi)$.
ii) For every p, there is a finite subtype $p_0 \subseteq p$ with the same rank.

These results are immediate from the definition. The following fundamental result is only slightly harder to prove.

1.14 Proposition. *Fix a formula ϕ. If for each countable A, $|S_\phi(A)| \leq \aleph_0$ then for every p, $R(p, \phi) < \omega$.*

Proof. By Exercise 1.11 unless $R(p, \phi)$ is bounded below ω, each $\Gamma_p(\phi, n)$ is consistent; so to, by Theorem 1.6, is $\Gamma_p(\phi, \aleph_0)$. But if $A = \{\overline{c}_s : s \in 2^{<\aleph_0}\}$, $|A| = \aleph_0$ while $|S_\phi(A)| = 2^{\aleph_0}$.

We come now to the critical notion of a definable type. We shall see that stable theories can be characterized by the definability of every type. Moreover, we will construct our freeness relation by generalizing the intuition that definable extensions of types are free.

1.15 Definition. Let $p(\overline{x})$ be an n-type over A. Let d be a map from formulas $\phi(\overline{x}; \overline{y})$ into formulas $d\phi(\overline{y})$ with parameters from B. Then d *defines* p over B if for each formula $\phi(\overline{x}; \overline{y})$ and each $\overline{a} \in A$:

$$\phi(\overline{x}; \overline{a}) \in p \text{ implies } \models d\phi(\overline{a}) \text{ and } \neg\phi(\overline{x}; \overline{a}) \in p \text{ implies } \models \neg d\phi(\overline{a}).$$

Note that if p is complete we can simplify this definition to $\phi(\overline{x}; \overline{a}) \in p$ if and only if $\models d\phi(\overline{a})$. In Corollary 1.23 and Lemma 1.26 we clarify the dependence of d on both p and ϕ and show that for A a model d is a Boolean homomorphism.

1.16 Exercise. Let M be a model of the theory of an equivalence relation with infinitely many infinite classes. Find a definition over M for each 1-type in $S(M)$.

1.17 Exercise. Show that if $\{\phi(x, \bar{a})\} \vdash p$ then p is definable over $\overline{a}$.

1.18 Exercise. Using the quantifier elimination results for modules discussed in Chapter I, verify that every 1-type, p, over an Abelian group A is definable as follows. Let $p = t(c; A)$. Every formula $\phi(x; \overline{y})$ is equivalent to a Boolean combination of formulas of the forms: $n|(kx + y)$ or $kx = y$. (Check that the replacement of $\overline{y}$ by y is legitimate.) In the first case, if for some $a \in A$, $n|kc + a$ let $d\phi$ be $n|y - a$; otherwise, let $d\phi$ be $y \neq y$. In the second case, if for some $a \in A$, $kc = a$, let $d\phi$ be $y = a$; otherwise, let $d\phi$ be $y \neq y$.

The following exercise uses the same trick explained before Exercise 1.3 to code information about a finite set of formulas in terms of one formula.

1.19 Exercise. Show that for any set A with $|A| \geq 2$ and any n formulas $\phi_0(\overline{x};\overline{y}), ..., \phi_{n-1}(\overline{x};\overline{y})$ there is a formula $\psi^*(\overline{x};\overline{y},\overline{z})$ such that for any χ, if a χ-type p is definable by one of the ϕ_i, then it is definable by an instance of ψ^*.

Now we can define the fundamental concept of this book.

1.20 Definition. The theory T is *stable* if for every $A \subseteq \mathcal{M}$ and every $p \in S(A)$, p is definable over A.

In accordance with our convention this means that for some unspecified n, every n-type is definable. After we relate the definability of n-types to the cardinality of $S^n(A)$, we will note (Exercise 1.36) that this condition holds for one n if and only if it holds for all n. In the meantime we do not hesitate, as in the next exercise, to verify that a theory is stable by showing only that all its 1-types are definable.

1.21 Exercise. Deduce from Exercise 1.18 that for any Abelian group A, $\mathrm{Th}(A)$ is stable.

Using the rank machinery we can immediately link the definability requirement with conditions on the cardinality of Stone space

1.22 Theorem. *The theory T is stable if and only if for each formula ϕ and for each set A, $|S_\phi(A)| \leq \sup(|A|, |T|)$.*

Proof. Clearly if T is stable the number of complete ϕ-types over A is bounded by the number of definitions over A. So, for each A, $|S_\phi(A)| \leq |A|$. If $|S_\phi(A)| \leq |A|$ for all A and ϕ, then, by Theorem 1.9 and Proposition 1.14, for every $p \in S(A)$, $R(p,\phi)$ is finite. Fix p with, say, $R(p,\phi) = n$. Choose by Proposition 1.13 a finite subset p_0 of p so that $R(p,\phi) = R(p_0,\phi)$. Note that $\phi(\overline{x};\overline{a}) \in p$ if and only if $R(p_0 \cup \{\phi(\overline{x};\overline{a})\},\phi) = n$. That is, $\phi(\overline{x};\overline{a})$ is in p if and only if $\Gamma_{p_0\cup\{\phi(\overline{x};\overline{a})\}}(\phi,n)$ is consistent. The consistency of $\Gamma_{p_0\cup\{\phi(\overline{x};\overline{a})\}}(\phi,n)$ can be expressed by a single first order formula $d\phi(\overline{y})$ whose parameters are those occurring in p_0. That is, $\models d\phi(\overline{a})$ just if $\Gamma_{p_0\cup\{\phi(\overline{x};\overline{a})\}}(\phi,n)$ is consistent.

We now explore several ways to uniformize the definability of types in a stable theory. The difference between Corollary 1.23 and Lemma 1.24 is clarified by the proof of Theorem 2.23. First we show that we can 'define' the collection of parameters which may serve to define p.

1.23 Corollary (Harnik). *If T is stable, for every formula $\phi(\overline{x};\overline{y})$ and each $p \in S(A)$ there exist L-formulas $\theta(\overline{x};\overline{z})$ and $\chi(\overline{y};\overline{z})$ such that:*

i) *For some $\overline{a} \in A$, $\theta(\overline{x};\overline{a}) \in p$.*
ii) *For every $\overline{a},\overline{b} \in A$, if $\theta(\overline{x};\overline{a}) \in p$ then $[\phi(\overline{x};\overline{b}) \in p$ iff $\models \chi(\overline{b};\overline{a})]$.*

Proof. We use the notation from the proof of the last theorem. Let $\theta_0(\overline{x};\overline{z})$ be the result of replacing the parameters in the conjunction of p_0 by the variables $\overline{z}$, let $\theta(\overline{x};\overline{z})$ be $\theta_0(\overline{x};\overline{z}) \wedge (\bigwedge \Gamma_{\theta_0(\overline{x};\overline{z})}(\phi,n))$ and let $\chi(\overline{y};\overline{z})$ express the consistency of $\Gamma_{\{\theta(\overline{x};\overline{z})\}\cup\{\phi(\overline{x};\overline{y})\}}(\phi,n)$.

Note that both θ and χ depend on p as well as ϕ. Yet our notation in Definition 1.15 indicates that d depends only on ϕ even though a formula $d\phi$ which satisfied the definition could depend on both ϕ and p. We now show that we can choose a formula $d\phi(\overline{y};\overline{z})$ (depending only on ϕ) and express the dependence on p by the choice of the parameter substituted for $\overline{z}$.

Technically, the following lemma only holds when any A under consideration has cardinality at least two. Since this is the only interesting case and can always be easily arranged we suppress that hypothesis to ease readability.

1.24 Lemma. *If for each A, every $p \in S(A)$ is definable over A then there is a map d taking formula $\phi(\overline{x};\overline{y})$ to formulas $d\phi(\overline{y};\overline{z})$ such that for each A and each $p \in S(A)$ there is a sequence $\overline{a}_{\phi,p} \in A$ such that $d\phi(\overline{y};\overline{a}_{\phi,p})$ defines p.*

Proof. If not, there is a formula ϕ, such that for each formula $\psi(\overline{y};\overline{z})$ there is a set A, and a type $p \in S(A)$ such that for each $\overline{a} \in A$, $\psi(\overline{y};\overline{a})$ does not define p_ϕ. Add to L a new unary predicate R and a new constant $\overline{c}$. Now the following set of sentences is consistent.

$$\{(\forall\overline{z})(R(\overline{z}) \to ((\exists\overline{y})(R(\overline{y}) \wedge \neg[\phi(\overline{c};\overline{y}) \leftrightarrow \psi(\overline{y};\overline{z})]))) : \psi(\overline{y};\overline{z}) \in L\}.$$

But, if B is the interpretation of R in a model of these sentences then $t(\overline{c};B)$ is not definable over B, contrary to the stability of T.

The argument for Lemma 1.24 seems rather nonconstructive. Note, however, that by the completeness theorem the function d is recursive (in T). Moreover, the most natural candidate for a constructive proof of the theorem fails. That is, we would like to argue that there is some uniform choice of the type p_0 which arises in the proof of Theorem 1.22. We couldn't hope to fix the type p_0 but since it is a conjunction of instances of ϕ and instances of $\neg\phi$, we might hope to fix the number of each. The following example shows that life is not that simple.

1.25 Exercise. If p is a finite ϕ-type, let $|p| = (n, k)$ assert that p contains n instances of ϕ and k instances of $\neg\phi$. Show that there is a stable theory and a formula ϕ such that for infinitely many distinct (n, k) there exist an A and a $p \in S(A)$ such that if p_0 is the least (where $|p|$ is ordered by the sum of n and k) subtype of p with $R(p, \phi) = R(p_0, \phi)$, then $|p_0| = (n, k)$. (Hint: Consider the theory of an equivalence relation with arbitrarily large finite classes.)

We will rely on the following easy but important fact about the map d which defines a type p.

1.26 Lemma. *Let T be stable. For any model M and $\overline{c}$, the map d which defines $t(\overline{c};M)$ is a Boolean homomorphism from $F(T)$ into $F(M)$.*

1.27 Exercise. Prove Lemma 1.26; give an example showing the necessity to assume that we work over a model M rather than an arbitrary set A.

The following definition provides an alternative view of definability. The duality between the roles of parameters and the sets they pick out which is exploited in the next few results recurs in the symmetry lemma in Section 3.

1.28 Definition. We say N is a *conservative extension* of M and write $M \prec_c N$ if for every formula $\phi(\overline{x};\overline{y})$ and every $\overline{n} \in N$, there is a formula $\phi^*(\overline{x};\overline{z})$ and an $\overline{m} \in M$ such that $\phi(N,\overline{n}) \cap M = \phi^*(M,\overline{m})$.

1.29 Exercise. Show that if $M \prec N$ are models of $\mathrm{Th}(Z,S)$ then N is a conservative extension of M.

1.30 Exercise. Show that as models of the theory of dense linear order without endpoints the reals are not a conservative extension of the rationals.

1.31 Lemma. *Let N be an elementary extension of M. Then N is a conservative elementary extension if and only if every type over M which is realized in N is definable over M.*

Proof. Suppose M is an elementary submodel of N and for each $\overline{n} \in N$, $t(\overline{n};M)$ is definable over M. Then for any formula $\phi(\overline{x};\overline{y})$,

$$\begin{aligned}\{\overline{m} \in M : N \models \phi(\overline{n};\overline{m})\} &= \{\overline{m} \in M : \phi(\overline{x};\overline{m}) \in t(\overline{n};M)\} \\ &= \{\overline{m} \in M : M \models d\phi(\overline{m})\}.\end{aligned}$$

Similarly, if N is a conservative extension of M,

$$\begin{aligned}\{\overline{m} : \phi(\overline{x};\overline{m}) \in t(\overline{n};M)\} &= \{\overline{m} : N \models \phi(\overline{n};\overline{m})\} \\ &= \{\overline{m} : M \models \phi^*(\overline{m})\}\end{aligned}$$

so we can use ϕ^* as $d\phi$ to define $t(\overline{n};M)$.

We can now easily prove the following proposition.

1.32 Proposition. *Suppose for every pair, M, N, of models of T, $M \prec N$ implies $M \prec_c N$. Then T is stable.*

1.33 Exercise. Prove Proposition 1.32.

1.34 Exercise. Show $\mathrm{Th}(Z,S)$ is stable. Show $\mathrm{Th}(Q,<)$ and $\mathrm{Th}(Z,<)$ are not.

The following theorem summarizes this section.

1.35 Theorem. *The theory T is stable iff it satisfies any of the following equivalent conditions.*

i) *For every A, $|S(A)| \le |A|^{|T|}$.*
ii) *For every A, if $|A|^{|T|} = |A|$ then $|S(A)| \le |A|$.*
iii) *For every A, if $|A| = \aleph_0$ then for each ϕ, $|S_\phi(A)| \le \aleph_0$.*
iv) *For every A and every ϕ $|S_\phi(A)| \le |A|$.*
v) *For every ϕ and p, $R(p,\phi) < \infty$.*
vi) *Every elementary extension of a model of T is a conservative elementary extension.*

The following exercise and condition ii) of Theorem 1.35 justify our failure to distinguish the various $S^n(A)$ when defining stability.

1.36 Exercise. Show that for each A with $|A| \leq \lambda$, $|S^1(A)| \leq \lambda$ if and only if for each A with $|A| \leq \lambda$ and each $n < \omega$, $|S^n(A)| \leq \lambda$. (It is essential that we quantify over A on both sides of the equivalence.)

1.37 Historical Notes. Morley began the investigation of cardinalities of families of Stone spaces $S(A)$ for A contained in a model of a countable complete theory T in [Morley 1965]. Ressayre [Ressayre 1969] and Rowbottom [Rowbottom 1964] extended this analysis to uncountable theories while working on the generalized Los conjecture: if T is categorical in some $\lambda > |T|$ then T is categorical in all powers greater than T. Shelah ([Shelah 1969a], [Shelah 1971]) discovered the importance of considering the restriction to instances of a single formula. Most of the results in this section come from those two papers. In particular, the definability lemma was first proved by Shelah [Shelah 1971]. A weaker version (for ω-stable T) was proved by Baldwin [Baldwin 1971] at about the same time. This formulation was applied directly in [Baldwin 1975] and, with more effect, in Lascar [Lascar 1973]. The definition of stability in terms of conservative extensions was first made explicitly in [Baldwin 1975]. It was suggested by analogy with the role of conservative extensions of models of Peano arithmetic. This analogy deserves further examination [Kirby & Pillay 1986]. The significance of varying Theorem 1.22 to obtain Corollary 1.23, by uniformizing the defining formula, was discovered by Harnik in 1980. Various applications of the strengthened form appear in [Harnik & Harrington 1984] and later in this book. In [Shelah 1974], Shelah provides another argument for Theorem 1.22 in the guise of a generalization of the Chang-Makkai Theorem. The examples of two 'obviously' independent transcendental numbers given in the first paragraph of this section require some acquaintance with number theory. (One cannot simply choose an apparently random pair of transcendental numbers. It is unknown whether e and π are algebraically independent or whether $e + \pi$ is rational; or even whether e and e^π are algebraically independent.) The example given follows from the Lindenbaum-Weierstrass theorem: If $b_1, \ldots, b_p$ are algebraic numbers which are linearly independent over the rationals then $e^{b_1}, \ldots, e^{b_p}$ are algebraically independent. Siegel [Siegel 1949] contains a clear introduction to this material.

2. *Types Over Models*

In this section we begin the proof that a freeness relation satisfying the axioms of Chapter II can be defined in any stable theory. We investigate the distinguished (i.e. definable) extensions of types over models. In the process, we consider the relation between definability and Poizat's fundamental order. We also introduce the notion of a strongly saturated model and prove an initial version of the exchange lemma. This chapter provides the groundwork for the extension to types over arbitrary sets in Section 3. For the remainder of this chapter the theory T is tacitly assumed to be stable.

We begin by noticing that if $p \in S(B)$ is definable over A, this definition induces a (possibly inconsistent) set of formulas over any set.

2.1 Definition. Let p be in $S(A)$ and B be arbitrary. If p is defined by d, the *d-extension of p* on B, denoted $d(p, B)$, is the collection of formulas $\phi(\overline{x}; \overline{b})$ with parameters from B which satisfy: $\phi(\overline{x}; \overline{b}) \in d(p, B)$ iff $\models d\phi(\overline{b})$.

We use the term d-extension because usually A will be a subset of B. The map d encodes all the information from p so we sometimes write just $d(B)$ for $d(p, B)$. This observation leads to the following definition.

2.2 Definition. i) A *preschema over A* is any map d taking formulas $\phi(\overline{x}; \overline{y}) \in F(\emptyset)$ to formulas $d\phi(\overline{y}) \in F(A)$.

ii) A *good schema* is a preschema which is also a Boolean algebra homomorphism.

Recall from Lemma 1.26 that if d defines $p \in S(M)$ then d is a good schema. Lascar [Lascar 1976] has an intermediate notion of schema.

2.3 Exercise. Let p be a type over a model M. Suppose d and d' are definitions of p. Show that for each formula $\phi(\overline{x}; \overline{y})$, $\models d\phi(\overline{y}) \leftrightarrow d'\phi(\overline{y})$.

2.4 Exercise. Show Exercise 2.3 fails if M is replaced by an arbitrary subset A.

2.5 Exercise. Give an example of $p \in S(A)$ such that $d(p, B)$ is not a (consistent) type for some B containing A.

In general, the d-extension of $p \in S(A)$ to a set may not even be consistent. However, if p is definable over a model, it is. We have the following slightly more general result.

2.6 Theorem. *Let M be a model and $p \in S(B)$. Suppose the preschema d defines over M an extension of p to an element of $S(M \cup B)$. Then for any A containing B, $d(p, A)$ is a consistent complete type.*

Proof. Consider any finite subset $\phi_1(\overline{x}; \overline{a}_1), \ldots \phi_n(\overline{x}; \overline{a}_n)$ of $d(p, A)$. Then $d\phi_i(\overline{a}_i)$ holds for each i. Now M satisfies the formula:

$$(\forall \overline{u}_1) \ldots (\forall \overline{u}_n)[\bigwedge_{i \in I} d\phi_i(\overline{u}_i) \rightarrow (\exists \overline{x}) \bigwedge_{i \in I} \phi_i(\overline{x}, \overline{u}_i)]$$

and M is a model of T so $(\exists \overline{x}) \bigwedge_{i \in I} \phi_i(\overline{x}; \overline{a}_i)$ holds in the monster model whence $d(p, A)$ is consistent. But clearly $d(p, A)$ is complete since for each $\overline{a} \in A$ and $\phi(\overline{x}; \overline{y})$, $d\phi(\overline{a}) \vee \neg d\phi(\overline{a})$ holds and $\neg d\phi$ is $d(\neg\phi)$.

2.7 Exercise. Show that if d is a good schema then for any A, $d(A)$ is a consistent complete type.

We introduce now some simple notions which turn out to be extremely useful in the context of stable theories. We begin with a simple but significant ordering relation on types.

2.8 Definition. The formula $\phi(\overline{x};\overline{y})$ is *represented* in the type p over A if for some $\overline{a}$ in A, $\phi(\overline{x};\overline{a})$ is in p.

Thus the formula $y = x^2$ is represented in $t(e^{1/2}; Q \cup \{e\})$ although not in $t(e^{1/2}; Q)$. Suppose E is an equivalence relation. If E has only finitely many classes then $E(x;y)$ is represented in every type whose domain is a model; this fails if E has infinitely many classes.

We now formalize the idea that one type satisfies more relations than another.

2.9 Definition (The Fundamental Order). Let p be in $S(A)$ and q in $S(B)$. Then $p \geq q$, if every formula which is represented in p is represented in q. We write $p \simeq q$ if $p \leq q$ and $q \leq p$. We denote the equivalence class of a type p under the resulting equivalence relation by $[p]$.

Note that the domain of this ordering relation is the collection of all types with domain a subset of the monster model. When $C \subseteq \operatorname{dom} p \cap \operatorname{dom} q$ we can naturally strengthen this notion to $p \geq_C q$ if every formula in $L(C)$ which is represented in p is represented in q. We call the resulting order the fundamental order over C. Observe that this ordering is 'upside down.' That is, $q \leq p$ if q represents more formulas than p. Nothing in this definition restricts us to types of finite sequences. Usually only a type whose domain is a model is considered in defining the fundamental order [Lascar & Poizat 1979].

2.10 Example. i) Consider the theory T of dense linear order without endpoints and suppose $a < b_1 < b_2 < \ldots$ are elements of a model of T. Let $B = \{b_1, b_2, \ldots\}$ and $C = \{b_2, b_3, \ldots\}$ and let $p = t(a, B)$ and $q = t(a, C)$. Recalling that T is quantifier eliminable, we see that (up to equivalence in T) the only formulas represented in p are equivalent to Boolean combination of $x < y_1, x < y_1 < y_2, \ldots$ and some equality formulas and these are all represented in q so $p \geq q$. Since $C \subseteq B$, $q \geq_C p$. However, the formula $x < y < b_2$ in $L(C)$ is represented in p but not in q thus $p \not\geq_C q$.

ii) Let T be a complete theory of Abelian groups which admits elimination of quantifiers (in the language of groups). Then every formula $\phi(\overline{x};\overline{y})$ is equivalent to a Boolean combination of equations of the form: $\sum_{i<n} p_i x_i + \sum_{j<m} q_j y_j = 0$. Now the position in the fundamental order of a 1-type $q \in S(A)$ is determined by the ideal, I_q, of those integers n such that if c realizes q, then nc is in A, the function f from I_q into A defined by $f(n) = nc$, and K_q, the ideal composed of integers which annihilate any realization of q. Note that K_q is determined from I_q and f by $K_q = ker(f)$.

2.11 Exercise. Let $p \in S(M)$. Show that if the type p is minimal in the fundamental order then p is realized in M. Show the converse is false. (Hint: Consider the formula $x = y$.)

2.12 Exercise. Construct the fundamental order for the theory T_1 of two refining equivalence relations and for the theory T_2 of two crosscutting equivalence relations. Assume each equivalence relation has infinitely many classes, all infinite. (These theories are considered in more detail in Section III.4.)

2.13 Exercise. Show that if p and q are comparable in the fundamental order, they have the same restriction to the empty set.

Certainly if q extends p, $p \geq q$. We distinguish those extensions q of p where the reverse inequality also holds.

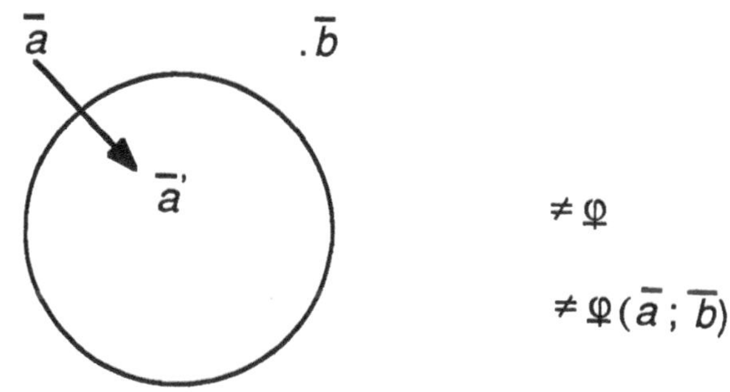

Fig. 2. $t(\bar{a}; M \cup \bar{b})$ is finitely satisfied in M.

2.14 Definition. (Fig. 2). Let M be a model of T and M be contained in A.

i) Let $p \in S(M)$; $q \in S(A)$ is an *heir* of p on A if and only if q extends p and $p \simeq_M q$.
ii) $t(\bar{a}; A)$ is a *coheir* of $t(\bar{a}; M)$ on A if and only if $t(\bar{a}; A)$ is finitely satisfiable in M.

The following exercise explains coheir in terms of heir.

2.15 Exercise. Suppose $M \subseteq A$. Let $\bar{a}$ realize $p \in S(M)$. Then $t(\bar{a}; A)$ is a coheir of p on A if $t(A; M \cup \bar{a})$ is an heir of $t(A; M)$.

Here is another characterization of a coheir.

2.16 Exercise. Show $t(\bar{a}; A)$ is a coheir of $t(\bar{a}; M)$ on A iff $t(\bar{a}; A)$ is an element of the topological closure of $\{t(\bar{c}; A) : \bar{c} \in M\}$.

We can derive the definability of types from the hypothesis that T is stable in some power λ using Beth's theorem as follows. From Beth's theorem one can deduce that each type which has a unique heir in $S(A)$ for every A containing $\operatorname{dom} p$ is definable. A non-trivial compactness argument shows that this result propagates to show that if $p \in S(M)$ is not definable (and thus has two heirs), then every heir $p_1 \in S(M_1)$ of p with $M \equiv M_1$ also has two heirs. Now a union of chains argument shows T is unstable in all sufficiently large λ. The details can be found in [Lascar & Poizat 1979] or [Pillay 1983a].

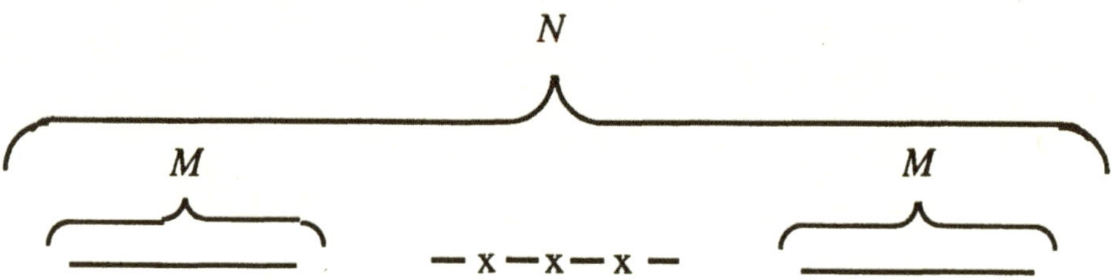

Fig. 3. Heirs

2.17 Example. Consider $(Q, <) = M$ as a dense linear order and let p be the 1-type of an irrational cut. Let N be an elementary extension of M with another copy of Q in the cut. Now p has continuum many heirs on N, namely one for each type over N extending p but not realized in N. (Fig. 3). However, p has only two coheirs on N, namely the types of points in the cuts between the models M and N. (Fig. 4). For, suppose there are elements a, b in $N - M$ such that $a < x < b$ is in q. Let c realize q. Then $x_1 < c < x_2$ is in $t(a\frown b; Q \cup \{c\})$ and $x_1 < y < x_2$ is not represented in $t(a\frown b; Q)$.

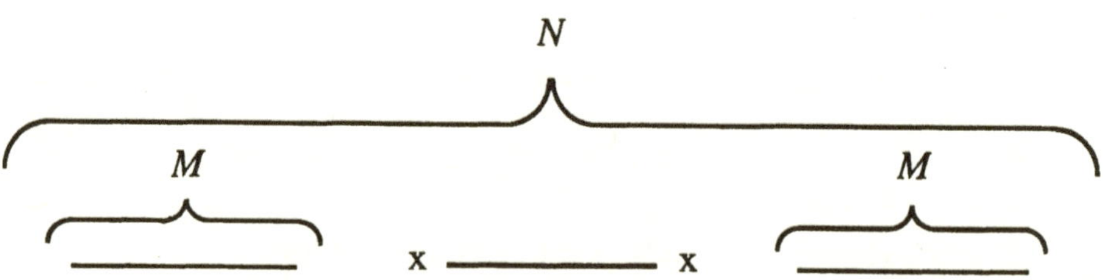

Fig. 4. Coheirs

2.18 Exercise. i) Let $A \subseteq B$ be Abelian groups and suppose $\operatorname{Th}(B)$ is quantifier eliminable, $p \in S(A)$, $q \in S(B)$. Show $p \geq q$ iff $K_p = K_q$,

$I_p \subseteq I_q$, $f_q|I_p = f_p$ and $f_q^{-1}(A) = I_p$. (K_p, I_p are defined in Exercise 2.10.)

ii) Let T be the theory of algebraically closed fields of characteristic zero. Let $k \models T$ and suppose L, K are superfields of k. Show $t(K; L)$ is an heir of $t(K; k)$ if and only if K and L are algebraically independent (equals linearly disjoint) over k ([Zariski & Samuel 1958]). What, if anything, can be said if the restriction that k is a model of T is removed?

We will consider the generalization of Exercise 2.18 i) to an arbitrary module in Section 5.

2.19 Definition. Let B be contained in A and $p \in S(A)$. Then p *splits* over B if for some $\overline{a}$, $\overline{b}$ in A, $\overline{a}$ and $\overline{b}$ realize the same type over B but for some formula $\phi(\overline{x}; \overline{y})$, $\phi(\overline{x}; \overline{a})$ is in p and $\neg\phi(\overline{x}; \overline{b})$ is in p.

2.20 Theorem. *Let M be contained in A, p an element of $S(M)$, and q a coheir of p on A. Then q does not split over M.*

Proof. Suppose $\overline{c}$ realizes q but $\phi(\overline{c}; \overline{a})$ holds while $\phi(\overline{c}; \overline{b})$ does not. Then, since $t(\overline{a}\frown\overline{b}; M \cup \overline{c})$ is the heir of $t(\overline{a}\frown\overline{b}; M)$ for some $\overline{m} \in M$, $\phi(\overline{m}; \overline{a})$ holds and $\phi(\overline{m}; \overline{b})$ does not. But then $\overline{a}$ and $\overline{b}$ do not realize the same type over M.

2.21 Corollary. *For any M a model of T and any p in $S(M)$ and any elementary extension N of M, the maximal number of coheirs of p on N depends only on the cardinality of M (not N).*

Note that we do not need to assume T is stable.

2.22 Exercise. Compute an upper bound in terms of $|\operatorname{dom} p|$ on the number of coheirs of p.

Now we examine the relation between the fundamental order and definability of types.

2.23 Theorem. *Let T be stable. If M is a model, $M \subseteq A$, and $p \in S(M)$ then $d(p, A)$ is the unique type q extending p such that $q \simeq p$. A fortiori, $d(p, A)$ is the unique heir of p on A.*

Proof. Note that $d(p, A)$ is an heir of p. For, $\phi(\overline{x}; \overline{a}, \overline{m}) \in d(p, A)$ (with $\overline{a} \in A - M$, $\overline{m} \in M$) implies $\models d\phi(\overline{a}, \overline{m})$ and thus $\models (\exists \overline{y}) d\phi(\overline{y}, \overline{m})$. Since M is an elementary submodel of the monste· model, this in turn implies $\models d\phi(\overline{m}', \overline{m})$ for some $\overline{m}' \in M$ which holds if and only if $\phi(\overline{x}, \overline{m}')$ is in p.

For the uniqueness, suppose $r \in S(A)$, $r \simeq p$ and $r \neq d(p, A)$. Fix $\theta(\overline{x}; \overline{z})$ and $\lambda(\overline{y}; \overline{z})$ as in Corollary 1.23. Since $r \neq d(p, A)$, the formula $\neg[\phi(\overline{x}; \overline{y}) \leftrightarrow \lambda(\overline{y}; \overline{z})] \wedge \theta(\overline{x}; \overline{z})$ is represented in r. But then it is represented in p, contrary to the definability of p.

If $M \subseteq A$ and $M \models T$, we can now speak of *the* definable extension of p without mentioning the particular definition d, since the last theorem asserts that all definitions of p yield the same extension to $S(A)$.

If $M \subseteq A$ and $p \in S(M)$ then p has a coheir on A. To see this, fix any $\overline{c}$ realizing p and let A' realize $d(t(A, M), M \cup \{\overline{c}\})$. Then, if f is an automorphism which fixes M and maps A' to A, $f(\overline{c})$ realizes a coheir of p on A. Unlike Theorem 2.23, this remark does not require T to be stable. With the stability assumption we could conclude the coheir is unique.

Note that Theorem 2.23 reduces the question of whether an extension of $p \in S(M)$ is an heir of p to properties of the fundamental order over the empty set rather than the fundamental order over M. More precisely, we have the following result.

2.24 Exercise. Let $M \subseteq N$, $p \in S(M)$ and $p \subseteq q \in S(N)$. Show $p \simeq q$ implies q is an heir of p.

2.25 Exercise. If $p \in S(M)$ and p is not realized in M then for any A containing M, $d(p, A)$ has infinitely many solutions in the monster model.

Now we will show the relation between heir and coheir in stable theories. This relationship leads ultimately to the exhange lemma for our notion of independence. We first indicate by a series of exercises how to prove the result using the basic definability theorem Theorem II.2.27. (This proof is due to Poizat.)

2.26 Exercise. Suppose $M \subseteq N$, N is $|M|^+$-saturated, $p \in S(M)$ and q is a coheir of p on N. Use Theorem 2.20 to show that q is definable not only over N but over M. Observe that this means that for any A with $M \subseteq A \subseteq N$, the coheir of p on A is definable over M.

2.27 Exercise. Use the previous exercise and Theorem 2.23 to show that if $p \in S(M)$ and $M \subseteq A$, the heir of p on A equals the coheir of p on A.

To introduce some useful techniques and show some further consequences of stability, we will give a different proof based on the following characterization of stability. This argument will be the only place in the development of the forking notion that we appeal to the characterization of a stable theory in terms of the cardinality of Stone spaces. The previous exercises show that this appeal is unnecessary.

2.28 Definition. The formula $\phi(\overline{x}; \overline{y})$ has the *order property* if in some model of T there exist infinite sequences $\{\overline{a}_i : i < \omega\}$ and $\{\overline{b}_i : i < \omega\}$ such that $\models \phi(\overline{a}_i; \overline{b}_j)$ iff $i < j$.

2.29 Theorem. *If some formula $\phi(\overline{x}; \overline{y})$ has the order property then T is not stable.*

Proof. Show that $\Gamma_{\langle\,\rangle}(\phi, n)$ is consistent for each n.

The converse to this theorem also holds ([Shelah 1978], Chapter II) but we will not need it here. In fact, the class of unstable theories can be divided into those which satisfy the strict order property and those which satisfy the independence property. These two notions mean that there is a formula $\phi(\overline{x}; \overline{y})$ and a sequence $\langle \overline{a}_i : i < \omega \rangle$ such that the sets $\{\phi(\mathcal{M}; \overline{a}_i) : i < \omega\}$ are, respectively, linearly ordered by inclusion or independent (i.e., every finite Boolean combination is nonempty.) We discuss these notions in more detail in Sections 4.42 and 4.43.

2.30 Theorem. *Let M be a model of the stable theory T. Suppose M is contained in C and p is in $S(M)$; then the heir of p on C equals the coheir of p on C.*

Proof. Let $\overline{a}$ realize q, a coheir of p on C. If q is not $d(p, C)$, the heir of p on C, then there is a formula $\phi(\overline{x}; \overline{y})$ and a sequence $\overline{c} \in C$ such that $\phi(\overline{x}; \overline{c}) \in q$ but $\neg d\phi(\overline{c})$. Let $r = t(\overline{c}; M)$. Since q is the coheir of p on C, $t(\overline{c}; M \cup \overline{a})$ is the heir, hence the definable extension, of r on $M \cup \overline{a}$. This means that, letting $\psi(\overline{y}; \overline{x}) = \phi(\overline{x}; \overline{y})$, $\models \phi(\overline{a}; \overline{c})$ iff $\models d'\psi(\overline{a})$ where d' defines $t(\overline{c}; M)$.

Now, choose $\overline{e}_i$, $\overline{f}_i$ for $i < \omega$, by induction, with $\overline{e}_0 = \overline{a}$ and $\overline{f}_0 = \overline{c}$ so that

i) $\overline{e}_i$ realizes the definable extension of p on $M \cup E_i \cup F_i$ and
ii) $\overline{f}_i$ realizes the definable extension of r on $M \cup E_{i+1} \cup F_i$.

Since $\overline{e}_i$ realizes p we have $d'\psi(\overline{e}_i)$ for all i. Similarly, since each $\overline{f}_i$ realizes r, we have $\neg d\phi(\overline{f}_i)$ for all i.

Now i) implies for $i > j, \phi(\overline{e}_i, \overline{f}_j)$ iff $d\phi(\overline{f}_j)$ while ii) implies for $i \leq j$ that $\models \phi(\overline{e}_i, \overline{f}_j)$ iff $\models d'\psi(\overline{e}_i)$. Thus $\models \phi(\overline{e}_i, \overline{f}_j)$ iff $i \leq j$ so T is unstable.

We can summarize the last few results in the following theorem.

2.31 Theorem. *Let M be a model of a stable theory T, $p \in S(M)$ and $M \subseteq C$; then p has a unique distinguished extension to a complete type over C which is its heir, coheir, and definable extension.*

In the next section we will define the notion of nonforking. Under this definition, for any model M and superset A, if $p \in S(A)$, p does not fork over M just if p is the definable extension of $p|M$. This notion satisfies the monotonicity and transitivity properties. ¿From Lemma II.2.10 and Theorem 2.31 we will deduce the symmetry theorem for nonforking.

We introduce now a strengthening of the notion of a saturated model which plays an important role in the investigation of stable theories. We show in Theorem XI.2.3 that this notion agrees with Shelah's notion of an F^a_κ-saturated model.

2.32 Definition. The model M is *strongly κ-saturated* if for each set A with $|A| < \kappa$, any type over A in fewer than κ free variables over A which is finitely satisfied in M is realized in M.

The crucial point in this definition is that A need not be a subset of M. The concept defined by replacing a type of less than κ elements by a type of any finite number of elements is apparently weaker. For, $t(\overline{b}_0 {}^\frown \overline{b}_1; A)$ finitely satisfied in M does not imply that $t(\overline{b}_1; A \cup \overline{b}_0)$ is finitely satisfied in M. Thus, there is no obvious way to obtain the analog to the fact that if a model is saturated for types over finite sets then it is saturated for ω-types over finite sets.

2.33 Exercise. Let $p \in S(B)$ and $A \subseteq B$. Suppose there is a strongly $|B|^+$-saturated model M with $B \subseteq M$ and there is an extension of p to $p' \in S(M)$ such that p' is definable over every $N \models T$ with $A \subseteq N \subseteq M$. Then the definable extension (coheir) of p' to a global type is definable over every model containing A.

On first reading of the proof of the following lemma, take κ to be ω. We explain after the proof the, primarily notational, adjustments which must be made for arbitrary κ. The key to the proof is the recognition that in a formula $\phi(\overline{x}, \overline{y}, \overline{z})$, we are free to decide which variables are to stand for parameters and which to be viewed as the free variables of a type. We reflect this in our notation by writing $(d_r\overline{y})\phi(\overline{x}, \overline{y}, \overline{z})$ to indicate that $\overline{y}$ is the variable of the type r which is being defined. This marvelous notation was introduced in [Hrushovski 1986]. The quantifier $(d_r\overline{y})$ is read 'For generic $\overline{y}$ realizing r.'

2.34 Lemma. *If M is a κ-saturated model of a stable theory and $\kappa > |T|$ then M is strongly κ-saturated.*

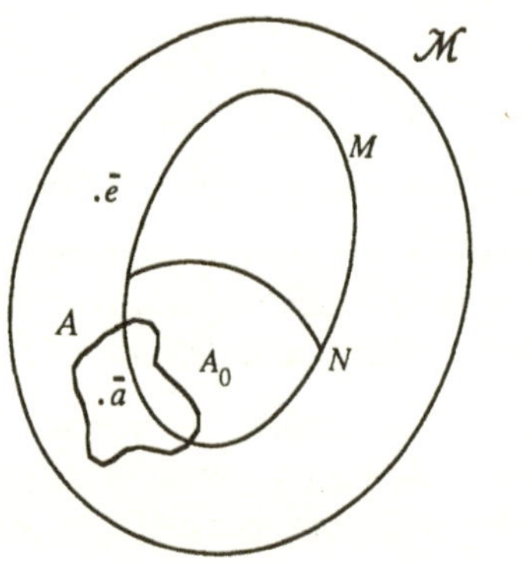

Fig. 5. Lemma 2.34

Proof. (Fig. 5). Suppose $|A| < \kappa$ and let $A \cap M = A_0$. Let $\overline{e}$ be a sequence of less than κ elements such that $p = t(\overline{e}; A)$ is finitely satisfiable in M. For any finite sequence $\overline{a}$ from $A - A_0$, we will show how to convert $t(\overline{e}; A_0 \cup \overline{a})$ into an equivalent type q over a subset N of M with $|N| < \kappa$ and $A_0 \subseteq N \subseteq M$. Given this conversion, the κ-saturation of M guarantees that $t(\overline{e}; A)$ is realized in M.

Let d define $r = t(\overline{a}; M)$ over M. By Lemma 1.24 we may suppose only $|T|$ parameters occur in the range of d. Let N be a submodel of M with A_0 and all these parameters contained in N and with $|N| < \kappa$. Now for any $\phi(\overline{x}, \overline{y}, \overline{z})$, $\overline{m} \in M$, $\overline{a} \in A - A_0$, and $\overline{b}$ from A_0,

$$\models \phi(\overline{m}, \overline{a}, \overline{b}) \text{ iff } \models (d_r\overline{y})\phi(\overline{m}, \overline{y}, \overline{b}).$$

So the required translation type is $\{(d_r\overline{y})\phi(\overline{m}, \overline{y}, \overline{b}) : \phi(\overline{x}; \overline{a}, \overline{b}) \in p\}$. To see this type is consistent, note that for any ϕ, $\models \phi(\overline{e}; \overline{a}, \overline{b})$ implies by finite satisfiability that there is an $\overline{e}' \in M$ such that $\models \phi(\overline{e}'; \overline{a}, \overline{b})$ and thus $\models (d_r\overline{y})\phi(\overline{e}', \overline{y}, \overline{b}')$.

If κ is uncountable, fix an enumeration of $\overline{e}$ (and thus of $\overline{x}$). Now the translation type is formed by applying the procedure above to each formula $\phi(\overline{x}', \overline{y}, \overline{z})$ where $\overline{x}'$ is a finite subsequence of $\overline{x}$.

2.35 Definition. Let $A \subseteq M$ and $p, q \in S(M)$. We say p is *conjugate* to q over A and write $p \sim_A q$ if there is an automorphism of M which fixes A and maps p to q.

We now relate this notion to the fundamental order. In the next argument we use the expression $t(B; A \cup \overline{c}) = t(B'; A \cup \overline{d})$. Recall from the discussion in Chapter I that this means there is an automorphism which fixes A, maps B to B' and $\overline{c}$ to $\overline{d}$.

2.36 Theorem. *Let T be stable. If M is saturated, $|M| = \kappa > |T|$, and $|A| < \kappa$ then for any $p, q \in S(M)$, if $p \simeq_A q$ then $p \sim_A q$.*

Proof. Let $\overline{c}$ realize p and $\overline{d}$ realize q. Choose for each formula $\phi(\overline{x}; \overline{y})$, formulas $\theta(\overline{x}; \overline{z})$ and $\lambda(\overline{y}; \overline{z})$ to define $t(\overline{c}; M)$ in the manner of Corollary 1.23. Choose $B \subseteq M$ such that $A \subseteq B$, $|B| \leq |A| + |T|$ and B contains all parameters necessary to define p. That is, for each θ chosen according to Corollary 1.23, there is a $\overline{b} \in B$ such that $\theta(\overline{x}; \overline{b}) \in p$ and for every $\overline{m} \in M$, $[\phi(\overline{x}; \overline{m}) \leftrightarrow \lambda(\overline{m}; \overline{b})] \in p$. Since $p \simeq_A q$, if $\beta \in \mathrm{Aut}(M)$ fixes A and maps $\overline{c}$ to $\overline{d}$, $t(\beta(B); A \cup \overline{d})$ is finitely satisfied in M. Now by strong saturation, choose $B' \subseteq M$ such that $t(B; A \cup \overline{c}) = t(B'; A \cup \overline{d})$. Let α be an automorphism of M fixing A and taking B to B'. Now for each ϕ, and each θ and λ depending on ϕ, if $\theta(\overline{x}; \overline{b}) \in p$ then $\theta(\overline{x}; \alpha\overline{b}) \in q$. Now towards a contradiction suppose that $\alpha p \neq q$. Therefore $\neg[\phi(\overline{x}; \overline{m}) \leftrightarrow \lambda(\overline{m}; \alpha\overline{b})] \in q$. Since $p \simeq_A q$, there exist $\overline{b}'$ and $\overline{m}'$ with $\theta(\overline{x}; \overline{b}') \in p$ and $\neg[\phi(\overline{x}'; \overline{m}') \leftrightarrow \lambda(\overline{m}'; \overline{b}')] \in p$, contrary to our choice of θ and λ. Thus $\alpha p = q$ as required.

We need to have M saturated to express our condition in terms of full automorphisms of the model M. A somewhat weaker version (expressing the result in terms of extending partial automorphisms) holds with only the hypothesis that M is $|A| + |T|$-saturated.

2.37 Exercise. Show the hypothesis of stability is necessary for Theorem 2.36. (Hint: Consider Example 2.10.)

Since $\mathcal{M}$ is extremely saturated and conjugate types certainly represent the same formulas we have

2.38 Corollary. *On $S(\mathcal{M})$, $\simeq_A$ and $\sim_A$ define the same equivalence relation.*

There is a further obvious corollary.

2.39 Corollary. *If T is stable in $|A|$ then the equivalence relation $\sim_A$ has at most $2^{|T|+|A|}$ equivalence classes.*

2.40 Historical Notes. There are several variants in the literature of the definition of a good schema. The definition here agrees with that in [Lascar 1976] and [Pillay 1983e] but not with the one in [Berline 1983]. The notions of coheir and fundamental order first appear in [Lascar & Poizat 1979]. The notion of splitting was introduced by Shelah. The order property characterization of stability is from [Shelah 1971]. The proof given for heir = coheir is adapted from [Lascar 1976]. The definition here of strong saturation is from [Baldwin 1984]. The relation to Shelah's notion of $F^a_{\kappa(T)}$-saturation is discussed in Chapter XI. This notion is entirely different from that called strong saturation by Shelah in [Shelah 198?c]. Theorem 2.36 is implicit in [Lascar & Poizat 1979]. There has been considerable work, [Lascar & Poizat 1979], [Harnik & Harrington 1984], [Pillay 198?] developing a local stability theory where one assumes only that a particular formula is stable rather than the entire theory. Since this project can only aid in the study of independence, as opposed to generation, we have not pursued it in this book.

3. Nonforking Types Over Sets

In Section 1 we saw that for $p \in S(M)$, M a model, the definable extension of p to a set A containing M is the kind of canonical extension we envisioned in Chapter II. Here, we find an analogous extension for types over arbitrary sets and verify that all the axioms of Section II.1 hold. This analog will be called a nonforking extension of p. Following a suggestion of M. Ziegler, we introduce nonforking by first defining the notion for types over the monster model (so called global types) and then extending it to types over arbitrary subsets in the natural manner to satisfy the monotonicity axioms.

3.1 Definition. i) Let $\hat{p} \in S(\mathcal{M})$. Then $\hat{p}$ *does not fork over A* iff for every $M \models T$ with $A \subseteq M$, $\hat{p}$ is definable over M.

ii) Let p be a type over B. Then p *does not fork over A* if for some $\hat{p} \in S(\mathcal{M})$ with $p \subseteq \hat{p}$, $\hat{p}$ does not fork over A. We say the formula $\phi(\overline{x};\overline{y})$ *does not fork over A* if the type $\phi(\overline{x};\overline{y})$ does not fork over A.

The following exercise gives a sufficient, but by no means necessary, condition for showing a type $\hat{p}$ does not fork over A.

3.2 Exercise. Show using Theorem 2.23 that if $\hat{p} \in S(\mathcal{M})$ and for some model $M \supseteq A$, $\hat{p}|M$ is definable over A then $\hat{p}$ does not fork over A.

There are actually three layers of complexity in defining nonforking: complete types over models, complete types over arbitrary sets, and incomplete types. For most purposes one can ignore the third case on a first reading. The cases where the extension to incomplete types gives the most trouble are pointed out.

3.3 Notation. We may write '$(p\downarrow B; A)$' for p does not fork over A where $B = \operatorname{dom} p$. We write $(\bar{c}\downarrow B; A)$ or $\bar{c}\downarrow_A B$ for $t(\bar{c}; B \cup A)$ does not fork over A. This notation naturally extends to $(C\downarrow B; A)$ as explained in Section II.2. We often read this as C is independent from B over A. Note that we do not require $A \subseteq B$ to write $t(C; B)$ does not fork over A, although this will frequently be the case. If so, $(C\downarrow B; A)$ is equivalent to $t(C; B)$ does not fork over A.

If $A \subseteq B$ we write $N(B, A)$ for $\{p \in S(B) : p \text{ does not fork over } A\}$.

If the type p with parameters from A is stationary and $A \subseteq B$ we write p^B for the, necessarily unique, nonforking extension of p to a complete type over B whose existence is guaranteed by Theorem 3.11. Other authors write $p|B$ for this extension; we reserve that notation for the restriction of a type.

3.4 Exercise. Show that $p \in S(B)$ does not fork over A iff for some strongly $(|T| + |A| + |B|)^+$-saturated model M_1 with $A, B \subseteq M_1$ there is an extension of p to $p' \in S(M_1)$ such that p' is definable over every $N \models T$ with $A \subseteq N \subseteq M_1$.

3.5 Exercise. If $\hat{p}$, $\hat{q} \in S(\mathcal{M})$, $\hat{p}$ does not fork over A and $\hat{p} \sim_A \hat{q}$ then $\hat{q}$ does not fork over A.

3.6 Exercise. Give an example of a theory T and elements a, b, c which are pairwise independent but not independent. (Hint: Almost any theory with an addition operation will do.)

We now want to show that the nonforking relation satisfies the axioms for freeness. Immediately from the definition, we see that nonforking satisfies the monotonicity axioms and the second extension axiom.

3.7 Lemma. *The nonforking relation satisfies the following axioms.*

M_1. *If $q \subseteq p$ and p does not fork over A then q does not fork over A.*
M_2. *If $A \subseteq B$ and p does not fork over A then p does not fork over B.*
E_2. *If p does not fork over A and $\operatorname{dom} p \subseteq B$, there is a $p_1 \in S(B)$ which extends p such that p_1 does not fork over A.*

To verify our remaining existence axiom, E_1, we show that if $p \in S(A)$ then p does not fork over A. We derive this from the fact that heir and coheir are the same for types over models and the following lemma. The lemma is proved by a compactness argument. The next exercise spells out the method for showing the consistency of each finite subset of the infinite set of sentences considered in the lemma.

3.8 Exercise. Suppose $p_1, \ldots, p_k$ is an increasing sequence of types in the fundamental order with each $p_i \in S(M_i)$. Show there is a model M containing $\operatorname{dom} p_i$ for each i and a type $p \in S(M)$ such that $p \geq p_i$ for each i. (Hint: Let M contain $\bigcup_i M_i$ and let p be the definable extension of p_k to M.)

3.9 The Extension Lemma. Let $p \in S(A)$. There exists a $\hat{p} \in S(\mathcal{M})$ with $p \subseteq \hat{p}$ such that for every model M with $A \subseteq M$, $\hat{p}$ is the heir of $\hat{p}|M$.

Proof. Let $Z = \{q$: There exists a model M, $A \subseteq M$, $q \in S(M)$ and $p \subseteq q\}$. Z is partially ordered by $\leq_A$. If p^* is a maximal element of Z and $\hat{p}$ is the heir of p^* on $\mathcal{M}$ then by Theorem 2.23, $\hat{p}$ is as required. To show such a p^* exists, we need, invoking Zorn's lemma, show only that if $P = \{p_i : i \in I\}$ is an increasing chain in Z then P has an upper bound p'. For this, let N be a model of the following set of sentences, Γ, which can easily be shown consistent using the previous exercise. Γ is in the language obtained by adding to L new constant symbols $c_0, \ldots, c_{n-1}$, names for the members of $\operatorname{dom} p_i = M_i$, and a new unary predicate symbol R. Let Γ be:

$$T \cup T|R \cup \bigcup_{i \in I} \operatorname{Diag}(M_i) \cup \{R(m) : m \in M_i, i \in I\}$$
$$\cup \{(\forall \overline{y})(R(\overline{y}) \rightarrow \neg\phi(\overline{c}; \overline{y})) : (\exists i)\phi(\overline{x}; \overline{y}) \text{ is not represented in } p_i\}.$$

There is little difficulty in extending Lemma 3.9 to allow the type p to have infinitely many variables and we will use the extended version without comment.

3.10 Theorem. *If p is a type over A then p does not fork over A.*

Proof. In view of the first monotonicity axiom we can assume $p \in S(A)$. By the previous lemma, we can find $\hat{p} \in S(\mathcal{M})$ which extends p and is the heir of each of its restrictions to a model containing A. By Theorem 2.31, (heir = coheir) it is the definable extension of each of these restrictions. Thus $\hat{p}$ and, a fortiori, p does not fork over A.

Note that the use of (heir = coheir = definable extension) is vital here; a direct proof of the existence of a $\hat{p}$ which is the coheir of each of its restrictions can be given but is much longer (see Paragraph 3.30).

We can now establish a useful equivalent to nonforking.

3.11 Definition. (Fig. 6).

i) The formula $\phi(\overline{x}; \overline{c})$, where the parameters $\overline{c}$ come from anywhere in the monster model, is *almost satisfied* in A if for every model M containing A, $\phi(\mathcal{M}; \overline{c}) \cap M \neq \emptyset$.
ii) The type p is *almost satisfied* in A if every finite conjunction of formulas in p is almost satisfied in A.

3.12 Lemma. *Let $A \subseteq B$ and let p be a type over B; p does not fork over A iff p is almost satisfied in A.*

Proof. If p does not fork over A then $p \subseteq \hat{p}$ for some global type $\hat{p}$ which is definable over M for each model M containing A (and thus by Theorem 2.31 finitely satisfied in M). For the converse, suppose p is almost satisfied in A. Choose by the Extension Lemma a strongly $|B|^+$-saturated model M such that $t(M; B)$ does not fork over A. Then p is finitely satisfied in M and thus realized in M by some $\overline{c}$. Now $p \subseteq t(\overline{c}; B)$ which does not fork over A. By monotonicity, p does not fork over A.

3.13 Corollary. *If p is a type over B which is closed under conjunction and p forks over A then there is a formula $\phi(\overline{x}; \overline{b}) \in p$ such that every type containing $\phi(\overline{x}; \overline{b})$ forks over A.*

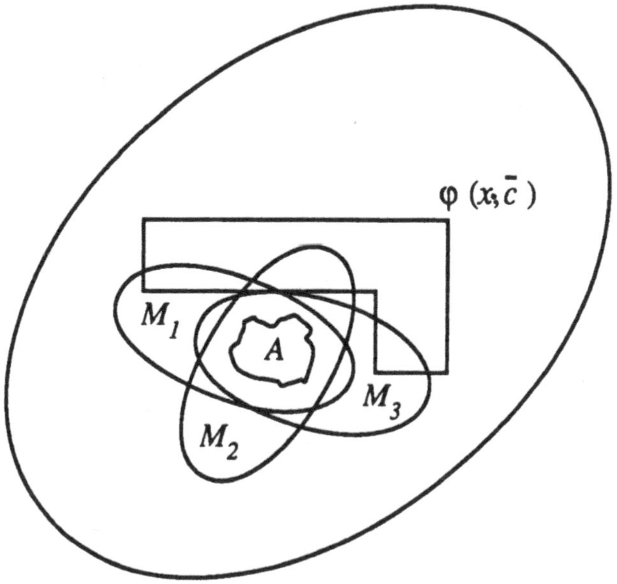

Fig. 6. $\phi(x;\overline{c})$ is almost satisfied in A.

Proof. The notion of almost satisfaction obviously has the required finite character.

The necessity of the assumption that p is closed under finite conjunction is demonstrated in Example V.3.10.

The notation disguises an important aspect of Corollary 3.13. The corollary does not say there is a formula $\phi(\overline{x};\overline{y})$ such that for any $\overline{b}'$, any type containing $\phi(\overline{x};\overline{b}')$ forks over A. This holds only for those '$\overline{b}'$' which realize $t(\overline{b};A)$. We will show in Corollary IV.2.2 a closely related result: if $A \subseteq B$, $q \in S(B)$ and q forks over A then there is a formula $\phi(\overline{x};\overline{y})$ such that for some $\overline{b}$, $\phi(\overline{x};\overline{b}) \in q$, and for any $\overline{c}$, $q|A \cup \phi(\overline{x};\overline{c})$ forks over A.

The following exercise gives Corollary 3.13 a useful topological form.

3.14 Exercise. Show that for any $A \subseteq B$, $N(B,A)$ (the collection of types in $S(B)$ which do not fork over A) is closed in $S(B)$.

The next few results connect the fundamental order with the notion of forking. Recall $[q]$ denotes an equivalence class in the fundamental order.

3.15 Definition. For any A and any $p \in S(A)$, the *bound of p*, $\beta(p)$ is the least upper bound of $W = \{[q] : p \subseteq q \text{ and } q \text{ is a type over a model}\}$.

We justify this definition by deducing from the following lemma that W does have a unique least upper bound.

3.16 Exercise. Show, using both the fact that the class of nonforking types is closed under isomorphism and the Extension Lemma that for any B and A we can find an N such that $t(B;N)$ does not fork over A.

3.17 Lemma. *Let $A \subseteq B$, $p \in S(B)$, and suppose p extends to $\hat{p} \in S(\mathcal{M})$ which does not fork over A. Let M be any model containing B and let $p|A \subseteq q \in S(M)$. Then $[q] \leq_A [\hat{p}|M]$.*

Proof. Choose, by Exercise 3.16, an N which is strongly $|M|^+$-saturated and so that $t(M;N)$ does not fork over A. Since $\hat{p}$ does not fork over A, $\hat{p}|M$ is realized in N by some $\overline{c}$. Let $\overline{d}$ realize q and let α be an automorphism of $\mathcal{M}$ which fixes A and maps $\overline{d}$ to $\overline{c}$. Let $\hat{q}$ denote the definable extension of q to a global type. Then, $\alpha t(\overline{d};M) = (\alpha\hat{q})|(\alpha M) = t(\overline{c};\alpha M)$. Now suppose $\phi(\overline{x};\overline{y}) \in F(A)$ is represented in $[\hat{p}]$. This implies that for some $\overline{m} \in M$, $\models \phi(\overline{c};\overline{m})$. Since $t(\overline{m};N)$ is almost satisfiable in A, for some $\overline{m}' \in \alpha M$, $\models \phi(\overline{c};\overline{m}')$. That is, $\phi(\overline{x};\overline{y})$ is represented in $\alpha\hat{q}$, hence in $\hat{q}$, hence in q.

Of course, in Lemma 3.17 we could replace $\leq_A$ by the weaker relation $\leq$. Now we can show rapidly that the nonforking relation satisfies the rest of the axioms from Section II.1. We note first that forking can now be described in terms of the fundamental order.

3.18 Corollary. *Fix any A and $p \in S(B)$ with $A \subseteq B$. If $q = p|A$ then p forks over A iff $\beta(p) < \beta(q)$.*

Proof. Let $\hat{p}$ and $\hat{q}$ be $\beta(p)$ and $\beta(p|A)$ respectively. Then $\hat{q}$ does not fork over A. If $\hat{p}$ does not fork over A then p does not fork over A so Lemma 3.17 implies $[\hat{p}] = [\hat{q}]$. For the converse, if p forks over A, then for every $\hat{p}$ extending p to S($\mathcal{M}$) there is an M containing A such that $\hat{p}$ is not the definable extension of $\hat{p}|M$ so by Theorem 2.23, $[\hat{p}] < [\hat{p}|M]$. But by Lemma 3.17, $[\hat{p}|M] \leq [\hat{q}|M]$ and hence $\beta(p) < \beta(p|A)$.

We now can deduce the transitivity axiom without effort. From Theorem III.2.31 and the reduction of the symmetry lemma to symmetry for types over models in Lemma II.2.12 we derive the symmetry lemma.

3.19 Corollary. i) (Transitivity) *Let $A \subseteq B \subseteq C$ and let $p \in S(C)$. If p does not fork over B and $p|B$ does not fork over A then p does not fork over A.*

ii) (Symmetry) *If $t(\overline{a};A \cup \overline{b})$ does not fork over A then $t(\overline{b};A \cup \overline{a})$ does not fork over A.*

3.20 Exercise. Show that $(\overline{a}^\frown\overline{b}\downarrow\overline{c}^\frown\overline{d};A)$ implies $(\overline{a}\downarrow\overline{d};A \cup \overline{b} \cup \overline{c})$.

Applying Lemma 3.17 once again we get the conjugacy lemma for non-forking extensions.

3.21 Corollary (The Conjugacy Lemma). *If $\hat{p}|A = \hat{q}|A$ and neither $\hat{p}$ nor $\hat{q}$ forks over A then $\hat{p} \sim_A \hat{q}$.*

Proof. By Lemma 3.17 we have $\hat{p} \simeq_A \hat{q}$ whence the result follows by Theorem 2.36.

The following exercise provides an extremely useful corollary to the conjugacy lemma.

3.22 Exercise. Let $\phi(\overline{x};\overline{b})$ be a formula. Show that the set of types q in $S(\mathcal{M})$ which do not fork over A and contain a conjugate of $\phi(\overline{x};\overline{b})$ is the same as the collection of q in $S(\mathcal{M})$ which do not fork over A and satisfy: $q|A$ has a nonforking extension to a type over $\mathcal{M}$ which contains $\phi(\overline{x};\overline{b})$. Note that the first of these collections is open in $N(\mathcal{M}, A)$.

The next result, the open mapping theorem, has enormously important consequences. These consequences and some generalizations of the result will be discussed in Chapter X. It is an easy consequence of the results at hand.

3.23 Theorem (The Open Mapping Theorem). *The restriction map r from $N(B, A)$ onto $S(A)$ is an open map.*

Proof. First note it suffices to prove the result for $B = \mathcal{M}$, since the restriction map from $\mathcal{M}$ to A is the composition of the map from $\mathcal{M}$ to B with the one from B to A. We use the following easy topological fact: if f is a continuous map between compact spaces and W is a saturated (i.e., $W = f^{-1}(f(W))$), open subset of the domain, then $f(W)$ is open. Thus, to show r is an open map it suffices to show that if U is an open subset of $N(\mathcal{M}, A)$ there is a saturated $\bar{U}$ with $r(U) = r(\bar{U})$. But this is exactly the content of the previous exercise.

Since it is clear that types over models are stationary we have established all the axioms except $\overline{\kappa}(T) < \infty$. This easily follows from Corollary 3.18 but we delay the derivation until Theorem 4.22 since Section 4 is devoted to an exhaustive analysis of $\kappa(T)$ and its variants.

The following exercise provides a useful reformulation of the open mapping theorem.

3.24 Exercise. Show that if $\phi(\overline{x};\overline{b})$ does not fork over M then there is a formula $\psi_{\phi}(\overline{x};\overline{a}) \in F(M)$ such that if $p \in N(M \cup \overline{b}, M)$ and $\psi_{\phi}(\overline{x};\overline{a}) \in p$ then $\phi(\overline{x};\overline{b}) \in p$. (Hint: Use the fact that *all* types over M are stationary.)

3.25 Exercise. Show the previous exercise fails if the model M is replaced by an arbitrary set A.

We can summarize the various characterizations of a nonforking extension as follows.

3.26 Theorem. *Let $A \subseteq B$ and $p \in S(B)$. The following are equivalent.*

i) *p does not fork over A.*
ii) *p is almost satisfied in A.*
iii) *There is a model M with $B \downarrow_A M$ and a good schema d over M with $p \subseteq d(\mathcal{M})$.*

Proof. By Lemma 3.12, i) implies ii). To see that ii) implies iii), note that if p is finitely satisfiable in M and M is strongly $|\operatorname{dom} p|^{+}$-saturated then p is contained in a type which is defined by a good schema d over M (taking $d\phi(\overline{x};\overline{y})$ to be $\phi(\overline{m};\overline{y})$ where $\overline{m}$ realizes p). Now choosing an appropriate

M we have ii) implies iii). Assuming iii), let $\bar{c}$ realize the extension of p to $M \cup B$ given by the schema d. Then $t(\bar{c}; M \cup B)$ does not fork over M. Since we have $t(B; M)$ does not fork over A, symmetry and transitivity yield $t(B; M \cup \bar{c})$ does not fork over A; by symmetry and monotonicity we conclude $t(\bar{c}; B)$ does not fork over A and we finish.

3.27 Exercise. Let $A \subseteq M$ and $p \in S(A)$. Show that $q \in S(M)$ does not fork over A if and only if $[q]$ is maximal among the $[r]$ for r an extension of p to a type over a model.

3.28 Exercise. Give an example of an $A \subset B$ and a $p \in S(B)$ such that p forks over A even though p is definable over A. (Hint: Consider an equivalence relation with infinitely many infinite classes. Let a be a member of one these classes and show the type over a which contains only the formula $E(x, a)$ is definable over $\emptyset$ but forks over $\emptyset$.)

3.29 Exercise. Show using the example in the previous exercise that the stronger form of Corollary 3.13 discussed in the paragraph after Corollary 3.13 is false.

3.30 Other Approaches to Forking. The definition here of forking is a variant on the definition in [Lascar & Poizat 1979]. We will briefly discuss another variation on the Lascar-Poizat approach due to Harnik and Harrington. Both the approach in the text above and the one outlined below differ from that in Lascar-Poizat primarily in the way they avoid certain arguments which depend on an extremely judicious choice of objects in general position. This theme is by no means eliminated from the theory, it is just played less often. In Section V.3 we discuss the original approach of Shelah and in Chapter VII we consider the definition of forking in terms of rank.

3.31 Definition. The type $p \in S(A)$ *needs* the formula $\phi(\bar{x}; \bar{y})$ if for each extension of p to a type p' over a model, $\phi(\bar{x}; \bar{y})$ is represented in p'. In particular, if $\phi(\bar{x}; \bar{a}) \in p$ then p needs $\phi(\bar{x}; \bar{y})$.

3.32 Exercise. The type p needs $\phi(\bar{x}; \bar{y})$ iff for some formula $\delta(\bar{v})$ over A and some finite set I with $\bar{v}_i \subseteq \bar{v}$ for $i \in I$:

$$(\exists \bar{v})\delta(\bar{v}) \wedge (\forall \bar{v})(\delta(\bar{v}) \rightarrow \bigvee_{i \in I} \phi(\bar{x}; \bar{v}_i)) \in p.$$

3.33 Definition. If $A \subseteq B$ and $p \in S(B)$ then p *does not fork* over A just if every formula needed by p is also needed by $p|A$.

With this definition it is easy to verify the transitivity axiom but the extension lemma requires a complicated compactness argument referring to the definition of stability in terms of the number of ϕ-types. The key step is to prove: If p needs $\phi \vee \psi$ then p needs ϕ or p needs ψ.

3.34 Exercise. Derive this assertion assuming that the two notions of forking are the same.

3.35 Exercise. Show that $p \in S(M)$ does not fork over A if and only if for every $\phi(\overline{x};\overline{y})$ which is represented in p, $p|A$ needs ϕ.

3.36 Exercise. Prove directly that if p needs $\phi \vee \psi$ then p needs ϕ or p needs ψ. (Hint: (Ziegler) The following five steps are one way to solve this exercise.)

1. Show using Exercise 3.32 that if p needs $\phi(\overline{x};\overline{u}) \vee \psi(\overline{x};\overline{u})$ then for some formula $\delta(\overline{v})$: $(\exists\overline{v})\delta(\overline{v}) \wedge (\forall\overline{v})[\delta(\overline{v}) \rightarrow \phi(\overline{x};\overline{v}) \vee \psi(\overline{x};\overline{v})] \in p$.
2. Now if p is over A and p needs $\phi \vee \psi$ but p needs neither ϕ nor ψ show that there is a $|T|^+$-saturated $M \supseteq A$ and sequences $\overline{b}_0$, $\overline{b}_1$ realizing p such that for every $\overline{m} \in M$, $\models \neg\phi(\overline{b}_1;\overline{m}) \wedge \neg\psi(\overline{b}_0;\overline{m})$.
3. By mapping $\overline{b}_0$ to $\overline{b}_1$ and considering both M and its image under this map, trade the picture of two realizations of p over one model, for a picture with one realization of p and two models. That is, construct an element $\overline{c}$ and two models M_0, M_1 contained in a third model N with all three saturated such that

   i) $\overline{c}$ realizes p.
   ii) $\phi(N;\overline{c}) \cap M_0 = \emptyset$; $\psi(N;\overline{c}) \cap M_1 = \emptyset$.
   iii) $\emptyset \neq \delta(N) \cap M_0 \subseteq \psi(N;\overline{c})$; $\emptyset \neq \delta(N) \cap M_1 \subseteq \phi(N;\overline{c})$.
4. Using the fact that $N \approx M_i$ for $i = 1, 2$, iterate this construction ω times to build a tree.
5. Deduce that $\phi(\overline{x};\overline{u})$ is unstable.

3.37 Historical Notes. The notion of forking was invented by Shelah [Shelah 1978]. The use of finite satisfiability as a characterization of a nonforking type over a model also appears in [Shelah 1978]. The fundamental order was discovered by Lascar and Poizat[Lascar & Poizat 1979]. They showed how to define forking in terms of the fundamental order, heirs, and coheirs. The approach in Definition 3.30 is described in [Harnik & Harrington 1984]. Some further refinements occur in Harnik [Harnik 1985]. Both of these approachs require an appeal to the characterization of stability in terms of cardinality of Stone spaces as well as the definability of types. Our approach, which avoids this appeal and which uses strong saturation to eliminate still more of the appeals to symmetry, first appeared inBaldwin [Baldwin 1984]. A further refinement of this approach is due to Rothmaler [Rothmaler 1983]. In fact, the precise definition of almost satisfaction (closing under conjunction) is taken from [Rothmaler 1983]. Another approach which avoids the appeal to uncountable cardinals was developed by Hodges [Hodges 1981].

4. $\kappa(T)$ and the Spectrum of Stability

One of the most remarkable features of the stability notion is that it provides a fruitful division of all first order theories into four classes: unstable, stable, superstable and ω-stable. Most of this book is devoted to developing the positive properties of the last three of these classes. In this section we define this classification and provide prototypic examples of stable and superstable theories. We also calculate the stability spectrum (the class of cardinals where T is stable) and the spectrum of saturation (the class of cardinals where T has a saturated model) of a theory T. We show, with the possible exception of a few small cardinals, these spectra are the same. We prove in fact that there are only four possible functions for the spectra of stability and these determine the classification.

We introduce three invariants of a theory T: $\kappa(T)$, $\lambda(T)$, and $\mu(T)$. The first two determine the stability classification. The main step in the classification is Theorem 4.25, whose proof provides a certain free tree to witness the non-superstability of a theory. This tree is used here to calculate the spectrum of saturation of T and will be used in Section IX.6 to show that a stable but not superstable theory has 2^λ models of power λ. We refine this tree to show the nonexistence of saturated models in certain cardinalities. We conclude by discussing unstable theories.

4.1 Definition. i) T is *stable in* μ if for every A with $|A| \leq \mu$, $|S(A)| \leq \mu$.

ii) T is *stable* if T is stable in some μ.

iii) The theory T is *superstable* if T is stable in all $\mu \geq 2^{|T|}$.

iv) The countable theory T is *ω-stable* (or totally transcendental) if T is stable in all infinite μ.

A theory T is *strictly stable* if it is stable but not superstable while T is *strictly superstable* if it is superstable but not ω-stable. Shelah [Shelah 1978] provides the appropriate generalization of ω-stability to uncountable theories.

One of the most natural ways of constructing theories which are stable but not ω-stable is through consideration of families of equivalence relations. There are several variations on this idea. These variations depend on three factors, how many equivalence relations there are, whether the different equivalence relations crosscut or refine, and whether there is any additive structure present. The following exercise is an important tool for investigating these examples.

4.2 Exercise. Show that if T is a quantifier eliminable theory of equivalence relations (i.e. T asserts each basic relation of T is an equivalence relation.) then T is stable.

4.3 Example (Refining Equivalence Relations with Finite Splitting). See Fig. 7. For each ordinal α let REI_α denote the theory of α equivalence relations E_i, $i < \alpha$, such that for $i < j$, E_j refines E_i and each E_i class is refined into infinitely many E_{i+1} classes.

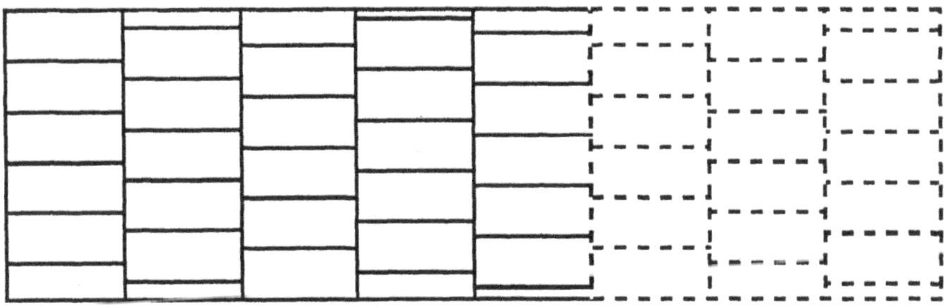

Fig. 7. REI_2

The following series of exercises outlines the crucial properties of a number of examples. Full solutions to these exercises, particularly Exercise 4.16, would be much longer than the solutions to the other exercises in this book.

4.4 Exercise. Show each REI_α is quantifier eliminable and stable.

4.5 Exercise. Show REI_α is not superstable if α is infinite.

4.6 Example (Refining Equivalence Relations with Infinite Splitting).

i) Let REF_α be the variant on Example 4.3 obtained by insisting that each E_i equivalence class is refined into two E_{i+1} equivalence classes.
ii) Let REF^+_α be the theory of the structure $(2^\alpha, +, E_i)_{i<\alpha}$ where the $E_i(\sigma, \tau)$ holds only if $\sigma|i = \tau|i$ The operation $+$ is interpreted as coordinatewise addition mod 2.

4.7 Exercise. Show that REF_α is stable. Show that if $\alpha < \omega$ then REF_α is superstable.

4.8 Exercise. Show that if $\alpha \geq \omega \cdot \omega$ then REF_α is not superstable.

4.9 Example (Crosscutting Equivalence Relations with Finite Splitting).

i) Let CEF_κ be the variant on Example 4.6 obtained by demanding that for $i < \kappa$ each equivalence class of E_{i+1} splits *each* equivalence class of E_i into two classes. Thus, each E_i has only two classes.
ii) Let CEF^+_ω be the theory obtained from CEF_ω as REF^+_ω is obtained from REF_ω, i.e. $\mathrm{Th}(2^\omega, E_i, +)$ where $E_i(\sigma, \tau)$ holds if and only if $\sigma|n = \tau|n$ and $+$ denotes coordinatewise addition.

4.10 Exercise. Show that if $\kappa \leq \omega$ then CEF_κ is superstable.

4.11 Exercise. Show that if T is CEF_ω or REF_ω, and A is an arbitrary subset of a model of T, then for any unrealized $p \in S(A)$, p does not fork over $\emptyset$.

4.12 Example (Crosscutting Equivalence Relations with Infinite Splitting). (Fig. 8). Let CEI_κ be the variant of Example 4.9 obtained by insisting that for $i < j$ each E_j equivalence class intersects *each* E_i class in infinitely many classes.

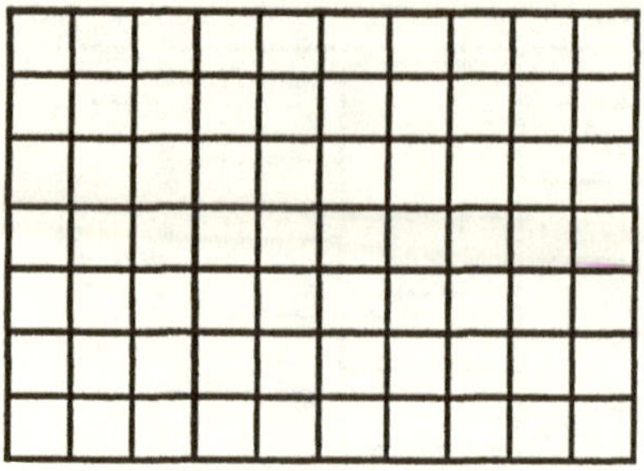

Fig. 8. CEI_2

Note that the theories of crosscutting equivalence relations are defined with respect to a cardinal while the refining equivalence relations are defined with respect to an ordinal.

4.13 Exercise. Formulate precisely the idea indicated in Example 4.12 and determine the spectrum of stability of the resulting theory.

4.14 Notation. When we refer to any of these examples without specifying κ, we mean κ to be ω.

4.15 Exercise. Show each of these examples is ω-stable if κ is finite.

4.16 Exercise. Compute the number of models of each of these theories in every power. (This is rather subtle for Example 4.6.)

4.17 Exercise. Find natural models for each of these theories. For $\kappa = \omega$, the universe of the natural model will be either 2^ω or ω^ω .

In Chapter II and Section III.2 we discussed the local (or finite) character of nonforking. That is, for any type p, there is a 'small' subset A of $\operatorname{dom} p$ such that p does not fork over A. We show here the size of this A can be uniformly bounded for all p by a cardinal $\kappa(T)$ and this cardinal largely determines the stability spectrum of T. There are several slight variations on the definition of the cardinal $\kappa(T)$. We begin by defining $\overline{\kappa}(T)$ as well and describing the relation between these two cardinals. If $|T|$ is singular we will have to consider still a third invariant $\kappa_r(T)$.

4.18 Definition. i) $\kappa(T)$ is the least infinite cardinal, if one exists, such that for any finite sequence $\overline{a}$ and each strictly ascending sequence of sets $\langle A_i : i < \kappa(T)\rangle$, for some i, $t(\overline{a}; A_{i+1})$ does not fork over A_i.

ii) $\overline{\kappa}(T)$ is the least infinite cardinal, if one exists, such that for every type p there is a set $A \subseteq \operatorname{dom} p$ with $|A| < \overline{\kappa}(T)$ and p does not fork over A.

iii) In the event that the cardinals described in i) and ii) do not exist we set $\overline{\kappa}(T)$, respectively $\kappa(T)$, equal to ∞.

iv) Let $\kappa_r(T) = \kappa(T)$ if $\kappa(T)$ is regular and $\kappa(T)^+$ if $\kappa(T)$ is singular.

While $\overline{\kappa}(T)$ is the bound that we will apply in practice; it is easier to evaluate $\kappa(T)$. Fortunately, the cardinals are almost the same. In particular, they are equal for countable T. The next two results showing the relations between these cardinals are of most interest for uncountable languages. We begin with a technical proposition.

4.19 Proposition. *For every* $\lambda < \kappa(T)$, $\mathrm{cf}(\lambda) < \overline{\kappa}(T)$.

Proof. Suppose $\langle A_i : i < \lambda \rangle$ and $\overline{a}$ are such that $t(\overline{a}; A_{i+1})$ forks over A_i for each $i < \lambda$ and the A_i are a continuous increasing chain. Then if $B \subseteq A_\lambda$ and $t(\overline{a}; A_\lambda)$ does not fork over B, $|B| \geq \mathrm{cf}(\lambda)$. So for every $\lambda < \kappa(T)$, $\mathrm{cf}(\lambda) < \overline{\kappa}(T)$.

4.20 Lemma. *For every* T, $\overline{\kappa}(T) \leq \kappa(T) \leq \overline{\kappa}(T)^+$. *If* $\overline{\kappa}(T)$ *is regular then* $\kappa(T) = \overline{\kappa}(T)$.

Proof. Choose $\lambda < \overline{\kappa}(T)$ and let p be a type such that for every $A \subseteq \mathrm{dom}\, p$ with $|A| < \lambda$, p forks over A. Define $\langle A_i : i < \lambda \rangle$ so that $A_{i+1} - A_i$ is finite and $p|A_{i+1}$ forks over A_i. (This is possible by the finite character of forking.) Then $\langle A_i : i < \lambda \rangle$ demonstrates $\kappa(T) > \lambda$. Thus, $\kappa(T) \geq \overline{\kappa}(T)$. The second inequality and the second assertion are immediate from the preceding proposition.

4.21 Examples. $\overline{\kappa}(\mathrm{REI}_{\aleph_\omega}) = \aleph_\omega$. $\kappa(\mathrm{REI}_{\aleph_\omega}) = \aleph_\omega^+$. Let T be the theory of a disjoint union of models of REI_α for $\alpha < \aleph_\omega$. Then $\overline{\kappa}(T) = \kappa(T) = \aleph_\omega$.

Now we bound $\kappa(T)$.

4.22 Theoren. *For every* T, $\kappa(T) \leq |T|^+$.

Proof. Suppose $\langle A_i : i < \lambda \rangle$ and $\overline{a}$ are such that $t(\overline{a}; A_{i+1})$ forks over A_i for each i and the A_i are a continuous increasing chain. Now, if $p_i = t(\overline{a}; A_i)$, by Corollary 3.20 we have $[\hat{p}_i] > [\hat{p}_j]$ if $i < j$. But it is clear that any decreasing sequence in the fundamental order has at most $|T|$ elements so $\lambda < |T|^+$ as required.

4.23 Exercise. What are the possible values of $\kappa(T)$ and $\overline{\kappa}(T)$ when T is countable?

4.24 Exercise. Define $\kappa^1(T)$ to be the cardinal one obtains in the definition of $\kappa(T)$ if the types p are required to be 1-types. Show $\kappa^1(T) = \kappa(T)$. (Hint: If $t(a^\frown b; A_{i+1})$ forks over A_i then either $t(a; A_{i+1})$ forks over A_i or $t(b; A_{i+1} \cup a)$ forks over $A_i \cup a$.)

We will now show that for stable T there is a cardinal $\lambda(T) \leq 2^{|T|}$ such that T is stable in μ if and only if $\mu = \mu^{<\kappa(T)}$ and $\mu \leq \lambda(T)$. Simple cardinal computations show that the conclusion of this assertion holds for exactly the same μ if $\kappa(T)$ is replaced by $\kappa_r(T)$.

This computation of the stability spectrum could be carried out for an arbitrary freeness relation satisfying the axioms discussed in Section II.1. Such an argument would show that any first order theory with a freeness relation satisfying the axioms is stable. We give such an abstract formulation in Section VII.1 and show the 'categoricity' of the freeness axioms. Note that in the next construction we use only the monotonicity, extension, symmetry, and transitivity axioms along with the definition of $\kappa(T)$. We could rephrase this theorem as saying that if $\kappa(T)$ is infinity then T is unstable.

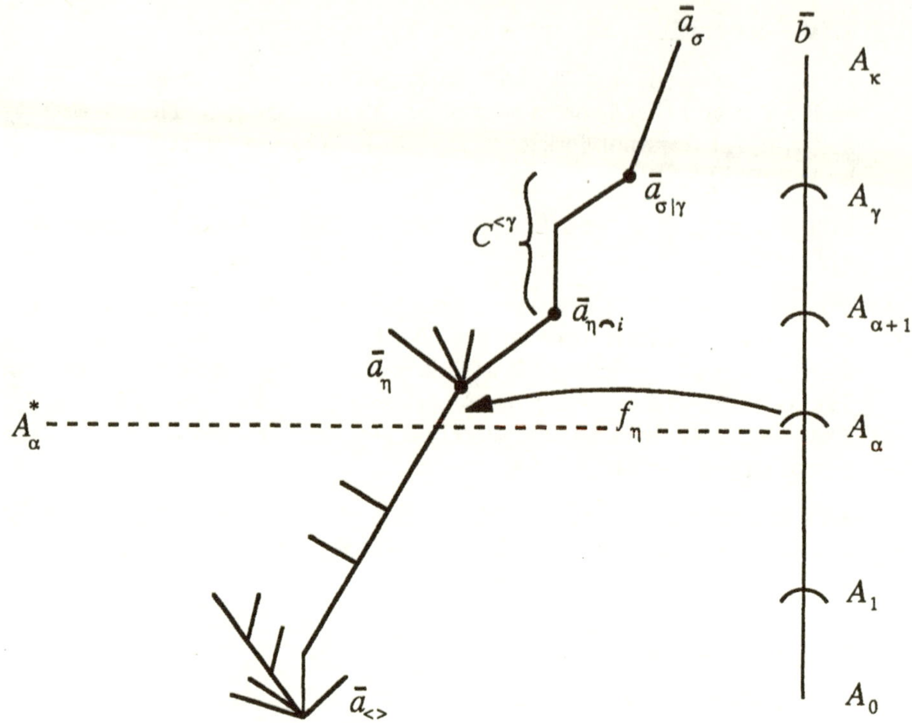

Fig. 9. Building a Shelah Tree

4.25 Theorem. *If T is stable and $\mu < \mu^{<\kappa(T)}$ then T is not stable in μ.*

Proof. (Fig. 9). Let κ be the least cardinal such that $\mu^\kappa > \mu$. Then there exists a sequence $\langle A_i : i < \kappa\rangle$ and a sequence $\bar{b}$ such that $t(\bar{b}; A_{i+1})$ forks over A_i (as κ is certainly less than $\kappa(T)$). Without loss of generality we may assume that $|A_i| \le |i| + \aleph_0$, the sequence is continuous and for each i, $A_{i+1} - A_i$ is a finite sequence $\bar{a}_i$. (This depends on the finite character of forking.) Now for each $\nu \in \mu^{\le\kappa}$ we will define an elementary map f_ν with $\operatorname{dom} f_\nu = A_{\lg(\nu)}$ and $\operatorname{rng} f_\nu \subseteq \mathcal{M}$. If $\lg(\nu) = \delta$, a limit ordinal, let f_ν be $\bigcup_{\alpha<\delta} f_{\nu|\alpha}$. If $\lg(\nu) = \alpha + 1$, first let $A^*_\alpha = \bigcup_{\eta\in\mu^\alpha} \operatorname{rng} f_\eta$. Now for $\nu = \eta\frown i$, define $f_{\eta\frown i}$ by induction on $i < \mu$ by choosing $\bar{a}_{\eta\frown i} = f_{\eta\frown i}(\bar{a}_\alpha)$ so that $t(\bar{a}_{\eta\frown i}; A^*_\alpha \cup \{\bar{a}_{\eta\frown j} : j < i\})$ does not fork over $A_\eta = f_\eta(A_\alpha)$ and extends $f_\eta(t(\bar{a}_\alpha; A_\alpha))$. For $\sigma \in \mu^\kappa$, let g_σ be an arbitrary extension to $A_\kappa \cup \{\bar{b}\}$ of f_σ and extend $t(g_\sigma(\bar{b}); A_\sigma)$ to a complete type over $A^* = \bigcup_{\alpha<\kappa} A^*_\alpha$ realized by $\bar{a}_\sigma$ such that $t(\bar{a}_\sigma; A^*)$ does not fork over A_σ.

Now, if $\sigma \ne \tau$, then $t(\bar{a}_\sigma; A^*) \ne t(\bar{a}_\tau; A^*)$. To see this, note first that since g_σ is an elementary map, $t(\bar{a}_\sigma; A_{\sigma|\beta} \cup \bar{a}_{\sigma|\beta+1})$ forks over $A_{\sigma|\beta}$ for any $\beta < \kappa$. Let $\alpha = \beta + 1$ be least such that $\bar{a}_{\sigma|\alpha} \ne \bar{a}_{\tau|\alpha}$; so $A_{\sigma|\beta} = A_{\tau|\beta}$. We will show $t(\bar{a}_\sigma; A_{\tau|\beta} \cup \bar{a}_{\tau|\alpha})$ does not fork over $A_{\tau|\beta}$. Let $C_\gamma = \{\bar{a}_{\sigma|\delta} : \beta < \delta \le \gamma\}$. We show by induction that for every γ, with $\beta < \gamma \le \kappa$:

$$(C_\gamma \downarrow \bar{a}_{\tau|\alpha}; A_{\sigma|\beta}). \qquad (*)$$

The induction step depends on the following exercise which is essentially Lemma II.2.11.

4.26 Exercise. Let $A \cap B = \emptyset$ and $A \cap C = \emptyset$. If $C \downarrow_A B$ and $D \downarrow_{A\cup C} B$ then $C \cup D \downarrow_A B$.

Returning to the proof of the theorem, for $\gamma = \beta + 1$, $(*)$ holds by the choice of $\overline{a}_{\tau|\alpha}$ (and the symmetry lemma). Now fix γ with $\beta < \gamma \leq \kappa$; let $C^{<\gamma} = \{\overline{a}_{\sigma|\delta} : \beta < \delta < \gamma\}$ and suppose by induction: $C^{<\gamma} \downarrow_{A_{\sigma|\beta}} \overline{a}_{\tau|\alpha}$ To apply the exercise, take A as $A_{\sigma|\beta}$, B as $\overline{a}_{\tau|\alpha}$, C as $C^{<\gamma}$, and D as $\overline{a}_{\sigma|(\gamma+1)}$; then $C \cup D$ is C_γ. The second hypothesis of the exercise follows from the construction by monotonicity. Applying the exercise we finish the induction.

Consider the case $\gamma = \kappa$. It asserts $C_\kappa \downarrow_{A_{\sigma|\beta}} \overline{a}_{\tau|\alpha}$. By monotonicity, as $\overline{a}_{\sigma|\kappa} = \overline{a}_\sigma$, $\overline{a}_\sigma \downarrow_{A_{\sigma|\beta}} \overline{a}_{\tau|\alpha}$. Thus, $t(\overline{a}_\sigma; A_{\sigma|\beta} \cup \overline{a}_{\tau|\alpha}) \neq t(\overline{a}_\tau; A_{\sigma|\beta} \cup \overline{a}_{\tau|\alpha})$ and we finish.

In this construction we only had to assure:

$$\overline{a}_{\eta\frown i} \underset{A_\eta}{\downarrow} A^*_\alpha \cup \{\overline{a}_{\eta\frown j} : j < i\}.$$

The relation of $\overline{a}_{\eta\frown i}$ to $\overline{a}_\nu$ for a ν with $\lg(\nu) = \alpha$ but $\nu|\alpha \neq \eta|\alpha$ is irrelevant. We can deduce a further consequence to the proof. In fact, as we see below, a_σ can realize at most $2^{\kappa(T)}$ distinct types of the form $t(a_\tau; A_\tau)$. But $|A_\sigma| \leq \kappa(T)$. This leads to a proof (see Exercise 4.40 below) of Corollary 4.27 ii) with the additional hypothesis that $\mu \leq 2^{(T)}$. However, we can exploit the indiscernibility of the tree constructed in Theorem 4.25 to obtain an even stronger result.

The tree described in Theorem 4.27 i) epitomises a strictly stable theory. It is sometimes referred to as a Shelah tree.

4.27 Corollary. *Let T be a countable stable, but not superstable, theory.*

i) *There exists a sequence of formulas $\phi_i(\overline{x};\overline{y})$ for $i < \omega$ and for each μ sequences $\overline{a}_\eta$ for each $\eta \in \mu^{<\omega}$ such that:*

  a) *For each $\eta \in \mu^{<\omega}$, $\{\overline{a}_{\eta\frown\alpha} : \alpha < \mu\}$ is a set of indiscernibles over $A_\eta = \{a_\nu : \nu \not\subseteq \eta\}$.*
  b) *For each $\sigma \in \mu^\omega$, $p_\sigma = \{\phi_i(\overline{x};\overline{a}_{\sigma|i}) : i < \omega\}$ is consistent.*
  c) *For each $\eta \in \mu^{<\omega}$ and $\alpha \neq \beta < \mu$, $\phi_i(\overline{x}, a_{\eta\frown\alpha}) \wedge \phi_i(\overline{x}; a_{\eta\frown\beta})$ is inconsistent.*

ii) *If $\aleph_1 \leq \mu < \mu^\omega$ then T does not have a saturated model of power μ.*

Proof. To prove i) we will construct a sequence of at most countably many approximations to the ϕ_i and $\overline{a}_\eta$. The kth approximation will satisfy conditions a) and b) (for $\overline{a}_\eta$ depending on k) and the following weaker form of c).

c′) For each $i < \omega$ and each $\eta \in \mu^i$ there is an m^k_i such that if $X \subset \mu$ and $|X| \geq m^k_i$, $\{\phi^k_i(\overline{x};\overline{a}_{\eta\frown\alpha}) : \alpha \in X\}$ is inconsistent.

The chief point of the construction is that for each $i < \omega$, $m_i^{k+1} \leq m_i^k$ and for some $i < \omega$, $m_i^{k+1} < m_i^k$. In fact, for each i, for all sufficiently large $k < \omega$, $m_i^k = 2$. Thus, we eventually satisfy condition c).

For $k = 0$, we take the $\overline{a}_\eta$ constructed in Theorem 4.25 and choose $\phi_i(\overline{x}; \overline{y})$ to witness that $t(\overline{b}; A_i \cup \overline{a}_i)$ forks over A_i. Thus $\phi(\overline{x}; \overline{a}_\eta)$ forks over A_η. Since the $\langle \overline{a}_{\eta \frown \alpha} : \alpha < \mu \rangle$ are independent over A_η, Theorem II.2.18 guarantees the existence of m_i^0 for each $i < \omega$.

We say that i requires attention at stage k if for each $j < i$, $m_j^k = 2$. We may suppose then that at a certain stage, i is the least ordinal which requires attention. Thus we have $\overline{a}_\eta$, ϕ_j^k and m_j^k for $j < \omega$ so that conditions a), b) and c′) are satisfied and $m_j^k = 2$ if $j < i$. We must choose new $\overline{a}_\eta$, ϕ_j^{k+1} and m_j^{k+1} so that $m_i^{k+1} < m_i^k$. Fix $\eta \in \mu^i$. There are two cases.

To describe the first case, let $\hat{j}$ denote a sequence of j 1's. Then we are in the first case if the following collection of formulas is consistent.

$$\Gamma = \{\phi_l^k(\overline{x}; \overline{a}_{\eta|l}) : l < i\} \cup \{\phi_{i+j}^k(\overline{x}; \overline{a}_{\eta \frown \hat{j}}) : j < \omega\} \cup \{\phi_i^k(\overline{x}; a_{\eta \frown 0})\}.$$

Let

$$\begin{aligned} \psi_j^{k+1} &= \psi_j^k \quad \text{if } j \neq i \\ \psi_i^{k+1}(\overline{x}; \overline{y}_1, \overline{y}_2) &= \psi_i^k(\overline{x}; \overline{y}_1) \wedge \psi_i^k(\overline{x}; \overline{y}_2) \\ m_j^{k+1} &= m_j^k \quad \text{if } j \neq i \\ m_i^{k+1} &= [(m_i^k + 1)/2]. \end{aligned}$$

The indiscernibility of $\{\overline{a}_{\eta \frown \alpha} : \alpha < \mu\}$ at the previous stage guarantees that we have satisfied the condition at $\overline{a}_\eta$ and thus at $\overline{a}_\nu$ for each $\nu \in \mu^i$.

If Γ is inconsistent, we are in the second case. That is, for some $n < \omega$

$$\Gamma_0 = \{\phi_l^k(\overline{x}; \overline{a}_{\eta|l}) : l < i\} \cup \{\phi_{i+j}^k(\overline{x}; \overline{a}_{\eta \frown \hat{j}}) : j < n\} \cup \{\phi_i^k(\overline{x}; a_{\eta \frown 0})\}$$

is inconsistent. In this case, let

$$\begin{aligned} \phi_l^{k+1} &= \phi_l^k \quad \text{if } l < i \\ \phi_i^{k+1} &= \textstyle\bigwedge_{l < i+n} \phi_l(\overline{x}; \overline{y}_l) \\ \phi_l^{k+1} &= \phi_{l+n}^k \quad \text{if } l > i \\ m_l^{k+1} &= m_l^k \quad \text{if } l < i \\ m_i^{k+1} &= 2 \\ m_l^{k+1} &= m_{l+n}^{k+1} \quad \text{if } l > i. \end{aligned}$$

For ii), let κ be the least cardinal such that $\mu^\kappa > \mu$. Suppose for contradiction that M is a saturated model of T with power μ. Since $\mu^{<\kappa} \leq \mu$ and M is μ-universal, the tree constructed in part i) can be embedded in M. Identify the original tree with its image in M. Now the $\{p_\sigma : \sigma \in \mu^\omega\}$ are pairwise inconsistent and $\mu^\omega < \mu$ so they cannot all be realised in M contrary to the assumption that M is saturated.

Now we introduce the invariant that was mentioned in the introduction to Part A, the bound on the number of free (i.e. nonforking) extensions of an independent set. We will denote this invariant by $\mu(T)$ but it will not figure prominently in the sequel since we will shortly (Lemma 4.32) bound it by a more visible invariant denoted by $\lambda(T)$. If we had proceeded totally abstractly we would have deduced stability of T from the assumption that $\kappa(T)$ and $\mu(T)$ are less than infinity (and our other axioms).

4.28 Definition. Let p be a type in finitely many variables.

i) Let $\mu(p)$, the *multiplicity* of p be the least cardinal μ such that p does not have μ^+ pairwise contradictory nonforking extensions.
ii) Let $\mu(T)$, the *global multiplicity of* T, be the least cardinal μ such that for any type p, $\mu(p) \leq \mu$.
iii) Let $\mu_\phi(p)$, the *ϕ-multiplicity of* p, be the least cardinal μ such that there do not exist μ^+ contradictory ϕ-types which do not fork over $\operatorname{dom} p$ but whose restriction to $\operatorname{dom} p$ is p_ϕ.

We can derive a sufficient condition for stability in χ in terms of the global multiplicity of T and one further invariant.

4.29 Definition. For a stable theory T, let $\lambda(T)$ denote the least cardinal in which T is stable.

4.30 Lemma. *For any cardinal χ, if $\mu(T)$ and $\lambda(T)$ are at most χ and $\chi = \chi^{<\kappa(T)}$ then T is stable in χ.*

Proof. Let $|B| = \chi$. For each $A \subseteq B$ with $|A| < \kappa(T)$, $|S(A)| \leq 2^{<\kappa(T)}$. Each element of $S(A)$ has at most $\mu(T)$ distinct extensions in $S(B)$ which do not fork over A. But if $p \in S(B)$, for some $A \subseteq B$ with $|A| < \overline{\kappa}(T) \leq \kappa(T)$, p does not fork over A. Thus, there are at most $\chi^{<\kappa(T)}$ (the number of subsets of B with power $\kappa(T)$) times $2^{<\kappa(T)}$ (the number of types over each A) times $\mu(T)$ (the number of nonforking extensions of a type over A) which equals χ types over B.

To discuss the properties of $\lambda(T)$ and $\mu(T)$ for uncountable theories we need the following lemma which is obvious when the language is countable or if $\chi > |T|$.

4.31 Lemma. *If T is stable in χ then every set of power χ is contained in a model of power χ.*

Proof. Let A be an arbitrary set of power χ. Define a sequence A_i for $i < \omega$ with $A_0 = A$ and such that $|A_i| = \chi$ for all i and A_{i+1} realizes all types over A_i. Then $A_\omega = \cup\{A_i : i < \omega\}$ is the required model of T.

Now we can bound $\mu(T)$ by $\lambda(T)$ and $\lambda(T)$ by $2^{|T|}$.

4.32 Lemma. *If T is stable (i.e. if $\lambda(T) < \infty$) then $\mu(T) \leq \lambda(T) \leq 2^{|T|}$.*

Proof. For the first inequality, let $p \in S(B)$. There exists A with $A \subseteq B$ and $|A| < \overline{\kappa}(T)$ such that p does not fork over A. We know from Lemma 4.25 (taking $\lambda(T)$ as μ) that $\lambda(T) \geq \kappa(T)$. So if $|A| < \kappa(T)$, we can choose by Lemma 4.31 a model, M, containing A with $|M| \leq \lambda(T)$ and thus $|S(M)| \leq \lambda(T)$. But types over models are stationary so the number of contradictory nonforking extensions of $p|A$ and, a fortiori, of p is bounded by $\lambda(T)$ as required.

Now we prove the second inequality. For any A and any $p \in S(A)$, p can be viewed as the product over $\phi \in L$ of p_ϕ. But, since T is stable for each ϕ, $|S_\phi(A)| \leq |A|$ so $|S(A)| \leq |A|^{|T|}$. In particular, T is stable in $2^{|T|}$.

4.33 Exercise. Show there is no relation between $\mu(T)$ and $\kappa(T)$. For example, if T is theory of countably many independent unary predicates $\kappa(T) = \omega$ and $\mu(T) = 1$. But, if T is CEF_ω, $\kappa(T) = \omega$ and $\mu(T) = 2^\omega$.

Now using Lemma 4.32 we calculate the class of cardinals in which a stable theory is stable.

4.34 Theorem (The Stability Spectrum Theorem). *If T is stable then T is stable in χ if and only if $\chi = \lambda(T) + \chi^{<\kappa(T)}$.*

Proof. By Theorem 4.25 and the definition of $\lambda(T)$, the equality holds if T is stable in χ. The converse is immediate from Lemma 4.29 and 4.32.

From Theorem 4.34, Exercise 1.8, and simple facts about cardinals (e.g. There are arbitrarily large cardinals of cofinality ω; $\chi^{\mathrm{cf}(\chi)} > \chi$.) one can easily deduce the following result.

4.35 Corollary. *T is superstable iff $\kappa(T)$ is ω.*

Theorem 4.34 gives a characterization of the stability spectrum above $2^{|T|}$. For countable T, $\lambda(T) = \aleph_0$ or $\lambda(T) = 2^{\aleph_0}$ and the following theorem describes the situation completely. If T is uncountable and $\lambda(T) < 2^{|T|}$ the more complicated situation is completely described in III.5 of [Shelah 1978].

4.36 Theorem. *For a countable theory T one of the following four mutually exclusive situtations holds.*

i) *T is stable in all χ.*
ii) *T is stable in all $\chi \geq 2^{\aleph_0}$.*
iii) *T is stable in χ iff $\chi^{\aleph_0} = \chi$.*
iv) *T is stable in no χ.*

Proof. This easily follows from our previous results noting that $\kappa(T)$ is either $\aleph_0$ or $\aleph_1$.

We can combine Theorems 4.25 and 4.34 to establish

4.37 Theorem. *Suppose T is a countable stable theory and $\chi \geq \kappa(T)$. Then T has a saturated model of power χ iff T is stable in χ.*

Proof. Suppose that T is not stable in χ. Then by Theorem 4.34, either $\lambda(T) > \chi$ (and certainly T does not have a saturated model of power χ) or $\chi < \chi^{<\kappa(T)}$. In the second case, T does not have a saturated model in power χ by Corollary 4.27.

We now prove that if T is stable in χ then T has a saturated model of power χ. Let M_0 be an arbitrary model of power χ and construct by induction a continuous chain of models $\{M_i : i < \chi\}$ such that $|M_i| = \chi$ for all i and every type over M_i is realized in M_{i+1}. Now let M be the union of the chain $\{M_i : i < \chi\}$. We will show M is saturated. Let A be a subset of M with $|A| < \chi$ and let p be in $S(A)$. If χ is regular, it is easy to see any such p is realized in M so for the rest of the proof we assume χ is singular. Extend p to r in $S(M)$. There exists A_0 contained in M such that r does not fork over A_0 and $|A_0| < \kappa(T)$. Thus, we may assume A_0 is contained in M_0 as, by Theorem 4.25, $\kappa(T) \leq \mathrm{cf}(\chi)$. Now, choose $E = \{\bar{e}_i : i < \chi\}$ such that $\bar{e}_i$ is in M_{i+1} and realizes $r|M_i$. Note that $r|M_0$ is stationary. As in the proof of Theorem II.1.30, choose $E_0 \cup M_0 \subseteq E$ such that $t(A; E \cup M_0)$ does not fork over $E_0 \cup M_0$ and $|E_0| \leq \kappa(T) + |A| < \chi$ (since χ is singular). Thus $E - E_0$ is not empty. Any element e of $E - E_0$ satisfies: $t(e; E_0 \cup A \cup M_0)$ does not fork over $E_0 \cup M_0$ (by the symmetry lemma). But also $t(e; E_0 \cup M_0)$ does not fork over M_0 so $t(e; E_0 \cup A \cup M_0)$ does not fork over M_0. Hence e realizes p.

This argument is very similar to that in Theorem II.1.30. We obtain the sharper result because we have computed $\kappa(T)$. Note that the existence of saturated models did not rely essentially on the hypothesis that T is countable.

The following exercises consider some variations on Corollary 4.27 ii) and Theorem 4.37.

4.38 Exercise. Show that if $\langle M_i : i < \delta\rangle$ is an increasing sequence of λ-saturated models and $\kappa(T) \leq \mathrm{cf}(\delta)$ then $\bigcup_{i<\delta} M_i$ is λ-saturated.

4.39 Exercise. Find examples of superstable countable theories which do and do not have countable saturated models.

4.40 Exercise. Prove the following weaker form of Corollary 4.27 ii) without relying on Theorem 4.27 i). If $2^{|T|} \leq \mu \leq \mu^{<\kappa(T)}$ then T has no saturated model of power μ. (Hint: Imbed the tree of Theorem 4.25 in a putatively saturated model and show no $\bar{b}$ can realize more than $2^{|T|}$ of the p_σ.)

4.41 Exercise. Prove the following generalization of Theorem 4.37 to uncountable languages. Suppose T is stable and $\chi > \kappa(T)$. Then T has a saturated model of power χ if and only if T is stable in χ. (Hint: Extend the proof of Theorem 4.27 i) to deal with trees contained in $\mu^{<\kappa(T)}$ for uncountable $\kappa(T)$. Then proceed as in Theorem 4.37.)

4.42 Unstable Theories. In this subsection we briefly summarise the properties of unstable theories. An introductory but more detailed account of this subject appears in [Pillay 1983a]; a full treatment appears in Chapter II of [Shelah 1978].

For any theory T let $\sigma(\kappa) = \sigma(T, \kappa) = \sup\{|S(A)| : |A| \leq \kappa\}$ (where A is a subset of the monster model of T). The stability spectrum theorem shows that for countable stable T there are only three possibilities for this function: $\sigma_1(\kappa) = \kappa$, $\sigma_2(\kappa) = \sup(\kappa, 2^\omega)$, $\sigma_3(\kappa) = \kappa^\omega$. In the presence of the generalized continuum hypothesis the only other possible such function is

$\tau(\kappa) = 2^\kappa$. With no hypothesis beyond ZFC it is provable that there are only three possible such functions for unstable theories. These functions depend on the value of $\mathrm{ded}(\kappa)$, the supremum of the cardinals λ such that there is a linear ordering of power κ with λ Dedekind cuts. Shelah [Shelah 1978] and Keisler ([Keisler 1976], [Keisler 1978]) have shown that the only other possibilities for σ are $\sigma_4(\kappa) = \mathrm{ded}(\kappa)$, $\sigma_5(\kappa) = \mathrm{ded}(\kappa)^\omega$, and $\sigma_6(\kappa) = 2^\kappa$. Each of these functions is determined by a syntactic property of T.

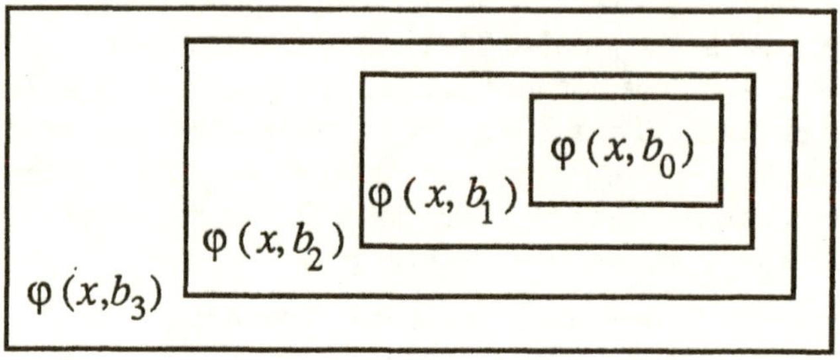

Fig. 10. $\phi(x; y)$ has the strict order property.

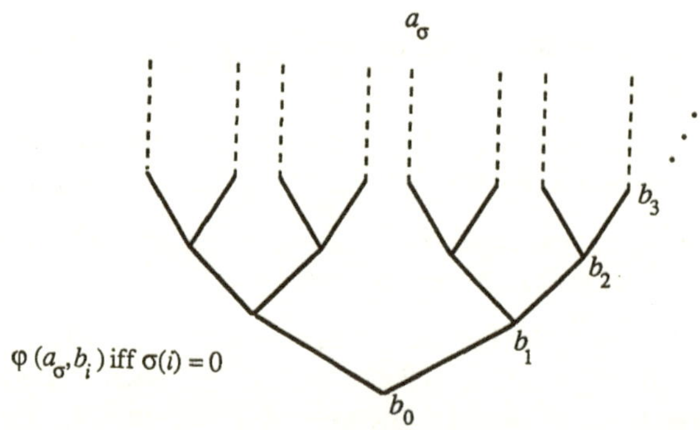

Fig. 11. $\phi(x; y)$ has the independence property.

4.43 Definition. i) T has the *strict order property* (Fig. 10) if there are a formula $\phi(\overline{x}; \overline{y})$ and sequences $\langle \overline{b}_i : i < \omega \rangle$, $\langle \overline{a}_i : i < \omega \rangle$ such that

$$\models \phi(\overline{a}_i; \overline{b}_j) \text{ iff } i < j.$$

ii) T is *multiply ordered* if there exist a family of formulas ϕ_i for $i < \omega$ and sequences $\overline{a}_{i,j}$ for $i, j < \omega$ and $\overline{c}_\sigma$ for $\sigma \in \omega^\omega$ such that

$$\models \phi_i(\overline{c}_\sigma; \overline{a}_{i,j}) \text{ iff } j < \sigma(i).$$

iii) T has the *independence property* (Fig. 11) if there are a formula $\phi(\overline{x}; \overline{y})$ and sequences $\langle \overline{a}_i : i < \omega \rangle$ and $\langle \overline{c}_\sigma : \sigma \in 2^\omega \rangle$ such that

$$\models \phi(\overline{a}_i; \overline{c}_\sigma) \text{ iff } \sigma(i) = 0.$$

Shelah [Shelah 1978] shows a theory has the order property (Fig. 12) if and only if it is unstable (Fig. 13) if and only if it has either the strict

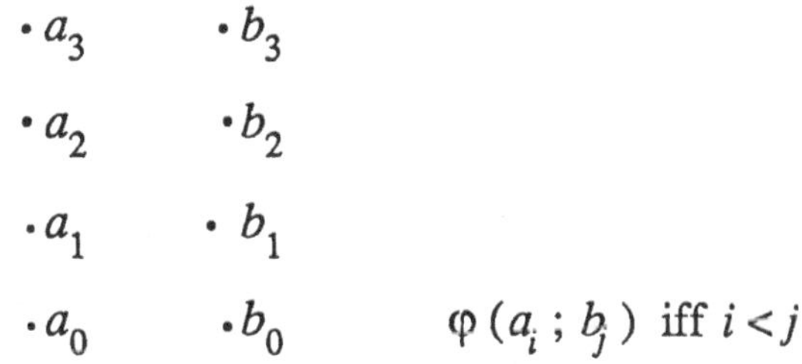

Fig. 12. $\phi(x;y)$ has the order property.

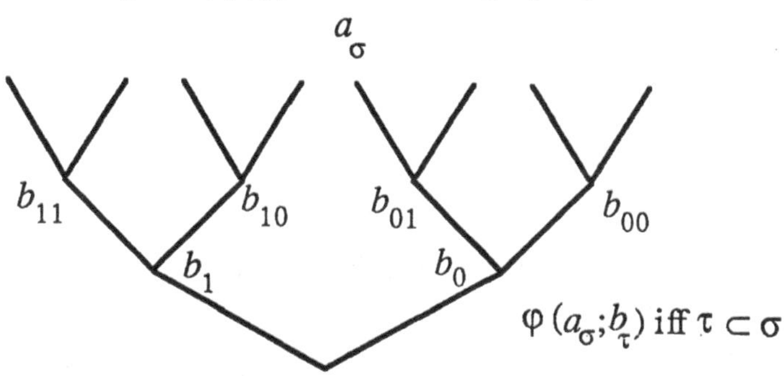

Fig. 13. $\phi(x;y)$ is unstable.

order property or the independence property ([Shelah 1978] II.2.25). Keisler shows the equivalence between $\sigma(T,\kappa) = \sigma_5$ and T is multiply ordered. It is not known whether it is consistent for σ_4 and σ_5 to be distinct functions. There has been considerable work on such questions as replacing $\phi(\overline{x};\overline{y})$ by $\phi(x;\overline{y})$ in the above definitions and in trying to extend results about stable theories to theories which are not too badly unstable.

The results discussed here illustrate a common phenomenon in stability theory which recurs for example in the solution of the spectrum problem for the number of models of an ω-stable theory. The spectrum functions that we have computed involve the exponential function and thus the computation of the functions is very much subject to axioms of set theory beyond those of ZFC. The properties which distinguish the cases, however, stability, strict order property etcetera are all very low in the Levy hierarchy (arithmetic in this case) and so are absolute between, e.g. V and L.

4.44 Historical Notes. Morley [Morley 1965] introduced the notion of an ω-stable countable theory and proved that an ω-stable theory is stable in all powers. Shelah, in considering uncountable languages, discovered the importance of considering single formulas and introduced the full stability hierarchy. The key Theorem 4.25 for establishing the stability spectrum theorem appears here with a new proof (also due to Shelah). This proof substitutes machinery from stability theory for the combinatorial argument in [Shelah 1971]. The particular argument given here for Theorem 4.37 was suggested to me by W.W. Tait. The theorem was first proved in various special cases by Harnik [Harnik 1975a] and in full generality by Shelah [Shelah 1978]. Poizat [Poizat 1985] gives a straightforward proof that an unstable theory has no saturated model of power λ if λ is singular.

5. *Definable Chain Conditions in Algebra*

In this section we apply the stability classification to the classification of such natural algebraic structures as groups, rings, modules and fields. Our primary interest is the identification of a family of examples for the general theory. Since the basic notions of stability theory are extremely general formulations of well known algebraic phenomena, they should and do acquire a concrete and natural interpretation when applied to the most common algebraic structures.

In general the calculation of the stability spectrum depends on finding the cardinality of the Stone space and in finding certain trees of formulas. But if a structure admits a multiplication then these trees can be replaced by chains.

5.1 Definition. i) A group G satisfies the *ω-stable descending chain condition* if there is no infinite properly descending chain of definable subgroups of G. (Fig. 14).

ii) A group G satisfies the *superstable descending chain condition* if there is no infinite properly descending chain of definable subgroups of G such that each subgroup has infinite index in its predecessor.

iii) A group G satisfies the *stable descending chain condition* if there is no infinite properly descending chain of definable subgroups of G each defined by an instance of a single formula $\phi(x;\overline{y})$.

iv) Let G be a group, $\phi(x;\overline{y})$ a formula, and Γ the collection of finite intersections of ϕ-definable subgroups of G. The group G satisfies the *full stable descending chain condition* if there is no infinite properly descending chain of elements of Γ.

There are two somewhat unusual usages in the statements of these definitions and the following theorem. First, when we say G is a group, we mean some formula in the language of G defines a binary operation under which G is a group. Thus, G may admit additional structure. For example, any field is a group in this sense. Secondly, in the following theorem we take properties such as stability, which are defined for theories, to be properties of algebraic structures, by saying, for example, G is stable when the theory of G is. With these caveats we can state the following theorem.

5.2 Theorem. i) *If G is ω-stable then G satisfies the ω-stable d.c.c.*

ii) *If G is superstable then G satisfies the superstable d.c.c.*

iii) *If G is stable then G satisfies the stable and even the full stable d.c.c.*

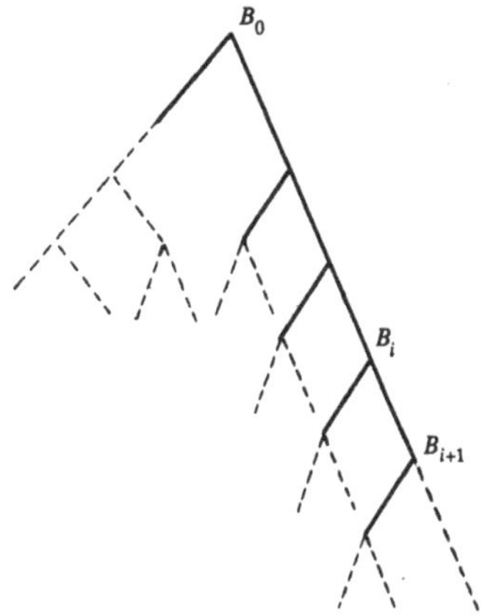

Fig. 14. The ω-stable d.c.c.

Proof. i) Let $\langle B_i : i < \omega \rangle$ be a properly decreasing chain of definable subgroups of G. Choose $b_i \in B_i - B_{i+1}$. For each $\eta \in 2^\omega$ define the type p_η which asserts that $x \in C_{\eta|i}$ where $C_{\eta|i}$ is the coset of B_{i+1} which contains $\prod_{j<i} \eta(j) b_j$. Now as the types p_η are pairwise inconsistent and can be extended to complete types over the countable set of elements necessary to define the B_i and their translates, there are $2^{\aleph_0}$ types over this set so G is not ω-stable.

The proof for ii) is similar, noting that since the index of B_{i+1} in B_i is infinite, one can construct an ω-branching tree and thus contradict superstability.

iii) It is immediate from the definition that a stable group satisfies the stable d.c.c. The following argument extends the result to the full stable descending chain condition. Let $\psi(x, y, \overline{z})$ denote $\phi(x \cdot y, \overline{z})$. Thus, each translate of a subgroup defined by an instance of ϕ is defined by an instance of ψ. So, if Γ contains an infinite descending chain the same type of argument as in i) allows us to constuct a full binary tree of instances of ψ. Thus, T is unstable.

This consideration of formulas which define the translations of sets within a group is the foundation for Poizat's *stratified order* [Poizat 1981]. This refinement of the notion of the fundamental order provides an elegant setting for much of the material discussed in this section. We survey this approach in Section VII.3.

5.3 Exercise. Show that if G is a stable group then G satisfies the descending chain condition on centralizers of elements. (Recall that the centralizer of $a \in G$ is $\{b \in G : ba = ab\}$.) We denote it by $C_G(a)$.)

5.4 Exercise. Show that if G is a stable group then for any subset X of G there is a finite $X_0 \subseteq X$ with $C_G(X) = C_G(X_0)$.

5.5 Exercise. Show that if G is an ω-stable Abelian group then G has the form $D \oplus H$ where D is a divisible group and every element of H

has order less than n for some integer n. In this situation H is said to have *bounded exponent.* (Hint: Consider the descending chain of definable subgroups $\langle n!G : n < \omega\rangle$.)

Macintyre [Macintyre 1971] proved the converse to the last exercise and noted that ω-stability is preserved by finite direct sums.

While the descending chain conditions are implied by the stability properties for any group, Garavaglia discovered the remarkable fact that for modules the converses hold. The statement of these converses is facilitated by the following useful notation.

5.6 Definition. Fix a theory T of modules. For any $p = t(\overline{a}; B)$, let p^+ denote $\{\phi(\overline{x}; \overline{b}) \in p \colon \phi(\overline{x}; \overline{y})$ a p.p. formula$\}$. Further, let p^- denote $\{\neg\phi(\overline{x}; \overline{b}) \in p \colon \phi(\overline{x}; \overline{y})$ a p.p. formula$\}$.

Note that p is axiomatized by $p^+ \cup p^-$ and p^- is determined by p^+. Thus the number of types over a set A is determined by the number of possible p^+ over A. The following exercise relies on the p.p. elimination theory for modules discussed in Section II.

5.7 Exercise. Prove that every theory of modules is stable.

5.8 Theorem. *For any module M, M is superstable, respectively, ω-stable just if M satisfies the superstable chain condition, respectively, the ω-stable chain condition on p.p.-definable subgroups.*

Proof. We treat the superstable case and leave the easier ω-stable version as an exercise. If M satisfies the superstable d.c.c. for p.p.-definable subgroups, then for every type p there is a formula $\phi(x; \overline{b}) \in p$ such that there is no $\psi(x, \overline{c}) \in p$ such that $\psi(\mathcal{M}; \overline{0})$ has infinite index in $\phi(\mathcal{M}; \overline{0})$. Now p^+ can be axiomatized by p.p. formulas $\psi(x; \overline{c})$ such that $\models \psi(x; \overline{c}) \to \phi(x; \overline{b})$. But since each $\psi(\mathcal{M}; \overline{0})$ has finite index in $\phi(\mathcal{M}; \overline{0})$, there are for each ψ only finitely many choices for inequivalent formulas $\psi(x; \overline{c})$. Thus we have only $|R| + |B| + \aleph_0$ choices for $\phi(x; \overline{b})$ and then only $2^{|R|+\aleph_0}$ choices for the sequence of $\psi(x; \overline{c})$. That is, the number of positive types over B and thus the number of types over B is bounded by $2^{|R|+\aleph_0} \times |B|$. Hence T is superstable.

The following argument provides a sufficient condition for a theory to be unstable.

5.9 Lemma. *Let M be a structure and $\phi(\overline{x}; \overline{y})$ an atomic formula. If there exist $\overline{a}, \overline{b} \in M$ such that*

$$\models \phi(\overline{a}; \overline{a}) \wedge \phi(\overline{b}; \overline{b}) \wedge \phi(\overline{a}; \overline{b}) \wedge \neg\phi(\overline{b}; \overline{a})$$

then M^ω is unstable.

Proof. Let $\langle \overline{c}_i : i < \omega\rangle$ be the sequence of elements of M^ω defined by: $\overline{c}_i(j) = \overline{a}$ if $j \geq i$ and $\overline{c}_i(j) = \overline{b}$ if $j < i$. Then $M^\omega \models \phi(\overline{c}_i; \overline{c}_k)$ if and only if $i \leq k$. Thus, ϕ has the order property (Definition 4.43) and so T is unstable.

This lemma has proved a key for applying stability theory to investigate varieties of algebras. [Baldwin & Lachlan 1973] contains a complete discussion of categoricity in power for varieties and more generally for universal Horn classes. [Baldwin &McKenzie 1982] shows that much of the spectrum problem for, at least, congruence modular varieties can be reduced to the study of varieties of modules. All of these studies are based on extending the definition of stability to include incomplete theories. An arbitrary theory T is said to be stable if every complete extension of T is stable. The following dichotomy summarises the value of this definition in the solution of the spectrum problem for incomplete theories. If some extension of T has the maximal number of models then so does T; the requirement that every complete extension of T be stable imposes very strong conditions on T.

5.10 Exercise. Show that if M is not ω-stable then M^ω is not superstable.

5.11 Exercise. Show that every non-Abelian variety of groups is unstable. (Hint: Apply Lemma 5.9 to the formula $\phi(x_0, x_1, y_0, y_1)$ which asserts $[x_0, y_1] = 1$. See [Baldwin & Lachlan 1971], [Baldwin & Saxl 1976].)

Many of the examples in the rest of the book come from the theory of modules. Thus, we need to develop some machinery for determining whether types in the theory of modules fork. The following definition is the first step.

5.12 Definition. Let T be a theory of modules, $A \subseteq \mathcal{M}$, and $p \in S(A)$.

i) Let $\mathcal{G}(p)$ be

$$\{\phi(\mathcal{M};\bar{0}) : \text{ for some } \bar{a} \in A,\ \phi(\bar{x};\bar{a}) \in p^+\}$$

and let $\mathcal{G}_0(p)$ be

$$\{\phi(\mathcal{M};\bar{0}) : \phi(\bar{x};\bar{y}) \text{ is p.p., } [\psi : \phi \wedge \psi] < \omega \text{ for some } \psi(\bar{x};\bar{0}) \in \mathcal{G}(p)\}.$$

ii) Let $G(p) = \bigcap \mathcal{G}(p)$ and $G_0(p) = \bigcap \mathcal{G}_0(p)$.

5.13 Exercise. Show that if $p \subseteq q$ then $G(p) \supseteq G(q) \supseteq G_0(q)$ and $G_0(p) \supseteq G_0(q)$.

We show that q is a nonforking extension of p just if the final containments of the last exercise can be reversed. Our argument relies on the characterization of nonforking extensions of a type in terms of the number of conjugates which is discussed in Section IV.1.

5.14 Theorem. *Let T be a theory of modules and $p \subseteq q$. Then q is a nonforking extension of p if and only if $G_0(q) = G_0(p)$.*

Proof. First observe that if $\hat{q}$ is a global type extending p and $G_0(\hat{q}) \supseteq G_0(p)$ then $\hat{q}$ is a nonforking extension of p. For, each conjugate of $\hat{q}$ is determined by a map which assigns to each p.p.-definable subgroup of $\mathcal{M}$, the coset of that subgroup which is in $\hat{q}$. But $G_0(\hat{q}) \supseteq G_0(p)$ implies that every such subgroup has finite index in some member of $\mathcal{G}(p)$. Thus, $\hat{q}$ has at most

$2^{|R|+|A|+\aleph_0}$ conjugates over A and thus does not fork over A (Lemma IV.1.14).

The following lemma shows that any q with $G_0(q) \subseteq G_0(p)$ can be extended to a $\hat{q}$ with the same property. By monotonicity, if $G_0(q) = G_0(p)$, q is a non-forking extension of p. But if q does not fork over p then q extends to a conjugate of a type $\hat{p}$ with $G_0(\hat{p}) = G_0(p)$. Since this last property is preserved by conjugacy over A we see $G_0(q) = G_0(p)$ and finish.

5.15 Lemma. *If $p \subseteq q$ and $G_0(q) \supseteq G_0(p)$ then there exists $\hat{q} \in S(\mathcal{M})$ with $p \subseteq \hat{q}$ and $G_0(\hat{q}) = G_0(p)$.*

Proof. It suffices to show that

$$p \cup \{\neg\psi(\overline{x};\overline{m}) : \overline{m} \in \mathcal{M}, \psi(\mathcal{M},\bar{0}) \notin \mathcal{G}_0(p)\}$$

is consistent. If not, there exist p.p. formulas ϕ, χ_i, ψ_i such that for some $\overline{a} \in A$, $\phi(\overline{x};\overline{a}) \in p$, $\neg\chi(\overline{x};\overline{a}) \in p$, and $\psi(\mathcal{M};\bar{0}) \notin G_0(p)$ but for some $\overline{m} \in \mathcal{M}$

$$\mathcal{M} \models \phi(\overline{x};\overline{a}) \rightarrow (\bigvee_{i<n} \chi_i(\overline{x};\overline{a}) \vee \bigvee_{j<m} \psi_j(\overline{x};\overline{m})).$$

Since $\mathcal{M} \not\models \phi(\overline{x};\overline{a}) \rightarrow (\bigvee_{i<n}\chi_i(\overline{x};\overline{a}))$, Lemma I.4.13 implies that for some j, $\phi(\mathcal{M};\bar{0}) \cap \psi_j(\overline{x};\bar{0})$ has finite index in $\phi(\mathcal{M};\bar{0})$. But this, in turn, implies that $\psi_j(\mathcal{M};\bar{0})$ has finite index in $\phi(\mathcal{M};\bar{0}) + \psi_j(\mathcal{M};\bar{0})$. Since the last group is in $\mathcal{G}(p)$ we have a contradiction.

5.16 Exercise. Show that if $G_0(p) = G(p)$ then p is stationary. Find a counterexample to the converse.

5.17 Exercise. Let T be the theory of $M = Z_4^{\aleph_0}$. For each p.p.-formula ϕ, let $\langle\phi\rangle$ denote the set of p.p. formulas which are deducible from ϕ. Let p be the type determined by $p^+ = \langle 2 \times v = 0\rangle$. Let $a \in M$ be an element of order 4 and let q be the type determined by $q^+ = \langle 2(x-a) = 0\rangle$. Then in $M \oplus Z_4$, p is realized by $\langle 0, 2\rangle$ and q is realized by $\langle a, 2\rangle$. Show that $G_0(p) = G_0(q) = \{x : 2 \times x = 0\}$ and thus that p does not fork over 0 but q does.

The definition of $G_0(p)$ reflects an idea that plays an important role in this area.

5.18 Definition. The group G is *connected* if it has no proper definable subgroup of finite index.

This name is suggested by analogy with the connected component of the identity in an algebraic group with the Zariski topology. The notion is important here both because of the relative ease with which connected groups can be found in this context and the consequences of assuming a group to be connected. In fact, if G is an affine algebraic group over an algebraically closed field F and we consider G in the language imposed by the imbedding of G in F^n, then G is connected in our sense exactly if G is connected in the sense of the Zariski topology.

The following terminology is convenient for discussing the structure of stable groups. The usage is standard in the study of infinite groups.

5.19 Definition. Let $\mathcal{H}$ and $\mathcal{K}$ be classes of groups. Then the group G is said to be $\mathcal{H}$ *by* $\mathcal{K}$ if there is a normal subgroup H of G which is in $\mathcal{H}$ such that $G/H \in \mathcal{K}$.

5.20 Exercise. Show that if G is an ω-stable group then G is connected by finite. (Hint: Apply the d.c.c.)

We prove a more delicate version of this result which is in [Baur, Cherlin & Macintyre 1979].

5.21 Theorem. *If G is $\aleph_0$-categorical and stable then G is connected by finite.*

Proof. Let Γ_n be the collection of subgroups of G which are definable with at most n parameters and have finite index in G. By the $\aleph_0$-categoricity and Ryll-Nardzewski's theorem (Theorem I.3.1), there are only a finite number of possible defining formulas for these groups. Thus by the full stable d.c.c. there is a minimal such group, H_n which is definable without parameters. If $\langle H_n : n < \omega \rangle$ forms an infinite decreasing chain, then G realizes infinitely many 1-types, again contradicting Ryll-Nardzewski's theorem.

The following results presage the proof in Chapter VIII that a superstable division ring is an algebraically closed field.

5.22 Theorem. *If F is an infinite stable division ring then the additive group F^+ of F is connected.*

Proof. Suppose for contradiction that A is a proper definable subgroup of F^+ with $[F^+ : A] < \omega$. Let Γ be the collection of subgroups aA for a a nonzero element of F. By the full stable d.c.c., there is a group $A_0 = \bigcap \Gamma$ with finite index in F. But then A_0 is an ideal in F and since F is infinite, $A_0 \neq (0)$; thus, $A_0 = F$.

A very similar argument solves the following exercise.

5.23 Exercise. Show that if N is a finite normal subgroup of a connected group G, then N is contained in the center of G. (Hint: Imbed G in the automorphism group of N via conjugation. Consider the kernel of this imbedding.)

5.24 Exercise. Show that if G is a group, f is an endomorphism of G such that $\ker f$ is finite, and $[G : f(G)]$ is infinite, then $[f(G) : f^2(G)]$ is infinite.

5.25 Theorem. *Let G be a connected superstable group and h a definable endomorphism of G with finite kernel. Then h is surjective.*

Proof. If f is not onto, G is connected implies $[G : f(G)]$ is infinite. Applying Exercise 5.24 and induction we see $[f^n(G) : f^{n+1}(G)]$ is infinite for each $n < \omega$. But this contradicts the superstable d.c.c.

5.26 Historical Notes. The reduction of stability to chains in structures admitting a multiplication is foreshadowed in [Macintyre 1971] and explicit in [Baldwin & Saxl 1976] and [Shelah 1975]. The proof that stable groups satisfy the full descending chain condition first appears in [Cherlin & Reineke 1976]. As remarked there the argument generalizes that for centralizers which Baldwin generalized from a lemma in [Baldwin & Saxl 1976]. The simple argument given here was pointed out by Ziegler. Poizat [Poizat 1981] gives still another proof.

There have been a number of powerful applications of the methods described in this section. It has been shown in varying degrees of generality in [Felgner 1975], [Cherlin & Reineke 1976], [Sabbagh 1975], and [Baldwin & Rose 1977] that any semisimple stable ring is a finite direct sum of matrix rings over division rings. This generalizes the classical proof of the Wedderburn-Artin theorem by showing that each use of the descending chain condition on all ideals can be replaced by a use of the stable descending chain condition.

The strongest known result concerning stable groups is that an $\aleph_0$-categorical stable group is nilpotent by finite; moreover, if the group is ω-stable then it is Abelian by finite. This profound result was proved in [Baur, Cherlin & Macintyre 1979].

In another direction, Cherlin [Cherlin 1979] has conjectured that the ω-stable simple groups with finite Morley rank are 'classical' simple groups and verified this result for groups with low Morley rank. Simon Thomas [Thomas 1983] has verified this conjecture for locally finite groups. Further important work on this line is contained in [Nesin 1985], [Berline & Lascar 1986], [Hrushovski 198?d], and [Hrushovski 1986].

Chapter IV

Finite Equivalence Relations, Definability, and Strong Types

We study in this chapter the nonforking extensions of a type $p \in S(A)$. In the first section we see that although two such types cannot be distinguished by formulas over A, they can be distinguished by formulas that are 'almost over' A. This leads to a straightforward proof of Lachlan's theorem that an $\aleph_0$-categorical superstable theory is ω-stable. In the second section we study in more detail the multiplicity of p - the number of nonforking extensions of p. We apply this analysis to reformulate the stability hierarchy in terms of definability. We introduce the important notions of the 'base' and the 'strong base' of a type. We show each type over a strongly $\kappa(T)$-saturated model is strongly based on a subset of power less than $\kappa(T)$. In Section 3 we summarise the fundamental properties of strong types. The strong type of a over A allows one to analyze the relation of a and A in terms of a stationary (i.e. multiplicity one) type. This is an essential tool for the further development of the theory. Finally we study the relation between the strong type (more generally, the multiplicity) of a pair and that of its components.

1. Finite Equivalence Relations

Unfortunately, the study of types will not provide as smooth a theory of independence as we would like. A simple example of the difficulty arises if we consider the theory of an equivalence relation with two infinite classes. Then the unique type, p, over the empty set has two extensions to a type over a model which do not fork over the empty set; namely, the type of a generic point in either equivalence class. Morley addressed this problem by saying 'the type has degree two' and the work of Lachlan [Lachlan 1975], [Lachlan 1978], successfully develops this approach. However, we find it more convenient to follow Shelah in introducing the notion of strong type. This approach describes the situation in the above example, by saying 'there are two strong types extending p', one for each equivalence class.

This approach also deals with another issue. Let A be a set and suppose $\bar{b}$ is not in A, but is algebraic over A. Consider a formula $\phi(\bar{x};\bar{b})$. Although $\phi(\bar{x};\bar{b})$ is not in $F(A)$, clearly it is 'close' to being so. We will formalize a generalization of this property by defining the notion that $\phi(\bar{x};\bar{b})$ is 'almost over' A.

Recall that for $X \subseteq \mathcal{M}$ we say X is definable over A (i.e. definable with parameters from A), or X is just over A if there is a formula $\phi(\bar{x};\bar{y})$ and a sequence $\bar{a}$ from A such that $X = \phi(\mathcal{M};\bar{a})$. Although all three expressions for this concept are commonly used we will eschew here this usage of 'definable over A' so as to distinguish this notion from the notion of a type being 'definable over A' considered in III.1 and below. We will say $X \subseteq \mathcal{M}$ is definable if it is definable with some parameters from $\mathcal{M}$ (i.e. if X is over $\mathcal{M}$). We recall from Chapter I a useful criterion for determining whether X is over A.

I.2.30 Lemma. *If X is a definable subset of $\mathcal{M}$ and X is fixed by every A-automorphism of $\mathcal{M}$ then X is over A.*

Recall that p is definable over A if for each $\phi(\bar{x};\bar{y})$ the set $\{\bar{c}:\phi(\bar{x};\bar{c}) \in p\}$ is over A. We distinguish the notions, 'p is over A' and 'p is definable over A' even though for stable T and $p \in S(A)$ they coincide. Generalizing the notion, 'X is over A', we replace the requirement that X be fixed by A-automorphisms with the requirement that it have 'few' conjugates. Since we are working in the monster model, a cardinal number is small or large depending on whether it is less than the cardinality of the monster model. In analogy with set theory, we say a collection (e.g. of formulas with parameters or of types) 'is a set' or 'has bounded cardinality' if its cardinality is less than that of the monster model and a class if its cardinality is the same as the cardinality of the monster model. Formally:

1.1 Definition. i) Let X be a definable subset of $\mathcal{M}$. Then X is *almost over A* if $|\{\alpha(X) : \alpha \in \mathrm{Aut}_A(\mathcal{M})\}| < \omega$.

ii) The formula $\phi(\bar{x};\bar{b})$ is *almost over A* if the set $\phi(\mathcal{M},\bar{b})$ is almost over A.

iii) The type p is *almost over A* if each formula in p is almost over A.

Note that if $\bar{b}$ is in the algebraic closure of A then for each formula $\phi(\bar{x};\bar{y})$, $\phi(\bar{x};\bar{b})$ is almost over A. Further, if T is the theory of an equivalence relation E with two classes, both infinite, then for any b, $E(x;b)$ is almost over the empty set. Virtually no complete type (not over A) is almost over A since if $b \notin \mathrm{cl}(A)$ the formula $x \neq b$ is not almost over A. However, the members of a very important class of incomplete types, the strong types over A (cf. Definition 3.1), are almost over A.

The following characterization (Theorem 1.3) underlines the significance of this notion.

1.2 Definition. i) A *finite equivalence relation* is a definable equivalence relation on $\mathcal{M}^m$ for some m which has only finitely many classes. We say E is over A if E is definable with parameters from A.

ii) The formula $\phi(\overline{x}; \overline{b})$ *depends on the finite equivalence relation* E if the solution set of ϕ is a union of equivalence classes of E.

We denote by $FE^m(A)$ the collection of finite equivalence relations (on m-tuples) which are over A. We write $FE(A)$ to refer ambiguously to some $FE^m(A)$.

1.3 Theorem. *The following are equivalent.*

i) $\phi(\overline{x}; \overline{c})$ *is almost over* A.
ii) $\phi(\mathcal{M}; \overline{c})$ *has a bounded number of conjugates over* A.
iii) *There is an* $E \in FE(A)$ *and* $\overline{b}_0, \ldots, \overline{b}_{k-1}$ *such that:*

$$\phi(\overline{x}; \overline{c}) \leftrightarrow \bigvee_{i<k} E(\overline{x}; \overline{b}_i).$$

iv) $\phi(\mathcal{M}; \overline{c})$ *is over* M *for every model* M *containing* A.

Proof. It is easiest to first argue that ii) and iii) are separately equivalent to i). Clearly i) $\rightarrow$ ii) and the converse is an easy compactness argument. If iii) holds, each conjugate of $\phi(\overline{x}; \overline{c})$ is equivalent to a union of equivalence classes of E so it is easy to bound the number of conjugates in terms of the number of classes of E. Now we show i) $\rightarrow$ iii). Define $E(\overline{x}; \overline{y})$ to hold just if $\phi(\overline{x}; c_i) \leftrightarrow \phi(\overline{y}; c_i)$ for each of the formulas $\phi(\overline{x}; c_i)$ representing the finitely many conjugates of $\phi(\overline{x}; \overline{c})$. Then E is a finite equivalence relation, definable with parameters from $\mathcal{M}$, and fixed under A-automorphisms. By Lemma I.2.30, $E \in FE(A)$. Now we show iii) implies iv) and iv) implies ii). By iii), the solution set in $\mathcal{M}$ of $\phi(\bar{x}; \bar{c})$ is the union of equivalence classes of a finite equivalence relation over A. Thus for any M containing A there are k sequences $\overline{m}_i$ from M such that: $\models \phi(\overline{x}; \overline{c}) \leftrightarrow \bigvee_{i<k} E(\overline{x}; \overline{m}_i)$ and iv) holds. On the other hand, if $\phi(\overline{x}; \overline{c})$ is over every M containing A, so are all the conjugates of $\phi(\overline{x}; \overline{c})$. That is, all conjugates of $\phi(\overline{x}; \overline{c})$ are over a single model M so $\phi(\overline{x}; \overline{c})$ has a bounded, hence finite, number of conjugates and is almost over A.

Thus ϕ is almost over A iff ϕ depends on some $E \in FE(A)$.

The fourth characterization in the previous theorem reinforces a second meaning for the word 'almost' which we have seen before. Recall from Definition III.3.13 that $\phi(\overline{x}; \overline{c})$ is said to be almost satisfiable (or almost satisfied) in A if for every M extending A, $\phi(\mathcal{M}; \overline{c}) \cap M \neq \emptyset$. A property P is almost true in A if it is true in every model containing A. We will consider several other uses of the word almost in this sense.

1.4 Exercise. Show that the collection of formulas almost over A is closed under the logical operations (Boolean and quantifiers), permutation of variables and substitution of dummy variables. (This is a routine application of Theorem 1.3. Details are in III.2.2 of [Shelah 1978].)

1.5 Corollary. *If the consistent type* p *is almost over* A *then each formula in* p *is almost satisfiable in* A *and so* p *does not fork over* A.

Proof. By Theorem 1.3 each $\phi(\overline{x};\overline{c})$ in p defines the union of some equivalence classes of a finite equivalence relation over A. But each class must intersect every model containing A so p is almost satisfiable in A.

1.6 Exercise. If $A \subseteq B$, $p \in S(B)$ does not fork over A, and r is almost over A, show that if $r \cup p$ is consistent then $r \cup p$ does not fork over A.

1.7 Exercise. Show that $p \in S(A)$ is stationary iff every $\phi(\overline{x};\overline{y})$ represented in $\hat{p}$, the bound of p, is represented in p in the following weak sense. If $q \in [\hat{p}]$ and $\phi(\overline{x};\overline{b}) \in q$ there is a formula $\psi(\overline{x};\overline{a}) \in p$ such that $\models \phi(\overline{x},\overline{b}) \leftrightarrow \psi(\overline{x};\overline{a})$ (Baldwin-Prest).

1.8 Exercise. Suppose $\phi(\overline{x};\overline{b})$ is almost over A and $\phi(\overline{x};\overline{b}_1),\ldots,\phi(\overline{x};\overline{b}_n)$ is a list of representatives of all the A-conjugates of $\phi(\overline{x};\overline{b})$. Show that $\bigwedge_{i\leq n}\phi(\overline{x};\overline{b}_i)$ and $\bigvee_{i\leq n}\phi(\overline{x};\overline{b}_i)$ are each equivalent to a formula over A.

1.9 Exercise. Show that if the finite equivalence relation E is almost over A, then there is another finite equivalence relation E' which is over A such that E depends on E'. (Hint: Consider the conjunction of the conjugates of E.)

It is immediate from Theorem 1.3 that there are at most $2^{(|T|+|A|)}$ distinct types almost over A. In Lemma 1.12 we will reduce this bound to $\sup(2^{|T|},|S(A)|)$. Now we combine the concepts 'defining a type' and 'almost over' by saying that p is definable almost over A if p is definable in the sense of III.1.23 but the formulas $d\phi$ are almost over A.

1.10 Definition. The type p is *definable almost over* A if for each $\phi(\overline{x};\overline{y})$, $\{\overline{c} : \phi(\overline{x};\overline{c}) \in p\}$ is almost over A.

It is immediate from the definitions and the characterization of almost over in Theorem 1.3 iv) that for any global type $\hat{p}$, $\hat{p}$ does not fork over A if and only if $\hat{p}$ is definable almost over A. We gave in Exercise III.3.28 an example of an incomplete type which was definable (and so certainly almost definable) over the empty set but still forked over the empty set. Thus this remark does not extend from global types to arbitrary types. Exercise 1.15 notes that in one direction this extension is possible; Exercise 1.16 points out a strengthening of the hypothesis which regains the converse.

1.11 Exercise. Show that $\hat{p}$ is definable almost over A if and only if for each formula $\phi(\overline{x};\overline{y})$, $d\phi$ is almost over A.

The following lemma yields another characterization of the nonforking extensions of a type p.

1.12 Lemma. *If $\hat{p}$ does not fork over A then $\hat{p}$ has at most $2^{|T|}$ conjugates over A.*

Proof. Since $\hat{p}$ does not fork over A, there is a model M and a map $d : F(L) \mapsto F(\mathcal{M})$ so that $\hat{p}$ is defined by d and each $d\phi$ is almost over A. Then each conjugate of $\hat{p}$ is defined by conjugates of the $d\phi$. But each of the $|T|$ formulas $d\phi$ has only finitely many conjugates over A, so $\hat{p}$ has at most $2^{|T|}$ conjugates over A.

1.13 Exercise. Give a proof of Lemma 1.12 without appealing to definability. (Hint: First assume without loss of generality that $|A| < \kappa(T)$, then choose a model of cardinality at most $|T|$ containing A and note that the nonforking types over A are determined by their restrictions to that model.)

We can show a strong form of the converse.

1.14 Lemma. *If $\hat{p}$ has a bounded number of conjugates over A then $\hat{p}$ does not fork over A.*

Proof. If $\hat{p}$ has a bounded number of conjugates over A then for each ϕ, $d\phi$ has a bounded number of conjugates over A. So $\hat{p}$ is definable almost over A and so does not fork over A.

For any p we have shown the extensions of p to types in $S(\mathcal{M})$ are partitioned into conjugacy classes by the fundamental order (Corollary III.2.38). There are at most $2^{(|T|+|A|)}$ classes. One of them, the class of nonforking extensions has at most $2^{|T|}$ members. The others have a class (i.e. $|\mathcal{M}|$) of members.

1.15 Exercise. Let $A \subseteq B$ and suppose p is a type over B such that $p|A$ is complete and p does not fork over A. Show that if d is any defining scheme for p then each formula $d\phi$ is almost over A.

1.16 Exercise. Show that if $p \in S(M)$ is definable almost over A then p does not fork over A. Give an example to show the necessity of assuming M is a model.

We have shown that if p doesn't fork over A then p is definable almost over A. We can strengthen this result if $p|A$ is stationary or even stationary relative to $\operatorname{dom} p$ in the following sense.

1.17 Definition. i) Let $A \subseteq B$ and $p \in S(A)$. Then we say p is *stationary inside* B if p has a unique extension in $S(B)$ which does not fork over A.

The remainder of this definition makes precise two ways in which a type can be founded on a 'small' set. These notions will play an increasingly important role later on. We discuss them more fully after Definition 2.4.

ii) Let $A \subseteq B$ and $p \in S(B)$. Then

a) p is *based* on A if p does not fork over A.
b) p is *strongly based* on A if p does not fork over A and $p|A$ is stationary.

WARNING: Our terminology here differs with that of Shelah. He only defines 'based' for a type which is the average type of a sequence of indiscernibles and does so in such a way that his 'based' is equivalent to our 'strongly based'.

Note that if the set B in Definition 1.17 i) is a model then the type p is, in fact, stationary. (See Theorem III.2.23.)

1.18 Exercise. Give an example of $A \subseteq B$ with $p \in S(A)$ which is stationary inside B but not stationary.

1.19 Exercise. Show that if $p \in S(A)$ is stationary then for every $B \supseteq A$, p is stationary inside B.

1.20 Lemma. i) *If $A \subseteq B$, $p \in S(B)$ does not fork over A, and $p|A$ is stationary inside B then p is definable over A.*

We will be most interested in the following special case.

ii) (Stationary types are definable). *Suppose $p \in S(A)$ is stationary. There is a definition d of p over A such that for every B containing A, if $p \subseteq q \in S(B)$ and q does not fork over A then $q = d(p, B)$.*

Proof. Let $\hat{p}$ be an extension of p to a global type which does not fork over B (and hence does not fork over A). Then $\hat{p}$ is definable almost over A by some map d taking ϕ to $d\phi(\overline{y};\overline{c})$. Let $d^*\phi(\overline{y})$ be $\bigvee_{\alpha \in \mathrm{Aut}_A(\mathcal{M})} d\phi(\overline{y};\alpha\overline{c})$. Then $d^*\phi(\overline{y})$ is invariant under automorphisms which fix A and so is equivalent by Theorem I.2.30 to a formula over A (thus justifying our omission of parameters). But since $d^*\phi(\mathcal{M}) \cap B$ is $\{\overline{b} : \phi(\overline{x};\overline{b}) \in p\}$ we have defined p over A as required.

Note that the $d^*\phi$ found in the proof of Lemma 1.20 is a positive Boolean combination of instances of $d\phi$. We replace $d\phi$ by ϕ in Lemma V.2.6.

1.21 Exercise (Berline). Show $d^*\phi(\overline{c})$ holds if and only if $\phi(\overline{x};\overline{c})$ is in every nonforking extension of p to a global type.

1.22 Exercise. Prove 1.20 ii) directly rather than deducing it from 1.20 i).

The following table summarises the relations among the notions considered in this section. In each column there are a number of statements equivalent to the assertion that a global type has the number of conjugates over A given at the head of the column. Each row represents a particular kind of description.

1	$\leq 2^{\lvert T\rvert}$	∞
definable over A	definable almost over A	
over A	does not fork over A	forks over A
strongly based on A	based on A	

A type may be almost over T — in that each formula in the type has only finitely many conjugates over A — but the type still has $2^{\lvert T\rvert}$ conjugates.

Suppose $A \subseteq B$, $p \in S(B)$, and p has no proper conjugates over A. Then by Theorem I.2.30 (The basic definability lemma) p is definable over A. We have just seen that if p has few conjugates then p does not fork over A and is definable almost over A. We now show that two distinct nonforking extensions of a type over A can be separated by a formula which is almost over A.

1.23 Theorem. *Let $A \subseteq M$, $p, q \in S(M)$. Suppose p and q do not fork over A and $p|A = q|A$. If $p \neq q$ then there is formula $\psi(\overline{x})$ almost over A with $\psi(\overline{x}) \in p$ and $\neg\psi(\overline{x}) \in q$.*

Proof. By Exercise II.2.6 we can choose $\overline{c}$ and $\overline{d}$ realizing p and q respectively with $\overline{c}^\frown\overline{d} \downarrow_A M$. Choose $\phi(\overline{m};\overline{x}) \in F(M)$ so that $\phi(\overline{m};\overline{c})$ and $\neg\phi(\overline{m};\overline{d})$. Let $\hat{r} \in S(\mathcal{M})$ be an nonforking extension of $r = t(\overline{m}; A \cup \overline{c} \cup \overline{d})$ to a global type. By symmetry and monotonicity, $\hat{r}$ does not fork over A. Thus there is a formula $d\phi(\overline{x})$ which is over M, almost over A, and defines $\hat{r}_\phi$. But then we have $\models d\phi(\overline{c}) \wedge \neg d\phi(\overline{d})$. So this formula is in p but not in q as required.

We now establish one of the principal tools of stability theory.

1.24 Theorem (The Finite Equivalence Relation Theorem). Let $A \subseteq B$ and $p, q \in S(B)$. Suppose $p|A = q|A$ and neither p nor q forks over A. If $p \neq q$ then there is an $E \in FE(A)$ such that: $p(\overline{x}) \cup q(\overline{y}) \vdash \neg E(\overline{x};\overline{y})$.

Proof. We know by Theorem 1.23 that there is a formula $\phi(\overline{x};\overline{b})$ in p but not q which is almost over A. Choose an $E \in FE(A)$ so that $\phi(\overline{x};\overline{b})$ depends on E. Now clearly $p(\overline{x}) \cup q(\overline{y}) \vdash \neg E(\overline{x},\overline{y})$.

For the following exercise consult the definition of Morley rank in Chapter VII.1.

1.25 Exercise (Relations With Morley Rank). Show that if $p \in S(M)$ then p has degree 1. Show that if the rank of M is α there is a finite equivalence relation definable over the empty set with $\deg(M)$ classes each with rank α.

One of the most important corollaries of the Finite Equivalence Relation Theorem is the following proof of Lachlan's theorem that a countable $\aleph_0$-categorical superstable theory is ω-stable. Recall from III.3.3 that $N(M, A)$ denotes the complete types over M which do not fork over A.

1.26 Theorem. *If T is a countable $\aleph_0$-categorical superstable theory then T is ω-stable.*

Proof. If not, there is a model M of T with $|S(M)| > |M|$. Without loss of generality we may fix an integer m such that the number of m-types over M is greater than $|M|$. Since T is superstable, for each p in $S(M)$ there is a finite subset of M over which p does not fork. Since there are only $|M|$ finite subsets of M, the theorem follows if we can show that for any finite A contained in M, $N(M, A)$ is also finite. By Ryll-Nardzewski's Theorem, $S(A)$ is finite so it suffices to show that any fixed member r of $S(A)$ has only finitely many extensions in $N(M, A)$. Thus, suppose $p(\overline{x})$ and $q(\overline{x})$ are distinct members of $N(M, A)$ extending $r \in S(A)$. Then by Theorem 1.24 there is an $E(\overline{x},\overline{y}) \in FE^m(A)$ such that $p(\overline{x}) \cup q(\overline{y})$ implies $\neg E(\overline{x};\overline{y})$. (Fig. 1). Thus $|N(M, A)|$ is bounded by the product of $\{n(E) : E \in FE^m(A)\}$ where $n(E)$ denotes the number of equivalence classes of E. But the $\aleph_0$-categoricity of T implies by Ryll-Nardzewski's theorem that the number of formulas with $|A| + 2m$ free variables is finite and this number certainly bounds $|FE^m(A)|$. Thus $N(M, A)$ is finite and the theorem follows.

Lachlan [Lachlan 1974] conjectured that the hypotheses of this theorem could be weakened by assuming only that T is stable. This conjecture

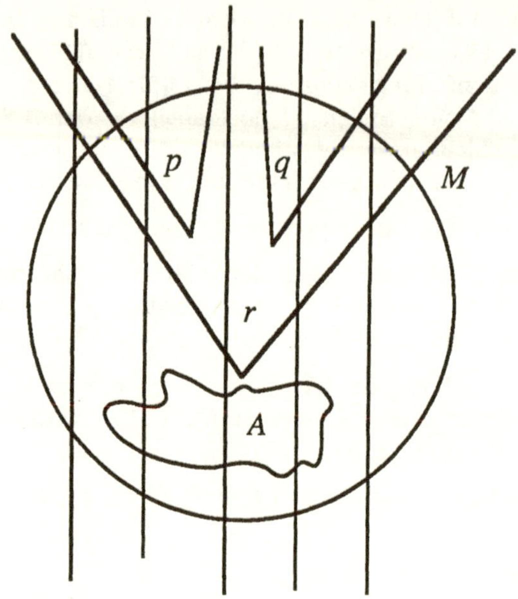

Fig. 1. Theorem 1.26. The vertical columns are the E-equivalence classes.

remains one of the critical unsolved problems in the area. This problem is the first of several which show that the investigation of stable but not superstable theories will require significant new tools. In his work on this problem Lachlan proved the normalization lemma which we discuss in detail in Chapter VIII. Moreover, he introduced the concept of pseudoplane which is also discussed in Chapter VIII. This notion has proved to be at the combinatorial core of the subject.

The following more algebraic version of the finite equivalence relation theorem is to due to John Vaughn [Vaughn 1985]. The terminology from Boolean algebras used in the next few paragraphs can be found in [Sikorski 1964]. To simplify the following proof we make the following definitions.

1.27 Definition. Let $B \subseteq \mathcal{M}$. Recall that $F(\mathcal{M})$ is the Lindenbaum algebra of formulas over $\mathcal{M}$. We fix the following notation. (Fig. 2).

i) F_B is the set of formulas in $F(\mathcal{M})$ which fork over B.
ii) A_B is the set of formulas in $F(\mathcal{M})$ which are almost over B.
iii) N_B is the set of formulas in $F(\mathcal{M})$ which do not fork over B.
iv) $\neg F_B$ is the set of formulas in $F(\mathcal{M})$ whose negations fork over B.

Recall that an ideal, I, in a Boolean algebra is a subset which is closed under joins and such that if $b \in I$ then for any a, $a \wedge b \in I$.

1.28 Exercise. Show that F_B is an ideal in $F(\mathcal{M})$.

We denote by $\phi \mathbin{\triangle} \psi$ the symmetric difference of the formulas ϕ and ψ. Recall that an ideal, I, in a Boolean algebra determines a congruence on

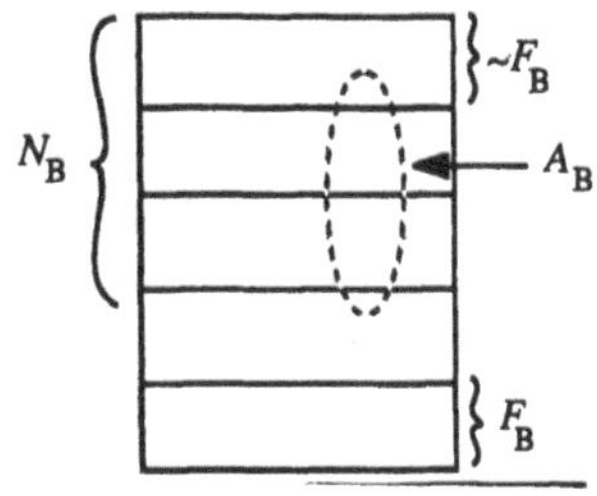

Fig. 2. Definition 1.27

the algebra where $\phi \simeq \psi$ if and only if $\phi \,\Delta\, \psi \in I$. Now we can reformulate the finite equivalence relation theorem as providing a normal form for $F(\mathcal{M})/F_B$.

1.29 Theorem. $F(\mathcal{M})/F_B \approx A_B$.

Proof. We show that for each $\phi(\overline{x};\overline{m}) \in F(\mathcal{M})$ there is a $\psi(\overline{x};\overline{b})$ which is almost over B such that $\phi(\overline{x};\overline{m}) \simeq \psi(\overline{x};\overline{b})$. As any formula in F_B is equivalent $\operatorname{mod} F_B$ to $x \neq x$ and and any formula in $\neg F_B$ is equivalent $\operatorname{mod} F_B$ to $x = x$, we can assume $\phi(\overline{x};\overline{m}) \in N_B - \neg F_B$. Thus, both $Y = U_{\phi(\mathcal{M},\overline{m})} \cap N(\mathcal{M},B)$ and $X = U_{\neg\phi(\mathcal{M},\overline{m})} \cap N(\mathcal{M},B)$ are nonempty closed, and thus compact sets. Now for each $q \in Y$ and $p \in X$, the finite equivalence relation theorem implies there is a clopen $C_{p,q} \subseteq S(\mathcal{M})$ which is defined by some $\psi_{p,q}(\overline{x};\overline{b}_{p,q})$ with $q \in C_{p,q}$ and $p \notin C_{p,q}$. By compactness, for each p the union, D_p, of some finite subset of the $C_{p,q}$ contains Y and is almost over B. By compactness again, some finite intersection of the D_p is defined by a formula $\psi(\overline{x};\overline{b})$ which is almost over B and such that $U_{\psi(\overline{x};\overline{b})} \cap N(\mathcal{M},B) = Y$. Thus, $\psi(\overline{x};\overline{b}) \,\Delta\, \phi(\overline{x};\overline{m}) \in F_B$ and we finish.

1.30 Historical Notes. Most of these notions appeared first in [Shelah 1978]. The importance of calculating the number of conjugates was emphasized by Lascar and Poizat. The treatment here was greatly influenced by conversations with Ziegler. The finite equivalence relation theorem is due to Shelah [Shelah 1978]. The improvement of it in Theorem 1.23 was conjectured by Harnik and proved by Ziegler. Theorem 1.26 was proved by Lachlan [Lachlan 1974] using rank. The argument used here is due to Shelah. Definition 1.27 and Theorem 1.29 are due to John Vaughn.

2. Definability and the Stability Hierarchy

In this section we use the notions of definability of types developed in section 1 to provide another characterization of the stability hierarchy. To begin with we need more 'local' information about the possible extensions of a single type. Then we can proceed to the 'global' discussion of the relation between the number of types over each model and the extent to which these types are definable. We assume T is stable throughout this section.

In Definition III.4.28 we defined the ϕ-multiplicity of a type p to be the maximal number of nonforking mutually contradictory ϕ-types which are consistent with p. We now use the fact (Exercise III.3.16) that the collection of nonforking extensions of p to complete types over $\mathcal{M}$ is closed to show this number must always be finite. Then we will use that result to refine the assertion that $N(\mathcal{M}, A)$ is closed.

2.1 Theorem. *For any A, any $p \in S(A)$, and any formula ϕ, $\mu_\phi(p)$ (the ϕ-multiplicity of p) is finite.*

Proof. If not, there is an infinite family of ϕ-types p_i such that $p \cup p_i$ does not fork over A but for each i and j there is an $\overline{a}_{i,j}$ such that $\phi(\overline{x}; \overline{a}_{i,j}) \in p_i$ and $\neg\phi(\overline{x}; \overline{a}_{i,j}) \in p_j$. Without loss of generality for some M containing A we may extend each $p \cup p_i$ to a type $q_i \in S(M)$ which does not fork over A. Now for any κ construct κ distinct nonforking extensions of p to a model as follows. Add to L a new unary predicate symbol P, constants $\{\overline{c}_i : i < \kappa\}$ and $\{\overline{a}_{i,j} : i, j < \kappa\}$. Now let Γ contain $\phi(\overline{c}_i, \overline{a}_{i,j})$ if and only if $i = j$ and assert that P is the universe of an elementary submodel of the universe, $A \subseteq P$, all $\overline{a}_{i,j}$ are in P, and $t(\overline{c}_i; P)$ does not fork over A. The last clause is obtained by including in Γ, $(\forall \overline{y})[P(\overline{y}) \to \neg\phi(\overline{c}_i; \overline{y})]$ if $\phi(\overline{x}, \overline{y}) \notin \beta(p)$. These κ types contradict Lemma III.4.32 which showed every type has at most $\lambda(T)$ nonforking extensions.

2.2 Exercise. Conclude from Theorem 2.1 that for any A and $p \in S(A)$ there are only finitely many types $q_i \in S_\phi(\mathrm{cl}(A))$ with $p \cup q_i$ consistent.

2.3 Exercise. Show that in Theorem 2.1 it is essential to assume p is complete.

Recall that Corollary III.3.15 asserted that if p is a type over B and p forks over $A \subseteq B$ then there is a formula $\phi(\overline{x}; \overline{y})$ and a type $q \in S(A)$ such that if $\overline{b}$ realizes q, then any type, r, containing $\phi(\overline{x}; \overline{b})$ forks over A. We will modify the result by removing the requirement that $\overline{b}$ realize a complete type but replacing the requirement r contains $\phi(\overline{x}; \overline{b})$ with the requirement that $p|A$ be complete and r contains $p|A \cup \phi(\overline{x}; \overline{b})$.

2.4 Corollary. *For every $p \in S(A)$ and every formula $\phi(\overline{x}; \overline{y})$ there is a formula $\theta(\overline{y}; \overline{a}) \in F(A)$ such that for any $\overline{b}$, $p \cup \phi(\overline{x}; \overline{b})$ does not fork over A if and only if $\models \theta(\overline{b}; \overline{a})$.*

Proof. Invoking Theorem 2.1, let $\{q_1, \ldots, q_n\} \subseteq S_\phi(\mathcal{M})$ with $n = \mu_\phi(p)$ be the nonforking complete ϕ-types over $\mathcal{M}$ such that $p \cup q_i$ does not fork over A. Let $\theta_l(\overline{y}; \overline{m}_l)$ define q_l for $l < n$ and let θ be the disjunction of the θ_l. Now $p \cup \phi(\overline{x}; \overline{b})$ does not fork over A if and only if for some $l < n$, $p \cup \phi(\overline{x}; \overline{b}) \subseteq q_l$ if and only if $\models \theta(\overline{b})$. It remains only to show that the formula θ is over A. But this follows from the fundamental definability lemma (Theorem I.2.30), since clearly it is fixed by every A-automorphism of $\mathcal{M}$.

We will now prove that in a countable theory the multiplicity of each complete type is either finite or $2^{\aleph_0}$. This yields as an immediate corollary that if T is ω-stable $u(p)$ is finite for each p. There are several ways to establish these results. Probably the most efficient is via ϕ-ω rank which we discuss briefly in Chapter VII (cf. II of [Shelah 1978]). We choose here to use techniques from Chapter III. The following lemma is the key step.

2.5 Lemma. *Let T be a stable theory and $p \in S(A)$ a complete type. Suppose that there are extensions p_i, p'_i of p for each $i < \omega$ such that $p_i \subseteq p_{i+1}$, $|\operatorname{dom} p_{i+1} - \operatorname{dom} p_i| < \omega$, $\operatorname{dom} p_{i+1} = \operatorname{dom} p'_{i+1}$, p_{i+1} and p'_{i+1} are distinct nonforking extensions of p_i. Then,*

i) $\mu(p) \geq 2^{\aleph_0}$.
ii) *It is impossible that all the p_i (p'_i) have the form $p \cup q_i$ ($p \cup q'_i$) where q_i (q'_i) is a ϕ-type.*

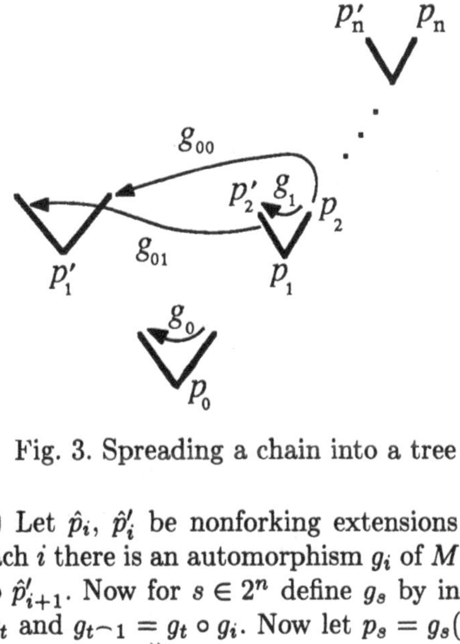

Fig. 3. Spreading a chain into a tree

Proof. (Fig. 3). i) Let $\hat{p}_i$, $\hat{p}'_i$ be nonforking extensions of p_i, p'_i to global types. Then for each i there is an automorphism g_i of M which fixes $\operatorname{dom} p_i$ and maps $\hat{p}_{i+1}$ to $\hat{p}'_{i+1}$. Now for $s \in 2^n$ define g_s by induction: $g_{\langle\,\rangle}$ is the identity, $g_{t\frown 0} = g_t$ and $g_{t\frown 1} = g_t \circ g_i$. Now let $p_s = g_s(p_i)$. For $\sigma \in 2^\omega$ let $p_\sigma = \bigcup_{n<\omega} p_{\sigma|n}$. The p_σ are $2^{\aleph_0}$ distinct nonforking extensions of p.

ii) If the p_i (p'_i) all have the form $p \cup q_i$ ($p \cup q'_i$) for ϕ-types q_i (q'_i) we have $2^{\aleph_0}$ ϕ-types over a countable set and contradict stability.

2.6 Theorem. *If T is a countable stable theory and p is a complete type $\mu(p)$ is finite or $2^{\aleph_0}$.*

Proof. Suppose $p \in S(A)$ and $A \subseteq M$. Since types over models are stationary it suffices to count the nonforking extensions of p to $S(M)$. Let Δ_i for $i < \omega$ be an increasing union of finite sets which exhaust $F(L)$. Now we define a finite branching tree of extensions of p. Let $\langle q^i_j : j < k_0\rangle$ list the complete Δ_i types over M which are consistent with p. Write $p_{<j>}$ for $p \cup q^0_j$. If p_s with $\lg(s) = n$ has been defined let $p_{s\frown j} = p_s \cup q^n_j$ if this type is consistent and do not define $p_{s\frown j}$ otherwise. If $\mu(p)$ is infinite, this tree contains infinitely many t such that for some $i \neq j$ both $p_{t\frown i}$ and $p_{t\frown j}$ are defined. By König's Lemma we have a chain satisfying Lemma 2.5 i) so $\mu(p) = 2^{\aleph_0}$.

Since Theorem III.4.34 showed $\mu(p)$ is countable in an ω-stable theory we have

2.7 Corollary. *If T is a countable ω-stable theory then for every p, $\mu(p)$ is finite.*

There are two parts to the proof of Theorem 2.6. One is to describe a finite branching tree and prove it has an infinite path; the second is to convert the infinite path to a complete binary tree. Pillay [Pillay 1983a] (p. 61) uses the finite equivalence relation theorem to construct the finite branching tree and constructs the binary tree somewhat more concretely than we have.

Corollary 2.7 can be proved more directly with the machinery of Morley rank and degree; we will see in Chapter VII that $\mu(p)$ is the Morley degree of p.

2.8 Exercise. Give examples of types with each possible multiplicity.

Corollary 2.4 allows us to define for each $p \in S(A)$ and each formula $\phi(\overline{x};\overline{y})$ a formula $d_p\phi(\overline{y})$ with parameters from A such that for each nonforking extension $\hat{p}$ of p to a global type, $\phi(\overline{x};\overline{b}) \in \hat{p}$ if and only if $d_p\phi(\overline{b})$. We use this notation in the following exercise.

2.9 Exercise (Berline). Suppose $A_0 \subseteq A$ and each $d_p\phi$ is a formula over A_0. Show p is the unique nonforking extension of $p|A_0$ in $S(A)$.

The virtue of having every type over a set A definable over a finite subset of A may not be immediately clear. We will see many applications of this fact in Part D; even more appear in [Shelah 198?]. For the moment, we content ourselves with singling out sets with this property.

2.10 Definition. The set A is *good* if for every $p \in S(A)$ there is a subset A_0 of A with $|A_0| < \kappa(T)$ such that p does not fork over A_0 and $p|A_0$ is stationary inside A.

Note that if A is the universe of model, M, we can shorten the definition of good by saying that every $p \in S(M)$ is strongly based on some $A_0 \subseteq M$ with $|A_0| < \kappa(T)$.

Intuitively, if a type p is based on a set A then A 'controls' the properties of p. Thus, we would like to be able to base each type on a small subset of its domain. This is not always possible however.

2.11 Exercise. Find a model M of the theory REI_ω and a type $p \in S(M)$ which is not based on any finite subset of M.

2.12 Exercise. Find a model M of the theory REF_ω and a type $p \in S(M)$ which is based on some finite subset of M but not strongly based on any finite subset of M.

2.13 Exercise. Show the prototypical model of REF_ω is not good.

2.14 Lemma. *If $M \models T$ and for every $p \in S(M)$ there is an $A \subseteq M$ with $|A| < \kappa(T)$ such that p is based on A and $\mu(p|A) < \omega$ then M is good.*

Proof. The finitely many nonforking extensions of $p|A$ to complete types over M can be distinguished by finitely many finite equivalence relations over A. Since each class of each of these relations is realized in M it is easy to extend A to A' such that $p|A'$ is stationary inside M.

Good sets play an extremely important role in the structure theory. Much of the work needed in the extensions of the material in this book to completely solve the spectrum problem involves finding large classes of substructures of models of T which are good. We show now that every model of an ω-stable theory is good. We show the same for every strongly $\kappa(T)$-saturated model of a stable theory in Theorem 3.22.

2.15 Corollary. *If T is a countable ω-stable theory then every model of T is good.*

Proof. This is immediate from Theorem 2.7 and Lemma 2.14.

With this result in hand, we recast the stability hierarchy in terms of definability. We require one more technical remark.

2.16 Lemma. *If $p, q \in S(M)$ are definable over $A \subseteq M$ and $p \neq q$, then $p|A \neq q|A$.*

Proof. By Definition III.3.1 p and q do not fork over A; whence by the finite equivalence relation theorem (Theorem 1.24) if $p \neq q$ there is a finite equivalence relation E over A such that: $p(\overline{x}) \cup q(\overline{y}) \to \neg E(\overline{x}; \overline{y})$. Now suppose that p is defined by d. Write $dE(\overline{v}, \overline{w})(\overline{w})$ as $dE(\overline{w})$. Let $r(\overline{w})$ denote $p(\overline{w})|A = q(\overline{w})|A$. Then $\neg dE(\overline{w})$ is in $q|A$. But, $dE(\overline{w})$ is in $p|A$. For, if not, choosing a sequence $\overline{b}_i$ for $i \in \omega$ such that $\overline{b}_i$ realizes $d(p; \{A \cup B_j : j < i\})$, we have $\neg E(\overline{b}_i, \overline{b}_j)$ if $i \neq j$ contradicting the assumption that E has only a finite number of classes. Thus, $p|A \neq q|A$.

We distinguish one further class of theories which play an important role in the study of countable models of countable theories. Recall that $S^n(T)$ denotes the collection of all n-types over the empty set.

2.17 Definition. We say T is a *small theory* if for each n, $|S^n(T)| \leq |T|$.

This class of theories is important because any theory with fewer than $2^{\aleph_0}$ countable models is small. Thus, attempts to solve Vaught's conjecture for stable theories or to show there is no stable theory with finitely many (but more than one) countable models easily reduce to investigations of small stable theories.

2.18 Exercise. Show REF_ω is small but CEF_ω is not.

2.19 Theorem. The countable small theory T is ω-stable if and only if for every model M of T and every p in $S(M)$, there is a finite subset B of M such that p is definable over B.

Proof. Suppose first that T is ω-stable. Let M be a model of T and p be in $S(M)$. Then by Corollary 2.15 there is a finite subset B of M such that p is strongly based on B. By Theorem 1.20 ii) p is definable over B. Thus for any M and any p in $S(M)$, there is a finite subset B of M such that p is definable over B. Conversely, since there are only $|M|$ finite subsets of M and since T is small, there are only countably many types in each $S(B)$. Mapping each element of $S(M)$ to its restriction to a finite set over which it is definable shows (by Lemma 2.16) that there are only $|M|$ types in $S(M)$.

This theorem requires the assumption that T is small. Consider the theory CEF_ω of crosscutting equivalence relations. Now CEF_ω is superstable but not ω-stable yet every 1-type over any model is definable over a singleton.

The next exercise is an easy step towards the following improvement of Theorem 2.19. If for every pair of countable models $M \subseteq N$ of a small stable theory, T, there is an element $a \in N - M$ with $t(a; M)$ definable over a finite subset of M then T is ω-stable. The general result appears in [Pillay & Steinhorn 1985]

2.20 Exercise (Pillay-Steinhorn). Show that if T is a countable small superstable theory and every type over a finite set has finite multiplicity then T is ω-stable.

2.21 Theorem. *The countable theory T is superstable if and only if for every A contained in a model of T and every $p \in S(A)$, there is a finite B contained in A such that p is defined almost over B.*

Proof. Suppose first that T is superstable. Let M be a model of T and p be in $S(M)$. Then, since $\kappa(T) = \omega$, there is a finite B contained in M such that p does not fork over B. Thus p is definable almost over B. Now let M be a model of T and suppose that for each p there is a finite subset B of M such that p is definable almost over B. But then p does not fork over B so $\overline{\kappa}(T) = \omega$. Since T is countable $\overline{\kappa}(T) = \kappa(T)$ and we finish.

2.22 Historical Notes. The definability characterization of the stability hierarchy is implicit in III.4 of [Shelah 1978]. The proof of Theorem 2.1 is adapted from [Pillay 1983a]. The argument given here for Corollary 2.4 is taken from [Shelah 198?]. The other arguments at the beginning of this section are adapted from [Baldwin 1981].

3. Strong Types and Multiplicity

In Section 1 of this chapter we proved the finite equivalence relation theorem and immediately applied it to show the enormous significance of finite equivalence relations in investigating the nonforking extensions of a type. In this section we introduce the notion of the strong type of an element, which encodes all the information about it which is provided by finite equivalence relations. We investigate closely the relations among types over A, strong types over A, and types which are almost over A. In particular, we consider what it means for a type to be implied by a type which is almost over A. Then we analyze the relation between the strong type of a pair and the strong type of its projections. We conclude by discussing the relation between the multiplicity of a pair and that of its projections.

We begin with the fundamental definition for a rigorous discussion of the types which are almost over a set.

3.1 Definition. For any $\overline{c}$ and A, the *strong type* of $\overline{c}$ over A, denoted by $stp(\overline{c}; A)$, is $\{E(\overline{x}; \overline{c}) : E \in FE(A)\}$.

Note that $\mathrm{dom}(stp(\overline{c}; A))$ is not A but $A \cup \overline{c}$. However, if M is any model containing A then $t(c; M)$ implies $stp(c; A)$ and even equals $stp(c; M)$. The meaning of $stp(\overline{a}; A) = stp(\overline{b}; A)$ is slightly mysterious, since the two types have different domains. This equality does not assert the equality of two sets of formulas (which do not have the same parameters). Rather it asserts that for any $\overline{c} \in \mathcal{M}$, $\overline{c}$ realizes $stp(\overline{a}; A)$ iff $\overline{c}$ realizes $stp(\overline{b}; A)$. The situation is reminiscent of Cantor's definition of addition for cardinal numbers without specifying a representative for each equivalence class under equinumerousity.

Strangely, $stp(\overline{c}; A)$ is not preserved by A-automorphisms of $\mathcal{M}$ but equality of strong types is. For example, consider the theory of an equivalence relation with two infinite classes. Then any two elements of any model are conjugate over the empty set but inequivalent elements realize distinct strong types.

3.2 Exercise. Show that if $stp(\overline{a}; A) = stp(\overline{b}; A)$ and α is an automorphism which fixes A then $stp(\alpha\overline{a}; A) = stp(\alpha\overline{b}; A)$. The last example shows $stp(\alpha\overline{a}; A)$ need not equal $stp(\overline{a}; A)$.

3.3 Exercise. Show $stp(\overline{a}; A) = stp(\overline{b}; A)$ iff for each finite equivalence relation, E, over A (of appropriate arity) $\models E(\overline{a}, \overline{b})$.

3.4 Exercise. Suppose $stp(a; B) = stp(c; B)$. Show $t(a; \mathrm{cl}(B)) = t(c; \mathrm{cl}(B))$ and even that $stp(a; \mathrm{cl}(B)) = stp(c; \mathrm{cl}(B))$.

3.5 Exercise. Show that $t(\overline{a}; B)$ does not imply $t(\overline{a}; \mathrm{cl}(B))$.

3.6 Exercise. Show that if $M \models T$ and $\phi(\overline{x}; \overline{a})$ is almost over M then $\phi(\overline{x}; \overline{a})$ is equivalent to a formula which is over M.

3.7 Exercise. Show that for any $\overline{c}$ and A, $stp(\overline{c}, A)$ is stationary.

3.8 Exercise. With the aid of strong types, give a direct (rather than axiomatic) proof that the strong extension property (Exercise II.2.8) holds for nonforking.

3.9 Definition. Let $A \subseteq \mathcal{M}$. Then $\mathrm{Saut}_A(\mathcal{M})$ denotes the set of automorphisms of $\mathcal{M}$ which fix the strong types over A. We refer to an automorphism which fixes all strong types over A as a *strong A-automorphism.* Naturally, strong automorphisms which fix A preserve formulas which are almost over A. We write $\mathrm{Saut}(\mathcal{M})$ if A is empty.

3.10 Exercise. Show Saut($\mathcal{M}$) is a normal subgroup of Aut($\mathcal{M}$).

Let $G = Z_3 \oplus Z_2^{\aleph_0}$ and let b be one of the two elements of order three and $2b$ the other. Now in the language of Abelian groups the formula which asserts the order of x is six generates a complete type over the empty set. But this type has two nonforking extensions to types over G, both asserting $x \neq g$ for each $g \in G$ but one saying $2x = b$ and the other $2x = 2b$. Each of the latter two types is almost over the empty set. The formula $(2x = 2y)$ defines a finite equivalence relation, $E(x,y)$. If c is an element with $2c = b$ but c is not in G, $stp(c;\emptyset)$ is implied by the formula $E(x,c)$.

3.11 Exercise. Investigate the rather different situation when the roles of Z_2 and Z_3 are reversed in the definition of the group G. Note however that if Z_3 is replaced by Z_p for any prime other than 2 the situation remains exactly the same as in the example.

We saw after Definition 3.1 that the notion of equality of strong types is somewhat slippery. The notion of implication can be understood somewhat better by passing from strong types over A to types which are almost over A.

3.12 Lemma. *Let $B \subseteq A$ and $p = t(c;A)$. The following are equivalent.*

i) *For some $q \subseteq p$, $q \vdash p$ and q is almost over B.*
ii) *For some q which is almost over B, $q \vdash p$.*
iii) *$stp(c;B) \vdash p$.*
iv) *$stp(c;B) \vdash stp(c;A)$.*

Proof. Clearly i) implies ii). To see that ii) implies iii), let $\overline{c}'$ realize $stp(c;B)$. Then c' realizes q and so by ii) c' realizes p. Suppose iii) holds; choose c' with $stp(\overline{c}';B) = stp(c;B)$. Then, c' realizes p so $t(\overline{c}';A)$ does not fork over B as $stp(c;B)$ does not fork over B. Since $stp(c;B)$ is stationary this implies c' realizes $stp(c;A)$ so iii) implies iv).

Obviously, iv) implies iii) so we complete the proof by showing iii) implies i). For this, let $q = \{\phi(x,b) \in p : \phi(x,b) \text{ almost over } B\}$. We will show $q \vdash p$. Since $FE^m(B)$ is closed under conjunction, for each $\psi(x,b) \in p$ there is an $E(x,y) \in FE^m(B)$ such that $E(x,c) \rightarrow \psi(x,b)$. Let $\chi(x)$ be $(\forall y)[E(x,y) \rightarrow \psi(y,b)]$. Now, the solutions of χ are a union of equivalence classes of E so $\chi \in q$. But $E(x,x)$ holds so $\chi(x) \rightarrow \psi(x,b)$. Thus, $q \vdash p$ as required.

3.13 Exercise. Show that the (standing) hypothesis that T is stable is required only for iii) implies iv) of Lemma 3.12.

Note that just because $c \downarrow_A B$ it does not follow that $stp(c;B) \vdash stp(c;A)$. However the following weakening of this idea holds.

3.14 Exercise. Suppose T is superstable. Let $\overline{c}$ realize $t(\overline{a};B)$ with $\overline{c} \downarrow_B A$. Show that for some finite B' with $B \subseteq B' \subseteq A$, $stp(\overline{c};B') \vdash t(\overline{c};A)$.

3.15 Exercise. Suppose $stp(\overline{d};B) \vdash t(\overline{d};A)$. Show $stp(\overline{d};B) \vdash t(\overline{d};\text{cl}(A))$.

The next result provides a dual to the last exercise; it considers taking the algebraic closure of the elements realizing a type. In addition, it will will play an important role in the proof of the transitivity axiom for the notion of S-isolation in chapter VIII.

3.16 Lemma. *If $stp(C; B) \vdash t(C; A)$ then*

$$stp(\mathrm{cl}(B \cup C); B) \vdash t(\mathrm{cl}(B \cup C); A).$$

Proof. If $\overline{e} \in \mathrm{cl}(B \cup C)$ then there is a $\overline{c} \in C$ such that $t(\overline{e}; B \cup \overline{c})$ is algebraic. We will prove $stp(\overline{e}\frown\overline{c}; B) \vdash t(\overline{e}\frown\overline{c}; A)$; from this the theorem is clear. It suffices to show all realizations of $t(\overline{c}\frown\overline{e}; B)$ are independent from A over B. Using only properties of their type over B we will show $\overline{c}\frown\overline{e} \downarrow_B A$ which will then establish the result. Let $\overline{f} = \langle \overline{e}_i : i < m \rangle$ be a complete list of the realizations of $t(\overline{e}; B \cup \overline{c})$. By the strong extension property (Exercise II.2.8) we can choose an $\overline{f}'$ with $t(\overline{c}\frown\overline{f}'; B) = t(\overline{c}\frown\overline{f}; B)$ but $\overline{c}\frown\overline{f}' \downarrow_B A$. Now $\overline{f}'$ is just a permutation of the sequence $\overline{f}$. Thus, since $\overline{c}'\frown\overline{f}' \downarrow_B A$, $\overline{c}\frown\overline{f} \downarrow_B A$ so $\overline{c}\frown\overline{e} \downarrow_B A$ as required.

The precise result we have proved is: If $stp(\overline{c}; B) \vdash t(\overline{c}; A)$ and $\overline{e} \in \mathrm{cl}(B \cup \overline{c})$ then $stp(\overline{c}\frown\overline{e}; B) \vdash t(\overline{c}\frown\overline{e}; A)$.

3.17 Example. A type p may be implied by a type q which is almost over B without p itself being almost over B. Let T be the theory of an equivalence relation, R, with two classes, one infinite and one with a single element a. Let $B = \emptyset$ and let A be an infinite set of equivalent elements. Now if $p = t(a; A)$, p is implied by the formula $(\exists! y)R(x, y)$ which is almost over (indeed over $\emptyset$) but p is not almost over $\emptyset$. For, p contains the formulas $x \neq c$ for each $c \in A$.

The description of implications from strong types in terms of automorphisms is sometimes helpful. It is easy to verify the following lemma.

3.18 Lemma. *Let $B \subseteq A$. The following are equivalent.*

i) $stp(\overline{d}; B) \vdash t(\overline{d}; A)$.

ii) *If there exists an $\alpha \in \mathrm{Saut}_B(\mathcal{M})$ with $\alpha(\overline{d}) = \overline{d}'$ then there exists an automorphism $\hat{\alpha} \in \mathrm{Saut}_A(\mathcal{M})$ with $\hat{\alpha}(\overline{d}) = \overline{d}'$.*

This trivially yields that if $stp(\overline{d}; B) \vdash stp(\overline{d}; A)$ and $\alpha \in \mathrm{Aut}_B(\mathcal{M})$ then $stp(\alpha(\overline{d}); B) \rightarrow stp(\alpha(\overline{d}); \alpha(A))$.

The next topic is the relation between the multiplicity (or more specifically) the strong type of a pair and that of each of its projections. It is clear that $t(\overline{a}\frown\overline{b}; A) = t(\overline{a}'\frown\overline{b}'; A)$ if and only if $t(\overline{a}; A) = t(\overline{a}'; A)$ and for some automorphism α fixing A with $\alpha(\overline{a}') = \overline{a}$, $t(\overline{b}; \overline{a} \cup A) = t(\alpha(\overline{b}'); \overline{a} \cup A)$. (We may write $t(\overline{b}; \overline{a} \cup A) = t(\overline{b}'; \overline{a}' \cup A)$.) The analogous result obtained by replacing t by stp fails. The next three lemmas establish exactly which parts of it are correct and which fail.

3.19 Lemma. i) *$stp(\overline{a}\frown\overline{b}; A)$ implies $stp(\overline{a}; A)$.*

ii) *$stp(\overline{a}\frown\overline{b}; A)$ need not imply $stp(\overline{b}; A \cup \overline{a})$.*

Proof. i) Suppose $stp(\overline{a}\frown\overline{b};A) = stp(\overline{a}'\frown\overline{b}';A)$. Then $stp(\overline{a};A) = stp(\overline{a}';A)$. For, suppose for contradiction that $\phi(\overline{x};\overline{y})$ defines a finite equivalence relation over A which separates $\overline{a}$ and $\overline{a}'$. Regard $\phi(\overline{x}_1;\overline{y}_1)$ as a formula $\phi'(\overline{x}_1\frown\overline{x}_2;\overline{y}_1\frown\overline{y}_2)$. Then $\phi'(\overline{x}_1\frown\overline{x}_2;\overline{y}_1\frown\overline{y}_2)$ defines a finite equivalence relation on pairs which separates $\overline{a}\frown\overline{b}$ from $\overline{a}'\frown\overline{b}'$.

For ii), consider the theory of the following structure. Let L contain a single binary relation, R, and let the universe of M be the set of two elements subsets of ω. Interpret R to hold of $\{a,b\},\{c,d\}$ if and only if $\{a,b\}\cap\{c,d\}\neq\emptyset$.

Let $a = \{1,2\}$, $a' = \{3,4\}$ and $b = \{1,3\}$. We will show $stp(a\frown b;\emptyset) = stp(a'\frown b;\emptyset)$ but $stp(a;b)\neq stp(a';b)$.

For the first assertion we show that any pair $\langle c,d\rangle$ with $c\neq d$ and such that $c\cap d\neq\emptyset$ satisfies the same strong type over the empty set as $\langle a,b\rangle$. To see this choose such a pair $\langle e,f\rangle$ but with $e\cup f$ disjoint from both $a\cup b$ and $c\cup d$. Now consider any definable equivalence relation on pairs (from M), F. If F separates $\langle a,b\rangle$ and $\langle c,d\rangle$ then we must have $\neg F(a,b,e,f)$ or $\neg F(c,d,e,f)$. Since F is definable from R, its truth on a quadruple depends only on the intersections between elements of the quadruple. But then we have $\neg F(e,f,e',f')$ for any other pair with $e'\cap f'\neq\emptyset$ but $(e'\cup f')\cap(e\cup f) = \emptyset$. But then F has infinitely many classes; so $stp(a\frown b;\emptyset) = stp(c\frown d;\emptyset)$.

Finally, we see $stp(a;b)\neq stp(a';b)$ since they differ on the following finite equivalence relation: $E(x,y)$ if $(\forall z)[R(b,z)\rightarrow(R(z,x)\leftrightarrow R(z,y))]$.

There exist a, b, a', b', such that $a\frown b$ and $a'\frown b'$ have the same strong type over $\emptyset$ but $stp(a;b)\neq stp(a';b')$ (i.e. take $b = b'$ in Lemma 3.19 ii) and there is no ambiguity in the notation). Note this cannot happen (with $b = b'$) for types.

Suppose we follow the procedure from the paragraph before Lemma 3.19 and write $stp(a;b) = stp(a';b')$ if there is an $\alpha\in \mathrm{Saut}_A(\mathcal{M})$ which takes b' to b and satisfies $stp(a;b) = stp(\alpha(a');b)$. Then a simple composition of automorphisms argument shows that $stp(a;b) = stp(\alpha(a');b)$ implies $stp(a\frown b;\emptyset) = stp(a'\frown b';\emptyset)$. But, suppose we follow a more syntactic course and say $stp(a;b) = stp(a';b')$ if $stp(a;b)$ is the set of formulas $\{E(x,\alpha(a'),b) : E(x,a',b')\in stp(b',a')\}$. The next two lemmas show we can still conclude $stp(a\frown b;\emptyset) = stp(a'\frown b';\emptyset)$.

3.20 Lemma. *If $\phi(\overline{x}\frown\overline{y};\overline{c})$ is almost over A, then for any $\overline{b}$, $\phi(\overline{x};\overline{b}\frown\overline{c})$ is almost over $A\cup\overline{b}$.*

Proof. Suppose $\{\phi(\overline{x};\overline{b},\alpha_i(\overline{c})) : i\in I\}$ is an infinite family of distinct conjugates over $A\cup\overline{b}$ of $\phi(\overline{x};\overline{b}\frown\overline{c})$. Then for some fixed j and an infinite subset of the i, $\phi(\overline{x},\overline{y},\alpha_i(\overline{c}))\leftrightarrow\phi(\overline{x},\overline{y};\alpha_j(\overline{c}))$. But this contradicts the choice of distinct $\{\phi(\overline{x};\overline{b},\alpha_i(\overline{c})) : i\in I\}$.

The following lemma gives a more algebraic formulation of the previous one and thus simplifies its application.

3.21 Lemma. *Let* $stp(\overline{a}; A) = stp(\overline{b}; A)$ *and let* $\alpha \in \mathrm{Saut}_A(\mathcal{M})$ *map* $\overline{a}$ *to* $\overline{b}$. *For any strong type* q *over* $A \cup \overline{a}$ *let* $\alpha(q)$ *be* $\{E(x, \alpha(c); \overline{b}) : E(x, c; \overline{a}) \in q\}$. *If* d *realizes* $\alpha(q)$ *then* $stp(\overline{a}^\frown c; A) = stp(\overline{b}^\frown d; A)$.

Proof. We have $stp(\overline{a}^\frown c; A) = stp(\overline{b}^\frown \alpha(c); A)$ so it suffices to show

$$stp(\overline{b}^\frown d; A) = stp(\overline{b}^\frown \alpha(c); A).$$

If not, for some $E \in FE(A)$, $\models \neg E(\overline{b}^\frown d, \overline{b}^\frown \alpha(c))$. Thus, by Lemma 3.20, $E(\overline{b}, x, \overline{b}, y) \in FE(A \cup \overline{b})$ and so $E(\overline{a}, x, \overline{a}, y) \in FE(A \cup \overline{a})$. But, $E(\overline{a}, \overline{x}, \overline{a}, c)$ is in $q = stp(c; A \cup \overline{a})$ so d does not realize $\alpha(q)$ contrary to hypothesis.

The following result illustrates another way to apply strong types.

3.22 Theorem. *Any strongly* $\kappa(T)$*-saturated model* M *of a stable theory is good.*

Proof. Let $\overline{a}$ realize $p \in S(M)$ and choose $A \subseteq M$ with $|A| < \kappa(T)$ and such that p does not fork over A. Since $stp(\overline{a}; A)$ does not fork over A, it is finitely satisfied in M and hence realized in M by some $\overline{b}$. Then p is strongly based on $A \cup \overline{b}$.

The preceding material concerned the relation between the strong types of pairs and their projections. We now generalize that concern slightly and consider the relation between the multiplicity of a pair and that of its projections.

Let T be the theory REF_ω of infinitely many refining equivalence relations, each with finitely many classes. Now, if a, b realize the same strong type over the empty set and $p = t(a; \emptyset)$, the multiplicity of p is $2^{\aleph_0}$ but $\mu(t(a; b))$ is one.

The following lemma of Saffe shows that this phenomenon (of the multiplicity decreasing from infinite to finite) could only occur because $t(b; \emptyset)$ also had infinite multiplicity.

3.23 Lemma. *Let* T *be stable. If* $t(\overline{b}; A)$ *has infinite multiplicity and* $t(\overline{c}; A)$ *has finite multiplicity then* $t(\overline{b}; A \cup \overline{c})$ *has infinite multiplicity.*

Proof. Choose $\langle \overline{b}_i : i < \omega \rangle$ all realizing $t(\overline{b}; A)$ but realizing distinct strong types over A. For each $i < \omega$, choose $\overline{c}_i$ with $t(\overline{b}_i{}^\frown \overline{c}_i; A) = t(\overline{b}^\frown \overline{c}; A)$. Since $t(\overline{c}; A)$ has finite multiplicity, we may assume that all the $\overline{c}_i$ realize the same strong type over A. Thus for each $i < \omega$ there is an automorphism of $\mathcal{M}$ which fixes strong types over A and maps $\overline{c}_i$ to $\overline{c}$. Denote $f_i(\overline{b}_i)$ by $\overline{b}'_i$. Then, the $\overline{b}'_i$ realize distinct strong types over A and, a fortiori, over $A \cup \overline{c}$. But for each i, $t(\overline{b}_i; A \cup \overline{c}) = t(\overline{b}; A \cup \overline{c})$. Thus $t(\overline{b}; A \cup \overline{c})$ has infinite multiplicity as required.

3.24 Historical Notes. This section primarily clarifies some notions that are not spelled out in the literature. Lemma 3.12 is a variant on Lemma IV.2.1 of [Shelah 1978]. The example for Lemma 3.18 is primarily due to Steve Buechler. Matt Kaufmann supplied the proof that the example actually worked. Lemma 3.23 is due to Jürgen Saffe [Saffe 1981]. The proof here is taken from [Pillay & Steinhorn 1985].

Chapter V

Indiscernibles In Stable Theories

Sets of indiscernibles play a number of important roles in model theory. They are used to realize types (as in the proof of the existence of saturated models); they are also used to 'blow up' models (without realizing new types). Moreover the cardinalities of maximal sets of indiscernibles can be used as invariants in classifying models. We begin this chapter by expounding the basic properties of indiscernibles in a stable theory and explaining the distinction between sequences of indiscernibles and independent sets. Section 2 begins the rather lengthy process of using sets of indiscernibles as bases for models of stable theories (cf. the introduction to Section 2). In Section 3 we apply the notion of indiscernibility to show the equivalence between the notion of forking as introduced here and the original version of Shelah [Shelah 1978].

1. Sets Of Indiscernibles

If X is a set of algebraically independent elements in an algebraically closed field then every permutation of X is in fact an elementary map. This indiscernibility of the elements of X is closely related to their independence. We explain this connection in Lemma 1.8 and Theorem 1.23. In this section we study in detail indiscernible elements in a model of a stable theory. We define such notions as indiscernible sequences (of sequences) and indiscernible sets (of sequences). That is, we deal with families $E = \{\overline{e}_i : i \in I\}$ where each $\overline{e}_i$ is a finite sequence. Very little intuition is lost by thinking of each $\overline{e}_i$ as a single individual but the added generality is necessary.

1.1 Definition. The ordered set $X = \{\overline{x}_i : i \in l\}$ is a *sequence of* (*order*) *indiscernibles* if every order preserving map f mapping into a finite subset of the linear order l induces a partial elementary monomorphism of X by taking $\overline{x}_i$ to $\overline{x}_{f(i)}$.

The index set will be well-ordered unless we explicitly assert otherwise. We distinguish now between a sequence and a set of indiscernibles. While this is an important distinction in general model theory, we will see that in our situations the concepts coalesce.

1.2 Definition. The set $X = \{\overline{x}_i : i \in I\}$ is a *set of (pure) indiscernible sequences* if every partial permutation of X is an elementary monomorphism.

In Definition 1.1 (1.2) we refer to X as a *sequence* (*set*) of indiscernibles. Naturally, both of these definitions can be made relative to a set A.

The next theorem emphasizes two important properties of indiscernible sequences in a stable theory.

1.3 Theorem. *Let M be a model of the stable theory T.*

i) *Every infinite sequence of indiscernibles in M is a set of indiscernibles.*
ii) *For any formula $\phi(\overline{x};\overline{y})$, there is an integer $n(\phi)$ such that for any $\overline{b}$, if $E = \{\overline{e}_i : i \in \omega\}$ is an infinite set of indiscernibles over the empty set either $|\{i : \phi(\overline{e}_i;\overline{b})\}| < n(\phi)$ or $|\{i : \neg\phi(\overline{e}_i;\overline{b})\}| < n(\phi)$.*

Proof. In showing i), let S_n denote the symmetric group on n elements and denote the application of a permutation $s \in S_n$ to an element i by si. Fix $i_1 < \ldots < i_n \in \omega$, and consider an arbitrary formula $\phi(\overline{x}_1, \ldots, \overline{x}_n)$. Let T_n be the set of $s \in S_n$ such that $\models \phi(\overline{e}_{i_{s1}}, \ldots, \overline{e}_{i_{sn}})$. If T_n is $\emptyset$ or S_n we finish. If not, fix $t \in T_n$ and $s \in S_n - T_n$. Then s and t differ by a product of transpositions so there exist $u \in T_n$ and $v \in S_n - T_n$ such that for some $k < n, v = (k, k+1)u$. (This requires a little argument with permutations.) Let $\psi(\overline{x}_1, \ldots, \overline{x}_n) = \phi(\overline{x}_{v1}, \ldots, \overline{x}_{vn})$. Now we have

a) $\models \psi(\overline{e}_{i_1}, \ldots, \overline{e}_{i_n})$ and
b) $\models \neg\psi(\overline{e}_{i_1}, \ldots, \overline{e}_{i_{k-1}}, \overline{e}_{i_{k+1}}, \overline{e}_{i_k}, \ldots, \overline{e}_{i_n})$.

Note that a) and b) hold when $i_1, \ldots, i_n$ are replaced by any other properly ordered n-tuple from ω (by order indiscernibility). Now by compactness, we can replace ω by an index set R with order type of the real numbers such that a) and b) hold whenever $i_1, \ldots, i_n$ are replaced by a properly ordered n-tuple $r_1, \ldots, r_n$ from R. If r and r' are arbitrary distinct members of R, there exist rational numbers $\{q_i : 1 \le i \le n \wedge i \ne k\}$ such that $q_1 < \ldots < q_{k-1} < r < q_{k+1} < r' < q_{k+2} < \ldots < q_n$. Hence

c) $\models \psi(\overline{e}_{q_1}, \ldots, \overline{e}_{q_{k-1}}, \overline{e}_r, \overline{e}_{q_{k+1}}, \ldots, \overline{e}_{q_n})$ and
d) $\models \neg\psi(\overline{e}_{q_1}, \ldots, \overline{e}_{q_{k-1}}, \overline{e}_{r'}, \overline{e}_{q_{k+1}}, \ldots, \overline{e}_{q_n})$.

Thus distinct $\overline{e}_r$ for $r \in R$ realize distinct ψ-types over the countable set of $\overline{e}_q$ indexed by rationals. By Theorem III.1.6, T is not stable.

For ii), suppose first there is an infinite and coinfinite subset F of E such that $\phi(\overline{e}, \overline{b})$ holds iff $\overline{e}$ is in F. For each infinite and coinfinite subset W of E, there is an elementary monomorphism f_w with $f(F) = W$. Since the monster model is so homogeneous, we can extend f to an automorphism of the monster model. Now, $\phi(\overline{a}_i, f(\overline{b}))$ holds iff $\overline{a}_i$ is in W. Thus there are continuum many ϕ-types over E, contrary to the stability of T.

This argument depended on the choice of F and b. However, if arbitrarily large finite F could be found for various choices of $\bar{b}$, by compactness we could obtain a single F and $\bar{b}$ to which the above proof applies.

Note that Theorem 1.3 ii), but not necessarily i), holds for any theory which does not have the independence property (cf. Section III.4.43).

1.4 Exercise. Find an example showing the necessity of the assumption that the sequence is infinite in Theorem 1.3 i).

By Theorem 1.3 i) in stable theories all infinite sequences of indiscernibles are sets of pure indiscernibles. In fact the converse is true, but we won't prove it here (see Chapter 2 in [Shelah 1978]). Using the second part of this theorem we can attach to each set of indiscernible n-tuples a complete n-type as follows.

1.5 Definition. Let $X = \{\bar{a}_i : i \in I\}$ be an infinite set of indiscernibles over the empty set. The *average type* of X over A, denoted $\mathrm{Av}(X;A)$, is $\{\phi(\bar{x};\bar{a}) : \phi(\bar{x};\bar{a})$ in $F(A)$ such that for all but finitely many i, $\models \phi(\bar{a}_i;\bar{a})\}$.

Thus, if X is an indiscernible set of n-tuples, $\mathrm{Av}(X,A) \in S^n(A)$. Note that if X is indiscernible over any set B it is certainly indiscernible over the empty set; X is usually not indiscernible over A when we apply this definition.

1.6 Exercise. Show that if $\bar{e}$ realizes $\mathrm{Av}(E;A)$ where E is an infinite set of indiscernibles over $B \subseteq A$ and $B \cup E \subseteq A$ then $E \cup \bar{e}$ is a set of indiscernibles over B.

The identification of indiscernibles with independent elements in vector spaces, algebraically closed fields, and free algebras is not a fluke. These examples are somewhat misleading, however. Here are some more revealing ones.

1.7 Examples. First we show a sequence may be indiscernible without being independent.

i) Let T be the theory of an equivalence relation with infinitely many infinite classes. If X is an infinite set of points in the same equivalence class then X is a set of indiscernibles but X is not an independent set.

Here is an independent set which is not a set of indiscernibles.

ii) In the theory of infinitely many refining equivalence relations with finite splitting (REF_ω) let $\langle a_i : i < \omega\rangle$ be a sequence such that $E_i(a_k, a_j)$ iff $k, j \geq i$. Then for each i, $t(a_i; A_i)$ does not fork over $\emptyset$ but the sequence $\langle a_i : i < \omega\rangle$ is not indiscernible over the empty set. Note also that $t(a_i; A_i) \subseteq t(a_j; A_i)$ if $i < j$.

These examples show that independence does not quite guarantee indiscernibility; however, only one ingredient is missing. Recall Definition II.2.19 declares that an independent sequence is strongly independent over A if all elements of the sequence realize the same stationary type over A.

1.8 Lemma. *If E is an infinite strongly independent set over A then E is a set of indiscernibles over A.*

Proof. Fix a well ordering $\langle \overline{a}_\alpha : \alpha < \kappa \rangle$ of E. We prove by induction on $\beta < \kappa$ that E_β is an indiscernible sequence over A; by Theorem 1.3 this suffices. The result is clear for limit ordinals, so suppose $\beta = \alpha + 1$ and E_α is a sequence of indiscernibles over A. Let $\overline{a}_{i_0}, \ldots, \overline{a}_{i_n}$ and $\overline{a}_{j_0}, \ldots, \overline{a}_{j_n}$ be in E_β with $i_0 < i_1 < \ldots < i_n$, $j_0 < j_1 < \ldots < j_n$, and without loss of generality $i_n < j_n$. We need only show

$$t(\overline{a}_\alpha; \{\overline{a}_{i_0}, \ldots, \overline{a}_{i_n}\} \cup A) = t(\overline{a}_\alpha; \{\overline{a}_{j_0}, \ldots, \overline{a}_{j_n}\} \cup A).$$

By induction, there exists a map f which fixes A and sends $\overline{a}_{i_l}$ to $\overline{a}_{j_l}$ for $l \leq n$. Both $t(\overline{a}_\alpha; \{\overline{a}_{j_0}, \ldots, \overline{a}_{j_n}\} \cup A)$ and $f(t(\overline{a}_\alpha; \{\overline{a}_{i_0}, \ldots, \overline{a}_{i_n}\} \cup A))$ are nonforking extensions of the same stationary type so they are equal. Applying f^{-1} we have the result.

Theorem 1.23 below is almost a converse to this result. We can weaken the hypothesis that E is strongly independent over A slightly as follows. The second exercise shows that this provides a real weakening of the hypothesis.

1.9 Exercise. Show that if the infinite set E is independent over A and all $\overline{e} \in E$ realize the same strong type over A then E is a set of indiscernibles over A. (Hint: First apply Lemma 1.8 to show $E - \{\overline{e}\}$ is a set of indiscernibles over $A \cup \{\overline{e}\}$ for each $\overline{e} \in E$. Then deduce that E is a set of indiscernibles over A.)

1.10 Exercise. Let M be a model of CEF_ω and suppose $\{c_i : i < \omega\}$ is an infinite set of elements realizing the same strong type over $\emptyset$. Show using the previous exercise (or by inspection) that the c_i are indiscernible. Show also that no type over the empty set is stationary.

The argument for Lemma 1.8 does not depend on the particular properties of stable theories but relies only on the definitions of stationary type and strongly independent sequence. We will use stability theory to construct a strongly independent sequence. We first describe a somewhat more syntactic way of finding an indiscernible sequence and indicate how to find indiscernibles in models of arbitrary theories. We devote the remainder of the section to elucidating those properties of sets of indiscernibles which depend on stability.

The following exercises are not necessary for our development but they put our uses of the hypothesis of stability into context. For this set of exercises we drop the standing assumption that T is stable.

1.11 Exercise. Show that if $\langle p_i : i < \alpha \rangle$ is an increasing sequence of types, each $p_i \in S(A \cup E_i)$, with $A \subseteq \operatorname{dom} p_0$, no p_i splits (Definition III.2.9) over A, and e_i realizes p_i then $\langle e_i : i < \alpha \rangle$ is a sequence of indiscernibles.

Using this observation and the fact that coheirs over models always exist solve the following exercises for an arbitrary theory T.

1.12 Exercise. If M is a model of T, find in the monster model an infinite set of indiscernibles over M.

1.13 Exercise. If $M \subseteq N$ and N is $|M|^+$-saturated find in N an infinite set of indiscernibles over M.

1.14 Exercise. If $A \subseteq N$ and N is $(|A| + |T|)^+$-saturated find in N an infinite set of indiscernibles over A.

The remaining results in this section depend, at least indirectly, on the hypothesis that types over models are stationary. The essential point is that indiscernible elements in stable theories are even more similar than is evident from the definition of indiscernibility. We first show that if E is a set of indiscernibles over A (i.e. the elements of E cannot be distinguished by formulas which are over A) then in fact the elements of E cannot be distinguished by formulas which are almost over A.

Note that in the following lemma E is a sequence of sequences. Thus, each variable $\overline{x}$ and each constant $\overline{e}$ refers to a sequence of sequences, e.g. $\overline{x} = \langle \overline{y}_0, \ldots, \overline{y}_m \rangle$, where $\lg(\overline{y}_i)$ is the common length of the sequences in E.

1.15 Lemma. *If E is an infinite set of sequences which is indiscernible over A, and $\phi(\overline{x}_1, \ldots, \overline{x}_n; \overline{c})$ is almost over A then every sequence $\langle \overline{e}_1, \ldots \overline{e}_n \rangle$ from E yields the same truth value for $\phi(\overline{e}_1, \ldots, \overline{e}_n; \overline{c})$.*

Proof. If the lemma fails, we can choose disjoint sequences (of sequences) $\overline{h}$ and $\overline{g}$ from E such that $stp(\overline{h}; A) \neq stp(\overline{g}; A)$. (If $\overline{h}$ and $\overline{g}$ are not disjoint replace one of them by a sequence disjoint from both of the original two.) There is a member R of $FE^m(A)$ such that $\neg R(\overline{g}, \overline{h})$. Thus any pair of disjoint sequences from E each having the same length as $\overline{g}$ are inequivalent under R so R has infinitely many equivalence classes, contrary to the choice of R in $FE^m(A)$.

1.16 Exercise. Show that if E is an infinite set of indiscernibles over A then E is an infinite set of indiscernibles over $\mathrm{cl}(A)$. Show that it is essential to assume that E is infinite.

1.17 Exercise. Suppose $\theta(\overline{x})$ is almost over $B \subseteq A$ and $E \subseteq A$ is an infinite set of indiscernibles over B such that all but finitely many $\overline{e} \in E$ satisfy θ. Show that if $\overline{e}'$ realizes $\mathrm{Av}(E, A)$ then $\models \theta(\overline{e}')$.

Now we apply Lemma 1.15 to show that if E is a set of indiscernibles over A then E remains indiscernible over supersets of A which are independent from E over A. The next theorem does all the work. This is one of the most important facts about the nonforking relation.

1.18 Theorem. *If E is an infinite set of indiscernibles over A and if $t(\overline{b}; E \cup A)$ does not fork over A then E is a set of indiscernibles over $A \cup \overline{b}$.*

Proof. It suffices (since E is indiscernible over A) to show that the type q of an arbitrary n-tuple of elements from E has a unique extension $q' \in S(A \cup \overline{b})$ which is realized in E. But if $r \in S(A \cup \overline{b})$ is the type of an n-tuple from

E, by the symmetry lemma r does not fork over A. Thus, if E is not a set of indiscernibles over $A \cup \bar{b}$, we have two extensions, r and r' of q which are realized in E and do not fork over A. By a form (Theorem IV.1.23) of the Finite Equivalence Relation Theorem, this implies that r and r' imply contradictory formulas which are almost over A. This contradicts Lemma 1.15.

The proof of Lemma 1.15 does not rely on stability. However, to make effective use of Lemma 1.15 and Theorem 1.18 we need to have $\kappa(T) < \infty$. Some simple cardinal computations and the definition of $\kappa(T)$ yield the following consequences.

1.19 Corollary. *Let E be indiscernible over A.*

i) *For any $\bar{b}$, there is an $E_0 \subseteq E$ with $|E_0| < \overline{\kappa}(T)$ such that $E - E_0$ is indiscernible over $A \cup E_0 \cup \{\bar{b}\}$.*
ii) *For any B there is a subset $E_0 \subseteq E$ such that $E - E_0$ is indiscernible over $E_0 \cup A \cup B$ with*

   a) $|E_0| \le \kappa(T) + |B|$ *and*
   b) *if* $|B| < \kappa(T)$ *then* $|E_0| < \kappa(T)$.

Proof. For i) use the definition of $\overline{\kappa}(T)$; for ii) we iterate the construction. For each finite sequence $\bar{b}$ from B choose $E(\bar{b})$ contained in E such that $|E(b)| < \kappa(T)$ and $E - E(\bar{b})$ is indiscernible over $A \cup \{\bar{b}\}$. Then, let E_0 be $\bigcup\{E(b) : b$ a finite sequence from $B\}$. It is easy to check E_0 has the required cardinality.

If we are willing to replace the $\overline{\kappa}(T)$ in i) by $|T|^+$, an argument both more naive and more general will suffice.

1.20 Exercise. Use Theorem 1.3 to show that if T does not have the independence property (cf. Paragraph III.4.42) and E is a set of indiscernibles over A then for any $\bar{b}$ there exists $E_0 \subseteq E$ with $|E_0| \le |T|$ such that $E - E_0$ is indiscernible over $A \cup E_0 \cup \bar{b}$.

Our next aim is to prove a suitable converse to Lemma 1.8. Recall the definition from II.2.14 of a coherent sequence.

1.21 Lemma. *Let $E = \{\bar{e}_i : i < \beta\}$ be a coherent sequence over A.*

i) *If $\beta \ge \omega$ then all $\bar{e}_m$ for $m \ge \omega$ realize the same strong type over A.*
ii) *If $|E| \ge \overline{\kappa}(T)$ then E contains a subset E_0 with $|E_0| < \kappa(T)$ such that $E - E_0$ is independent over $A \cup E_0$.*

Proof. i) For each finite equivalence relation R over A, there is an n (one more than the number of classes of R) such that for all $\alpha \ge n$, $\bar{e}_\alpha$ is R-equivalent to some (the same) member of E_n. Thus all $\bar{e}_\alpha$ for $\alpha \ge \omega$ satisfy the same strong type over A.

ii) There exists $E_0 \subseteq E$ with $|E_0| < \overline{\kappa}(T)$ such that $\mathrm{Av}(E; E \cup A)$ does not fork over $E_0 \cup A$. Since E is coherent, for any $\bar{e} \in E - E_0$, $t(\bar{e}_\alpha; E_\alpha \cup A)$ is a restriction of $\mathrm{Av}(E; E \cup A)$ so we finish.

The property described in Lemma 1.21 i) depends on the order type rather than the cardinality of the sequence E. Note that Lemma 1.21 i) does not depend on the cardinality of L. It holds for a sequence E, even if $|L|$ is much greater than $|E|$.

1.22 Exercise. Show that even for countable T, there can be coherent sequences over a set A of arbitrary finite length $\langle e_i : i < n\rangle$ such that not all the e_i realize the same strong type over A.

Since any sequence of indiscernibles is a coherent sequence we could apply Lemma 1.21 directly to any indiscernible set E to obtain a proper subset E_0 of E over which $E - E_0$ is strongly independent and with $|E_0| < \kappa(T) + \aleph_1$. However, we can use Lemma 1.15 to improve the bound on $|E_0|$. In particular, if T is superstable we choose E_0 in the following theorem to be finite.

1.23 Theorem. *If E is a set of indiscernibles over A and $|E| \geq \overline{\kappa}(T)$ then there is a set $E_0 \subseteq E$ with $|E_0| < \overline{\kappa}(T)$ such that $E - E_0$ is strongly independent over $E_0 \cup A$.*

Proof. Since E is a coherent sequence, we can choose by Lemma 1.21 ii) a set E_0' with $|E_0'| < \overline{\kappa}(T)$ such that $E - E_0'$ is independent over $A \cup E_0'$. Let $E_0 = E_0' \cup \overline{e}$ for any $\overline{e}$ in $E - E_0'$. By Lemma 1.15 all elements of $E - E_0'$ realize the same strong type over $A \cup E_0'$ and this strong type is specified by the choice of $\overline{e}$.

We have discussed the notions: a) a strongly independent sequence b) an independent sequence and c) a sequence of indiscernibles. We have shown that a) implies both b) and c). Neither converse holds. But, Theorem 1.23 asserts that each set satisfying c) 'eventually' satisfies a).

From Lemma 1.15 we see that if E is a set of indiscernibles over A then all elements of E realize the same strong type over A. Moreover, for any k, any k-tuple of sequences from E realizes the same strong type over A. For applications, we need a somewhat stronger result. Namely, we would like the type over A to actually specify which strong type is realized by the sequence of indiscernibles. To arrange this for 1-tuples from E is easy; we just add one element of E to A and work over $A \cup \overline{e}$. It would seem that to deal with k-tuples from E we would have to add k members of E. This would be unfortunate if E had only $\aleph_0$ members. However, here we are able to use the strength of independence instead of mere indiscernibility. The case of the following theorem with $\lg(\overline{h}) = 2$ is Exercise II.2.1. The argument here requires one crucial observation about automorphisms and the equality of types: If α is an automorphism fixing B and taking $\overline{a}$ to $\overline{a}'$ and $t(\alpha\overline{b}; \overline{a}' \cup B) = t(\overline{b}'; \overline{a}' \cup B)$ then $t(\overline{b}^\frown\overline{a}; B) = t(\overline{b}'^\frown\overline{a}'; B)$.

1.24 Theorem. *If E is strongly independent over A then for any sequence $\overline{h}$ of sequences from E (i.e. $\overline{h} = (\overline{e}_{i_0}, \ldots, \overline{e}_{i_n})$), $t(\overline{h}; A)$ is stationary.*

Proof. The proof is by induction on the length of $\overline{h}$. If $\lg(\overline{h}) = 1$, this is the definition of strong independence. Now suppose $\overline{h} = \overline{g}^\frown\overline{e}$ and $t(\overline{g}; A)$

is stationary. Suppose that there exist $B \supseteq A$, $\overline{g}'$, $\overline{g}''$, $\overline{e}'$, $\overline{e}''$ such that $t(\overline{g}'\frown\overline{e}'; A) = t(\overline{g}''\frown\overline{e}''; A) = t(\overline{g}\frown\overline{e}; A)$ and $t(\overline{g}'\frown\overline{e}'; B)$ does not fork over A and $t(\overline{g}''\frown\overline{e}''; B)$ does not fork over A. Since $t(\overline{g}'\frown\overline{e}'; A) = t(\overline{g}\frown\overline{e}; A)$, the independence of E implies $t(\overline{g}'; A \cup \overline{e}')$ does not fork over A. But then, by Corollary II.2.10, $t(\overline{g}'; B \cup \overline{e}')$ does not fork over B. Since $t(\overline{e}'; B)$ does not fork over A and $t(\overline{e}''; B)$ does not fork over A and their common restriction to A is stationary, $t(\overline{e}'; B) = t(\overline{e}''; B)$. Hence, there is an automorphism α which fixes B and maps $\overline{e}''$ to $\overline{e}'$. We can apply to $\alpha\overline{g}''$ the same argument we applied to $\overline{g}'$ and conclude $t(\alpha\overline{g}''; B \cup \overline{e}')$ does not fork over A. Thus by the stationarity of $t(\overline{g}; A)$, $t(\overline{g}'; B \cup \overline{e}') = t(\alpha\overline{g}''; B \cup \overline{e}')$. Hence we can conclude $t(\overline{g}'\frown\overline{e}'; B) = t(\overline{g}''\frown\overline{e}''; B)$ as required.

1.25 Exercise. Produce an example where $t(\overline{a}; A \cup \overline{b})$ is stationary and $t(\overline{b}; A \cup \overline{a})$ is not. Refine the example to make $\overline{a} \downarrow_A A \cup \overline{b}$.

For simplicity of notation, we state the following theorem for a set of indiscernibles over the empty set.

1.26 Corollary. *If T is a small superstable theory and I is a countable set of indiscernibles over $\emptyset$, then $|S(I)| = \aleph_0$.*

Proof. Suppose $S(I)$ is uncountable. For every $p \in S(I)$ there is a finite $I_p \subseteq I$ such that p does not fork over I_p. Thus, there is an $I_0 \subseteq I$ with $2^{\aleph_0}$ types $p \in S(I)$ which do not fork over I_0. Without loss of generality, $I - I_0$ is strongly independent over I_0. Let, then, $\overline{a}$ and $\overline{b}$ realize the same type over I_0 but distinct types over I with $a \downarrow_{I_0} I$ and $b \downarrow_{I_0} I$. Thus, for some finite sequence $\overline{e} \in I$ we have $t(\overline{e}; I_0 \cup \overline{a}) \neq t(\alpha(\overline{e}); I_0 \cup \overline{a})$ where α is an automorphism of $\mathcal{M}$ which fixes I_0 and maps $\overline{b}$ to $\overline{a}$. But both $\overline{e} \downarrow_{I_0} \overline{a}$ and $\alpha(\overline{e}) \downarrow_{I_0} \overline{a}$ so this contradicts Theorem 1.24, which implies that $t(\overline{e}; I_0)$ is stationary.

1.27 Exercise. Suppose $\langle p_i : i < \alpha \rangle$ is an increasing sequence of types with the ordinal $\alpha > \omega$ and $\text{cf}(|\alpha|) > \kappa(T)$. If $E = \{e_i : i < \alpha\}$ where e_i realizes p_i then E contains an infinite indiscernible subsequence.

Thus far in this section we have dealt with indiscernibles without generally guaranteeing their existence or their location. In Exercises 1.13 and 1.14 we showed the existence of indiscernibles somewhere in the monster model; now, we employ a variant on an argument due originally to Morley to construct indiscernibles within a given set. The specific argument here was suggested to me by Lascar.

1.28 Theorem. *Let T be λ-stable, $\lambda \geq |T| + \kappa(T)$, and A a subset of B with $|B| > \lambda \geq |A|$. Then B contains a subset E with $|E| > \lambda$ which is strongly independent over some M containing A and thus indiscernible over A.*

Proof. (Fig. 1). Choose by induction sets $E_i \subseteq B$ such that E_{i+1} is maximal with respect to being independent over $A \cup E_i$ (and taking unions at limits). If for each i, $|E_i| \leq \lambda$ this construction can be continued at least λ times.

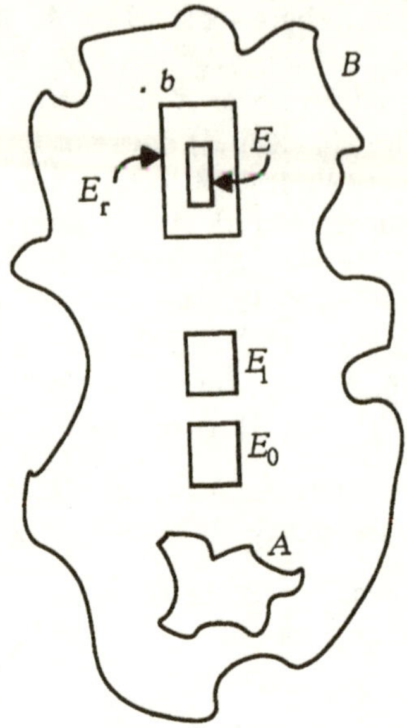

Fig. 1. Theorem 1.28. Constructing Indiscernibles

Choose any $b \in B - \bigcup_{i<\lambda} E_i$. Then, for each i, by the definition of E_i, $t(b; A \cup \bigcup_{j \leq i} E_j)$ forks over $\bigcup_{j<i} E_j \cup A$. Since $\lambda \geq \kappa(T)$, this is impossible. Thus for some $k < \lambda$, $|E_k| > \lambda$. Choose M containing A such that $|M| = \lambda$ and $t(M; B)$ does not fork over $\bigcup_{j<k} E_j \cup A$. Since T is stable in λ there is a subset E of E_k with $|E| > \lambda$ such that all elements of E realize the same, necessarily stationary, type over M. By the symmetry axiom and Corollary II.2.10, E is independent over M, whence by Theorem 1.3 indiscernible over M and a fortiori indiscernible over A.

1.29 Exercise. Derive Morley's theorem which states that if T is a countable ω-stable theory and A is an uncountable set then for any countable set $B \subseteq A$, A contains an uncountable set of elements indiscernible over B.

These ideas can be applied along with König's lemma to give the following results of [Zilber 1980b]. For the following three exercises, assume T is $\aleph_0$-categorical and superstable and so ω-stable. Let X be a finite subset of a model of T.

1.30 Exercise. Any infinite subset of a model of T contains an infinite set of indiscernibles over X. (Hint: use König's lemma and Lemma 1.21.)

Now using Theorem 1.18 deduce the following.

1.31 Exercise. If $X \subseteq A$, Y is infinite and $t(Y; A)$ does not fork over X then Y contains an infinite set of indiscernibles over A.

1.32 Exercise. If X is a subset of A, M is a minimal prime model over A (cf. Chapter IX), and Y is a subset of M which satisfies '$t(Y; A)$ does not fork over X', then Y is finite.

1.33 Historical Notes. The vital notion of a set of indiscernibles was introduced by Ehrenfeucht and Mostowski [Ehrenfeucht & Mostowski 1956] who used it to prove that a first order theory has models of arbitrary cardinality κ whose automorphism groups have cardinality 2^{κ}. In [Morley 1965], Morley discovered the important distinction between order and pure indiscernibles and introduced the inductive construction of sequences of indiscernibles. Shelah's recognition that this construction could be carried out with a more general technique than the minimization of Morley rank was one of the impulses towards the development of the nonforking notion. Shelah introduced the notion of an average type in [Shelah 1978]. Precursors of the important fact (Theorem 1.18) that nonforking extensions preserve indiscernibility are in [Harnik & Ressayre 1971] and [Shelah 1972].

2. Comparing Sets of Indiscernibles

We will eventually assign as invariants of a model the cardinalities of various sets of indiscernibles. While the structure of a vector space is determined by one magnitude or dimension, models of other theories will depend on more than one dimension. Consider first the theory of an infinite set and second the theory of an equivalence relation with two infinite classes. A model of the first is specified by one magnitude, its cardinality; a model of the second requires two magnitudes to be specified, the cardinality of each equivalence class. Once we agree to count more than one set of indiscernibles, we face a problem of redundant information. Of which sets of indiscernibles do we need to know the cardinality? For example, consider the theory of an equivalence relation with two infinite classes, each of which is a model of the theory of the integers under successor. Clearly, to determine a model we need to count the number of components in each class. In a particular class, a set of indiscernibles containing one element from each component and a second set consisting of the successors of the elements in the first set give the same information. (Fig. 2).

Fig. 2. The x's and o's give the same information.

Our aim, which can be achieved only in some cases, is to define an equivalence relation on sets of indiscernibles in a model, assign a dimension to each class, and have these dimensions determine the model up to isomorphism. For this procedure to be useful we need to check two things. First, that the obvious way of assigning such a dimension as the cardinality of a maximal set of indiscernibles in the equivalence class is well-defined. Second, that in fact the cardinalities of these maximal sets of indiscernibles determine the model up to isomorphism.

There are several difficulties in this program. The first condition is easily met for equivalence classes whose dimension is greater than $|T|$ and even without too much difficulty for those with cardinality above $\kappa(T)$. It is much more difficult for small (e.g. finite) dimensions. Unfortunately , it is just the finite dimensions which are crucial in investigating countable models and this is one reason for the difficulty of settling Vaught's conjecture. It is to meet this difficulty that the apparatus of regular types is developed in Chapter XII.

There is however a second difficulty with redundancy. It turns out that the equivalence relation suggested by the example beginning this section is not sufficiently coarse. If we were to expand that example by adding a 1-1 function between the two equivalence classes then we would need only one cardinal to specify a model instead of two. But, the indiscernibles in distinct classes are not equivalent. We will need to describe a second equivalence relation to allow for this second level of redundancy. This equivalence relation is defined in Chapter XI.

2.1 Definition. Let X_1 and X_2 be infinite sets of indiscernibles over A. We write $X_1 \sim_A X_2$ and say X_1 and X_2 are *equivalent* if there is an infinite set, Y, of indiscernibles over A such that $X_1 \cup Y$ and $X_2 \cup Y$ are sets of indiscernibles over A. We omit the A if it is the empty set.

Note that E_0 and E_1 can be equivalent without $E_0 \cup E_1$ being a set of indiscernibles. Just choose two distinct sets of representatives for the equivalence classes in the example considered in the paragraph before the definition. For another example, in the theory of the integers under successor let E_0 and E_1 each select a different element from each of infinitely many components of a model.

If $X_1 \sim_A X_2$ and $|X_1| = |X_2|$ then $t(X_1; A) = t(X_2; A)$ so there is an automorphism fixing A taking X_1 to X_2. Clearly $\sim_A$ is symmetric and reflexive; but to establish transitivity we prove the following characterization.

2.2 Theorem. *If X_1 and X_2 are infinite sets of indiscernibles over the empty set, the following are equivalent.*

i) $X_1 \sim X_2$.
ii) *For every A,* $\mathrm{Av}(X_1; A) = \mathrm{Av}(X_2; A)$.
iii) *For some model M containing $X_1 \cup X_2$,* $\mathrm{Av}(X_1; M) = \mathrm{Av}(X_2; M)$.

Proof. Suppose i) holds and $X_1 \cup Y$ and $X_2 \cup Y$ are both sets of indiscernibles over the empty set. Then for any A, since X_1, X_2, and Y are infinite, $\mathrm{Av}(X_1; A) = \mathrm{Av}(X_1 \cup Y; A) = \mathrm{Av}(Y; A) = \mathrm{Av}(X_2 \cup Y; A) = \mathrm{Av}(X_2; A)$. Thus ii) holds.

Now we show that ii) implies i). For this, first define by induction a set $C = \{\bar{c}_i : i \in \omega\}$ so that if D_n denotes $A \cup X_1 \cup X_2 \cup C_n$, $\bar{c}_n$ realizes $\mathrm{Av}(X_1; D_n)$. Observe that if the construction were repeated replacing $\mathrm{Av}(X_1; D_n)$ with $\mathrm{Av}(X_2; D_n)$ the resulting set C' would be isomorphic to C over $X_1 \cup X_2 \cup A$. Now, applying the fact that if F is a set of indiscernibles over an arbitrary set A and $\bar{c}$ realizes $\mathrm{Av}(F; A \cup F)$ then $F \cup \{\bar{c}\}$ is a set of indiscernibles over A, it is easy to show by induction that C is the required set of indiscernibles for the theorem.

Now we show ii) and iii) are equivalent. It is obvious that ii) implies iii). If for some A, $\mathrm{Av}(X_1; A) \neq \mathrm{Av}(X_2; A)$ then there is a formula $\phi(\bar{x}; \bar{a})$ such that $\phi(\bar{x}; \bar{a})$ is in $\mathrm{Av}(X_1, A)$ but $\neg\phi(\bar{x}; \bar{a})$ is in $\mathrm{Av}(X_2, A)$. Let $n = n(\phi)$ from Theorem 1.3 ii). Now choose $\bar{b}_1, \ldots, \bar{b}_n$ from X_1 and $\bar{c}_1, \ldots, \bar{c}_n$ from X_2 such that $\models \phi(\bar{b}_i; \bar{a}) \wedge \neg\phi(\bar{c}_i; \bar{a})$ for all $i < n+1$. Then since M is a model of T which contains both X_1 and X_2, for some a' in M,

$$\bigwedge_{i<n+1} \phi(\bar{b}_i; \bar{a}') \wedge \neg\phi(\bar{c}_i; \bar{a}')$$

is true. By Theorem 1.3, $\mathrm{Av}(X_1; M) \neq \mathrm{Av}(X_2; M)$.

Since ii) is obviously transitive, so is i) as required.

We want now to define the dimension of an equivalence class in this relation.

2.3 Definition. i) Let $A \subseteq B$, $E \subseteq B$ and suppose E is a set of indiscernibles over A. Then $\dim(E, A, B)$, the dimension of E in B over A, is the minimal cardinality of a maximal set of indiscernibles $F \subseteq B$ such that $E \sim_A F$.

We can extend this notion to types as follows.

ii) Let $p \in S(A)$ be a stationary type. If $A \subseteq B$, then $\dim(p, B)$ is the cardinality of a maximal independent coherent sequence over A in B of realizations of p.

The next lemma shows that for sufficiently large values of the dimension, it can be calculated as the size of *any* maximal independent set. This is analogous to the situation in linear algebra where a simple cardinality argument shows that the dimension of a vector space is well defined if it is infinite but a much more complicated argument is necessary for the case of finite dimension. In our situation, we will be unable to handle the case of finite dimension until Chapter XII when we restrict the class of sets of indiscernibles that we discuss.

2.4 Theorem. *Let E be a maximal set of indiscernibles over A with $A \cup E$ contained in B. If $\dim(E, A, B) \geq \kappa(T)$ then for any maximal indiscernible set $F \subseteq B$ with $F \sim_A E$, $|E| = |F|$.*

Proof. Without loss of generality, suppose for contradiction that $|F| > |E|$. Then by Corollary 1.19 we can find a subset F' of F with $|F'| > |E|$ such that F' is a set of indiscernibles over E. We will show $F' \cup E$ is a set of indiscernibles, contradicting the maximality of E. Since E and F are equivalent, there is an infinite set E_0 such that $F \cup E_0$ and $E \cup E_0$ are both indiscernible sets. Thus $F' \cup E_0$ is a set of indiscernibles and, possibly first extending E_0 and applying Corollary 1.19 again, we can assume $F' \cup E_0$ is indiscernible over E. But now it is easy to see that $F' \cup E$ is indiscernible over A.

Let E be a set of indiscernibles over A. The remainder of this section is devoted to the search for a syntactic description of E which preserves equivalence over A.

One description of E is $t(E; A)$. This description does not contain enough information to specify E up to equivalence. To see this, consider a theory T with two unary predicates R and S which partition the universe of each model. Further, let f be a function from R onto S such that each element has infinitely many preimages. Now consider sets X and Y of elements realizing R such that all elements of X map to a single a satisfying S; all elements of Y map to a single element b in S; $a \neq b$. Then each of X and Y is a set of indiscernibles and $t(X; \emptyset) = t(Y; \emptyset)$ but X is not equivalent to Y. For, suppose Z were an infinite set such that both $X \cup Z$ and $Y \cup Z$ were indiscernible. For any $z \in Z$, we must have $f(z) = a$ from the viewpoint of X and $f(z) = b$ from the viewpoint of Y.

We will describe E in terms of the type of a single element in the sequence E. To describe the cases in which this is possible, we require some futher definitions. These notions extend, but do not conflict with, the definition of a type being based or strongly based on a set (Definition IV.2.4).

2.5 Definition. i) The infinite set of indiscernibles E is *based* on A if for every B which contains A, $\mathrm{Av}(E, B)$ does not fork over A.

ii) If, in addition $\mathrm{Av}(E, A)$ is stationary then E is *strongly based* on A. If $q = \mathrm{Av}(E, A)$, we may write E is based on q.

iii) If $p \in S(A)$ and for some $\overline{c}$ realizing p, E is strongly based on q, we also say E is based on p.

Of course, either of the conditions in Definition 2.5 can be checked by considering $\mathrm{Av}(E, \mathcal{M})$ rather than all B containing A.

2.6 Lemma. *If E is an infinite indiscernible set then there is a definition d of* $\mathrm{Av}(E, E)$ *with parameters from E such that d defines* $\mathrm{Av}(E, \mathcal{M})$.

Proof. For each formula $\phi(\overline{x}; \overline{y})$ there is by Theorem 1.3 an $n(\phi)$ such that for any $\overline{c} \in \mathcal{M}$, $\phi(\overline{x}; \overline{c}) \in \mathrm{Av}(E, \mathcal{M})$ iff $\models \phi(\overline{e}_i; \overline{c})$ for at least $n = n(\phi)$ sequences $\overline{e}_i$ from E. Thus we can take as $d\phi$ the formula

$$\bigvee_{\substack{I \subseteq E_{n+1} \\ |I| = n}} \bigwedge_{i \in I} \phi(\overline{e}_i; \overline{y}).$$

We have, in fact, that $\mathrm{Av}(E, E)$ is defined by a positive Boolean combination of instances of ϕ. This result also holds for any type over a model (cf. [Baldwin & Shelah 1985] IV.2.15). This observation is also closely connected to the normalization lemma (cf. [Harnik & Harrington 1984] [Pillay

198?]). This similarity in role between a set of indiscernibles and a model appears again in Section X.4.

2.7 Theorem. *Let E be an infinite set which is indiscernible over A. Then E is strongly based on E.*

Proof. By Lemma 2.6, $\mathrm{Av}(E; \mathcal{M})$ is definable over E so $\mathrm{Av}(E; \mathcal{M})$ does not fork over E. It remains to show that $\mathrm{Av}(E; E)$ is stationary. For this, it suffices to show that if $\overline{a}$ realizes $\mathrm{Av}(E, E)$ then for any finite equivalence relation R over E, for some $\overline{e} \in E$, $R(\overline{x}; \overline{e}) \in t(\overline{a}; E)$. But this is immediate from the definition of average type.

The next exercise illustrates another aspect of the analogy between indiscernible sets and models.

2.8 Exercise. Show directly from the definition that if $E \subset B$ is an infinite set of indiscernibles then $\mathrm{Av}(E, B)$ is finitely satisfied in E and thus $\mathrm{Av}(E, B)$ does not fork over E.

2.9 Exercise. Let M be $|A|^+$-saturated and A a subset of M. Show that if $p \in S(M)$ does not fork over A and the sequence $\langle \overline{e}_n : n < \omega \rangle$ is chosen so that $E_0 = A$ and $\overline{e}_{n+1} \in M$ realizes $p|E_n$ then $p = \mathrm{Av}(E; M)$.

With this in hand, we see that $\mathrm{Av}(E; E)$ is the desired syntactic description of E. The next definition makes more precise the requirement we would like this syntactic definition to meet. We proceed by formalizing the solution to the first problem of redundancy discussed at the beginning of this section.

2.10 Definition. Let p be a type over A and q a type over B. The types p and q are *parallel*, written $p \parallel q$, if there is a global type $\hat{r}$ extending $p \cup q$ which does not fork over either $\operatorname{dom} p$ or $\operatorname{dom} q$.

If p is stationary, p has a unique nonforking extension to a global type. Thus, parallelism is an equivalence relation on stationary types. In fact, parallelism is a congruence relation with respect to all the major relations among stationary types and this will be one of the first properties we prove as we introduce each new notion.

We could have replaced 'global type' in the definition of parallel by 'type over some model containing $\operatorname{dom} p \cup \operatorname{dom} q$'. This is easily verified using transitivity of independence in one direction and monotonicity in the other.

2.11 Theorem. *Suppose E and F are infinite sets of indiscernibles. Then $E \sim F$ if and only if $\mathrm{Av}(E; E) \parallel \mathrm{Av}(F; F)$.*

Proof. Let p (resp. q) be the unique global type which is a nonforking extension of $\mathrm{Av}(E; E)$ (resp. $\mathrm{Av}(F; F)$). If $p = q$, it is easy to restrict p and q to some sufficiently saturated model M containing $E \cup F$ and then define a strongly independent sequence realizing $p|M$ which will witness the equivalence of E and F.

Conversely, if E_0 witnesses the equivalence of E and F and r is the unique nonforking extension of $\mathrm{Av}(E_0, E_0)$ to a global type, it is easy to see that $p = r = q$.

In general we can restrict the size of the set E over which we average by invoking Theorem 1.23 and the following easy remark.

2.12 Lemma. *If E is a set of indiscernibles over A and an infinite subset of E is strongly independent over $A \cup E_0$ (for some subset E_0 of E) then E is strongly based on $E_0 \cup A$.*

2.13 Exercise. Let T be the theory of an equivalence relation E with two classes. Let p be the unique type over the empty set. Suppose $\neg E(a, b)$, p_a, p_b are non-forking extensions of p in $S(a)$, $S(b)$. Show $p_a \parallel p$, $p_b \parallel p$ but p_a is not parallel to p_b.

2.14 Historical Notes. This discussion of the average of a type is primarily taken from Chapter III of [Shelah 1978]. There are two complementary techniques for discussing the eventual behavior of a stationary type. Shelah builds a set of indiscernibles based on this type and works directly with this set of indiscernibles. Lascar emphasizes the role of the base type. We have incorporated both viewpoints; this section contains the fundamental properties of the indiscernible sequences. The definition of dimension appears in Chapter III of [Shelah 1978].

3. Forking and Dividing

In this section we study the combinatorial core of forking. We have defined the notion of forking in terms of definability of types in a stable theory. The concept we call implicit division here is Shelah's original notion of forking, which is described by specific properties of formulas. This formulation makes sense in an arbitrary theory and there have been some uses of it in unstable theories [Shelah 1980a]. We will not rely on results from this chapter except for examples. It is useful for reading Shelah's book and it does make the notion of forking much more concrete. We could have introduced implicit division at any time, but the proof of the equivalence of this notion with the one we introduced in Chapter III requires some of the properties of indiscernibles discussed in Section V.1.

We begin by defining some combinatorial properties of formulas and types. Note that types may be incomplete unless they are explicitly asserted to be complete.

3.1 Definition. For $n > 1$ and ϕ an L-formula, the collection of formulas $\{\phi(\overline{x}; \overline{e}) : \overline{e} \in E\}$ is *n-inconsistent* if for any subset E_0 of E with n or more elements, $\{\phi(\overline{x}; \overline{e}) : \overline{e} \in E_0\}$ is inconsistent but for each $\overline{e}$, $\models (\exists \overline{x})\phi(\overline{x}; \overline{e})$.

Thus, 2-inconsistency means the sets defined by the formulas $\phi(\overline{x}; \overline{e})$ are non-empty and disjoint for distinct choices of $\overline{e}$. For a slightly more complicated example, consider the set E of natural numbers which are divisible by at most 7 distinct primes. Define a single binary relation $R(x, y)$ which holds of n and p just if p is a prime and $p|n$. Then $\{R(x, e) : e \in E\}$ is 8-inconsistent.

3.2 Definition. The formula $\phi(\overline{x};\overline{y})$ *divides* over A with respect to the type $q(\overline{y}) \in S(A)$ if i) $(\exists\overline{x})\phi(\overline{x};\overline{y})$ is in $q(\overline{y})$ and ii) there exists an infinite set E of sequences which realize q such that $\{\phi(\overline{x}, \overline{e}) : \overline{e} \in E\}$ is n-inconsistent for some n.

We frequently write $\phi(\overline{x};\overline{c})$ divides over A to mean, '$\phi(\overline{x};\overline{y})$ divides over A with respect to $t(\overline{c}; A)$.' We may also write '$\phi(\overline{x};\overline{y})$ and q divide over A' to indicate that ϕ divides with respect to q. However, the assertion, 'p divides over A' means there is a formula $\phi(\overline{x};\overline{c}) \in p$ such that $\phi(\overline{x};\overline{y})$ divides over A with respect to $t(\overline{c}; A)$. Note that if $t(\overline{c}; A)$ is algebraic, in particular if $\overline{c} \in A$, then $\phi(\overline{x};\overline{c})$ does not divide over A.

We would like to use the notion of dividing as our combinatorial equivalent of forking and eventually for stable theories we will be able to do so (see Lemma 3.9). For the moment, however, we introduce a notion of implicit dividing and show it is equivalent to forking. We must do this as it is not obvious that if p is a type over B and p does not divide over A, then p can be extended to a complete type over B which does not divide over A.

3.3 Definition. The type $p(\overline{x})$ *implicitly divides* over A, if there exist formulas $\phi_0(\overline{x};\overline{a}_0), \ldots, \phi_n(\overline{x};\overline{a}_n)$ such that p implies the disjunction of the ϕ_i and each ϕ_i divides over A.

There is no assumption that the $\overline{a}_i$ realize the same type over A or even have the same length. We want first to establish some simple properties of implicit division and then show that it satisfies the extension requirement.

3.4 Lemma. i) *If $\overline{y}$ is a subsequence of $\overline{x}$, $\phi(\overline{x};\overline{c})$ divides over A, and $\models (\exists\overline{y})\psi(\overline{y};\overline{c})$, $\models \psi(\overline{y};\overline{c}) \to \phi(\overline{x};\overline{c})$, then $\psi(\overline{y};\overline{c})$ divides over A.*

ii) *If $\phi(\overline{x};\overline{a})$ divides over A then $\phi(\overline{x};\overline{a})$ implicitly divides over A.*

iii) *The type p implicitly divides over A iff some finite subtype of p implicitly divides over A.*

iv) *(monotonicity) If A is contained in B and p, q are types such that q implies p (in particular if q extends p) then if p implicitly divides over B it follows that q implicitly divides over A.*

v) *If for each $i < n$, $p \cup \{\phi_i\}$ implicitly divides over A then $p \cup \bigvee_{i<n} \phi_i$ implicitly divides over A.*

Proof. All of these results follow immediately from the definitions.

We now consider a slightly more complicated condition which is equivalent to dividing. The proof is just a variant on the Ehrenfeucht-Mostowski method of constructing indiscernibles; such variants will abound in this section.

3.5 Lemma. *The formula $\phi(\overline{x};\overline{c})$ divides over A iff there is an infinite set $E = \{\overline{e}_i : i \in I\}$ of indiscernibles over A such that $\{\phi(\overline{x};\overline{e}) : \overline{e} \in E\}$ is n-inconsistent for some n and $\overline{c} = \overline{e}_0$.*

Proof. Clearly the latter condition implies that $\phi(\overline{x};\overline{c})$ divides over A. For the converse, suppose that F is a set of sequences such that $\{\phi(\overline{x};\overline{f}) : \overline{f} \in F\}$ is n-inconsistent and for each $\overline{f} \in F$, $t(\overline{f};A) = t(\overline{c};A)$. By Ramsey's theorem and compactness, there is a set E of indiscernibles over A such that each member of E realizes $t(\overline{c};A)$ and $\{\phi(\overline{x};\overline{e}_i) : \overline{e}_i \in E\}$ is n-inconsistent. Now choose an automorphism of the monster model which maps $\overline{e}_0$ to $\overline{c}$. The image of E under this automorphism satisfies the lemma.

Now by Lemma 1.15 we can rephrase Lemma 3.5 as follows.

3.6 Theorem. *If $\phi(\overline{x};\overline{y})$ and $t(\overline{c};A)$ divide over A, then there is an infinite set E with $\overline{c} \in E$ such that*

i) *any two sequences of the same length from E satisfy the same strong type over A and*
ii) *$\{\phi(\overline{x};\overline{e}) : \overline{e} \in E\}$ is n-inconsistent.*

Our next step is to show that for a stable theory, forking is the same as implicitly dividing.

3.7 Lemma. *If p implicitly divides over A then p forks over A.*

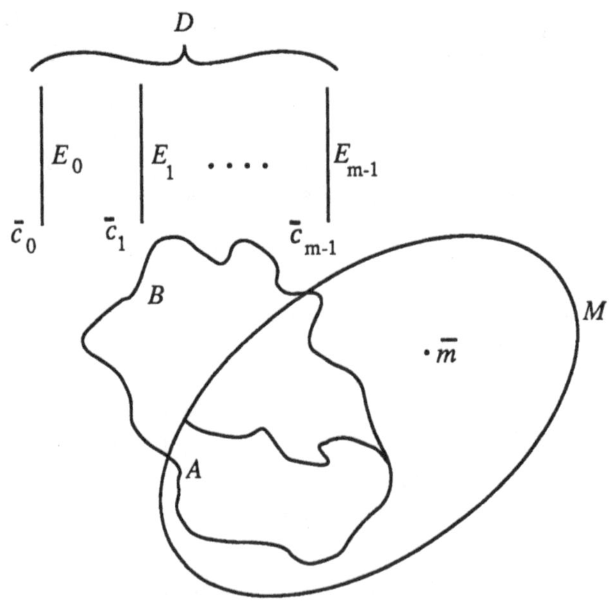

Fig. 3. Lemma V.3.7.

Proof. (Fig. 3). Let $B = \operatorname{dom} p$. If p does not fork over A, we can extend p to p' in $S(B)$ such that p' does not fork over A. If p implicitly divides over A, then by Lemma 3.4 iv) p' implicitly divides over A. Thus, if there is any counterexample p to the theorem, we may without loss of generality assume that $p \in S(B)$.

Now p implies a disjunction of, say m, formulas $\phi_i(\overline{x}; \overline{c}_i)$ which divide over A. That is, for each $i < m$ there is a set E_i of indiscernibles over A with $\overline{c}_i$ in E_i such that $\{\phi_i(\overline{x}, \overline{c}) : \overline{c} \in E_i\}$ is n-inconsistent for some n. Let D denote the union of the E_i. Choose M containing A such that $(M \downarrow D; A)$. Now p is definable over M (by Theorem III.3.26) since p does not fork over A. So we can extend p to $p_1 \in S(M \cup B \cup D)$ which is definable over M. Then for each formula $\phi(\overline{x}; \overline{y})$, there is a formula $d\phi(\overline{y}; \overline{m})$ in $F(M)$ such that for $\overline{e} \in M \cup B \cup D$, $\phi(\overline{x}; \overline{e})$ is in p_1 iff $d\phi(\overline{e}; \overline{m})$. By Lemma 1.18, since $t(\overline{m}; A \cup D)$ does not fork over A, E_i is a set of indiscernibles over $A \cup \{\overline{m}\}$. Now p implies $\bigvee_{i<m} \phi_i(\overline{x}; \overline{c}_i)$ and p is contained in the complete type p_1 so, for some $\phi_i(\overline{x}; \overline{c}_i)$, call it $\phi(\overline{x}, \overline{c})$, $\phi(\overline{x}, \overline{c}) \in p_1$, so $d\phi(\overline{c}; \overline{m})$ holds. But then, by indiscernibility, $d\phi(\overline{e}; \overline{m})$ holds for all $\overline{e}$ in E_i. Thus $\phi(\overline{x}, \overline{e})$ is in p_1 for all $\overline{e}$ in E_i, contradicting n-inconsistency and proving the lemma.

The next lemma justifies the extension from divides to implicitly divides. Note that it does not require stability.

3.8 Lemma. *If p is a type over B which does not implicitly divide over A, then p can be extended to a complete type q in $S(B)$ which does not implicitly divide over A.*

Proof. Let $W = \{\phi(\overline{x}) \in F(B) : \phi$ implicitly divides over $A\}$. Let q' be $p \cup \{\neg\phi(\overline{x}) : \phi \in W\}$. We first show q' is consistent. If not, there is a finite subset p' of p and a finite subset W_0 of W such that $p' \cup \{\neg\phi : \phi \in W_0\}$ is inconsistent. But then p implies the disjunction of the formulas $\phi(\overline{x})$, in W_0. Since each $\phi(\overline{x})$ implicitly divides over A, by Lemma 3.4 iv) and v) p implicitly divides over A.

Let q be any extension of q' in $S(B)$; we now show that q does not implicitly divide over A. If it does, then q implies the disjunction of a finite number of formulas, each of which implicitly divides over A. Then some finite subset, q_0, of q implies this disjunction so if $\theta(\overline{x})$ denotes the conjunction of q_0, $\theta(\overline{x})$ implicitly divides over A. So $\neg\theta$ is in q' which is contained in q, contradiction.

We have shown that 'implicit dividing' implies 'forking'. We now show a result somewhat stronger than the converse. Namely we show 'forking' implies 'dividing' (for a stable theory). The original proof of the following result was considerably shortened by a suggestion of Jürgen Saffe.

3.9 Lemma. *Let T be a stable theory $A \subseteq B$, and p a type over B which is closed under finite conjunction. If p forks over A then p divides over A.*

Proof. If p forks over A then by Corollary III.3.13, there is a formula $\phi(\overline{x}; \overline{b}) \in p$ such that $\{\phi(\overline{x}; \overline{b}')\}$ forks over A for any $\overline{b}'$ realizing $q = t(\overline{b}; A)$. It suffices to show that $\phi(\overline{x}; \overline{b})$ and q divides over A. Note that $t(\overline{b}; A)$ cannot be algebraic, since if it were $\phi(\overline{x}; \overline{b})$ would be almost over A which would contradict Corollary IV.1.5. Let E, with $|E| \geq \kappa(T)$, be an independent set of sequences realizing q. Choose E to be strongly based on A. If $\phi(\overline{x};\overline{b})$ and q do not divide over A then $\{\phi(\overline{x};\overline{e}) : \overline{e} \in E\}$ is n-consistent for every n. But then for some $\overline{c} \in \mathcal{M}$, $\models \phi(\overline{c};\overline{e})$ for all $\overline{e} \in E$. But, $\overline{c} \not\downarrow_A \overline{e}$ for each $\overline{e} \in E$, which contradicts Theorem II.2.18.

The following example shows the necessity both here and in Corollary III.3.13 of assuming that the type p is closed under conjunction.

3.10 Example. Let the language L contain three unary relation symbols, U_1, U_2, U_3 and a binary relation symbol E. The theory T asserts that the U_i partition the universe into infinite sets and that E is an equivalence relation splitting U_2 into infinitely many infinite classes but each element of U_1 and U_3 is related only to itself. Let b be an element of U_2. Then neither of the formulas, $\phi_1 \equiv E(x,b) \vee U_1(x)$, nor $\phi_2 \equiv E(x,b) \vee U_3(x)$ forks over the empty set but the formula $\phi_1(x,b) \wedge \phi_2(x,b)$ does.

Even without assuming that p is closed under conjunction we easily obtain the following.

3.11 Lemma. *If p is a type over B and p forks over A then p implicitly divides over A.*

Proof. If p does not implicitly divide over A we can, by Lemma 3.8, extend p to a complete type over B which does not implicitly divide over A. By the first monotonicity axiom this complete type forks over A so without loss of generality we may assume we have a counterexample p, to the lemma which is in $S(B)$. Now, by Lemma 3.9 we have the result.

The following example due to Charles Steinhorn shows the necessity of assuming T is stable for the results in this section.

3.12 Example. We construct a theory T and a formula ϕ such that ϕ implicitly divides over the empty set but ϕ does not divide, nor even fork, over the empty set. Let T have a language with two unary relations U_1 and U_2 and one binary relation R. The prototypical model of T interprets U_1 as an infinite set X, U_2 as the power set of X, and R as the membership relation. Suppose a and b are elements of U_2 such that a names the complement of the set named by b. Then the formula $R(x,a) \vee R(x,b)$ does not divide over the empty set since it is equivalent to $U_1(x)$. But each of the disjuncts divides.

We can summarize this chapter as follows.

3.13 Corollary. *Let T be a stable theory, $A \subseteq B$ and $p \in S(B)$. The following are equivalent.*

i) *p forks over A.*
ii) *p divides over A.*
iii) *p implicitly divides over A.*

3.14 Historical Notes. The term 'implicitly divides' is new here. The equivalence of dividing and implicitly dividing for stable theories was first shown by Shelah. Further work on dividing can be found in [Shelah 1980a].

Chapter VI

Orthogonality

In this chapter we investigate two extensions of the nonforking notion in an attempt to find the right notion of freeness. In the first section we discuss triples (A, B, C) such that A and B are not only independent over C but are persistently so. We say $t(A; C)$ is orthogonal to $t(B; C)$. In the second section we discuss a way in which A can be even more independent from B over C, namely $t(A; C)$ is orthogonal to every type over B. Our goal is this chapter is to describe the principal properties of these relations so we can use them as tools later. We first consider orthogonality of two types as a freeness relation in the sense of Chapter II. It turns out that, while regarding nonorthogonality of two types as a dependence relation is very natural; viewing nonorthogonality of a type and a set as a dependence relation is somewhat forced. A more natural approach to this latter relation is considered in Section 3 where we introduce an important partial (pre)order on the types over a set: the dominance order.

1. Orthogonality Of Types

In this section we discuss an important extension of the notion of independence: orthogonality. This concept has both a local and global form. The local form describes the orthogonality of two types over subsets of the monster model; the global form describes the orthogonality of two global types (i.e. types over the monster model). The notion of parallelism bridges the gap between these two forms of orthogonality. We begin by defining the local form.

1.1 Definition. Let p and q be complete types over C. We say p is *orthogonal* to q (over C) and write $p \perp q$ if the following holds. For every E containing C and every $\overline{a}$ realizing p, $\overline{b}$ realizing q, if $\overline{a} \downarrow_C E$ and $\overline{b} \downarrow_C E$ then $\overline{a} \downarrow_E \overline{b}$.

To simplify notation, we may write $(\overline{a} \perp \overline{b}; C)$. The role of C turns out to be unimportant; we usually just write $p \perp q$.

We can immediately extend this definition to types with different domains by the following device which we will use for other concepts later.

1.2 Definition. Let p,q be types over C, D, respectively. Then $p \perp q$ if for any nonforking extension p' of p to $S(C \cup D)$ and any nonforking extension q' of q to $S(C \cup D)$, $p' \perp q'$.

Definition 1.2 justifies speaking of the orthogonality of strong types. That is, $stp(c; B) \perp q$ if and only if the unique nonforking extension of $stp(c; B)$ to a complete type over $B \cup c$ is orthogonal to q.

We will investigate orthogonality as a notion of freeness in the sense of Chapter II. There we defined a relation $\mathcal{F}$ between a type p and a set C. This was in fact a relation among a realization $\overline{a}$ of p, the domain B of p and the set C. Much of our discussion focused on the triple $(\overline{a}, B, C)$ but the relation was actually a property of $t(\overline{a}^\frown B; C)$. Here our relation is among a similar triple $(\overline{a}, \overline{b}, C)$ but (and this is the essence of orthogonality) the relation depends not on the entire type $t(\overline{a}^\frown \overline{b}; C)$ but only on its two projections $t(\overline{a}; C)$ and $t(\overline{b}; C)$.

We observed after Axiom II.1.15 that although we defined the notion of freeness in terms of finite sequences it really was a property of the sets which are the ranges of the relevant sequences. Accordingly, this same observation applies to the derived relation, orthogonality, and we can write $(A \perp B; C)$. Similarly, as after Definition II.1.17 we can remove the restriction that A and B be finite because nonorthogonality inherits the finite character of nonforking.

1.3 Proposition. *If $(A \not\perp B; C)$ then for some finite A_0 and B_0 contained in A, B respectively, $(A_0 \not\perp B_0; C)$.*

Exercises 1.4 and 1.6 just describe the translation of properties of orthogonality from set notation to type notation.

1.4 Exercise. If for each finite $\overline{n} \in N$, $p \perp t(\overline{n}; N_0)$ show $p \perp t(N; N_0)$.

This differs from the finite character discussed in Chapter II because we have only the algebraic, as opposed to the logical, form. We have identified finite sets but not specific formulas as a cause of nonorthogonality . We give a more syntactic characterization in Exercise 2.19.

It is easy to see that orthogonality is strictly stronger that nonforking since we can use the base set as the test in the definition of orthogonality. Formally, we have

1.5 Lemma. $(A \perp B; C)$ *implies* $(A \downarrow B; C)$.

1.6 Exercise. Let $p = t(\overline{a}; C)$ and $q = t(\overline{b}; D)$. Then p is orthogonal to q iff for every $\overline{a}'$ realizing p, every $\overline{b}'$ realizing q and every E containing $C \cup D$, if $\overline{a}' \downarrow_C E$ and $\overline{b}' \downarrow_D E$ then $\overline{a}' \downarrow_E \overline{b}'$.

1.7 Exercise. Show that if $p \in S(A)$ is algebraic then no consistent extension of p forks over A.

1.8 Exercise. Conclude that if $p \in S(A)$ is algebraic then every q is orthogonal to p.

There is an important simplification available when checking whether two types are orthogonal. We can replace the quantifiers 'for all $\overline{a}'$, $\overline{b}'$, $\overline{e}'$ by 'there exists an $\overline{a}'$ such that for all $\overline{b}'$ and $\overline{e}'$. This is easily established by an automorphism argument.

1.9 Exercise. Show that $t(A;C) \perp t(B;C)$ iff there exists an A' realizing $t(A;C)$ such that for every E and every B' realizing $t(B;C)$, if $A' \downarrow_C E$ and $B' \downarrow_C E$ then $A' \downarrow_C B'$.

This illustrates an important principle which we will invoke repeatedly in trying to parse our definitions. Because of the homogeneity of the monster model, 'One "for all" is "for free".' That is, in many definitions the first occurrence of an assertion, 'for every element realizing a type p such that ...' can be replaced by the assertion,'there is an element realizing a type p such that ...'.

1.10 Examples. i) The simplest example of two orthogonal types occurs in the theory with one unary predicate U such that both U and $\neg U$ are infinite. Then for any model M, the type p(x), which asserts $x \notin M$ and $U(x)$, and the type $q(x)$, which asserts $\neg U(x)$ and $x \notin M$, are orthogonal over M. Only slightly more complicated is the similar situation of an equivalence relation with two infinite classes.

ii) The next level of complexity arises by considering an equivalence relation with infinitely many infinite classes. Then the type, p, over a model M which fixes an equivalence class but is not realized in M is orthogonal over M to the type, q, which is realized by an element in a 'new' equivalence class.

iii) A much more interesting example arises by considering the theory of $Z_2^{\aleph_0} \oplus Z_4^{\aleph_0}$. Let M be a model of T and consider the types p, q, and r defined as follows. Let p be the type of an element of order four satisfying $2x \neq m$ for each $m \in M$. Let q, r be types which are not realized in M but are satisfied by elements of order two. Let any realization of q be divisible by two but $(2m \neq x) \in q$ for each $m \in M$. Let no realization of r be divisible by two. Then, $p \perp r$, $p \not\perp q$ (since twice some realization of p may realize q), and $q \perp r$. Thus, orthogonality need not be transitive.

We now consider which of the other axioms for freeness hold of orthogonality. It is easy to see that orthogonality is symmetric. But the monotonicity axioms are more problematical.

1.11 Lemma. i) *If* $(A \perp B; C)$ *then* $(B \perp A; C)$.

ii) *Orthogonality satisfies the first monotonicity axiom. If* $(A \perp B; C)$ *and* $A_0 \subseteq A$ *then* $(A_0 \perp B; C)$.

iii) *The following stronger version of ii) is more useful. Suppose that* $t(A; B) \perp t(C; D)$ *and* $A_0 \subseteq A$. *Then* $t(A_0; B) \perp t(C; D)$.

Proof. Check i) immediately from the definition. For ii), fix A_0', B' and E so that A_0' realizes $t(A_0;C)$, B' realizes $t(B;C)$, $A_0' \downarrow_C E$, and $B' \downarrow_C E$. Let A'' be the image of A under an automorphism which takes A_0 to A_0' while fixing C. Apply the strong extension property (Exercise II.2.8) to choose A' so that $(A' \downarrow E; C)$ and $t(A'; A_0' \cup C) = t(A''; A_0' \cup C)$. By symmetry, applying $A_0' \downarrow_C E$, we have $(A \perp B; C)$ implies $(A' \downarrow B; E)$ so by monotonicity of nonforking $(A_0' \downarrow B; E)$ as required.

To derive iii) from ii) let A_0', C' realize $t(A_0;B)$, $t(C,D)$, respectively, with $A_0' \downarrow_B B \cup D$ and $C' \downarrow_D B \cup D$. We must show $t(A_0'; B \cup D) \perp t(C'; B \cup D)$. Apply the strong extension principle to choose $A' \supseteq A_0'$ with $A' \downarrow_B B \cup D$. By the definition of orthogonality for types over different sets, $t(A; B \cup D) \perp t(C'; B \cup D)$. By part ii) $t(A_0'; B \cup D) \perp t(C'; B \cup D)$. Since this argument holds for any choice of C' which does not depend on $B \cup D$ over D, we conclude $t(A_0; B) \perp t(C; D)$.

In this proof we have carefully traced the argument for each nonforking extension of $t(A;B)$. The second two sentences of the proof of ii) justify the assertion 'without loss of generality $A_0 = A_0'$.' We will not hesitate to make such assumptions later.

The following example shows that upward monotonicity in the third coordinate fails, i.e. $(A \perp B; C)$ does not imply $(A \perp B; C')$ for arbitrary C' containing C. In fact the remedy from Section III.3 (requiring C' to be contained in B) does not work either since in the example $t(C' - C; C) \subseteq t(B; C)$.

1.12 Example. Let T be the theory of $Z_2^{\aleph_0} \oplus Z_4^{\aleph_0}$. Let r be the type over the empty set asserting x has order four. Let $q(y)$ be the type over the empty set asserting y has order two and two does not divide y. Let M be a countable model of T and let N be $M \oplus Z_2 \oplus Z_4$. Consider the following three elements of N

	M	Z_2	Z_4
a	$\bar{0}$	0	1
b	$\bar{0}$	1	0
b'	$\bar{0}$	1	2

Now a realizes the nonforking extension of r to M while B and b$'$ realize the nonforking extension of q to M. We have $(a \perp b; M)$ but $(a \not\perp b; M \cup b')$. For, $t(a; M \cup b, b')$ forks over $M \cup b'$ since $2a = b + b'$.

Using the transitivity of nonforking, it is easy to show the 'correct' form of monotonicity in the third coordinate. Namely, orthogonality is preserved by nonforking extensions. Hrushovski has made important applications ([Hrushovski 1986], [Hrushovski 198?]) of the concept of *hereditary orthogonality* (p is hereditarily orthogonal to q if *every* extension of p is orthogonal to q.).

1.13 Lemma. *If $p \perp q$ and p', q' are nonforking extensions of p and q then $p' \perp q'$.*

The following exercise translates this result into the set notation.

1.14 Exercise. If $(A \perp B; C)$ and $C \subseteq C'$ with $A \downarrow_C C'$ and $B \downarrow_C C'$ then $(A \perp B; C')$.

We show now that if we view nonorthogonality as a dependence relation then a suitable modification of the transitivity of independence axiom holds for orthogonality. The clearest argument for the next proposition relies on Theorem 1.37 which asserts that if the nonforking extensions p', q' of the stationary types p and q are orthogonal then so are p and q. Lemma 1.15 is not used in the proof of Theorem 1.37.

1.15 Lemma. *If $B \subseteq C \subseteq D$ and $(A \perp D; C)$ and $(A \perp C; D)$ then $(A \perp D; B)$.*

Proof. Let A' realize $t(A;B)$, D' realize $t(D;B)$, and suppose $(A' \downarrow E; B)$ and $(D' \downarrow E; B)$ for some E containing B. By Exercise 1.9, we may choose A' to realize $stp(A;B)$. We must show $(A' \downarrow D'; E)$. Choose $C' \subseteq D'$ with $t(C';B) = t(C;B)$. By monotonicity, $(C' \downarrow E; B)$. Now applying the orthogonality of A and C over B we have $(A' \downarrow C'; B \cup E)$. By monotonicity and the independence of A' from E over B, this implies $(A' \downarrow C' \cup E; B)$. ¿From $D' \downarrow_B E$ we deduce by monotonicity and symmetry that $D' \downarrow_{C'} E \cup C'$. Suppose for the moment that $stp(D';C') \perp stp(A';C')$. Then $D' \downarrow_{C' \cup E} A'$. Since $A' \downarrow_B C' \cup E$, we deduce $D' \downarrow_B A'$ as required.

The argument that $stp(D';C') \perp stp(A';C')$ is somewhat circuitous. Since $stp(D;C) \perp stp(A;C)$ and $A \downarrow_B C$, Theorem 1.39 implies $stp(D;C) \perp stp(A;B)$. There is an automorphism fixing B and mapping D to D' and C to C'. Thus, since orthogonality is preserved under automorphisms, $stp(D';C') \perp stp(A;C)$. But $stp(A;B) = stp(A';B)$ and $A' \downarrow_B C'$. So by Lemma 1.13 $stp(D';C') \perp stp(A';C')$.

We can drop the assumption that $C \subseteq D$.

1.16 Exercise. Show that if B is a subset of C and $(A \perp C; B)$, and $(A \perp D; C)$ then $(A \perp D; B)$. (Hint: Note that $t(D;C) \perp t(D \cup C; C)$ and then apply Lemma 1.15.

1.17 Example (Stationarity Fails). Let T be the theory of an equivalence relation with infinitely many infinite classes. Then the unique 1-type over the empty set has an unbounded number of pairwise orthogonal non-algebraic extensions.

We will discuss later theories in which each type has a bounded number of mutually orthogonal non-algebraic extensions (i.e. which satisfy for orthogonality the intent of Axiom II.1.23) and will see that they form a particuarly well behaved class. (Such theories are said to have bounded width or be non-multidimensional).

There is one way in which orthogonality is a much stronger relation than independence.

1.18 Definition. We call a notion of freeness $\mathcal{F}$ *trivial* when $(A \mathcal{F} B_0; C)$, $(A \mathcal{F} B_1; C)$, and $(B_0 \mathcal{F} B_1; C)$ together imply $(A \mathcal{F} B_0 \cup B_1; C)$.

For example, if T is any completion of the theory of a single 1-1 unary function then nonforking is a trivial dependence relation. Intuitively, orthogonality should be trivial. For, if $t(A;C) \perp t(B;C)$ then A and C are persistently independent. That is, they remain independent over any set which is independent from each of them; in particular, they should be independent over any realization of a type which is orthogonal to each of $t(A;C)$ and $t(B;C)$. More formally, we show that orthogonality is trivial by showing an even stronger result which we call *strong triviality.*

1.19 Theorem. *Orthogonality is a trivial dependence relation. In fact, let E be an independent set over A. Suppose that for any $\bar{b}$ and each $\bar{e} \in E$, $(\bar{b} \perp \bar{e}; A)$ then $(\bar{b} \perp E; A)$.*

Proof. Without loss of generality, by Proposition 1.3, we assume E is finite. The proof is by induction on $n = |E|$ and the result is obvious if $n = 0$. So assume the result for $n = m$ and we will establish it for $n = m + 1$. Fix any D with $A \subseteq D$ such that $\bar{b} \downarrow_A D$ and $E_{m+1} \downarrow_A D$ Since nonforking preserves independence and since E_{m+1} is independent over A, $\bar{e}_m \downarrow_D E_m$. By induction $(\bar{b} \perp E_m; A)$ so by the definition of orthogonality $\bar{b} \downarrow_D E_m$. Now since $(\bar{b} \perp \bar{e}_m; A)$, we have $(\bar{b} \downarrow \bar{e}_m; D \cup E_m)$. Thus $\bar{b} \downarrow_D E_{m+1}$ as required.

The following exercise provides a useful rephrasing of Theorem 1.19.

1.20 Exercise. Show that if $E = \langle \bar{e}_i : i < \alpha \rangle$ is independent over A and for each i, $p \perp t(\bar{e}_i; A)$ then $p \perp t(E; A)$.

1.21 Exercise. Find a counterexample to the following assertion. If for each $e \in E$, $p \perp t(e; A)$ then $p \perp t(E; A)$.

We now will discuss orthogonality from a more syntactic viewpoint, one which relies heavily on model theory as opposed to the abstract theory of dependence relations.

1.22 Definition. i) Let $p(\bar{x}), q(\bar{y}) \in S(A)$. p is *weakly orthogonal* to q, $p \perp^w q$, if $p(\bar{x}) \cup q(\bar{y})$ is a complete type.

ii) Let p and q be types (almost) over A. p and q are *almost orthogonal,* $p \perp^a q$, if for any a, b realizing p, q, respectively, $a \downarrow_A b$.

The following exercises will be applied repeatedly without reference. They are all proved by permuting Definition 1.22.

1.23 Exercise. Show $t(\bar{a}_1; A) \perp^w t(\bar{a}_2; A)$ iff $t(\bar{a}_1; A) \vdash t(\bar{a}_1; A \cup \{\bar{a}_2\})$ iff $t(\bar{a}_2; A) \vdash t(\bar{a}_2; A \cup \{\bar{a}_1\})$. Hence if $t(\bar{a}_1; A) \perp^w t(\bar{a}_2; A)$ we can conclude $\bar{a}_1 \downarrow_A \bar{a}_2$.

1.24 Exercise. Show that $p \perp^w q$ implies $p \perp^a q$.

1.25 Exercise. Show that $p \perp q$ implies $p \perp^a q$.

1.26 Exercise. Show that if p is weakly orthogonal to $q(\bar{x})^\frown r(\bar{y})$ then p is weakly orthogonal to each of q and r.

1.27 Exercise. If one of p and q is stationary, $p \perp^a q$ implies $p \perp^w q$.

1.28 Exercise. $p \perp^a q$ iff for some a realizing p and any $\bar{b}$ realizing q, $t(\bar{b}; A \cup \{\bar{a}\})$ does not fork over A.

1.29 Exercise. If for some model $M, p, q \in S(M)$ and $p \perp^a q$ then $p \perp^w q$.

We extend by convention the definition of weak orthogonality and almost orthogonality to include types with different domains. Recall the notation $N(A \cup B, A)$ from Section III.3.

1.30 Definition. Let $p \in S(A)$ and $q \in S(B)$ and suppose p', q' extend p and q respectively.

i) $p \perp^w q$ iff for each $p' \in N(A \cup B, A)$ and $q' \in N(A \cup B, B)$, $p' \perp^w q'$.
ii) $p \perp^a q$ iff for each $p' \in N(A \cup B, A)$ and $q' \in N(A \cup B, B)$, $p' \perp^a q'$.

By the preceding exercises, for stationary p and q we have $p \perp^w q$ iff $p \perp^a q$. Thus $\perp^a$ is the weak version of orthogonality where instead of extending p and q freely to an arbitrary extension of their domains, we extend only to $\operatorname{dom} p \cup \operatorname{dom} q$.

The subtle difference between the following two exercises plays an important role in problems relating to the axiomatizability of totally categorical theories [Ahlbrandt 1984].

1.31 Exercise. Show that if $t(a; A)$ is algebraic then it is stationary if and only if a is in the definable closure of A. Conclude that if $t(\bar{a}; C)$ and $t(\bar{b}; C)$ are both algebraic and stationary then $t(\bar{a}; C) \perp^w t(\bar{b}; C)$.

1.32 Exercise. Show the previous result fails if the hypothesis that the types are stationary is dropped.

The following two examples show that, strictly speaking, orthogonality and weak orthogonality are incomparable concepts. Of course, we have already noted that if one of p, q is stationary then $p \perp q$ implies $p \perp^w q$.

1.33 Example (Weak orthogonality need not imply orthogonality). i) We first show that two types over a set A can be weakly orthogonal but not orthogonal.

Let T be the theory of the $Z/(4)$ module $Z_4^{\aleph_0}$. Denote by p the type over the empty set of an element of order 2. Fix a set A of realizations of p, $a \in A$ and q the type of an element of order 4 which satisfies $2x = a$. Letting p' be the nonforking extension of p to A, we will show $p' \perp^w q$ but $p' \not\perp q$.

To see that $p' \perp^w q$, note that if $b \neq c \in \mathcal{M}$ (the monster model) satisfy q there is an automorphism of $\mathcal{M}$ which fixes $p(\mathcal{M})$ pointwise and maps b to c. Exercise 1.23 implies this is more than enough. (For this, regard $\mathcal{M}$ as a free $Z/(4)$-module with basis $\langle b_i : i < \kappa \rangle$. The basis can be chosen so that $b = b_0$ and $c = b_0 + 2b_1$. Then, $\{b, c\} \cup \{b_i : 1 < i < \kappa\}$ is another basis for $\mathcal{M}$. Interchanging b and c and fixing the rest of the basis yields an automorphism of $\mathcal{M}$ which fixes $p(\mathcal{M})$ pointwise.)

To see that $p' \not\perp q$, fix a realization c of q and let q' (p'') be the nonforking extensions q (p') to $A \cup c$. Now, $p' \not\perp q$ since if c' realizes q' and a' realizes p'' we may or may not have $c' = c + a'$.

This counterexample is intrinsic to the situation. That is, $Z_4^{\aleph_0}$ is the prototype of an $\aleph_1$-categorical theory which is not almost strongly minimal. Tsuboi showed that if T is such a theory, A is an infinite dimensional subset of a strongly minimal set D, and p is the unique nonalgebraic type over A which contains $D(x)$ then for some nonalgebraic $q \in S(A)$, $q \perp^w p$. The $\aleph_1$-categoricity of T guarantees that $q \not\perp p$.

ii) We now show that there are examples of types over models which are weakly orthogonal but not orthogonal. Lascar has proved (cf. [Lascar 1982a] and XIII.3.7) that this is impossible in an ω-stable theory.

Let T be the theory CEF_ω^+ (III.4.9) of the structure $(2^\omega, E_i, +)$. Let M be the model whose universe is the collection of eventually constant functions, and let σ, τ be two sequences which do not differ by an eventually constant function. We claim that $p = t(\sigma; M)$ and $q = t(\tau; M)$ are weakly orthogonal but not orthogonal. To see $p \perp^w q$, we need only show that all formulas about σ, τ, and $\sigma + \tau$ with parameters from M are implied by $p \cup q$. This is straightforward using the quantifier eliminability of T. To see that $p \not\perp q$, note that $\tau \downarrow_M \tau + \sigma$ and $\sigma \downarrow_M \tau + \sigma$ but $(\sigma \not\downarrow \tau; M \cup \sigma + \tau)$.

Note that in this example forking reduces to algebraic dependence.

1.34 Exercise. Why are the equivalence relations E_i necessary for the preceding example?

Example 1.33 ii) arose from the following more complicated, but less contrived, example due to Mike Prest.

1.35 Exercise. Let $Z_{(2)}$ denote the additive group of those rational numbers whose denominators are not divisible by two (the integers localized at two). Show there are two types over $Z_{(2)}$ which are weakly orthogonal but not orthogonal.

A more abstract, and so perhaps more believable explanation of the phenomenon in Paragraphs 1.33 through 1.35 is given on page 504 of [Poizat 1985].

1.36 Example (For nonstationary types orthogonality does not imply weak orthogonality). Let T be the theory of two unary predicates P and Q which partition the universe and an equivalence relation with two classes each of which intersects both P and Q infinitely. Then if $p \in S(\emptyset)$ is generated by $P(x)$ and $q \in S(\emptyset)$ is generated by $Q(y)$ it is easy to see that $p \perp q$ but $p \not\perp^w q$.

The following theorem gives the relationship between orthogonality and weak orthogonality.

1.37 Theorem. *Let $p \in S(A)$ and $q \in S(B)$, then $p \perp q$ if and only if for any nonforking extensions of p and q to global types $\hat{p}$ and $\hat{q}$, $\hat{p} \perp^w \hat{q}$.*

Proof. If $p \perp q$ then, by the definition of orthogonality, for any $\overline{a}'$ realizing $\hat{p}$ and any $\overline{b}'$ realizing $\hat{q}$, $\overline{a}' \downarrow_M \overline{b}'$. Since $\hat{p}$ and $\hat{q}$ are stationary, Exercise 1.27 shows $\hat{p} \perp^w \hat{q}$. For the converse, suppose there exist $\overline{a}$,$\overline{b}$ realizing p,q respectively and a set D such that $\overline{a} \downarrow_A D$ and $\overline{b} \downarrow_B D$ but $\overline{a} \not\downarrow_D \overline{b}$. Let $\overline{a}'\frown\overline{b}'$ realize an extension of $t(\overline{a}\frown\overline{b}; D)$ to the monster model which does not fork over D. Then by Corollary II.2.10, $\overline{a}' \not\downarrow_{\mathcal{M}} \overline{b}'$. Since types over models are stationary this contradicts the weak orthogonality of $t(\overline{a}'; \mathcal{M})$ and $t(\overline{b}'; \mathcal{M})$. (Clearly, these types do not fork over A, B, respectively.)

This argument violates at least the spirit of the monster model idea by appealing to $\overline{a}'$ and $\overline{b}'$. This appeal can be avoided by a somewhat more complicated formulation.

1.38 Theorem. *Let $p \in S(A)$ and $q \in S(B)$ with $|A \cup B| < \lambda$, then $p \perp q$ if and only if for any nonforking extensions of p and q to types p' and q' in $S(M)$, where M is a λ-saturated model containing $A \cup B$, $p' \perp^w q'$.*

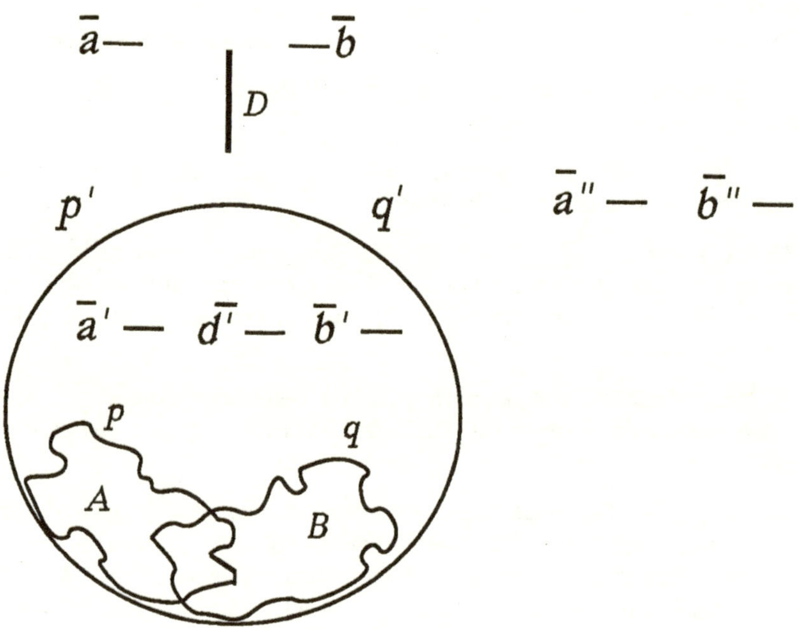

Fig. 1. Theorem VI.1.38

Proof. (Fig. 1). If $p \not\perp q$ then there exists D containing $A \cup B$ with $\overline{a} \downarrow_A D$ and $\overline{b} \downarrow_B D$ but $\overline{a} \not\downarrow_D \overline{b}$ for some $\overline{a}$ and $\overline{b}$ realizing p, q respectively. Without loss of generality, we may assume $D - (A \cup B)$ is a sequence $\overline{d}$ with $|\overline{d}| < \omega$. Now, by the saturation of M, choose $\overline{a}'\frown\overline{b}'\frown\overline{d}' \in M$ realizing $t(\overline{a}\frown\overline{b}\frown\overline{d}; A \cup B)$. Then $\overline{a}' \downarrow_A \overline{d}'$ and $\overline{b}' \downarrow_B \overline{d}'$. Choose $\overline{a}''\frown\overline{b}''$ realizing $t(\overline{a}'\frown\overline{b}'; A \cup B \cup \overline{d}')$ so that $t(\overline{a}''\frown\overline{b}''; M)$ does not fork over $A \cup B \cup \overline{d}'$. Now $(\overline{a}' \not\downarrow \overline{b}'; A \cup B \cup \overline{d}')$ so by Corollary II.2.10 $\overline{a}'' \not\downarrow_M \overline{b}''$. But $\overline{a}'' \downarrow_A M$ and $\overline{b}'' \downarrow_B M$ so this contradicts the hypothesis.

We next prove that parallelism repects orthogonality. Using this observation we are able to improve the formulation of Theorem 1.38. We noted in Lemma 1.13 the routine observation that orthogonality is preserved by nonforking extensions (orthogonality goes up). The fact that orthogonality also goes down is the content of the current lemma.

1.39 Lemma. *Let p,q, and r be arbitrary stationary types. If $p \parallel r$ then $p \perp q$ iff $r \perp q$.*

Proof. It suffices to show $p \perp q$ implies $q \perp r$. Since $p \parallel r$ there exists a global type $\hat{p}$ which is a common nonforking extension of p and r. If $\hat{q}$ is any nonforking extension of q to a global type then $p \perp q$ and Theorem 1.37 imply $\hat{p} \perp^w \hat{q}$. Applying Theorem 1.37 in the other direction, we conclude $r \perp q$.

1.40 Theorem. *If $p, q \in S(M)$ and M is strongly $\kappa(T)$-saturated then $p \perp^w q$ implies $p \perp q$.*

Proof. Since M is strongly $\kappa(T)$-saturated, by Corollary IV.2.8 there are subsets A and B of M with cardinality less than $\kappa(T)$ such that p is strongly based on A and q is strongly based on B. By Theorem 1.38, $p|A \perp q|B$. But $p|A \parallel p$ and $q|B \parallel q$ so $p \perp q$ by Lemma 1.39.

Some conditions on the domain of p and q are necessary to have weak orthogonality imply orthogonality. Example 1.33 showed that we must at least assume the domain is a model. Lascar [Lascar 1982a] has shown that for ω-stable theories this suffices.

There are two parts to the proof of Theorem 1.40. We need to know that the model M is good in the sense of Section IV.2. This part of course transfers immediately to any model of an ω-stable theory. Secondly, we need to know that the nonorthogonality of p and q can be passed to the nonorthogonality of $p|A$ and $q|B$. The above proof accomplishes this by appealing to the $\kappa(T)$-saturation of M. Lascar's argument [Lascar 1982a] substitutes an analysis of regular types.

Although $p \perp q$ was defined without assuming p or q was stationary many of the results in this section depended on the hypothesis that a type was stationary. The natural way to convert a type $p = t(\overline{c}; A)$ to a stationary type is to consider $stp(\overline{c}; A)$. The relation between orthogonality to p and the various strong types extending p is stated in the following corollary which follows easily from Theorem 1.39.

1.41 Corollary. *If $p \in S(A)$ and $q \in S(B)$ then $p \perp q$ if and only if for each $\overline{c}$ realizing p, $stp(\overline{c}; A) \perp q$.*

The following two exercises emphasize, first, the importance of assuming that p and q are stationary in Theorem 1.39 and second the necessity of assuming $stp(\overline{c}; A) \perp q$ for each $\overline{c}$ realizing p in Corollary 1.41.

1.42 Exercise. Let T be the theory of an equivalence relation with two infinite classes. Show that the unique type over the empty set has a pair of nonforking extensions which are orthogonal. Conclude that the hypothesis that p and q are stationary is necessary for Theorem 1.39.

1.43 Exercise. Find an example of types $p, q \in S(\emptyset)$ so that $p \not\perp q$ but for some $\bar{c}$ realizing p, $stp(\bar{c}; \emptyset) \perp q$.

1.44 Historical Notes. The notion of orthogonality was introduced in Chapter V.1 of [Shelah 1978]. Shelah begins with the model theoretic notion of weak orthogonality. Then he defines orthogonality in terms of our Theorem 1.37. Spurred by some observations of Saffe, we noted that orthogonality can be derived in purely algebraic terms from any dependence relation satisfying the axioms of Chapter II. Thus we introduce weak orthogonality as a useful variant which arises in the context of first order theories. The third variant, almost orthogonality, was suggested by Makkai [Makkai 1984].

2. *Orthogonality of a Type and a Set*

The next two sections are devoted to the discussion of several closely related but subtly different notions: orthogonality of a type to a set, dominance, eventual dominance, and several minor variants of these. All of these notions can be derived from an arbitrary freeness relation. That is, just as in Definition 1.1, the extended property is defined from the nonforking relation without the intervention of any other stability theoretic concept. We will explore their full significance in Chapter X when we consider their relations with various notions of isolation.

We first consider the notion 'p is orthogonal to B' as meaning p is orthogonal to every type over B. If $p = t(A; C)$ we can think of this relation as meaning 'A is free from B over C' and so we are studying the kind of relation discussed in Chapter II.

2.1 Definition. i) Let $p \in S(C)$ and B be a set. Then p is *orthogonal* to B (written $p \dashv B$) iff $p \perp q$ for each $q \in S(B)$.

ii) If $p \dashv \emptyset$ and p is not algebraic, p is said to be *unbounded.*

Note that $t(A; C) \dashv B$ is actually an assertion about $t(A; C)$ and $t(B; C)$.

2.2 Examples. i) Let T be the theory of an equivalence relation with infinitely many infinite classes and let $M \models T$. Suppose $M \prec N$ and $a \in N$ is in an equivalence class which is not represented in M. Then, a type p over N which asserts $E(x, a)$ is orthogonal to M. The only difficulty in seeing this is to observe that as well as being orthogonal to any type which fixes an equivalence class different from $[a]$, p is orthogonal to the type which asserts of an element that it is not in any class which is represented in M. (Fig. 2).

ii) A similar situation arises if we consider the third example from Example 1.10. Any type which specifies a new coset of a subgroup with infinite index will be orthogonal to the ground model.

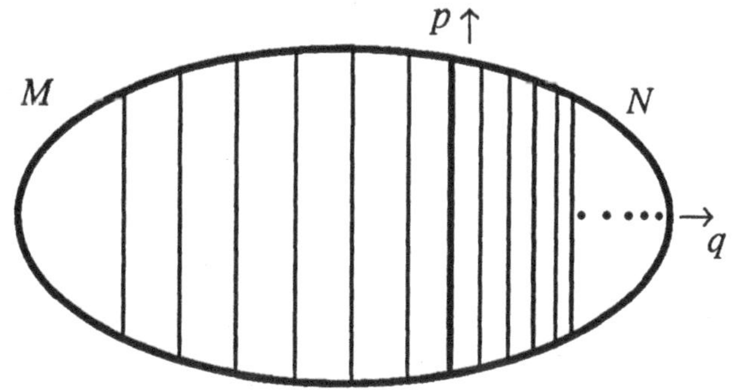

Fig. 2. $p \dashv M$

The following easy exercises will provide some familiarity with the notion.

2.3 Exercise. Show that $p \dashv A$ implies $p \perp stp(c; A)$ for every c. But, if $p \in S(B)$, $p \dashv A$ if and only if for each $\overline{c}$ realizing p, $stp(\overline{c}; B) \dashv A$.

2.4 Exercise. Show that if $p \dashv A$ then $p \dashv \text{cl}(A)$.

It is an easy exercise to show that if $t(A; C) \dashv B$ this relation continues to hold if we shrink A or B. The situation is more complicated if we increase C. A third monotonicity result, replacing B by a B' with $B' \downarrow_B C$, requires much more work and appears as Theorem 2.21.

2.5 Proposition. i) If $t(A; C) \dashv B$, $A' \subseteq A$ and $B' \subseteq B$, it follows that $t(A'; C) \dashv B'$.

ii) If $t(A; C) \dashv B$ and $C \subseteq C'$ with $C' \downarrow_C A$ then $t(A; C') \dashv B$.

iii) If $t(A; C) \not\dashv B$ then there is a $B' \subseteq B$ with $|B'| < \kappa(T)$ such that $t(A; C) \not\dashv B'$.

Part ii) of Proposition 2.5 just says that $p \dashv A$ is preserved by taking a nonforking extension of p.

2.6 Exercise. Show that for any A and B if there is a C with $t(A; B) \dashv C$ then $t(A; B) \dashv \emptyset$.

The following three exercises illustrate some arguments involving trees of types orthogonal to sets. These arguments play an important role in Chapter XVI. The second exercise will usually be applied to $C \subseteq A_0 \subseteq B_0$ and $C \subseteq A_1 \subseteq B_1$.

2.7 Exercise. Show that if $A \downarrow_C B$ and $t(D; A) \dashv C$ then $D \downarrow_C B$.

2.8 Exercise. Let $A_0 \downarrow_C A_1$ and suppose that, for $i = 1, 2$, $t(B_i; A_i) \dashv C$ and $t(D_i; B_i) \dashv A_i$. Conclude $B_0 \downarrow_C B_1$ and $t(D_0; B_0) \perp t(D_1; B_1)$.

2.9 Exercise. Let $D \subseteq A \subseteq (B \cap C)$ and $B \downarrow_A C$. Show that if $t(B; A) \dashv D$ and $t(C; A) \dashv D$ then $t(B \cup C; A) \dashv D$.

At this point the analysis in terms of a freeness relation stops making much sense. The relation does not seem to satisfy the symmetry axiom in general. A literal translation of the transitivity axiom produces nonsense. We will prove a stronger version of monotonicity and then introduce another viewpoint on orthogonality which is more fruitful.

We want to prove that if $t(A; C) \dashv B$ and $C \downarrow_B D$ then $t(A; C) \dashv D$. In some sense we ought to be able to derive this result on the basis of the properties of orthogonality and freeness which we already have in hand. However, such a proof is unknown. Rather, we must introduce here several more complicated notions in order to obtain the result.

The following concept specializes the idea of a product of types which is described in Chapter XIII.2.

2.10 Notation. If $p \in S(A)$ is a stationary type and α is an ordinal, we denote by p^α the type $t(I; A)$ where I is a strongly independent sequence of length α based on p.

2.11 Exercise. Since p is stationary, p^α is well defined.

2.12 Exercise. Find a model N, a submodel M and a $q \in S(N)$ such that $q^\omega | M \neq (q|M)^\omega$. (Hint: consider Example 2.2 i)).

2.13 Exercise. Show that since p is stationary, p^α is stationary for every α.

2.14 Exercise. Show that $p \perp q$ if and only if $p^\omega \perp q$.

The following technical result is notable in two ways. First, it reduces the problem of deciding whether two types p and q are orthogonal to the question of weak orthogonality of associated types. Second, these associated types ($p^\omega|A$ and $q^\omega|A$) need not even have the same domain as p and q but are over an arbitrary subset of their common domain.

2.15 Lemma. *Let $p, q \in S(B)$ be stationary types and suppose $A \subseteq B$. If $(p^\omega|A) \perp^w (q^\omega|A)$ then $p \perp q$.*

Proof. If $p \not\perp q$, there exist $\bar{a}$, $\bar{a}'$, $\bar{b}$, $\bar{b}'$, and $\bar{c}$ such that $\bar{a}$ and $\bar{a}'$ realize p, $\bar{b}$ and $\bar{b}'$ realize q, and each of $\bar{a}$, $\bar{a}'$, $\bar{b}$, and $\bar{b}'$ is independent from $B \cup \bar{c}$ over B but $t(\bar{a}\frown\bar{b}; B \cup \bar{c}) \neq t(\bar{a}'\frown\bar{b}'; B \cup \bar{c})$. Let $\alpha = \sup(\bar{\kappa}(T), |B|^+)$. Now choose a sequence $\bar{e}_i = \bar{a}_i\frown\bar{b}_i$ for $i < \alpha$ such that $\bar{a}_i \downarrow_B A_i$ and $\bar{b}_i \downarrow_B B_i$ but $t(\bar{e}_i; E_i \cup B \cup \bar{c})$ extends $t(\bar{a}\frown\bar{b}; E_i \cup B \cup \bar{c})$ if i is even and $t(\bar{a}'\frown\bar{b}'; E_i \cup B \cup \bar{c})$ if i is odd. Now E realizes the type $p^\alpha|A\frown q^\alpha|A$. Since $(p^\omega|A) \perp^w (q^\omega|A)$, there is only one type over A of a sequence $\langle \bar{e}'_i : i < \alpha\rangle$ with $\bar{e}'_i = \bar{a}'_i\frown\bar{b}'_i$ which projects to a realization of $p^\alpha|A$ and to a realization of $q^\alpha|A$. Moreover, such a sequence can be chosen with $\bar{e}'_i \downarrow_B E'_i$ and all the $\bar{e}'_i$ realizing the same strong type over B. The resulting sequence, E', is a sequence of indiscernibles over B. Hence, E is a sequence of indiscernibles over B. Since $|E| > |B| + \kappa(T)$, by Corollary V.1.19 there is a final subsequence of E which is indiscernible over $B \cup \overline{c}$. But this contradicts the choice of E, so $p \perp q$.

2.16 Exercise. Show the hypothesis of the last lemma can be replaced by: for each $n < \omega$, $p^n \perp^w q^n$.

2.17 Exercise. Show that if $p, q \in S(A)$, $p \not\perp q$, and $A \subseteq B$ then

i) there exists $\overline{a}$ and $\overline{b}$ realizing p and q with $\overline{a} \downarrow_A B$, $\overline{b} \downarrow_A B$ and $\overline{a} \downarrow_B \overline{b}$.
ii) For some $n < \omega$ there exists $\overline{a}', \overline{b}'$ realizing p^n and q^n with $\overline{a}' \downarrow_A B$, $\overline{b}' \downarrow_A B$ and $\overline{a}' \not\downarrow_B \overline{b}'$.

It is easy to deduce the following summary of the last few lemmas and exercises.

2.18 Corollary. *Let p, $q \in S(B)$ be stationary. The following are equivalent.*

i) $p \perp q$.
ii) $p^\omega \perp^w q^\omega$.
iii) $p^\omega \perp^a q^\omega$.
iv) *For each natural number n, $p^n \perp^w q^n$.*

We noted, after Proposition 1.3, that the finite character we had established for nonorthogonality lacked the syntactic character that the finite character of forking has (Axiom II.1.15, Corollary III.3.15). Of course, weak orthogonality has this syntactic character. Thus, we can now give a syntactic expression for $p(\overline{x}) \not\perp q(\overline{y})$, but only by introducing more variables.

2.19 Exercise. Assume p and q are stationary. Show that if $p, q \in S(A)$ and $p \not\perp q$ then for some integer n and formula $\phi(\overline{x}_0, \ldots, \overline{x}_{n-1}; \overline{y}_0, \ldots, \overline{y}_{n-1})$ and for any $r \in S(A)$, if $\langle \overline{a}_0, \ldots, \overline{a}_{n-1} \rangle$ satisfies p^n, $\langle \overline{b}_0, \ldots, \overline{b}_{n-1} \rangle$ realizes r^n, and $\models \phi(\overline{a}_0, \ldots, \overline{a}_{n-1}; \overline{b}_0, \ldots, \overline{b}_{n-1})$ then $r \not\perp p$.

2.20 Exercise. Show that a fifth equivalent to the conditions in Corollary 2.18 is: For some (any) $\alpha \geq \omega$ $p^\alpha \perp^w q^\alpha$.

Now we can obtain the promised strong monotonicity result.

2.21 Theorem. *Let $p \in S(C)$. Suppose $p \dashv B$ and $C \downarrow_B D$. Then $p \dashv D$.*

Proof. First, we can assume that p is stationary. For, $p \dashv B$ implies that for each $\overline{d}$ realizing p, $stp(\overline{d}; C) \dashv B$. Thus it suffices to prove the theorem for $stp(\overline{d}; C)$ for an arbitrary $\overline{d}$ realizing p. Without loss of generality, invoking Proposition 2.5, $B \subseteq D$ and $B \subseteq C$. Since orthogonality is respected by parallelism, we may also assume without loss of generality that C and D (though not B) are universes of models. Thus if $q \in S(D)$, q is stationary. Let $E = C \cup D$ and let p', q' be nonforking extensions of p, q respectively to $S(E)$. Then $p^\omega = (p'|C)^\omega = (p')^\omega|C$. We establish $p \perp q$ by showing

$p^\omega \perp^w (q')^\omega|C$ and invoking Lemma 2.15. For this, let I realize $(q')^\omega|C$. Since 'one for all is for free' we may assume I realizes $(q')^\omega$. As for stationary types $\perp^a$ implies $\perp^w$ (Exercises 1.9 and 1.27), it suffices to show that for any J realizing p^ω, $J \downarrow_C I$. But $I \downarrow_D E$ implies by monotonicity that $I \downarrow_D C$. By hypothesis, $C \downarrow_B D$ so transitivity yields $I \downarrow_B C$. Now, $p \dashv B$ implies $p^\omega \dashv B$ so applying the definition of orthogonality to p^ω and $t(I; B)$ we have $I \downarrow_C J$ as required.

The following exercise is an easy application of Theorems 1.19, II.2.18, and 1.40.

2.22 Exercise. Show that for any p there does not exist a sequence of types $\langle p_i : i < \kappa(T)\rangle$ with $p \not\perp p_i$ for each i.

As a corollary to Theorem 2.21 and Exercise 2.22 we obtain a convenient test for $p \dashv B$.

2.23 Corollary. *Suppose $B \subseteq C$, $\operatorname{dom} p = C$ and f is any elementary map with domain C which fixes B pointwise. Suppose further that $stp(C; B) = stp(f(C); B)$ and $C \downarrow_B f(C)$. Then, $p \dashv B$ if and only if $p \perp f(p)$.*

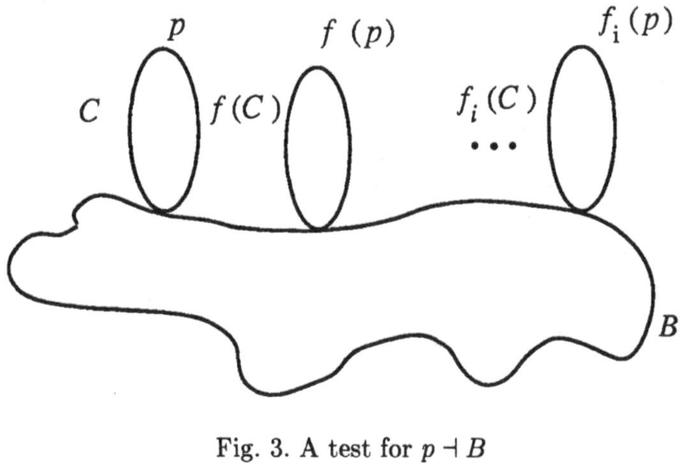

Fig. 3. A test for $p \dashv B$

Proof. (Fig. 3). If $p \dashv B$ then by Theorem 2.21, $p \dashv f(C)$ and, a fortiori, $p \perp f(p)$. Conversely, suppose that $p \not\dashv B$. For $i < |(S(B)| + |T|)^+$, define elementary mappings f_i such that $stp(f_i(C), B) = stp(C, B)$ and $t(f_i(C); \bigcup_{j<i} f_j(C))$ does not fork over B. Without loss, we may assume $f_0 = id|C$ and $f_1 = f$. If $p \perp f(p)$, then the $f_i(p)$ are a sequence of greater than $\kappa(T)$ pairwise orthogonal types. Since $p \not\dashv B$, each $p_i \not\dashv B$. Thus, for some d realizing p, $stp(d; B) \not\perp f_i(p)$ for each i. This contradicts Exercise 2.22 and yields the theorem.

It is essential for this argument that the strong types and not merely the types of B and $f(B)$ are equal. A simple counterexample can be found by considering the theory of an equivalence relation with two infinite classes.

2.24 Exercise. Let E be set of independent sequences and suppose that for each $\bar{e} \in E$ there is a $p_{\bar{e}} \in S(\bar{e})$ with $p_{\bar{e}} \dashv \emptyset$. Let $A_{\bar{e}}$ be a set of independent realizations of $p_{\bar{e}}$. Show $A = \{A_{\bar{e}} : \bar{e} \in E\}$ is an independent set. Show that if $p'_{\bar{e}}$ is the nonforking extension of $p_{\bar{e}}$ to $A_{\bar{e}} \cup \bar{e}$ then i) $p_{\bar{e}} \dashv A - A_{\bar{e}}$ and ii) $p_{\bar{e}} \perp t(A - A_{\bar{e}}; A_{\bar{e}})$.

2.25 Historical Notes. The notion of a type being orthogonal to a set arises in [Shelah 1978]. Our treatment was influenced by that of Lascar and Makkai. The important Corollary 2.18 first appears in [Shelah 1982]. We have tried to avoid arguments involving the construction of an infinite set of independent realizations of a type in favor of arguments directly about the type. They represent a technique which Shelah uses much more often in developing this material and which seems essential for Theorem 2.21. Hrushovski ([Hrushovski 1986], [Hrushovski 198?d]) has proved the remarkable result that in Corollary 2.18 iv) 'each natural number n' can be replaced by '$n \leq 3$'. More precisely, if m is minimal such that for some n $p^n \not\perp^a q^m$ then there is a $k \leq 3$ such that $p^k \not\perp^a q^m$.

3. *The Dominance Order*

In this section we consider a variant on the notion $t(A; B) \dashv C$. Rather than consider this variant as an independence relation, it is more useful to regard it in Definition 3.2 as a partial ordering relation on types over C.

3.1 Definition. Let $C \subseteq B$ and $p \in S(B)$. Then p is *almost orthogonal* to C if for each $q \in S(C)$, $p \perp^a q$. We write $p \dashv^a C$.

The restriction that $C \subseteq B$ can easily be seen to be superfluous. In fact replacing B by $B - C$ leads to another form of this definition which will be used more frequently.

3.2 Definition. i) For any $\bar{a}$, $\bar{b}$, and C we say $\bar{a}$ *dominates* $\bar{b}$ over C and write $(\bar{a} \rhd \bar{b}; C)$ or $\bar{a} \rhd_C \bar{b}$ if for every $\bar{d}$, $\bar{d} \downarrow_C \bar{a}$ implies $\bar{d} \downarrow_C \bar{b}$.

ii) For any p, $q \in S(C)$, we say p *dominates* q over C and we write $(p \rhd q; C)$ or $p \rhd_C q$ or simply $p \rhd q$ if there exists an $\bar{a}$ realizing p and $\bar{b}$ realizing q such that $\bar{a} \rhd_C \bar{b}$.

iii) If $p \in S(A)$ and $q \in S(B)$, we make the standard extension and say p *dominates* q and write $p \rhd q$ if for any nonforking extensions p' and q' to $A \cup B$, $p' \rhd_{A \cup B} q'$.

iv) We say B and C are *bidominant* or *domination equivalent* and write $B \sqsupseteq_A C$ if $B \rhd_A C$ and $C \rhd_A B$. Similarly if $p \rhd q$ and $q \rhd p$ we write $p \sqsupseteq q$.

bidominance

While $p \perp^a q$ means for *each* a realizing p and b realizing q $a \downarrow b$, $p \rhd q$ means for *some* a realizing p and b realizing q $a \rhd b$.

The following exercise shows that dominance is a strong negation to orthogonality. In particular, if $t(a; B \cup c) \dashv B$ and $t(a; B \cup c)$ is not algebraic then $c \triangleright_B a$.

3.3 Exercise. If $\overline{a} \downarrow_A \overline{b}$ and $t(\overline{b}; A)$ is not algebraic show $(\overline{a} \not\triangleright \overline{b}; A)$.

The following is proved by checking through the definitions.

3.4 Lemma. *Let B, $\overline{a}$, and $\overline{c}$ be arbitrary: $t(\overline{a}; B \cup \overline{c}) \dashv^a B$ if and only if $t(\overline{c}; B) \triangleright t(\overline{a}{}^\frown\overline{c}; B)$.*

3.5 Exercise. Show using the properties of independence developed in Section II.2 that we can extend the notion $\overline{a} \triangleright_C \overline{b}$ of domination of sequences to a notion of domination of sets $A \triangleright_C B$.

3.6 Exercise. Show that replacing 'there exists an $\overline{a}$ and a $\overline{b}$' by 'for every $\overline{b}$ there exists an $\overline{a}$' or by 'for every $\overline{a}$ there exists a $\overline{b}$' in Definition 3.2 iii) yields an equivalent definition of p dominates q.

3.7 Exercise. Show that the attempt to strengthen the notion of domination by replacing 'there exists an $\overline{a}$ and a $\overline{b}$' by 'for every $\overline{a}$ and for every $\overline{b}$' yields a relation which is never satisfied, except possibly by algebraic types.

3.8 Exercise. Let T be the theory of an equivalence relation with two infinite classes and let M be a model of T. Show that if a and c are equivalent elements which are not in M then $t(c; M \cup a) \dashv^a M$

3.9 Exercise. Show $p \triangleright p$.

The next lemma shows that dominance determines a preorder on the types over a set B. Moreover, we get a useful strong form of transitivity of dominance with varying base sets. The proofs are entirely routine.

3.10 Lemma. i) *The relation $p \triangleright_B q$ on types in $S(B)$ is reflexive and transitive.*

ii) *If $A \subseteq B \subseteq C$, $B \triangleright_A C$, and $C \triangleright_B D$ then $B \triangleright_A D$. In particular, if $B \triangleright_A C$ and $C \triangleright_A D$ then $B \triangleright_A D$.*

The following exercise provides a useful reformulation of Lemma 3.10 ii).

3.11 Exercise. Show that if $A \subseteq B \subseteq C$, $t(C; B) \dashv^a A$, and $t(D; C) \dashv^a B$ then $t(D; B) \dashv^a A$.

So far the relation of domination is entirely local. We next explore how it behaves under parallelism so we can extend to a relation on global types. The next lemma is a routine consequence of Corollary II.2.11. It provides the independence conditions necessary to show dominance is preserved when 'going up' and when 'going down'.

3.12 Lemma. *Suppose $A_0 \subseteq A$ and $B \downarrow_{A_0} A$.*

i) *$(B \triangleright C; A_0)$ implies $(B \triangleright C; A)$.*
ii) *$(B{}^\frown C \downarrow A; A_0)$ implies if $(B \triangleright C; A)$ then $(B \triangleright C; A_0)$.*

Proof. i) Let $D \downarrow_A B$. By monotonicity, $D^\frown A \downarrow_A B$ and thence by transitivity of independence, $D^\frown A \downarrow_{A_0} B$. The hypothesis yields $D^\frown A \downarrow_{A_0} C$ and monotonicity of independence again gives $D \downarrow_A C$.

ii) Let $D \downarrow_{A_0} B$. We must show $D \downarrow_{A_0} C$. Choose A' realizing $stp(A; A_0)$ and with $(A' \downarrow A \cup B \cup C \cup D; A_0)$. Note $t(A'; A_0 \cup B \cup C) = t(A; A_0 \cup B \cup C)$. Thus $(B \rhd C; A')$. Now by Corollary II.2.10 $D \downarrow_{A'} B$; whence by hypothesis $D \downarrow_{A'} C$ and so by Corollary II.2.10 again $D \downarrow_{A_0} C$.

We easily deduce the following corollary.

3.13 Corollary. *Let p', $q' \in S(B)$ be nonforking extensions of types p, q which are in $S(A)$. If $p \rhd q$ then $p' \rhd q'$.*

However, we can not conclude from $p' \rhd q'$ that $p \rhd q$. Consider Example 1.12. Using the notation from that example but renaming the types, let $p = t(a; M)$, $q = t(b; M)$, $p' = t(a; M \cup b')$, and $q' = t(b; M \cup b')$. Check that p' (q') is a nonforking extension of p (q). But $(p \not\rhd q; M)$ while $(p' \rhd q'; M \cup b')$ since b is in the prime model over $M \cup \{a, b'\}$. The last assertion follows from X.1.21.

Thus, in order to obtain a parallelism invariant definition we must modify the definition of domination slightly. Since domination is a strong negation to orthogonality and the orthogonality of two types in determined by considering all (appropriate) extensions, it suffices here to find one extension witnessing domination.

3.14 Definition. For $p \in S(A)$ and $q \in S(B)$ we say p *eventually dominates* q and write $p \rhd^e q$ if for some $C \supseteq A \cup B$ and some p', q', nonforking extensions of p and q to $S(C)$, $p' \rhd q'$.

In the natural way we obtain the notion of *eventual domination equivalence* of p and q, written $p \trianglerighteq\!\trianglelefteq^e q$.

Clearly $\rhd^e$ is a property of global types and is preserved by parallelism of stationary types. Thus we have established a preorder on the global types. This notion will be exploited extensively later in the book.

The following exercise will be a small but important step in some proofs in Part D.

3.15 Exercise. Let for $i = 1, 2$, $X_i \subseteq M_i$, and suppose $X_i \rhd_{M_0} M_i$ and $M_1 \downarrow_{M_0} M_2$. Show $X_1 \cup X_2 \rhd_{M_0} M_1 \cup M_2$.

3.16 Historical Notes. The treatment of domination presented here follows that of [Makkai 1984]. His account was largely based on [Lascar 1984] and [Lascar 1982a].

Chapter VII

Rank

Stability theory arose from Morley's investigation of $\aleph_1$-categorical theories, where he introduced the notion of the rank of a type in an ω-stable theory. From this beginning, Shelah and others developed a number of different rank notions to investigate more complicated (e.g. stable and superstable) theories. In this chapter we will investigate these various rank functions and their relationships to nonforking. It will become clear that our entire machine could have been constructed on the basis of rank. There are two reasons that we did not do this. In the first place rank is actually a finer tool than nonforking. We will show that for a rank function R satisfying certain axioms (in a superstable or ω-stable theory), p is a nonforking extension of q exactly if $R(p) = R(q)$. However, the rank codes additional information that is lost by the dichotomy 'forks or doesn't fork.' That is, if p is a forking extension of q we can ask how much less is $R(p)$ than $R(q)$. This additional information plays an important role in some branches of stability theory (e.g. [Cherlin, Harrington, & Lachlan 1985]). But, in the basic study of free extensions the additional information obscures the issues. More importantly, no equally simple definition of nonforking in terms of rank holds for stable theories. Rank also provides more information because it allows one to compare two types neither of which extends the other.

In Section 1 we exhibit some axioms for rank and show that any rank which satisfies these axioms induces the nonforking relation. In Section 2 we describe a number of the important rank functions. In Section 3 we prove some theorems illustrating the greater power of the rank notion.

1. *Ranks and Forking*

We describe here the properties that the notion of free extension satisfies when we define $p\mathcal{F}q$ if $q \subseteq p$ and $R(p) = R(q)$, where R satisfies the axioms given below. We show that these properties of the resulting freeness relation completely characterize the nonforking relation on stable theories.

1.1 Axioms For Rank. Let R be a map which associates with each p in $S(A)$ (for A a subset of $\mathcal{M}$) an ordinal $R(p)$ which is called its *rank*.

i) If f is an elementary map then $R(p) = R(fp)$.
ii) If $p \subseteq q$ then $R(q) \leq R(p)$.
iii) For each B containing $\operatorname{dom} p$ there is at least one $q \in S(B)$ with $R(q) = R(p)$.
iv) If $p \in S(A)$ there is a finite subset $A_0 \subseteq \operatorname{dom} p$ with $R(p) = R(p|A_0)$.
v) There is a cardinal λ such that any type p has at most λ mutually contradictory extensions with the same rank as p.

1.2 Notation. If R is a rank satisfying the axioms of 1.1, we denote by $\mathcal{F}_R$ the notion of freeness given by: p is a free extension of q if p extends q and $R(p) = R(q)$.

$\mathcal{F}_R$ is defined on complete types but we extend it to incomplete types by saying q is free over A if for some complete q' extending q, $q' \mathcal{F}_R q'|A$. It is now easy to verify the following lemma.

1.3 Lemma. *If R satisfies the axioms in 1.1 then $\mathcal{F}_R$ is a freeness relation obeying the isomorphism convention (II.1.2), the first monotonicity axiom (II.1.5 M_1), the extension axiom (II.1.14 E_1), and the full local character of freeness axiom (II.1.21). Moreover, there is a cardinal λ such that no p has more than λ mutually contradictory free extensions.*

Note that the condition on the number of free extensions is implied by the axiom asserting that types over models are stationary (in the presence of the second monotonicity axiom).

Now we show that the only relation satisfying these axioms is the relation of nonforking for stable theories. We have shown that nonforking satisfies these axioms (and others) in Sections III.3 and III.4.

1.4 Theorem. *If $\mathcal{F}$ satisfies the properties listed in Lemma 1.3 then T is superstable and $\mathcal{F}$ is the nonforking relation.*

Proof. Clearly the bound on the number of extensions in rank and the local character of freeness imply that T is stable in μ for all $\mu \geq \lambda + |T|$ (by counting the types over A). Now suppose $q \mathcal{F} p$ with $p \in S(A)$. Extend q to a global type $\hat{q}$ with $\hat{q} \mathcal{F} p$. Since all conjugates over A of $\hat{q}$ are free over p there at most λ of them. By Lemma IV.1.14, $\hat{q}$ and, a fortiori, q do not fork over A.

Conversely, suppose q is a nonforking extension of $p \in S(A)$. Extend q to $\hat{q}$ a global type which does not fork over A. Also extend p to $\hat{p}$ a global type with $\hat{p} \mathcal{F} p$. By the first part of the proof, $\hat{p}$ does not fork over A. Thus by the conjugacy lemma $\hat{p}$ and $\hat{q}$ are conjugate over A. Thus $\hat{q} \mathcal{F} p$ and by monotonicity $q \mathcal{F} A$.

Note that the symmetry and transitivity properties now follow from the axioms listed in Lemma 1.3 since we know they hold of the only relation which satisfies those axioms.

There is no ordinal valued rank function which is defined on all types exactly if T is a stable theory. However, a slight relaxation of the conditions in Lemma 1.3 describes the properties of a freeness relation which imply that the underlying theory is stable and the relation is nonforking.

1.5 Exercise. Show that if iv) of 1.1 is weakened by requiring the cardinality of the set A to be less than $|T|$, in Theorem 1.4 we can still conclude that T is stable and $\mathcal{F}$ is the nonforking relation. (rng R is a linear order.)

1.6 Historical Notes. The first axiomatic approaches to rank were in [Baldwin & Blass 1974] and [Lascar 1976]. Theorem 1.4 is derived from [Lascar 1976]. Another useful account is [Harnik & Harrington 1984].

2. A Plethora of Ranks

This section is devoted to the definitions of a number of rank functions and the verification that they satisfy the axioms discussed in Section VII.1. There are a number of minor variants of the definitions but we forgo the tedious discussion of equivalences among these variants. For details see [Shelah 1978].

The most important rank functions are Morley rank, U-rank, and continuous rank (Shelah-degree). The Morley rank of each type in a countable theory is defined if and only if the theory is ω-stable. The U-rank and continuous rank are defined for all complete types exactly if T is superstable. We describe below the distinctions among these three ranks. In short, continuous ranks and Morley ranks are defined on formulas; the rank can be extended to complete types. U-rank is defined only on complete types but it enjoys an additivity property which facilitates calculating the rank of the type of a pair of elements. One reason for the more satisfactory analyses of theories which are categorical in some infinite power is that on such theories Morley rank and U-rank coincide.

Most of the ranks we discuss can be defined either directly on types or on formulas. We will define the rank on formulas when possible and then extend to types by setting $R(p) = \inf\{R(\phi) : \phi \in p\}$. This automatically guarantees that each type has a finite subtype of the same rank, thus establishing the local character of dependence for the associated freeness relation.

Morley originally defined rank for a complete type over a set by an extension of the Cantor-Bendixson rank of the Stone space. In fact, if M is ω-saturated and $p \in S(M)$ then $R_M(p)$ is exactly the Cantor-Bendixson rank of p in $S(M)$. However, we prefer to define the rank directly on formulas. We give first a definition which specializes to both the original Morley rank and to the important local version considered in Section III.1. We require some further notation.

2.1 Notation. Let Δ be a collection of formulas of the form $\phi(\overline{x};\overline{y})$. We denote by $C(\Delta)$ the collection of finite conjunctions of substitution instances $\phi(\overline{x},\overline{a})$ of formulas in Δ.

The two most important uses of this notation occur when Δ is the collection of all n-ary formulas for some n and when $\Delta = \{\phi(\overline{x};\overline{y}), \neg\phi(\overline{x};\overline{y})\}$ for some single formula ϕ. By Exercise III.1.3 this second case is the same as considering an arbitrary finite set of formulas (instead of just one).

2.2 Definition ((Δ,μ)-Rank). Let $\phi(\overline{x};\overline{a})$ be a formula, μ a cardinal, Δ a set of formulas and α an ordinal. The *(Δ,μ)-rank*, R_Δ, of a formula $\phi(\overline{x};\overline{a})$ is defined as follows.

- $R_\Delta(\phi(\overline{x};\overline{a})) = -1$ iff $\models \neg(\exists\overline{x})\phi(\overline{x};\overline{a})$.
- $R_\Delta(\phi(\overline{x};\overline{a})) \geq 0$ iff $\models (\exists\overline{x})\phi(\overline{x};\overline{a})$
- $R_\Delta(\phi(\overline{x};\overline{a}) \geq \alpha + 1$ iff for each $k < \mu$, $*_{(\alpha,k)}$ holds:

  There exists a sequence of formulas $\theta_i(\overline{x};\overline{a}_i)$ for $i \leq k$ with $\theta_i(\overline{x};\overline{a}_i) \in C(\Delta)$ such that $R_\Delta(\phi(\overline{x};\overline{a}) \wedge \theta_i(\overline{x};\overline{a}_i)) \geq \alpha$ and where the formulas $\{\phi(\overline{x};\overline{a}) \wedge \theta_i(\overline{x};\overline{a}_i) : i < k\}$ are pairwise contradictory. $\quad *_{(\alpha,k)}$

- $R_\Delta(\phi(\overline{x};\overline{a})) \geq \delta$ where δ is a limit ordinal iff $R_\Delta(\phi(\overline{x};\overline{a})) \geq \beta$ for each $\beta < \delta$.
- $R_\Delta(\phi(\overline{x};\overline{a})) = \gamma$ if γ is the least ordinal β with $R_\Delta(\phi(\overline{x};\overline{a})) \not\geq \beta$.

Note that if $R_\Delta(\phi(\overline{x};\overline{a})) = \alpha$ then there is a least $m < \mu$ so that there is no sequence of formulas $\theta_i(\overline{x};\overline{a}_i)$ for $i < m+1$ satisfying $^*(\alpha, m+1)$. Such an m is called the *degree* of $\phi(\overline{x};\overline{a})$ and denoted $D_\Delta(\phi(\overline{x};\overline{a}))$. This notion is denoted $R^n(p,\Delta,\mu)$ in [Shelah 1978], where the n indicates that only formulas with n free variables are ranked. We have suppressed the explicit mention of μ assuming that it will be clear in context.

2.3 Definition. i) (Morley rank) If Δ is the collection of all formulas and $\mu = \omega$ we call the resulting rank *Morley rank* and denote it by $R_M(\phi(\overline{x};\overline{a}))$.

ii) (local rank) If Δ is a finite set of formulas and μ is 2, R_Δ is the local rank R studied in Section III.1. We often code Δ by a single formula ϕ and speak of ϕ-rank.

iii) If Δ is a finite set of formulas and μ is ω, R_Δ is called simply (Δ,ω) rank.

In this chapter for finite Δ, R_Δ denotes local rank unless we explicitly say otherwise.

It is now straightforward to verify the following result by induction.

2.4 Lemma. *Let T be stable. For any finite Δ and any p, $R_\Delta(p)$ is defined. If T is ω-stable then $R_M(p)$ is defined for any p. If μ is infinite and condition v) is restricted to Δ-types then each R_Δ satisfies conditions i) through v) of Axiom 1.1.*

The next exercise asks for an example showing the necessity of assuming μ is infinite to satisfy condition iii). Exercise 2.6 illustrates another aspect of the same point.

2.5 Exercise. Show that condition iii) may fail for local rank.

2.6 Exercise. Show that $R_M(\phi \vee \psi) = \max(R_M(\phi), R_M(\psi))$.

2.7 Exercise. Compute $R_M(x{=}x)$ for the three theories REF_ω, CEF_ω, and ACF_0.

Exercise 2.8 provides a naive illustration of the difficulty in satisfying Axiom 1.1 v) for all types with a Δ-rank for a Δ which does not contain all formulas.

2.8 Exercise. Show that if T is the theory of a single equivalence relation with infinitely many infinite classes, p is the unique 1-type over the empty set, and Δ is $\{x = y\}$ then p has unboundedly many incompatible extensions q_i with $R_\Delta(q_i) = R_\Delta(p)$.

The following definition and theorem provide a rank for stable theories. Note however that the values of the rank function are sequences of integers rather than ordinals.

2.9 Notation. If $p \subseteq q$, q is a free extension of p for finite sets of formulas (written $q\mathcal{F}_f p$) if for every finite Δ, $R_\Delta(q) = R_\Delta(p)$.

The proof of the following theorem approximates that of Theorem 1.4. We must replace Axiom 1.1 iv) by the observation that if we define

$$R_f(p) = \langle R_\Delta(p) : \Delta \text{ a finite subset of } F(L)\rangle$$

then every type has a subtype with the same rank which is over a set with cardinality at most $|T|$.

2.10 Theorem. *Let T be a stable theory. If $p \subseteq q$ then q does not fork over* $\operatorname{dom} p$ *iff $q\mathcal{F}_f p$.*

Proof. By Lemma IV.1.14 it suffices to show that for some λ, p has at most λ extensions q_i with $q_i\mathcal{F}_f p$. Since T is stable, for each finite Δ there are only finitely many Δ-types of extensions with the same Δ-rank. Thus there are at most $2^{|T|} \times |A|^{|T|}$ extensions of p to complete types with the same rank.

Now we consider a measure of exactly how much a type can be made to fork. This rank was introduced by Lascar who originally defined it as the least connected (see below) rank function. It differs significantly from the other ranks we discuss because although every type has a subtype over a finite set with the same rank, there may not be a single formula with the same rank. Accordingly, we define this notion for complete types over some subset of the monster model.

2.11 Definition. For $p \in S(A)$ we define the *U-rank* (also called the Lascar rank) of a type p, $U(p)$, by induction.

- $U(p) \geq 0$ iff p is consistent.
- $U(p) \geq \beta + 1$ iff there exists B containing $\operatorname{dom} p$ and $q \in S(B)$ such that $p \subseteq q$ and $U(q) \geq \beta$ and q forks over A.
- $U(p) \geq \delta$ for a limit ordinal δ iff $U(p) \geq \beta$ for each $\beta < \delta$.
- $U(p) = \alpha$ if α is the least ordinal γ such that $U(p) \not\geq \gamma$.

2.12 Exercise. Show that if T is superstable the U-rank of each type is defined.

The following notions distinguish Morley and Lascar rank. We say an ordinal-valued rank function is connected if the range of its restriction to complete types forms a connected subset of the ordinals under the order topology. Thus, we have

2.13 Definition. i) The rank function R is *connected* if for every type p with $R(p) > \alpha$, there is a complete type, p', extending p with $R(p') = \alpha$.

ii) The rank function R is *continuous* if for every set A and ordinal α, $\{p \in S(A) : R(p) \geq \alpha\}$ is a closed subset of $S(A)$.

2.14 Exercise. i) Show Morley rank is continuous and U-rank is connected.

ii) Show that Morley rank may not be connected and U-rank may not be continuous. (Hint: Example 2.17 is germane.)

Now we define another notion of rank in a completely different way from U-rank. These ranks share the property that if T is stable they are defined for all types precisely when T is superstable. The next notion was called *infinity rank* in [Shelah 1978] and *degree* (not to be confused with the Morley degree) when Shelah introduced it in [Shelah 1971]. The crucial fact about this continuous rank is that it assigns a rank to formulas rather than to types. The advantage of U-rank is the additivity properties discussed below.

2.15 Definition. The *continuous rank*, R_C, of a formula $\phi(\overline{x};\overline{a})$ is defined by induction as follows.

- $R_C(\phi(\overline{x};\overline{a})) \geq 0$ iff $(\exists\overline{x})\phi(\overline{x};\overline{a})$.
- $R_C(\phi(\overline{x};\overline{a})) \geq \beta + 1$ iff there exists a formula $\psi(\overline{x};\overline{y})$ and sequences $\overline{c}_i$ for $i < (2^{|T|})^+$ such that $R_C(\phi(\overline{x};\overline{a}) \wedge \psi(\overline{x};\overline{c}_i)) \geq \beta$ for each i and the set of formulas $\{\psi(\overline{x};\overline{c}_i)\}$ is n-inconsistent for some n.
- $R_C(\phi(\overline{x};\overline{a})) \geq \delta$ for a limit ordinal δ just if $R_C(\phi(\overline{x};\overline{a})) \geq \beta$ for each $\beta < \delta$.

$R_C(\phi(\overline{x};\overline{a})) = \gamma$ if γ is the least ordinal β such that $R_C(\phi(\overline{x};\overline{a})) \not\geq \beta$.

Shelah actually makes the bound in ii) to be $|T|^+$. The present definition yields the same result with fewer combinatorial difficulties. The delicate relation between various versions of this notion is discussed in Chapter II of [Shelah 1978]. The next theorem indicates the relative value of R_C and U. In one direction we are restricted to finite ordinals.

2.16 Theorem. *For each formula $\phi(\overline{x};\overline{y})$, each $\overline{a}$, and each natural number n, $R_C(\phi(\overline{x};\overline{a})) \geq n$ iff there exists a $p \in S(\overline{a})$ with $\phi(\overline{x};\overline{a}) \in p$ and $U(p) \geq n$. For any ordinal α, if there exists such a p with $\phi(\overline{x};\overline{a}) \in p$ and $U(p) \geq \alpha$ then $R_C(\phi(\overline{x};\overline{a})) \geq \alpha$.*

Proof. Suppose $R_C(\phi(\overline{x};\overline{a})) \geq n+1$. Then there are a formula $\psi(\overline{x};\overline{y})$ and sequences $\overline{c}_i$ for $i < (2^{|T|})^+$ such that $R_C(\phi(\overline{x};\overline{a}) \wedge \psi(\overline{x};\overline{c}_i)) \geq n$ and with $\{\psi(\overline{x};\overline{c}_i) : i < (2^{|T|})^+\}$ n-inconsistent. By the pigeonhole principle there is an infinite subset of the $\overline{c}_i$ which realize the same type over $\overline{a}$. That is, for some $\overline{c}_i$ (wlog $\overline{c}_0$)$\psi(\overline{x};\overline{c}_0)$ divides over $\overline{a}$. But then if q is any completion of $\{\phi(\overline{x},\overline{a}) \wedge \psi(\overline{x};\overline{c}_0)\}$ to a complete type over $\overline{a} \cup \overline{c}_0$, q forks over $\overline{a}$ and by induction $U(q) \geq n$. Thus $q|\overline{a}$ is the required member of $S(\overline{a})$ with $U(q|\overline{a}) \geq n+1$.

Since we have shown in Lemma V.3.9 that a complete type which forks contains a formula which divides the converse is easy.

This lemma is best possible in a sense which illustrates the difference between U being based on complete types and R_C being based on formulas.

2.17 Example. For any $\alpha < \aleph_1$ there is a theory T_α with $R_C(x = x) = \alpha$ but for every p, $U(p)$ is finite. We construct the T_α by induction on α. If α is finite there is nothing to prove. If we have T_α, let $T_{\alpha+1}$ be the theory of an equivalence relation with infinitely many infinite classes, each of which is a model of T_α. If δ is a limit ordinal let T_δ be the theory of the disjoint union of models of the T_α for $\alpha < \delta$.

The next theorem describes two important properties of continuous rank which we will exploit in the next section.

2.18 Theorem. i) *R_C is the least continuous rank function.*

ii) *If $[p] \geq [q]$ (in the fundamental order) then $R_C(p) \geq R_C(q)$.*

Theorem 2.18 i) is proved by a routine induction; Theorem 2.18 ii) follows directly from the definition (using Section V.3).

2.19 Exercise. Show that if R is a continuous rank function and T is superstable (in particular, if $R = R_C$) then there is a complete type with maximal R-rank.

As we have seen on many occasions one of the major problems in this subject is to relate the behavior of singletons to the behavior of pairs. This problem arises again for rank. The behavior of Morley rank in this situation is extremely complicated ([Lachlan 1980]). Lascar rank is better behaved.

2.20 Theorem. *For any $\overline{b},\overline{c}$ in a model of a superstable theory*

$$U(t(\overline{c};A\cup\overline{b}) + U(t(\overline{b};A)) \leq U(t(\overline{b}^\frown\overline{c};A)) \leq U(t(\overline{c};A\cup\overline{b})) \oplus U(t(\overline{b};A)).$$

Here $\oplus$ denotes the natural sum of ordinals. That is, to add two ordinals put them in Cantor normal form and add them as polynomials in ω. Note that natural sum agrees with ordinary arithmetic on finite ordinals. If $\overline{b} \downarrow_A \overline{c}$

then the second inequality becomes an equality. For the proof of this theorem see [Lascar 1976]. This inequality has been extensively exploited. See, for example, [Berline & Lascar 1986] and [Cherlin, Harrington, & Lachlan 1985].

2.21 Historical Notes. For further information see [Lascar 1976], [Lachlan 1980], [Shelah 1978], or [Pillay 1983a]. Pillay's book has a more detailed introductory survey which includes a fuller proof of Theorem 2.16. Morley rank was introduced in [Morley 1965]. If T is categorical in either $\aleph_1$ or $\aleph_0$, Morley rank and U-rank agree [Lascar 1976]. Moreover, in such theories the rank of every type is finite. For $\aleph_1$-categorical theories this was first proved in [Baldwin 1973]. Later, more accessible, proofs appear in [Poizat 1978] and [Zilber 1974]. The proof for $\aleph_0$-categorical theories is in [Cherlin, Harrington, & Lachlan 1985]. In fact, if T is unidimensional then R_C and U are equal and finite [Saffe 1984].

3. Ranks and Stable Groups

The major goal of this section is to prove the theorem of Cherlin, Shelah, and Macintyre that a superstable field is algebraically closed. The proof given here rests ultimately on the ability to compare the ranks of two types over the same set. Thus, we exploit one of the advantages of rank over forking which we mentioned at the beginning of this chapter. We will somewhat disguise this fact by expounding Poizat's notion of the stratified order which provides a nice framework for formulating the result.

The proof of the main theorem requires three major steps. We already know from Section III.5 that the additive group of a superstable division ring is connected. The model theoretic task in this section will be to show that this implies that the multiplicative group of the division ring is connected. Knowing that both groups are connected, an algebraic proof due to Macintyre yields the theorem.

The fundamental insight is that the action of a group on itself by (left) multiplication preserves rank. To make this idea precise we define the translate and the inverse of a type. As in Section III.5, by a 'group' we mean any structure which includes a group operation among its definable operations.

3.1 Definition. Let M be a group, $a \in M$, and let $c \in \mathcal{M}$ realize the type $p \in S(M)$. Then ap, the *left translate of p by a* is $t(ac; M)$. That is, $ap = \{\phi(a^{-1}x; \overline{m}) : \phi(x; \overline{m}) \in p\}$. Similarly, $p^{-1} = t(c^{-1}; M)$.

Clearly, the operation of left translation by elements of M induces an equivalence relation on $S^1(M)$ which preserves any of the rank functions we have discussed (except the local ranks). But the relation is too fine; we also want to identify types which are only potentially translates of each other. The following notion captures this idea.

3.2 Definition. If $M, N \models T, p \in S^1(M), q \in S^1(N)$ then the *stratum* of p is at least that of q, written $s(p) \geq s(q)$, if for every formula of the form $\phi(yx; \overline{z})$, whenever ϕ is represented in $p(x)$ then ϕ is represented in $q(x)$. That is, if for some a and $\overline{b}$, $\phi(ax; \overline{b}) \in p$ then for some a' and $\overline{b}'$, $\phi(a'x; \overline{b}') \in q$.

The partial order which results by identifying types p and q if $s(p) \geq s(q)$ and $s(q) \geq s(p)$ is called the *stratified order.* We call the equivalence class of p under this relation the *stratum* of p.

3.3 Exercise. Show that if $q \leq p$ in the sense of the fundamental order then $s(q) \leq s(p)$. Show the converse fails.

3.4 Exercise. Show that if $\phi(x; \overline{y})$ is represented in p and $s(p) \geq s(q)$ then $\phi(zx; \overline{y})$ is represented in q. (Remember $\phi(x; \overline{y})$ is $\phi(1x; \overline{y})$.) In particular, all translates of a type are in the same stratum.

3.5 Exercise. Show that there are at most $2^{|T|}$ elements in the stratified order.

3.6 Theorem. *Suppose* $p \in S^1(M), q \in S^1(N)$ *and* $s(p) \geq s(q)$. *Then there is an* $\alpha \in \mathrm{Aut}(\mathcal{M})$ *and an* $a \in \mathcal{M}$ *such that* $p \subseteq a\alpha(q^{\mathcal{M}})$.

Proof. Extend L by adding a unary predicate P, names for the elements of M and new constants a and c. We first show that there is a type q_1 such that $q_1 \simeq q$ and $p \geq aq_1$. Let Γ be the following set of sentences:

$$\begin{aligned} & T \cup T|P \cup \{P(a)\} \\ & \quad \cup \{(\exists \overline{y})[P(\overline{y}) \wedge \phi(ac; \overline{y})] : \phi(x; \overline{y}) \text{ is represented in } p\} \\ & \quad \cup \{(\exists \overline{y})[P(\overline{y}) \wedge \phi(c; \overline{y})] : \phi(x; \overline{y}) \text{ is represented in } q\} \\ & \quad \cup \{(\forall \overline{y})[P(\overline{y}) \rightarrow \neg\phi(c; \overline{y})] : \text{if } \phi(x; \overline{y}) \text{ is not represented in } q\}. \end{aligned}$$

Note that if $\phi(x; \overline{y})$ is represented in p then $\phi(zx; \overline{w})$ is represented in q. With this in mind it is easy to see that any finite subset of Γ is satisfiable by interpreting P as the universe of N, a as an appropriate element of N, and c as a realization of q. Then the required q_1 is $t(c; P(M_1))$ if $M_1 \models \Gamma$.

A similar compactness argument shows that for any r and p, if $p \geq r$ there is an extension r_1 of p with $r_1 \simeq r$. Apply this argument to p and aq_1 to get a model $N_1 \supseteq M$ and $r_2 \in S(N_1)$ such that $r_2 \simeq aq_1$ and $p \subseteq r_2$. Let c realize $a^{-1}r_2$ and let $q_2 = t(c; N_1)$. Then $q_2 \simeq q$.

Let $\hat{p}$ be a nonforking extension of p to $\mathcal{M}$, $\hat{q}$ a nonforking extension of q to $\mathcal{M}$, and $\hat{q}_2$ a nonforking extension of q_2 to $\mathcal{M}$. Then $\hat{q} \simeq \hat{q}_2$ so by Theorem III.2.36 there is an automorphism α of $\mathcal{M}$ with $\alpha(\hat{q}) = \hat{q}_2$ and so $p \subseteq a\alpha(q^{\mathcal{M}})$ as required.

The following exercise restates the omitted compactness argument in the last proof.

3.7 Exercise. Show that $p \geq q$ implies that p is contained in a conjugate of a nonforking extension of q.

We deduce the following crucial property from this result.

3.8 Corollary. *Suppose $s(p) = s(q)$; then if R is any one of R_M, U, or R_C, $R(p) = R(q)$. Moreover, $R(q) = R(q^{-1})$.*

Proof. All of these ranks are preserved by left translation.

Note that this result does not hold for all local ranks because ap_ϕ is not always a ϕ-type. Our aim now is to identify types of maximal stratum with types of maximal rank. This will have a number of useful corollaries, the first being the existence of types with maximal U-rank. To save space we give the most direct proofs for superstable theories. Some of the results hold for the stable case.

3.9 Lemma. *Let G be a group and suppose $T = \mathrm{Th}(G)$ is superstable. There is exactly one maximal stratum.*

Proof. Fix a strongly ω-saturated model of T. Since T is superstable, there is a $p \in S_1(M)$ with maximal continuous rank. As $s(q) > s(p)$ implies $q > p$ (in the fundamental order), which implies $R_C(q) > R_C(p)$ by Corollary 2.18, we have that the stratum of p is maximal. We want to show that if $R_C(q) = R_C(p)$ then $s(q) = s(p)$. We first show that if $R_C(q) = R_C(p)$ then $s(q^{-1}) = s(p)$. Applying this observation with $p = q$ yields $s(q) = s(q^{-1})$ and then applying it to p and q^{-1} yields the theorem.

To see that if $R_C(q) = R_C(p)$ then $s(q^{-1}) = s(p)$, let a realize p and b realize q with $a \downarrow_M b$. We first claim that $a \downarrow_M ab$. For this, choose $N \supset M \cup a$ with $b \downarrow_{M \cup a} N$. Now $R_C(t(ab; N)) = R_C(t(b; N))$ since ab is definable from $N \cup b$. But $R_C(t(b; N))$ is maximal so $ab \downarrow_M N$. In particular, $ab \downarrow_M a$ as claimed. This allows us to choose $M' \supset M \cup ab$ with $M' \downarrow_M a$. Denote $t(a; M')$ by p_1. Now, $b^{-1}a^{-1} \in M'$ and $b^{-1}a^{-1}p_1 = t(b^{-1}; M')$. Thus $s(t(b^{-1}; M')) = s(p_1) = s(p)$ which is maximal. But $q^{-1} \subseteq t(b^{-1}; M')$ implies $s(t(b^{-1}; M')) \leq s(q^{-1})$ implies $s(q^{-1}) = s(p)$ and we finish.

By emulating the proof of the existence of maximal elements in the fundamental order, Lemma III.3.11, and then refining the previous argument one can prove the following stronger result of Poizat.

3.10 Exercise. Show that if T is the theory of a stable group then there is a unique maximal stratum.

3.11 Exercise. Show that if T is the theory of a superstable group then the types in the maximal stratum do not fork over the empty set.

3.12 Corollary. *Let G be a group and suppose $T = \mathrm{Th}(G)$ is superstable. There are types of maximal U-rank.*

Proof. The types in the maximal stratum must have maximal U-rank. For, $s(p) \leq s(q)$ implies $U(p) \leq U(q)$.

Our next step is to link the connectivity property discussed in Section III.5 with the number of types of maximal stratum. The following exercise leads to the easy half of the connection.

3.13 Exercise. Show that if $\phi(x; \overline{m})$ defines a subgroup of M with finite index then $\phi(zx; \overline{y})$ is represented in every type over M.

3.14 Lemma. *If M contains a proper definable subgroup of finite index then there is more than one type in the maximal stratum.*

Proof. Let $\phi(x;\overline{m})$ define a proper subgroup H with finite index in M. Now, if e denotes the identity of the group and $a \in M - H$, $\phi(ax;\overline{m})$ and $\phi(ex;\overline{m})$ occur in distinct types of maximal stratum.

To show the converse of this lemma we require a little more notation.

3.15 Definition. i) Let $M \models T$ where T is a theory of groups and let p be a type (not necessarily complete) over M. The *stabilizer of* p is $F_p = \{a \in M : ap = p\}$.

ii) For each formula $\phi(x;\overline{y})$, let $\hat{\phi}(x,z,\overline{y})$ denote $\phi(zx;\overline{y})$.

3.16 Lemma. *For any $p \in S(M)$ and any formula ϕ, $F_{p_{\hat{\phi}}}$ is a definable subgroup of M. If p has maximal stratum then $[M : F_{p_{\hat{\phi}}}] < \omega$.*

Proof. The required definition is:

$$\chi(u) = (\forall\overline{y})(\forall z)[d\hat{\phi}(z;\overline{y}) \leftrightarrow d\hat{\phi}(uz;\overline{y})].$$

Now, $[M : F_{p_{\hat{\phi}}}] = |\{ap_{\hat{\phi}} : a \in M\}|$. But since p and thus ap is in a maximal strata, each $ap_{\hat{\phi}}$ extends to a global type which does not fork over the empty set. Each of the types is definable over every model of T. Thus, the number of translates of p is bounded by $2^{|T|}$. But then by compactness the number of translates of $p_{\hat{\phi}}$ must be finite.

3.17 Corollary. *If M is connected there is a unique type of maximal stratum.*

Proof. By Lemma 3.16, if p has maximal stratum, then for each ϕ, $F_{p_{\hat{\phi}}} = M$. Thus all types of maximal stratum are equal (since we can extend to a model where they are translates of each other).

3.18 Exercise. Show that if $p \in S(M)$, then F_p is defined by an infinite family of formulas. (That is, by an intersection of infinitely many definable sets.)

The concept of an infinitely definable subgroup has been exploited extensively ([Berline & Lascar 1986] and [Hrushovski 1986]).

3.19 Theorem. *If T is the theory of a superstable structure which admits two group structures then one is connected if and only if the other is connected.*

Proof. We have shown that the group is connected if and only if there is only one type in the associated maximal stratum. But by Corollary 3.17 the types in the maximal stratum for the two operations are the same, namely, the types of maximal U-rank.

We showed in III.5 that the additive group of a superstable division ring is connected. Now we can conclude:

3.20 Corollary. *If M is a superstable division ring, both the additive and multiplicative groups of M are connected.*

We turn now to the algebraic information necessary to conclude the proof of the main theorem. The following theorem can be found, for example, in [Lang 1971] or Chapter 23 of [Adamson 1982].

3.21 Theorem. *Let K be an extension of prime degree, q, of the field F such that F is the fixed field of the automorphism group of K and $x^q - 1$ splits in K. Let p be the characteristic of K.*

i) *If $q = p$ then K is generated over F by the solution of an equation $x^p - x = a$ for some $a \in F$.*
ii) *If $q \neq p$ then K is generated over F by the solution of an equation $x^q = a$ for some $a \in F$.*

In the first case K is called an *Artin-Schreier extension* of F and in the second a *Kummer* extension. Recall that a field F with characteristic $p \neq 0$ is *perfect* if every element of F has a pth root in F and a *Galois extension* of a field is an algebraic extension which is both normal and separable.

3.22 Lemma. *If F is a superstable field then F is perfect and has no Artin-Schreier or Kummer extensions.*

Proof. Let $h(x)$ be either of the following two maps.

i) $x \mapsto x^p - x$ (if the characteristic of F is $p > 0$).
ii) $x \mapsto x^n$ for $x \neq 0$ an element of F where $n \geq 1$ is an arbitrary natural number.

Now, h is a definable endomorphism of the additive group of F in the first case, and of the multiplicative group of F in the second. In each case the kernel of h is finite. Since we have just shown that both groups are connected, Theorem III.5.25 shows that in each case h is surjective. The surjectivity of the first map implies that F has no Artin-Schreier extension and the surjectivity of the second that F is perfect and has no Kummer extension. This yields the lemma.

3.23 Theorem. *If F is a superstable field then F is algebraically closed or finite.*

Proof. Assume for contradiction that F is an infinite superstable field which is not algebraically closed. By the previous lemma, F is perfect and so it has a Galois extension of some finite degree n. Consider all pairs of fields (F, K) such that (*) F is an infinite superstable field and K is a proper finite dimensional Galois extension of F. Choose such a pair with the degree, q, of K over F minimal.

We claim q is prime and $x^q - 1$ splits over F. If q is not prime, choose an r which divides q and let F_1 be the fixed field of an element of order r in $\mathrm{Gal}(K/F)$. Since F_1 is a finite dimensional extension of F, F_1 is also superstable and so the pair (F_1, K) contradicts the minimal choice of q.

Thus q is prime. Now, let K_1 be the splitting extension of $x^q - 1$ over F. Then the degree of K_1 over F divides $q - 1$, so by the minimality of q, $K_1 = K$, as required.

Now Lemmas 3.21 and 3.22 yield a contradiction unless $q = 1$ and we finish.

Theorem 3.23 has been strengthened by Cherlin to remove the hypothesis that F is commutative. We will not give the full argument for this result here. The basic theme of the proof is to show that the division ring D is finite dimensional over a subfield, F. One algebraic result and an application of Theorem 3.23 then yield the result. Many of the basic ideas in finding a large Abelian subgroup of a stable group are contained in the following theorem of Reineke.

We rely in the next proof on two standard group theoretic facts. If all elements of a group have order two then it is Abelian. There is a 1-1 correspondence between the conjugates of an element a and the cosets of $C_G(a)$. We write $[a, b]$ for the commutator, $a^{-1}b^{-1}ab$ of a and b.

3.24 Theorem. *If the theory of the group G is strongly minimal then G is Abelian.*

Proof. We will apply at various points in the proof the following strong consequences of assuming that G is strongly minimal. First, G can have no infinite definable proper subgroup, H (as H and any coset of H would be disjoint infinite sets). In particular, G is connected. Since G is superstable, we conclude from Exercise III.5.23 that any finite normal subgroup of G is contained in $Z(G)$. Since there are no infinite definable subgroups, we have, in fact, that every definable normal subgroup of G is contained in $Z(G)$. Thus, $Z^2(G) = \{a : \forall b[a, b] \in Z(G)\} = Z(G)$. It is now easy to see that $\bar{G} = G/Z(G)$ is centerless and strongly minimal.

Let $a \neq 1 \in \bar{G}$. Then, $C_{\bar{G}}(a)$ is finite. This easily implies that a has finite order (since all powers of a commute with a) and that C_a, the conjugacy class of $a \in \bar{G}$ is infinite (by the second remark before the theorem). Since $\{C_a : a \neq 1\}$ partition $\bar{G} - \{1\}$, all nonidentity elements of $\bar{G}$ are conjugate. This implies that they all have the same order, l, which is prime. For otherwise, we have elements of different orders that are conjugate. Since we finish if l is two, we may assume that l is odd.

For some b, $bab^{-1} = a^{-1}$. Thus, $(bab^{-1})^{-1} = ba^{-1}b^{-1} = a$. It is now easy using this fact to show by induction that $b^k ab^{-k}$ equals a if k is even and a^{-1} if k is odd. Thus, $b^l ab^{-l} = a^{-1}$. Since, the order of b is l, we conclude that $a = a^{-1}$ and thus the theorem.

We omit the proofs of the long string of generalizations of this result, [Baur, Cherlin & Macintyre 1979], [Cherlin 1978], [Poizat 1981], and just state the strongest result which is due to Lascar and Berline [Berline & Lascar 1986].

3.25 Theorem. *If G is a superstable group and $U(G) \geq \omega^\alpha$, then G contains a definable Abelian subgroup H with $U(H) \geq \omega^\alpha$.*

Assuming this result we finish the proof of Cherlin's commutativity theorem.

3.26 Theorem. *An infinite superstable division ring D is an algebraically closed field.*

Proof. Without loss of generality, we may assume D is $\aleph_1$-saturated. Let $U(D) = \omega^\alpha k + \beta$ with $\beta < \omega^\alpha$. Let H be a definable Abelian subgroup of D^* with $U(H) \geq \omega^\alpha$ whose existence is guaranteed by Theorem 3.25. Let $D_1 = C_D(H)$ and let F be the center of D_1.

We next show that D is finite dimensional over F. Since $H \subseteq F^*$, $U(F) \geq \omega^\alpha$. We claim that if $d_1, \ldots, d_m$ are F-linearly independent elements of D then the U-rank of $d_1F + \cdots + d_mF$ is $U(F)m$ which is at least $\omega^\alpha m$. Thus, m is at most k. Let D_0 be any model containing $\{d_1, \ldots, d_m\}$. For any m-tuple $\overline{a} = \langle a_1, \ldots, a_m\rangle$ from F, $\overline{a}$ is algebraic over $D_0 \cup \{d_1a_1 + \cdots + d_ma_m\}$ and $d_1a_1 + \cdots + d_ma_m$ is algebraic over $D_0 \cup \overline{a}$. So the types of $d_1a_1 + \ldots d_ma_m$ and $\overline{a}$ over D_0 have the same U-rank. But

$$U(t(\overline{a}; D_0)) \leq \sum_i U(t(a_i; D_0)) \leq m \cdot U(F).$$

Moreover, if the a_i are independent realizations of the type of maximal U-rank ($= U(F)$) then equality holds and we conclude the claim.

To conclude the argument we require two facts from algebra. The first is an elementary property of tensor products; the second appears in [Herstein 1975].

i) Let $Z \subseteq F \subseteq D$ be division rings (with Z a field). As vector spaces, $\dim_Z D = \dim_F(D \otimes_Z F)$. (cf. [Lang 1971])

ii) If the division ring D is finite dimensional over the maximal subfield F and $Z = Z(D)$ then $F \otimes_Z D \approx M_n(F)$ for some n. (This follows from the proof of the Corollary to Theorem 4.2.1 in [Herstein 1975].)

From the claim, we have that D is finite dimensional over F. Without loss of generality, we may assume that F is a maximal subfield of D. By ii), we have $D \otimes_Z F \approx M_n(F)$ is finite dimensional over F. But then, by i), and since $D \approx D \otimes_Z Z$, $\dim_Z D$ is finite. Thus Z is infinite. Since Z is a definable subfield of D, Z inherits the superstability of D. This implies Z is algebraically closed by Theorem 3.23. Now any element $a \in D - Z$ satisfies a polynomial over Z and since a centralizes Z, Z and a generate a subfield of D. As Z has no finite algebraic extensions we have the theorem.

3.27 Exercise. Show that for any of the rank functions considered if $a \downarrow_G b$ then $R(t(a;G)) \leq R(t(ab;G))$.

3.28 Historical Notes. This section comes primarily from [Cherlin & Shelah 1980] and [Poizat 1981]. Macintyre [Macintyre 1971a] had earlier proved that every ω-stable field is algebraically closed. Further important developments in this direction can be found in [Berline 1983] and [Berline & Lascar 1986] The account of Cherlin's result that a superstable division ring is commutative, is taken largely from [Berline & Lascar 1986]

Chapter VIII

Normalization and T^{eq}

In this chapter we give a brief introduction to some important topics in stability theory which do not play a major role in the part of stability theory emphasized in this book. These notions are extremely significant in the study of ω-stable, ω-categorical theories and their importance for the fine structure of the spectrum of models is becoming evident.

The first of these is the introduction of 'imaginary elements' to represent classes of equivalence relations which are definable in a theory T. This notion could have been used far more systematically in this book. Instead, we have placed a greater reliance on strong types. The second section is concerned with normalization, a topic which is closely related to the first and which leads to the introduction of geometric notions into the theory. We briefly discuss the role of these notions in the study of totally categorical theories in Section 3.

1. T^{eq}

The discussion of finite equivalence relations in Chapter IV suggests the desirability of giving special attention to the classes of definable equivalence relations. This idea is also suggested by the usual practice of forming quotient structures in algebra. Accordingly, we introduce an expanded language in which the equivalence classes of a definable equivalence relation can be considered as points. Some authors [Cherlin, Harrington, & Lachlan 1985] have proposed adding such equivalence relations piecemeal as they are needed. We follow Shelah in describing one large expansion, L^{eq}, which will encompass all the constructions we need to make. However, following a suggestion of Makkai we treat the expanded structure as a many-sorted model.

The properties of theories in this expanded language are 'conservative' for the non-technical results in stability theory. That is, T^{eq} provides a shortcut to prove theorems which would be much more difficult to conceptualize or prove without this notion. Shelah called the elements in the expanded models imaginary elements in analogy with the same conservative role the imaginary numbers play with respect to the reals. This analogy is fairly precise; just as there is an interpretation of the theory of the complex numbers into that of the reals, the theory T^{eq} can be interpreted into T.

We work below with many-sorted languages. In a many-sorted language each constant symbol and variable has a fixed sort. The relations (functions) are required to relate (respectively, map between) objects of specified sorts. The quantifiers are restricted to specific sorts. Each model is a disjoint union of sets of objects of the various sorts. The usual theorems of model theory, especially compactness and completeness, can be easily translated into the many-sorted situation (cf. [Enderton 1972]).

1.1 Definition. i) Let T be a theory in a first order language L. Then L^{eq} denotes the many sorted language which has a sort S_E for each equivalence relation (of arbitrarily many variables) which is definable without parameters in T. The symbols of L are considered as relations and functions on objects of sort $S_=$. For each n-ary equivalence relation E there is a function symbol F_E which maps n-tuples of sort $S_=$ to elements of sort S_E.

ii) Each L-structure M is expanded to an L^{eq}-structure denoted M^{eq} by taking the elements of M as the elements of sort $S_=$ and retaining the interpretations of the relations in L. The objects of sort S_E are the equivalence classes of the equivalence relation E. The interpretation of F_E maps a sequence $\overline{a} \in M$ to $\overline{a}/E$.

iii) T^{eq} denotes the theory of M^{eq}.

1.2 Exercise. Show the last part of Definition 1.1 is permissible; that is, if $M \equiv N$ then $M^{\mathrm{eq}} \equiv N^{\mathrm{eq}}$ (and conversely).

1.3 Exercise. Prove the map ^{eq} (from M to M^{eq}) is a functor from the category of models of T and elementary embeddings to the category of models of T^{eq} and elementary embeddings.

Although L^{eq} is a proper expansion of L, the formulas of L^{eq} which involve the new sorts can be easily translated into the original language. The following lemma, which is proved by induction on formulas, accomplishes this translation.

1.4 Lemma. *Let $\phi(x_0, \ldots, x_n)$ be an L^{eq}-formula where each x_i has sort S_{E_i}. There is an L-formula $\phi^*(\overline{x}_0, \ldots, \overline{x}_n)$ where each $\overline{x}_i$ has length m if E_i is an equivalence relation on m-tuples such that:*

$$\mathcal{M}^{\mathrm{eq}} \models \phi(\overline{a}_0/E_0, \ldots, \overline{a}_n/E_n) \leftrightarrow \phi^*(\overline{a}_0, \ldots, \overline{a}_n).$$

Note that if ϕ is an L-formula then ϕ^* is ϕ. The formulas ϕ^* are invariant in the sense that

$$\mathcal{M} \models \phi^*(\overline{a}_0, \ldots, \overline{a}_n) \leftrightarrow \phi^*(\overline{b}_0, \ldots, \overline{b}_n)$$

if $E_i(\overline{a}_i, \overline{b}_i)$ for each $1 \leq i \leq n$. With this result in hand, it is easy to see that most of the properties discussed in this book transfer from T to T^{eq}.

1.5 Exercise. Show that T is superstable, stable, ω-stable, etc. if and only if T^{eq} is.

The preservation of categoricity from T to T^{eq} is a more subtle matter. Because T^{eq} has infinitely many sorts, when one makes the natural translation of the many sorted language into a single-sorted one there will be elements in some models which realize 'non-standard' sorts. Naturally, such elements wreak havoc with categoricity. In the many-sorted version, these elements may be ignored. Formally, one can consider the reducts of models of T^{eq} to the standard sorts. This approach to many sorted structures is worked out in a different context in [Baldwin & Berman 1981]. Such fine notions as forking are also preserved. To see this we must consider the relation between types in T and types in T^{eq}.

1.6 Definition. Let $p \in S(A)$ and suppose that $\overline{b} \in \mathcal{M}$ realizes p. Then $p^{eq} = t^{eq}(\overline{b}; A)$ denotes the L^{eq}-type over A realized by $\overline{b}$.

Similarly we can define stp^{eq}. When we write $t^{eq}(\overline{b}; \mathrm{cl}(A))$, the algebraic closure is taken in T^{eq}. The following lemma justifies this definition, by showing p^{eq} does not depend on the choice of $\overline{b}$, and yields an important corollary.

1.7 Lemma. *For any $A \subseteq \mathcal{M}$ and $p \in S(A)$, $p \vdash p^{eq}$.*

Proof. If $\phi(\overline{b}; \overline{a}) \in p^{eq}$ then the L-formula $\phi^*(\overline{b}; \overline{a}) \in p$. Apply Lemma 1.4.

1.8 Corollary. *Let $A \subseteq B \subseteq \mathcal{M}$ and $p \in S(B)$. Then p forks over A if and only if p^{eq} forks over A.*

Proof. Since $p \subseteq p^{eq}$ one direction is trivial; but from $p \vdash p^{eq}$ we easily deduce the other.

The following exercise is immediate.

1.9 Exercise. For any type p, p is stationary if and only if p^{eq} is stationary.

All the results we have discussed so far are direct translations between T and T^{eq}. If this state of affairs held for all properties there would be no reason to introduce the notion. We now discuss several of the ways in which the theories differ.

To shorten notation, if we have a sequence $\overline{a}$ containing elements of various sorts, we write $F_{\overline{G}}(\overline{w}) = \overline{a}$ rather than the conjunction over specific equivalence relations G', specific subsequences $\overline{w}'$ of $\overline{w}$ and specific $a' \in \overline{a}$ of the formulas $F_{G'}(\overline{w}') = a'$.

1.10 Theorem. i) *Let A be an algebraically closed subset of $\mathcal{M}^{eq}$. If the formula ϕ is almost over A then ϕ is over A.*
ii) *$stp(\overline{c}; B) \vdash t(\overline{c}; A)$ if and only if $stp^{eq}(\overline{c}; B) \vdash t^{eq}(\overline{c}; \mathrm{cl}(A))$.*

Proof. i) Let $E(\overline{x}, \overline{y}; \overline{a})$ be a finite equivalence relation over A. We will show that each equivalence class of E corresponds to an element of $\mathcal{M}^{eq}$. (This is nontrivial because E is defined with parameters.) Then we will show each of these classes is over A and the theorem follows immediately by considering the finite equivalence relation on which ϕ depends.

Let $E^*(\overline{u}, \overline{v}; \overline{w})$ be the L-formula Lemma 1.4 associates with $E(\overline{x}, \overline{y}; \overline{z})$. Now let $\hat{E}$ be the equivalence relation on tuples with length $\lg(\overline{u}) + \lg(\overline{w})$ whose classes are as follows. If $E^*(\overline{u}, \overline{v}; \overline{a})$ is an equivalence relation then for each $\overline{b}$, $\{\overline{c} : E^*(\overline{c}, \overline{b}, \overline{a})\} \times \overline{a}$ is a class. A final class is the class of all tuples $\{\overline{c}\frown\overline{a}\}$ for those $\overline{a}$ such that $E^*(\overline{u}, \overline{v}; \overline{a})$ is not an equivalence relation. This equivalence relation is defined by the formula $\hat{E}(\overline{x}\frown\overline{w}_1; \overline{y}\frown\overline{w}_2)$:

$$(\neg E_1(\overline{w}_1) \wedge \neg E_1(\overline{w}_2)) \vee (\overline{w}_1 = \overline{w}_2 \wedge E^*(\overline{x}, \overline{y}, \overline{w}_1) \wedge E_1(\overline{w}_1))$$

where $E_1(\overline{a})$ holds if $E^*(\overline{x}, \overline{y}; \overline{a})$ is an equivalence relation. Let $d_1, \ldots, d_n$ in $\mathcal{M}^{\mathrm{eq}}$ be the equivalence classes of $\hat{E}$ which are associated with the equivalence classes of $E(\overline{x}, \overline{y}; \overline{c})$. If we find an L^{eq} formula over A with these as its solution we finish. For some $\overline{c}'$, $\overline{c} = \bar{F}_{\bar{G}}(\overline{c}')$, so such a formula is

$$\bigwedge_{1 \le i \le n} S_{\hat{E}}(z_i) \wedge (\exists \overline{x})(\exists \overline{w})[\bar{F}_{\bar{G}}(\overline{w}) = \overline{c} \wedge F_{\hat{E}}(\overline{x}\frown\overline{w}) = \overline{z}].$$

ii) This is immediate from the definitions of strong type and algebraic closure.

1.11 Corollary. *If A is an algebraically closed subset of $\mathcal{M}^{\mathrm{eq}}$ then every type over A is stationary.*

Proof. Let p and q be complete types over $B \supset A$ which are distinct, do not fork over A, and have a common restriction to A. Then by the finite equivalence relation theorem (Theorem IV.2.2) there is a formula ϕ which is almost over A and is in p but not in q. But by Theorem 1.10, ϕ is over A which contradicts the assumption that p and q have a common restriction to A.

Next we examine some notions which are primarily useful in $\mathcal{M}^{\mathrm{eq}}$.

1.12 Definition. For any set A, the *definable closure* of A, written $dcl(A)$, is the collection of points definable over A, i.e. the solutions of formulas with parameters from A which have unique solutions. If $dcl(A) = A$ we say A is definably closed.

If the global type $\hat{p}$ is strongly based on the set A, then every automorphism of $\mathcal{M}$ which fixes A also fixes $\hat{p}$. In general there are many such A.

1.13 Definition. i) Let $G_{\hat{p}}$ denote the *stabilizer* of $\hat{p}$, the subgroup of all automorphisms of $\mathcal{M}$ which fix $\hat{p}$.

ii) Let $F_{\hat{p}}$ denote the fixed set of $G_{\hat{p}}$, that is, the set of elements of $\mathcal{M}$ which are fixed by every automorphism of $\mathcal{M}$ which fixes $\hat{p}$.

iii) A *canonical base* of $\hat{p}$, written $\mathrm{cb}(\hat{p})$, is the (necessarily unique) set A such that A is definably closed and the automorphism f is in $G_{\hat{p}}$ if and only if f fixes A pointwise.

1.14 Exercise. Show that each $\hat{p}$ has at most one canonical base.

In general canonical bases need not exist. For example, consider the theory of an equivalence relation with two infinite classes. The type which is realized by a 'new' member of one of the classes has no canonical base. It is defined over any element of that class, but any such element can be moved by automorphisms which fix the type. Passing to T^{eq} we see that the canonical base is the class itself. This phenomenon generally holds; in $\mathcal{M}^{eq}$ every global type has a canonical base. One further notion is necessary to establish this result.

The definition of $\mathcal{M}^{eq}$ assigns an element to each equivalence class of an equivalence relation which is definable without parameters. In fact, each definable subset of $\mathcal{M}$ becomes a point of $\mathcal{M}^{eq}$ and if b is the element assigned to $\phi(\overline{x};\overline{a})$ then $b \in \mathcal{M}^{eq}$ is in the definable closure of $\overline{a}$. More formally, we have

1.15 Theorem. *For every formula $\phi(\overline{x};\overline{y})$ and every $\overline{a} \in M$ there is a point $b_\phi = b_{\phi(\overline{x};\overline{a})} \in \mathcal{M}^{eq}$ and a function F^ϕ, which is definable without parameters, such that $F^\phi(\overline{a}) = b_{\phi(\overline{x};\overline{a})}$.*

Proof. Define without parameters the equivalence relation R_ϕ by:

$$R_\phi(\overline{y},\overline{z}) \leftrightarrow [(\forall\overline{x})(\phi(\overline{x};\overline{y}) \leftrightarrow \phi(\overline{x};\overline{z}))].$$

Now $b = \overline{a}/R_\phi$ and F^ϕ can be taken as F_{R_ϕ}.

Note that if two formulas, say $\phi(\overline{x};\overline{a})$ and $\psi(\overline{x};\overline{b})$, define the same set in $\mathcal{M}$ then they give rise to distinct points in $\mathcal{M}^{eq}$; but all such points are in the definable closure of $\overline{a}$.

The following exercise provides another perspective for the last result.

1.16 Exercise. Show that for every definable (with parameters) subset X of M there is an element $X^* \in \mathcal{M}^{eq}$ such that an automorphism of M^{eq} leaves X invariant if and only if it fixes X^*.

We apply this exercise to prove each global type has a canonical base.

1.17 Theorem (Canonical Base). *If $\hat{p}$ is a type over $\mathcal{M}^{eq}$ then* $\mathrm{cb}(\hat{p})$ *exists.*

Proof. Let d define $\hat{p}$. That is, $\phi(\overline{x};\overline{a}) \in \hat{p}$ iff $\models d\phi(\overline{a};\overline{c}_\phi)$. Let A denote $\{\overline{c}_\phi/R_{d\phi} : \phi \in L\}$. Clearly, $F_{\hat{p}} = dcl(A)$ and every automorphism in $G_{\hat{p}}$ fixes A pointwise.

The definition of canonical base easily extends to any stationary type.

1.18 Definition. For any stationary type p, the *canonical base* of p, denoted $\mathrm{cb}(p)$, is the canonical base of the unique nonforking extension of p to a global type.

1.19 Exercise. If $p \parallel q$ then $\mathrm{cb}(p) = \mathrm{cb}(q)$ (in M^{eq}).

1.20 Exercise. There is a $B \subseteq \mathrm{cb}(p)$ with $|B| < \kappa(T)$ such that if $q \parallel p$ then q does not fork over B.

A slightly more involved argument shows that if p does not fork over A then $\text{cb}(p) \subseteq \text{cl}(A)$. Thus p is based on A in the sense of Definition IV.1.17 ii) if and only if $\text{cb}(p) \subseteq A$. In this manner of speaking, p is strongly based on A if $\text{cb}(p) \subseteq A$. A glance at the proof of Theorem 1.17 shows the following.

1.21 Lemma. *For any $\hat{p}$, $|\text{cb}(\hat{p})| \leq |T|$.*

1.22 Exercise. If T is superstable then for every global type $\hat{p}$ there is a finite $A \subseteq \text{cb}(\hat{p})$ such that $\text{cb}(\hat{p}) \subseteq \text{cl}(A)$. ($T^{\text{eq}}$)

The following extension of the notion of canonical base is due to Anand Pillay. It provides a useful tool for the proof of the normalization lemma in Section 2. Note that the proof makes essential use of T^{eq}.

1.23 Definition. Let P be a set of (possibly incomplete) types over $\mathcal{M}$.

i) G_P denotes the group of automorphisms of $\mathcal{M}$ which permute P.
ii) A *canonical base* for P is a set A such that A is definably closed and an automorphism α of $\mathcal{M}$ is in G_P if and only if it fixes A pointwise.

The following theorem is stated for a sufficiently closed subset Δ of formulas. On first reading, Δ can be taken to be all L-formulas.

1.24 Theorem. *Let T be a stable theory and let $\Delta(\overline{x})$ be a set of formulas closed under Boolean combination and substitution for parameter variables. Let P be a closed subset of $S_\Delta(\mathcal{M})$ with $|P| < |\mathcal{M}|$. The following are equivalent:*

i) *P has a canonical base.*
ii) *For each $p \in S_\Delta(\mathcal{M})$ and each $\phi(\overline{x};\overline{y}) \in \Delta$ the orbit of p_ϕ under G_P is finite.*

Proof. Suppose P has a canonical base, A, but for some $\phi \in \Delta$ and some $p \in P$, the orbit of p_ϕ under G_P is infinite. We will show $|P|$ is unbounded.

Let $\beta(\overline{y};\overline{c})$ define p_ϕ. Fix $\kappa > |P|$ and let $\overline{c}_i$ for $i < \kappa$ be new constant symbols. Let Σ assert that the $\overline{c}_i$ are distinct realizations of $t(\overline{c};A)$ and that if $i \neq j$ then $\beta(\overline{y};\overline{c}_i) \not\leftrightarrow \beta(\overline{y};\overline{c}_j)$. Since the orbit of p_ϕ under G_P is infinite, Σ is consistent. But the formulas $\beta(\overline{y};\overline{c}_i)$ thus define more than κ distinct conjugates of p_ϕ. Since all can be extended to members of P, this contradicts the choice of κ.

Now suppose ii) holds. Let $P = \bigcup_{i \in I} P_i$ where each P_i is an orbit of some $p \in P$ under G_P. Fix $\phi \in \Delta$ and $i \in I$. Let $\beta_1, \ldots, \beta_k$ define the finitely many ϕ-types in P_i. For each $j \leq k$, let $b_j = b_{\beta_j}$ be the point in M^{eq} attached to β_j using Theorem 1.15. Now $B_i = \{b_1, \ldots, b_k\}$ is a finite and thus definable subset of M^{eq}. Let $a_{\phi,i}$ be the point of M^{eq} attached to (some definition of) B_i by Theorem 1.15. An automorphism, f, of M^{eq} fixes $a_{\phi,i}$ if and only if f permutes B_i if and only if f permutes the ϕ-types in P_i.

We now show that $A = dcl(\{a_{\phi,i} : \phi \in \Delta, i \in I\})$ is the canonical base for P. If $f \in G_P$, f permutes the ϕ-types in P_i for each i, so f fixes A pointwise. Conversely, if f fixes A pointwise and $p \in P_i$ then for each ϕ,

$f(p_\phi) = q_\phi$ for some $q \in P_i$. For, p_ϕ is defined by some β_j and $f(\beta_j)$ defines q_ϕ for some $q \in P_i$. This shows $f(p)_\phi$ has an extension in P for each $\phi \in \Delta$. Since P is a closed subset of $S_\Delta(\mathcal{M})$, this implies $f(p) \in P$ as required.

Pillay [Pillay 198?] proves Theorem 1.24 with the hypothesis that P is closed replaced by a weaker but more complicated hypothesis.

1.25 Historical Notes. The advantage of considering equivalence classes as elements in models was seen at least as early as the Rabin-Scott-Ershov decision procedure [Eršov 1974]. Shelah initiated the systematic use of T^{eq} in stability theory in Chapter III.6 of [Shelah 1978] where all the results here (except Theorem 1.24) and more are obtained. The idea of viewing L^{eq} as a multisorted language was expounded by Makkai [Makkai 1984]. It has the advantage that $\mathcal{M}^{eq}$ is saturated and that categoricity as well as the stability hierarchy is preserved. The disadvantage, which we have chosen to suppress, is that stability theory has not been developed in many-sorted languages and so our use of it in the study of T^{eq} is not formally justified. Shelah's development goes directly to the step one takes in interpreting many-sorted logics into single sorted logic. He just expands L by adding a new unary predicate P_E whose intended interpretation is the set of E-equivalence classes. In this approach the formulas are somewhat more difficult to write since the range of the variables must be explicitly delimited, rather than thought of as built into the name of the variable. Definition 1.23 and Theorem 1.24 are from [Pillay 198?]. Poizat [Poizat 1985], [Poizat 1983a] provides an important view of the role (nonrole?) of T^{eq} in the theory of algebraically closed fields. In addition to the published expositions our treatment benefitted from discussions with Gisela Ahlbrandt and Anand Pillay.

2. *Normalization*

In this section we consider the normalization of a formula. This notion, which was introduced by Lachlan, is closely connected both with the finite equivalence relation theorem and with the concept of a canonical base discussed in Section 1.

One reason to consider normalization is the following situation. We want to investigate the relation between a uniformly definable family of sets $\langle \phi(\overline{x};\overline{a}) : \overline{a} \in \mathcal{M} \rangle$ and the set defined by the formula $(\exists \overline{y})\phi(\overline{x};\overline{y})$. Often the quantification may be restricted to a definable set. If, for example, T is ω-stable and $\phi(\mathcal{M};\overline{a})$ has Morley degree 1 there is a natural equivalence relation on the parameters given by $\overline{a} \simeq \overline{b}$ if

$$R_M(\phi(\overline{x};\overline{a}) \wedge \phi(\overline{x};\overline{b})) = R_M(\phi(\overline{x};\overline{a})).$$

We would like to find natural representatives for the equivalence classes under this equivalence relation. The application of the Normalization Lemma in this situation is discussed in Theorem 3.4 (cf. [Lachlan 1980]).

The situation is more easily understood from a slightly more abstract viewpoint.

We write $\triangle$ for the symmetric difference of two elements in a Boolean algebra.

2.1 Notation. Fix a stable theory T. Let $A \subseteq \mathcal{M}$ be invariant under automorphisms of $\mathcal{M}$. Let $\mathcal{F} = F_\Gamma(A)$, the set of A-substitution instances of a set Γ of parameter free formulas which is closed under Boolean combinations. Then $\mathcal{F}$ is a subalgebra of the Boolean algebra $F(\mathcal{M})$ which is invariant under automorphisms of $\mathcal{M}$. Let $\mathcal{I}$ be an ideal in $\mathcal{F}$ which is invariant under automorphisms of $\mathcal{M}$.

i) Two formulas, $\phi, \psi \in \mathcal{F}$ are *equivalent* relative to $\mathcal{I}$ if they are in the same congruence class modulo $\mathcal{I}$. That is, if $\phi \triangle \psi \in \mathcal{I}$. We write $\phi \simeq_{\mathcal{I}} \psi$, omitting the subscript if it is clear from context.

ii) Let $\psi(\overline{x};\overline{a})$ be in Γ. The set of *distinguished $\mathcal{I}$-extensions* of $\psi(\overline{x};\overline{a})$, denoted by $P_{\psi(\overline{x};\overline{a})}$, is the set of complete Γ types over $\mathcal{M}$ which contain $\psi(\overline{x};\overline{a})$ and which contain no formula from $\mathcal{I}$. When ψ is clear we write $P_{\overline{a}}$ for $P_{\psi(\overline{x};\overline{a})}$. We will not write $\mathcal{I}$ when $\mathcal{I}$ is fixed in context.

2.2 Examples. i) Let T be a stable theory, let $\mathcal{F} = F(\mathcal{M})$, and let $\mathcal{I}$ be the ideal of formulas which fork over $\emptyset$.

Of course, this example can be generalized by replacing the empty set by a fixed subset A of $\mathcal{M}$.

ii) Let T be an ω-stable theory. Let $\mathcal{F} = \{\phi(\overline{x};\overline{a}) : R_M(\phi(\overline{x};\overline{a})) \leq n\}$. Let $\mathcal{I}$ be the ideal of formulas with Morley rank $< n$.

By defining our notion in terms of symmetric difference, we do not need in Example 2.2 ii) the restriction, which is usually made in discussions of the normalization lemma, that $\phi(\overline{x};\overline{a})$ have Morley degree one.

2.3 Exercise. Show in Example 2.2 ii) that if $R_M(\phi(\overline{x};\overline{a})) = n$, $\overline{a}'$ is a conjugate of $\overline{a}$, and $\phi(\overline{x};\overline{a})$ has Morley degree one, then $\phi(\overline{x};\overline{a}) \simeq \phi(\overline{x};\overline{a}')$ if and only if $R_M(\phi(\overline{x};\overline{a}) \wedge \phi(\overline{x};\overline{a}')) = n$.

2.4 Exercise. Show that $\phi(\overline{x};\overline{a}) \simeq \phi(\overline{x};\overline{a}')$ if and only if $P_{\overline{a}} = P_{\overline{a}'}$.

2.5 Exercise. Show that the equivalence classes under $\simeq$ are closed under positive Boolean combinations.

2.6 Definition. A formula $\phi(\overline{x};\overline{a}) \in \mathcal{F}$ is said to be *normal* with respect to $\mathcal{I}$ if for any conjugate $\phi(\overline{x};\overline{a}')$ of $\phi(\overline{x};\overline{a})$, if $\phi(\overline{x};\overline{a}) \simeq_{\mathcal{I}} \phi(\overline{x};\overline{a}')$ then $\phi(\mathcal{M};\overline{a}) = \phi(\mathcal{M};\overline{a}')$.

With some further conditions on the pair $(\mathcal{F}, \mathcal{I})$, we will show that every formula in $\mathcal{F}$ can be normalized. That is, to each formula ϕ, there corresponds a formula ϕ^* such that $\phi \simeq_{\mathcal{I}} \phi^*$ and ϕ^* is normal. We prove this result here for the context we described in Paragraph 2.1. We distinguish now some conditions which play an important role in our proof. In fact, as we discuss below, varying the context slightly these conditions suffice to prove the normalization Lemma.

2.7 Conditions. Consider the following conditions on $\phi(\overline{x};\overline{a}) \in \mathcal{F}$ and $P_{\overline{a}}$.

N1. For each $q \in P_{\overline{a}}$, $\phi(\overline{x};\overline{a}) \in q$.
N2. $P_{\overline{a}}$ is closed under automorphisms of $\mathcal{M}$ which fix $\overline{a}$.
N3. $|P_{\overline{a}}| < |\mathcal{M}|$.
N4. If $t(\overline{a};\emptyset) = t(\overline{b};\emptyset)$, $P_{\overline{a}} = P_{\overline{b}}$ if and only if for each $q \in P_{\overline{a}}$, $\phi(x;\overline{b}) \in q$ and for each $r \in P_{\overline{b}}$, $\phi(x;\overline{a}) \in r$.

2.8 Exercise. Prove that in the situation of Paragraph 2.1, each of N1, N2, and N4 holds.

Now we can state the major result of this section.

2.9 Theorem (The Normalization Lemma). *Let $\langle \mathcal{F}, I\rangle$ satisfy the conditions of Paragraph 2.1. Suppose in addition that each $|P_{\overline{a}}| < |\mathcal{M}|$ (N3). For each formula $\phi(\overline{x};\overline{a}) \in \mathcal{F} - I$, there exists a formula $\phi^*(\overline{x},\overline{a})$ such that*

i) $\phi^*(\overline{x},\overline{a}) \simeq \phi(\overline{x},\overline{a})$.
ii) *ϕ^* is equivalent to a positive Boolean combination of conjugates of $\phi(\overline{x};\overline{a})$.*
iii) *$\phi^*(\overline{x};\overline{a})$ is normal with respect to I.*

Note that statement ii) of the theorem is made only 'up to equivalence'. There is no claim that the same formula both contains only the parameter $\overline{a}$ and is a Boolean combination of conjugates of $\phi(\overline{x};\overline{a})$. When reading papers in this area the reader should beware of an identification of properties of the set defined by a formula with properties of the formula.

Before proving the theorem we note that our examples satisfy the conditions of 2.7. Of course, in light of the exercise following their statement we need only check N3. We then derive some general properties of $\langle \mathcal{F}, I\rangle$ which satisfy N3.

2.10 Examples. We continue the examples begun in 2.2.

i) Fix a set $A \subseteq \mathcal{M}$. Then $P_{\overline{a}} = \{p \in S(\mathcal{M}) : \phi(x;\overline{a}) \in p$ and p does not fork over $A\}$.
ii) Let $P_{\overline{a}} = \{p \in S(\mathcal{M}) : \phi(x;\overline{a}) \in p$ and $R_M(p) = n\}$.

2.11 Exercise. Show that the two examples satisfy N3.

We now collect some properties which hold whenever Conditions 2.7 do. It is obvious from N2 and N3 that:

2.12 Lemma. *If $q \in P_{\overline{a}}$ then q does not fork over $\overline{a}$.*

Thus, if $q \in P_{\overline{a}}$, q is definable almost over $\overline{a}$. Let $\beta_i(\overline{y})$ list the formulas over $\mathrm{cl}(\overline{a})$ which define the types p_ϕ (the restriction of p to ϕ) for $p \in P_{\overline{a}}$.

2.13 Lemma. *There is a type $s(\overline{u},\overline{v})$ over $\emptyset$ so that $P_{\overline{a}} = P_{\overline{b}}$ if and only if* $\models s(\overline{a},\overline{b})$.

Proof. Since $P_{\overline{a}}$ is closed under automorphisms which fix $\overline{a}$, for each i and for each of the finitely many $\overline{a}$-conjugates β_i' of β_i, $\models \beta_i'(\overline{a})$. Since the conjunction of the β_i' is invariant under $\overline{a}$-automorphisms we may take it as a formula $\gamma_i(\overline{x};\overline{a})$. Now for each i, we have $\models \gamma_i(\overline{a};\overline{a})$ so $s(\overline{u},\overline{v}) = \{\gamma_i(\overline{u};\overline{v}) \wedge \gamma_i(\overline{v};\overline{u}) : i < \omega\}$ is a consistent type over the empty set. Now, $\models s(\overline{a},\overline{b})$ if and only if for each $p_i \in P_{\overline{a}}$, $\models \gamma_i(\overline{a},\overline{b})$ iff and only if $\phi(\overline{x};\overline{a}) \in p_i$. Switching the roles of $\overline{a}$ and $\overline{b}$ and applying N4 twice, we see $\models s(\overline{a},\overline{b})$ if and only if $P_{\overline{a}} = P_{\overline{b}}$. Thus, we have the Lemma.

The following proposition is the key to the proof of the normalization lemma. We defined the canonical base of a set of types in Definition 1.23.

2.14 Proposition. *With the above notation $P_{\overline{a}}$ has a canonical base.*

Proof. Clearly, $P_{\overline{a}}$ is a closed set and its cardinality is properly less than that of $|\mathcal{M}|$. So, it suffices by Theorem 1.24 to show that for each $p \in P_{\overline{a}}$ and each $\phi \in \Gamma$, the orbit of p_ϕ under $G_{P_{\overline{a}}}$ is finite. Let $\beta(\overline{y};\overline{c})$ be a definition of q_ϕ for some nonforking extension, q, of p to $S(\mathcal{M})$. Add new constants $\overline{a}_i, \overline{c}_i$ to the language for $i < \kappa = |P_{\overline{a}}|^+$. Let Σ be the set of sentences which assert:

$$t(\overline{a}_i{}^\frown\overline{c}_i;\emptyset) = t(\overline{a}{}^\frown\overline{c};\emptyset),$$

$$\models s(\overline{a}_i,\overline{a}),$$

$$\beta(\overline{y},\overline{c}_i) \not\leftrightarrow \beta(\overline{y};\overline{c}_j) \text{ if } i < j < \kappa \,.$$

Since $f \in G_{P_{\overline{a}}}$ if and only if $\models s(\overline{a}, f(\overline{a}))$, Σ is consistent. But the interpretations of Σ yield κ distinct conjugates of $p|\phi$ in $P_{\overline{a}}$ contrary to the choice of κ. Thus, by Theorem 1.24, we have the proposition.

With this in hand we can quickly prove the normalization lemma.

Proof of 2.9. We want to normalize $\phi(\overline{x};\overline{a})$. Choose a canonical base, A, for $P_{\overline{a}}$ by Proposition 2.14. Let $q \in S(\mathcal{M})$ be a nonforking extension of $t(\overline{a};A)$. By the saturation of $\mathcal{M}$ we can find a set of indiscernibles, E, realizing $t(\overline{a};A)$ so that $q = \mathrm{Av}(E,\mathcal{M})$. By Lemma V.2.6, there is a formula $\chi(\overline{e};\overline{y})$ which is a positive Boolean combination of A-conjugates of $\phi(\overline{x};\overline{a})$ such that for any $\overline{c} \in \mathcal{M}$, $\chi(\overline{e};\overline{c})$ if and only if $\phi(\overline{x};\overline{c}) \in q$. Since q does not fork over A, χ is almost over A. Let χ^* be a formula over A which is equivalent to the disjunction of the (finitely many) A-conjugates of χ. Since χ^* is equivalent to a positive Boolean combination of A-conjugates of $\phi(\overline{x};\overline{a})$ and $\simeq$ classes are closed under such combinations, $\chi^* \simeq \phi(\overline{x};\overline{a})$. It remains to show that χ^* is normal. Let χ' be a conjugate of χ by an automorphism f. Now $\chi' \simeq \chi^*$ if and only if $P_{\chi'} = P_{\chi^*}$. But, $P_{\chi^*} = P_{\overline{a}}$. Thus f permutes $P_{\overline{a}}$. But then f fixes A pointwise and so χ' is an A-conjugate of χ^*. Since χ^* is over A, this implies χ^* is normal.

Since A is a canonical base for $P_{\overline{a}}$, each $\overline{a}$-automorphism fixes A pointwise and thus leaves χ^* invariant. Thus, χ^* is equivalent to a formula $\phi^*(\overline{x};\overline{a})$ as required.

Steve Buechler gave another account of this result which is more ad hoc but gives some additional information. One can use the conditions of Proposition 2.7 as the hypotheses without assuming the exact context described in Paragraphs 2.1 through 2.6. In outline his argument proceeds as follows.

He proves that there is a family $\langle E_i : i < \omega\rangle$ of 0-definable (i.e. definable without parameters) equivalence relations satisfying the following condition. If $\overline{b}, \overline{b}'$ are conjugate to $\overline{a}$ then $\models \bigwedge_{i<\omega} E_i(\overline{b}, \overline{b}')$ if and only if $P_{\overline{b}} = P_{\overline{b}'}$. Now, let $A = \{\overline{a}/E_i : i < \omega\} \subseteq M^{eq}$. Moreover $A \subseteq dcl(\overline{a})$. Note that if $\overline{b}, \overline{b}'$ are conjugates of $\overline{a}$, then $t(\overline{b}; A) = t(\overline{b}'; A)$ if and only if $\overline{b}/E_i = \overline{b}'/E_i$ for each $i < \omega$. One of these implications is obvious. For the other, note that if $\overline{b}/E_i = \overline{b}'/E_i$ for each i, any automorphism which maps $\overline{b}$ to $\overline{b}'$ fixes A pointwise. With this observation we have the following reformulation of Lemma 2.13. $P_{\overline{b}} = P_{\overline{b}'}$ if and only if $t(\overline{b}; A) = t(\overline{b}'; A)$. Now an argument very similar to that for Theorem 2.9 derives a normalizing formula for $\phi(\overline{x};\overline{a})$.

2.15 Historical Notes. The normalization theorem was originally proved by Lachlan [Lachlan 1974] using a complicated rank argument. The more abstract versions were initiated by [Harnik & Harrington 1984]. The particular version chosen here is due to Anand Pillay [Pillay 1983]. Other proofs have been given by [Buechler 1984c], [Saffe 198?a], and, for the forking case, [Vaughn 1985].

3. *'Geometric' Stability Theory*

In this section we summarize recent developments in 'geometric' stability theory. By this we mean the investigation of the fine structure of models through the combination of techniques from combinatorial geometry and group theory with those of stability theory.

There are two major sources for this development. Our discussion is primarily organized around the first, Morley's question, "Can a theory which is categorical in all infinite powers and has no finite models (a totally categorical theory) be finitely axiomatizable?" A major step in the investigation of this problem was Zilber's isolation of the class of *strictly minimal* sets. A set is strictly minimal in the theory T if it is not only strongly minimal but admits no nontrivial definable equivalence relation. A geometry can be associated with any strictly minimal set as follows. The elements of the strictly minimal set are the points of the geometry; the algebraic closure of two points is called a line; the algebraic closure of three noncollinear points is called a plane, etc. The following theorem states the basic fact.

3.1 Theorem. *Each $\aleph_0$-categorical strictly minimal set is associated with one of the following geometries.*

i) *An infinite set with no structure (disintegrated).*
ii) *An infinite dimensional projective space over a finite field.*
iii) *An infinite dimensional affine space over a finite field.*

Cherlin [Cherlin, Harrington, & Lachlan 1985] and Mills (unpublished) independently proved this result by an analysis of the automorphism groups of the finite approximations to the strictly minimal set. This analysis depends upon the classification of the finite simple groups. Zilber [Zilber 1981] proved this result directly with no reliance on a high powered technique imported into the field.

Reflecting the modularity of the associated lattice of closed subsets, a strictly minimal set which is either disintegrated or projective is called *modular*. Otherwise, it is called *affine*.

By a variant on the usual construction of projective space, any affine space can be converted into a modular one in a language obtained *by naming a point*. The effect of such an extension by constants in this situation is so critical that it has spawned some further terminology. A set H is A-definable in M if it is defined by a formula with parameters from A. If $A = \emptyset$, H is said to be 0-definable.

If T is $\aleph_0$-categorical we naturally extend the notion of orthogonality to sets by saying the strongly minimal sets H_1 and H_2 are orthogonal if the unique non-algebraic types of elements in H_1 and H_2 are orthogonal. The crucial step in moving from the 'local' behavior of a particular strictly minimal set to the 'global' analysis of a model is

3.2 Theorem. *If T is ω-stable and $\aleph_0$-categorical, H_0 and H_1 are non-orthogonal* modular *strictly minimal sets then there is a unique* 0*-definable bijection between H_1 and H_2.*

This result leads to the two basic structural properties of an ω-stable, $\aleph_0$-categorical theory, the coordinatization theorem and the fundamental rank inequality. To state these we require some more terminology.

Recall that a group G acts n-transitively on a set X if for each pair of distinct n-tuples, $\overline{a}, \overline{b}$ from X there is an element of G which takes $\overline{a}$ to $\overline{b}$. If an $\aleph_0$-categorical countable structure, M, realizes only one 1-type, the automorphism group of M acts 1-transitively on M since M is homogenous. When the automorphism group of a structure acts 1-transitively on it, M is called *transitive.*

Let P and A be 0-definable subsets of an $\aleph_0$-categorical structure M. We say *A coordinatizes P*, if A generates a complete type over $\emptyset$ and for each $x \in P$, $\mathrm{cl}(x) \cap A \neq \emptyset$. Thus, each element of P is coordinatized by the finite set of elements in A which are in the algebraic closure of X. Here is the fundamental result.

3.3 Theorem (Coordinatization Theorem). *Let M be transitive, ω-stable and $\aleph_0$-categorical. There is a finite extension by definitions, M^* of M*

which is 0-interpretable into M in which there is a rank one set A which coordinatizes M.

M^* is just an appropriate approximation to M^{eq} by adding only finitely many sorts. This result leads to an important result whose statement is somewhat confusing. In the following theorem (and throughout this subject) we are able to discuss families of definable sets as though the sets are elements of the model and the family is a subset. Of course, this is accomplished by passing to (a suitable finite reduct of) T^{eq}. In particular, the rank of a family of definable sets is the rank of the associated set in $\mathcal{M}^{eq}$.

3.4 Theorem (The Fundamental Rank Inequality). *Let F be a definable collection of definable subsets of the $\aleph_0$-categorical, ω-stable structure N with $R_M(N) = n$. Suppose each element of F has the same Morley rank r and the intersection of any two has rank less than r. Then $r + R_M(F) \leq n$.*

The almost disjoint family F needed to apply Theorem 3.4 is obtained from the Normalization Theorem. The Coordinatization Theorem is first proved under the hypothesis that the rank of T is finite. It then leads to the fundamental rank inequality. Then a clever argument of Cherlin turns the situation on its head to prove

3.5 Theorem. *If T is ω-stable and $\aleph_0$-categorical then T has finite Morley rank.*

A second impetus for 'geometric' stability theory came from Lachlan's analyses of finite homogeneous structures [Lachlan 1984], [Lachlan 1986]. This approach coalesces with the results described so far in this section in

3.6 Theorem. *If T is $\aleph_0$-categorical and ω-stable then every sentence which is consistent with T has a finite homogeneous model. Thus, T is not finitely axiomatizable.*

Noting that all known ω-stable, $\aleph_0$-categorical stuctures are constructed from finite structures in fairly simple ways, e.g. a direct sum of finite abelian groups, leads to the conjecture that this is always true. Given the difficulty of describing a 'simple' construction without having it in hand, the following test question is formulated. Is every totally categorical theory *quasi-finitely axiomatizable*, that is, finitely axiomatizable relative to an axiom schema of infinity? To indicate the progress made on this question we need two more notions. A theory T is *almost strongly minimal* [Baldwin 1972] if every model of some finite inessential extension of T is contained in the algebraic closure of a strongly minimal set. In the early 1970's these theories were thought of as the 'simple' $\aleph_1$-categorical theories. For example, [Makowsky 1984] had proved that no such theory was finitely axiomatizable. Ahlbrandt connected this notion with the local study of strictly minimal sets via the following definition and theorem. The totally categorical theory T is *almost of modular type* if for some finite principal extension T^* of T, every strictly minimal set in $T^{*^{eq}}$ is modular.

3.7 Theorem. *The totally categorical theory T is almost strongly minimal if and only if it is almost of modular type.*

Finally Ahlbrandt and Ziegler proved

3.8 Theorem. *If the totally categorical theory T is almost of modular type then T is quasi-finitely axiomatizable.*

Another entry of geometry into stability theory came from Lachlan's proof [Lachlan 1974] that the question of the existence of a countable $\aleph_0$-categorical theory which is stable but not ω-stable could be given the following form. A *pseudoplane* is a two sorted structure of points and lines such that i) each line contains infinitely many points (and dually each point is on infinitely many lines) but ii) a pair of lines intersect in only finitely many points (and dually only finitely many lines pass through any fixed pair of points).

3.9 Theorem. *[Lachlan 1974] If there is a countable, $\aleph_0$-categorical and stable but not ω-stable theory then there is such a pseudoplane.*

Pseudoplanes have since showed up in a number of contexts in the theory. For example, Zilber's extension of the classification of strictly minimal sets to the $\aleph_1$ but not $\aleph_0$-categorical case provides the following theorem.

3.10 Theorem (The Trichotomy Theorem). *If M is an $\aleph_1$-categorical structure then exactly one of the following three conditions holds.*

i) *There is a pseudoplane definable in every strongly minimal structure definable in M*
ii) *Every strongly minimal structure definable in M is locally projective.*
iii) *Every strongly minimal structure definable in M is disintegrated.*

Zilber calls a strongly minimal set, D, *locally projective* if the geometry associated with D over a non-algebraic point is a projective geometry over a division ring. Intuitively, this divides $\aleph_1$-categorical theories into those which are field-like, module-like, or disintegrated. In particular, Zilber conjectures every $\aleph_1$-categorical pseudoplane is biinterpretable with an algebraically closed field.

Buechler [Buechler 1985b] made a significant application of the 'geometric' theory by using it to not only compute the spectrum of an ω-stable $\aleph_0$-categorical theory but to assign invariants to models in a more exact way than the general assignment for models of an arbitrary ω-stable theory. There is more detail on this in Chapter XVIII.

When dealing with theories which are not $\aleph_0$-categorical, certain properties of formulas in $\aleph_0$-categorical theories must be replaced by consideration of types. Thus, the notion of a type-interpretable pseudoplane, a pseudoplane whose incidence relation and sets of points and lines are defined by types, arises. Buechler [Buechler 1984] has shown that a trichotomy similar to Zilber's trichotomy holds for 'simple' types in arbitrary superstable theories. That is, if a stationary type contains a formula with continuous rank one then it defines either a strongly minimal, a locally projective, or a disintegrated set. Using this result he generalizes the coordinatization theorem and the fundamental rank inequality to superstable locally modular theories [Buechler 1986] with finite U-rank.

Pillay has defined a class of *weakly normal* theories. He shows that the stable theory T is weakly normal if and only if T does not type-interpret a pseudoplane. Moreover, in [Pillay 1986], he generalizes the Baur-Cherlin-Macintyre theorem that every stable $\aleph_0$-categorical group is Abelian by finite by showing that an ω-stable group which does not type-interpret a pseudoplane is Abelian by finite. In [Hrushovski & Pillay 1986] every weakly normal group is shown to be Abelian by finite.

3.11 Historical Notes. Since this section was a history there is little to say. Key papers include [Cherlin, Harrington, & Lachlan 1985], [Zilber 1981], [Zilber 1980b], [Zilber 1980a],[Zilber 1984],[Zilber 1984a], [Buechler 1985b], and [Ahlbrandt & Ziegler 1986]. Another interesting line is pursued by Srour and Pillay [Pillay 1983], [Pillay & Srour 1984], and [Pillay 198?a]. Another important tack is the investigation of finite homogeneous structures. Lachlan has written a series of papers which are listed in the bibliography. See also [Cherlin & Lachlan 1986] and [Lachlan & Shelah 1984]. In still another direction a detailed analysis of categorical varieties and quasivarieties has been given ([Givant 1973], [Givant 1976], [Palyutin 1973], [Palyutin 1976]). The representation theorems proved there are remarkably similar to Theorem 3.1.

Part B

Dependence and Prime Models

In this part of the book we focus on ways to 'generate' a model of a theory. Of course, the most natural notion of generation is closure under functions. However, we are interested in constructing elementary submodels and it is easy to see that the addition of Skolem functions destroys many model theoretic properties including, in particular, the stability hierarchy. For example, there exist ω-stable theories with no complete ω-stable Skolemization. Thus we turn to more general notions. One such idea is known to model theorists as the 'algebraic closure' of a set A; it consists of those points which lie in finite sets defined by formulas with parameters from A. Except in the extreme case of a strongly minimal set, this notion lacks the exchange (symmetry) property of a good dependence relation. A next attempt is to adjoin those points which realize a principal type over A. Here too, the full symmetry property is lacking but a reasonable substitute (cf. Section IX.3) holds. However, the requirement that every set have a closure with respect to this relation which is a model of T is extremely strong. In fact, the only known nontrivial sufficient condition for every set to have a closure with respect to this relation is the assumption that the theory is countable and ω-stable. It is necessary to consider some further variants in order to find a notion of closure which is both strong enough to be useful and which applies to a wide class of theories. In this part of the book we give a general treatment of several such properties.

Chapter IX
Atomic and Prime Models

The first four chapters of this book discussed generalizations of the notion of independence. We remarked in Chapter II that one feature of this generalization is that we can no longer define 'dependent' as merely 'not independent'. We must instead develop an apparatus to study 'generation'. Since the early 1960's it has been clear that the proper notion of 'submodel generated by A in the category of structures and elementary embeddings' is 'model prime over A'. It turns out, however, that in classifying the models of a first order theory, it does not suffice to investigate arbitrary models of the theory; the detailed study of saturated models of several sorts is essential. For each of these classes of models an appropriate notion of generation (of 'prime over') must be developed. This chapter is devoted to developing such notions.

We will discuss various concepts of prime and atomic models in this book. 'Prime' is a category theoretic concept: M is prime over A in the category K if every K-embedding of A into a member N of K can be extended to a K-embedding of M into N. 'Atomic' is a logical notion. Let $\mathbf{I}$ be a collection of types. Then, M is $\mathbf{I}$-atomic over A if for every $\overline{m} \in M$, there is some $q \in I$ with $q \subseteq t(\overline{m}; A)$ such that $q \vdash t(\overline{m}; A)$. Let K be the category of models of a countable first order theory T with elementary embeddings as morphisms and let $\mathbf{AT}$ be the collection of single formulas. It was realized in the early 1960's that the concepts, 'K-prime over $\emptyset$' and 'countable and $\mathbf{AT}$-atomic over $\emptyset$' are identical. Vaught used this equivalence to prove theorems characterizing prime models and proved that any two prime models of a countable theory are isomorphic. Morley noticed that the atomic component of the equivalence is useful in omitting types and applied this observation in proving his categoricity theorem.

A third aspect of 'primality' turns out to play an important role. Morley's construction of atomic models produces not only a structure which is atomic but one which has been built up in a specific way. This construction of the model corresponds to what we call a 'strictly prime' model. The proof of the uniqueness of strictly prime models is comparatively straightforward. If the countable theory T is stable and admits prime models over all sets we will prove in Chapter X the more difficult assertion that prime and strictly prime coincide and thus deduce the uniqueness of prime models.

In the first section below we sketch the results for the 'classical' case. Then we discuss the situation from a more abstract point of view. In Sections 2 and 3 we develop a set of axioms for primality (or generation) to complement the earlier axioms for independence. In Section 4 we prove the uniqueness of strictly prime models for all our notions simultaneously. In Chapter X we use the properties of independence to show the uniqueness of prime models for these notions in a countable stable theory.

The last two sections of this chapter are devoted to two notions of isolation which do fit precisely into the general scheme of isolation notion we developed in Section 2. In one sense, the failure of these notions to fit the more general rubric is purely technical; they do fit in the scheme of [Shelah 1978]. But there is a thematic reason for the difference. Neither of these notions of 'isolation' admit a natural notion of 'prime'. Section 5 concerns locally atomic models. We show that one can find a locally atomic model over any set if T is a stable theory. We apply this concept in Section 5 to the proof of two cardinal theorems. In Section 6, we describe the notion of $\mathbf{F}$-isolation and use it to show that a stable but not superstable theory has 2^λ models of power λ in each uncountable cardinality λ.

1. *Elementarily Prime Models*

We deal in this section with prime models over sets for countable theories. We first develop Vaught's characterization of prime models over the empty set as atomic models. Then we turn to the existence question for prime models over arbitrary sets. There are two ways to find prime models over the empty set (or indeed over arbitrary countable sets). The first, which appears in [Vaught 1961], uses the omitting types theorem to construct a model which omits the nonprincipal types and thus is atomic. This method depends essentially on the countability hypothesis. The second method, which is used almost exclusively here, is to build up the model guaranteeing that it is atomic during the construction. This method was introduced by Morley [Morley 1965] and this entire chapter is devoted to amplifications of the idea.

1.1 Definition. i) The model M is *prime* over A if every elementary embedding of A into a model N of T can be extended to an elementary embedding of M into N.

ii) The model M is a *prime model* of T if M is prime over the empty set.

iii) The theory T *admits prime models* if there is a prime model over every set A.

For example, the algebraic numbers are the prime model of the theory of algebraically closed fields; the natural numbers are the prime model of 'true arithmetic'. This purely categorical definition is expressed by the following diagram (Fig. 1). However, there are equivalent definitions of a

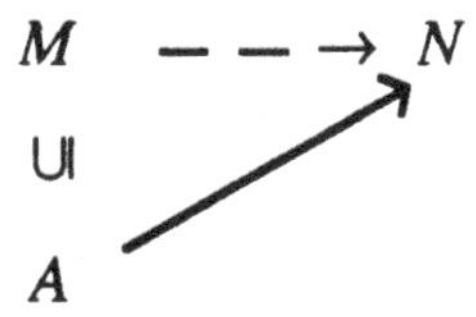

Fig. 1. M is prime over A.

logical character. We need some further notions in order to express them. Recall from Chapter I the notion of an isolated type.

1.2 Definition. Let $p \in S(A)$. Then p is *principal* or *isolated* if there is a single formula $\phi(x;\overline{a}) \in p$ such that $\phi(x;\overline{a}) \vdash p$. That is, for all $\psi \in p$, $\mathrm{Th}(A) \vdash (\forall \overline{x})[\phi(\overline{x};\overline{a}) \rightarrow \psi(\overline{x})]$.

Note that this is exactly the same as saying p is an isolated point in the topology on $S(A)$ induced by Stone duality, that is, the topology where each basic open set is of the form $U_\phi = \{p : \phi \in p\}$.

We denote by **AT** the set of all single first order formulas (with parameters anywhere in $\mathcal{M}$).

1.3 Definition. The model M is *atomic* over A if for each finite sequence $\overline{m} \in M$, $t(\overline{m}; A)$ is an isolated type. M is *atomic* if M is atomic over the empty set.

The immediate significance of these definitions is given by

1.4 Theorem. *M is a prime model of T if and only if M is atomic and countable. Moreover, any two countable atomic models are isomorphic so if T has a prime model it is unique.*

Proof. For one half of the theorem, we rely on the omitting types theorem. If M realizes a nonprincipal type then M is not prime as there is also a model N of T which omits p. For the converse, we are given a countable atomic model M and a model N; we want to elementarily embed M in N. Let $M = \{m_i : i \in \omega\}$. Now construct the embedding f by induction. Say $\phi_0(x)$ generates $t(m_0; \emptyset)$. Since T is complete $T \models (\exists x)\phi_0(x)$. Let $f(m_0)$ be any element of N such that $N \models \phi_0(m_0)$. Suppose we have defined f on $\{m_i : i < n\}$ so that $(M, m_0, \ldots, m_{n-1}) \equiv (M, f(m_0), \ldots, f(m_{n-1}))$. Let $\phi(x_0, \ldots, x_n)$ generate $t(m_0, \ldots, m_n; \emptyset)$. By the induction hypothesis, $N \models (\exists y)\phi(f(m_0), \ldots, f(m_{n-1}), y)$. Choose any witness for this formula as $f(m_n)$. To establish the 'moreover' clause, replace this 'forth' argument by a 'back and forth' argument.

1.5 Exercise. Prove the moreover clause in the last theorem.

1.6 Exercise. Show that if M_1 and M_2 are prime models over the countable set A then $M_1 \approx M_2$ by an isomorphism which fixes A.

We have shown that any theory T has at most one prime model over the empty set. The attempt to describe those theories which have prime models over the empty set leads easily to a sufficient condition (Definition 1.13) for a theory to have a prime model over any set; the uniqueness question is more difficult. We now consider how to construct atomic models over sets. The actual construction plays a vital role. That is, we will show in Section 5 that any two prime constructible models are isomorphic, while leaving open the the possibility that there are prime models which are not constructible. An argument like that for Theorem 1.4 shows that for any set A (even uncountable) two atomic models over A are $L_{\infty,\omega}$-equivalent over A. No such general argument can show the models are isomorphic (even assuming they have the same cardinality). In Chapter X, invoking the notions of freeness as well as isolation, we will show that if T is stable and has prime models over all sets then prime models over sets are unique.

1.7 Definition. A *construction* C over a set A is a sequence $\langle c_j : j \in J\rangle$ such that for each j, $t(c_j; A \cup C_j)$ is isolated. The set B is *constructible* over A if $B = A \cup \langle c_j : j \in J\rangle$ for some construction C.

It is somewhat anomalous that the set B which is constructible over A contains A while the actual range of the constructing sequence may not. The following exercise shows this is only an apparent difficulty.

1.8 Exercise. Show that if B is constructible over A by C, there is a construction C' over A such that $\{c'_j : j \in J\} = B$.

It is essential to remember that a construction is a sequence and not just a set. For example, any embedding of a sequence of order type ω into a model of the theory of dense linear order (without endpoints) is a construction but an embeddding of a sequence of order type $\omega + 1$ is not.

1.9 Lemma. *If B is constructible over A then any elementary monomorphism of A into a model of T can be extended to an elementary embedding of B into that model. Thus, if M is a model which is constructible over A then M is prime over A.*

1.10 Exercise. Prove Lemma 1.9.

Our next goal is to show that if B is constructible over A then B is atomic over A. This relies on the following key property of isolation.

1.11 Lemma (Transitivity Property). *Let $B \subseteq A$. If $t(\overline{d}; A)$ is isolated and for each $\overline{c} \in A - B$, $t(\overline{c}; B)$ is isolated then for each $\overline{c}_1 \in A - B$, $t(\overline{d}^\frown\overline{c}_1; B)$ is isolated. In particular, $t(\overline{d}; B)$ is isolated.*

Proof. For some $\overline{c} \in A - B$, let $\phi(\overline{x}; \overline{c}) \in F(B)$ generate $t(\overline{d}; A)$. Fix $\overline{c}_1 \in A$ and suppose $\psi(\overline{y}_1{}^\frown\overline{y}_2)$ generates $t(\overline{c}_1{}^\frown\overline{c}; B)$. Now, the formula

$$(\exists \overline{y}_2)(\phi(\overline{x}; \overline{y}_2) \wedge \psi(\overline{y}_1; \overline{y}_2))$$

generates $t(\overline{d}^\frown\overline{c}_1; B)$. For, suppose $\chi \in F(B)$ and $\models \chi(\overline{d}^\frown\overline{c}_1)$. Then,

$$\models \psi(\overline{y}_1; \overline{y}_2) \rightarrow (\overline{x})(\phi(\overline{x}; \overline{y}_2) \rightarrow \chi(\overline{x}; \overline{y}_1)).$$

Since it is clear that a union of a chain of structures atomic over A is atomic over A, we can rephrase this as

1.12 Theorem. *If C is constructible over A then C is atomic over A.*

We have seen how to extend a set A to a set B which is atomic over A. Can we find a B which is the universe of a model and with $|B|$ not much greater than $|A|$? The following property of a theory guarantees that such a model B exists with $|B| = |A| + |T|$.

1.13 Definition. The *isolated types are dense* in $S(A)$ if for each formula $\phi(x; \overline{a})$ there is a principal type $p \in S(A)$ with $\phi(x; \overline{a}) \in p$. The theory T is *atomistic* if the isolated types are dense in $S(A)$ for every subset A.

Thus, the theory T is atomistic just if for each $A \subseteq \mathcal{M}$, the Boolean algebra $F(T)$ is atomic.

The following argument is similar to the proof in Lemma III.1.7 and Exercise III.1.8 that a theory is ω-stable if and only if it is stable in all infinite powers.

1.14 Theorem. *If T is a countable ω-stable theory then T is atomistic.*

Proof. If T is not atomistic there is a formula $\phi(\overline{x}; \overline{a})$ such that no atom intersects $\phi(\mathcal{M}; \overline{a})$. Choose by induction $\psi_s(\overline{x}; \overline{a}_s)$ for $s \in 2^{<\omega}$ so that both

$$(\exists \overline{x})[\psi_s(\overline{x}; \overline{a}_s) \wedge \psi_{s\frown 0}(\overline{x}; \overline{a}_{s\frown 0})]$$

and

$$(\exists \overline{x})[\psi_s(\overline{x}; \overline{a}_s) \wedge \neg\psi_{s\frown 0}(\overline{x}; \overline{a}_{s\frown 0})]$$

are true in $\mathcal{M}$. We can find such a $\psi_{s\frown 0}$ because $\psi_s(\overline{a}; \overline{a}_s)$ is not an atom. Let $\psi_{s\frown 1}$ be $\neg\psi_{s\frown 0}$. There are $2^{\aleph_0}$ types over the countable set $\{\overline{a}_s : s \in 2^{<\omega}\}$ so T is not ω-stable.

However, ω-stability is by no means a necessary condition for atomicity. For example, the theories of dense linear order without endpoints and REF_ω are atomistic. On the other hand, CEF_ω is a glaring example of a stable theory which is not atomistic (since it doesn't even have a prime model over any singleton). It is possible to construct examples of $\aleph_0$-categorical theories which are not atomistic.

1.15 Exercise. Show that if T is an $\aleph_0$-categorical theory then T admits prime models over finite sets. That is, for every finite subset, A, of $\mathcal{M}$ there is a prime model of T over A.

1.16 Exercise. Let T be the theory of a dense linear order without endpoints but with a unary predicate P which picks out a dense codense subset. Show that T is $\aleph_0$-categorical but that if $A = P(M)$ is an infinite set there is no prime model over A. (This example is due to Steinhorn.)

1.17 Theorem. *If T is atomistic, then for every A there is a model M of T with $|M| = |A| + |T|$ which is constructible over A.*

Proof. Suppose that C is a substructure of a model of T and enumerate the consistent members of $F_1(C)$ as $\langle \phi_i : i < |C| + |T| \rangle$. Choose a sequence $\langle a_i : i < |C| + |T| \rangle$ such that $t(a_i; C \cup C_i)$ is isolated in $S(C \cup C_i)$ and $\phi_i \in t(a_i; C \cup C_i)$. This is possible because T is atomistic. For any C, let $G(C) = C \cup \{a_i : i < |C| + |T|\}$. Now for the A of the theorem, for each n let $G^{n+1}(A) = G(G^n(A))$. Letting $M = \bigcup\{G^n(A)\}$ completes the construction.

1.18 Exercise. Show that if the countable theory T is atomistic then for every $A \subseteq N \models T$ there is a prime model over A.

1.19 Exercise. Show that if $S(\emptyset)$ is countable then T has a prime model.

1.20 Exercise. Show that if a countable theory T has prime models over every countable set then it has prime models over all sets.

1.21 Exercise. Show that if T is a small countable theory then T admits prime models over all finite sets.

1.22 Historical Notes. Call a model M *algebraically prime* over A if every embedding (not necessarily elementary) of A into a model, N, of T extends to an embedding of M into N. Abraham Robinson introduced the notion of prime models in [Robinson 1956]. However, he worked in the context of model complete theories so the distinction between algebraically and elementarily prime models was obscured. Most of this section is taken from Vaught's beautiful paper [Vaught 1961]. The extension to prime models over sets was made by [Morley 1965]. The contrast between algebraically prime and elementarily prime is explored in [Baldwin & Kueker 1981].

2. The General Notion of Isolation

We give here a semi-abstract treatment of the notion of generation. The key to such a treatment is to describe what it means for a sequence to be in the model 'generated' by a set. The most primitive notion of this sort requires the sequence to be given by a term. Increasingly more general notions require the sequence to be in a finite set definable with parameters from the given set or to realize a principal type generated by a formula with parameters from the given set. Shelah gives in Chapter IV of [Shelah 1978] an axiomatic treatment to include many such notions. Our approach here is to extend directly from the last idea. We will deal with several extensions of the word 'formula' but consider a sequence to be 'generated' by a set if all its relations to that set are determined by a 'formula' over the set.

Our approach to isolation is more concrete than that in [Shelah 1978]. We proceed by discussing three families **I** of 'sets of formulas'. We express the idea that $\overline{a}$ is generated by B by saying $t(\overline{a}; B)$ is implied by some member of **I** whose domain is contained in B. At this level of generality we are able to handle the most important among the notions Shelah discusses. We can generalize somewhat more by replacing 'p is implied by a member of **I**' by some other relation. (Thus, in Section 5 we study the notion given by 'p is locally implied by a member of **I**'.)

As in Chapter II, for each of the exercises in this section all of the preceding axioms may be used unless the exercise expressly says otherwise. We require the following convention.

2.1 Notation. If q has the form $t(\overline{c}; A)$ then $\operatorname{dom}^* q = \operatorname{dom} q = A$; if q has the form $stp(\overline{c}, A)$ then $\operatorname{dom}^* q = A$.

2.2 Definition. A collection of **I**-formulas is one of the following collections:

i) $\mathbf{AT}_\lambda$ is the collection of all types containing fewer than λ first order formulas.
ii) $\mathbf{SET}_\lambda$ is the collection of all types which are over a set of power less than λ.
iii) $\mathbf{S}_\lambda$ is the collection of all types which are almost over a set of power less than λ.

We let **I** vary over $\mathbf{SET}_\lambda$, $\mathbf{AT}_\lambda$, and $\mathbf{S}_\lambda$. Each of these sets of types is given with a cardinal parameter λ which is referred to as $\lambda(\mathbf{I})$. We often abbreviate $\mathbf{AT}_{\aleph_0}$ by **AT** and $\mathbf{S}_{\kappa(T)}$ by **S**. Thus, if **I** is $\mathbf{AT}_\lambda$, an **I**-formula is a type p with $|p| < \lambda$. In particular, if **I** is $\mathbf{AT}_{\aleph_0}$, which we often abbreviate as **AT**, then an **I**-formula is an ordinary first order formula.

These notions correspond to those introduced by Shelah as follows: $\mathbf{AT}_\lambda$ to F^t_λ, $\mathbf{SET}_\lambda$ to F^s_λ, and $\mathbf{S}_\lambda$ to F^a_λ. To obtain some of Shelah's more exotic notions in this framework we must generalize the notion of implication. Thus, the set F^f_λ on page 155 of [Shelah 1978] would come from $\mathbf{SET}_\lambda$ by replacing 'implies' by 'does not fork over'. We have not investigated exactly what properties of the implication notion are needed to make such generalizations work.

In his more abstract treatment Shelah takes the following property as the definition of $\lambda(\mathbf{I})$.

2.3 Exercise. For each collection of formulas **I**, $\lambda(\mathbf{I})$ is the least cardinal such that for each $p \in I$, $|\operatorname{dom}^* p| < \lambda(\mathbf{I})$.

2.4 Exercise. $\mathbf{AT}_{\aleph_0}$ is the notion discussed in Section IX.1.

The following properties are easily seen to hold for each **I**.

2.5 Proposition. *Every set* **I** *of formulas satisfies the following closure conditions.*

i) **I** *is closed under isomorphism.*
ii) *If* $q \in \mathbf{I}$ *then any type (or strong type) obtained by changing the order or names of variables in* q *is also in* **I**.
iii) *Any subtype of a type (strong type) in* **I** *is in* **I** .
iv) *Every type consisting of a single first order formula is in* **I**.
v) *If* $\lambda(\mathbf{I})$ *is regular then* **I** *is closed under increasing unions of length less than* $\lambda(\mathbf{I})$. *More generally, if the consistent type* $p = \bigcup p_i$, *with each* $p_i \in \mathbf{I}$ *for* $i < \operatorname{cf}(\lambda(\mathbf{I}))$, *then* $p \in \mathbf{I}$.

2.6 Exercise. Show every algebraic type is in **I**.

Note that $\mathbf{AT}_\lambda \subseteq \mathbf{SET}_\lambda$ and if $\lambda > |T|$, $\mathbf{AT}_\lambda = \mathbf{SET}_\lambda$.

The distinctions between the several variations on the term 'isolated' made in the next definition are extremely important. Pay particular attention to the distinction between 'isolated over' and 'isolated by'.

2.7 Definition. i) The type p is **I** *-isolated* by $q \in \mathbf{I}$ if $q \vdash p$.

ii) The type p is **I**-*isolated over* A if for some $q \in \mathbf{I}$, $q \vdash p$ and dom* q is contained in A.

iii) The type p is **I**-*isolated by* A if for some $q \in \mathbf{I}$, $q \vdash p$ and dom* $q = A$.

iv) The type p is **I** *-isolated* if for some $q \in I$, p is **I**-isolated by q.

If $p = t(\bar{c}; B)$ is **I**-isolated over A and $|A| < \lambda(\mathbf{I})$, note that $stp(\bar{c}; A) \vdash p$. (If $\mathbf{I} = \mathbf{SET}_\lambda$ or $\mathbf{AT}_\lambda$ we can replace $stp(\bar{c}; A)$ by $t(\bar{c}; A)$.) Moreover if p is **I**-isolated by A then $|A| < \lambda(\mathbf{I})$. This last conclusion fails if 'by A' is replaced by 'over A'.

2.8 Exercise. Use Lemma IV.3.12 to show that $t(a; B)$ is **S**-isolated by $C \subseteq B$ if and only if $stp(a; C) \vdash stp(a; B)$.

There are several monotonicity and continuity properties which are easily seen to hold for any notion of isolation.

2.9 Exercise (Montonicity). If $p \subseteq q \subseteq r$ and r is **I**-isolated by p then q is **I**-isolated by p.

2.10 Exercise. (First Continuity Property) If $\langle p_i : i \in \alpha \rangle$ is an increasing sequence of types, each p_i is **I**-isolated, and $\alpha < \mathrm{cf}(\lambda(\mathbf{I}))$ then $\bigcup p_i$ is **I**-isolated.

This result easily yields the following more useful formulation.

2.11 Exercise. If $|F| < \mathrm{cf}(\lambda(\mathbf{I}))$ and for each $\bar{f} \in F$, $t(\bar{a}; B \cup \bar{f})$ is **I**-isolated then $t(\bar{a}; B \cup F)$ is **I**-isolated.

2.12 Exercise. (Second Continuity Property) If $\langle p_i : i \in \alpha \rangle$ is an increasing sequence of types, p_0 is complete, each p_i is **I**-isolated over B, and $|B| < \lambda(\mathbf{I})$ then $\bigcup p_i$ is **I**-isolated over B.

2.13 Exercise. Show the necessity of restricting the cardinality of J or B in the continuity properties. (Hint: consider $t(\sqrt{2}; Q)$ in the theory of dense linear order.)

2.14 Exercise. Suppose $t(\bar{a}; B)$ is **I**-isolated and $\bar{b} \in B$. Show $t(\bar{a}^\frown \bar{b}; B)$ is **I**-isolated.

Now we extend the notion of a tuple being isolated over a set to embrace 'the set B is isolated over the set A.' Since each set of **I**-formulas is closed under permutations of the variables, whether $t(\bar{b}; A)$ is isolated over A depends on the set enumerated by $\bar{b}$, not on the particular enumeration. As usual we extend the notion to infinite sets by imposing a requirement of finite character. Note, however, that just because every finite sequence from B is isolated over A, it does not mean that $t(B; A)$ is actually isolated over A. (Consider the case when $|A|$ and $|B|$ are both much greater than $\lambda(\mathbf{I})$.) Accordingly, we introduce a new term: **I**-atomic.

2.15 Definition. Let $A \subset B$. If each finite sequence $\bar{b}$ from B is **I**-isolated over A then B is **I**-*atomic* over A. If, in addition, B is the universe of a model M, we say M is an **I**-*atomic model* over A.

Although the following notion is introduced as a technical tool, it has a life of its own. A number of notions in stability theory can be profitably investigated by looking at the construction rather than the result. We will see this distinction at several stages in the proof of the uniqueness of prime models. Another example is Shelah's use of F^f_λ as a way of describing independent families of models (cf. Section 6 and [Shelah 1978]).

2.16 Definition. An **I**-*construction* over A is a sequence $E = \{e_j : j \in J\}$ such that for each $j \in J$, $p_j = t(e_j; A \cup E_j)$ is **I**-isolated. When **I** is understood we frequently simply say construction.

We sometimes write an **I**-construction as $\{\langle e_j, B_j\rangle : j \in J\}$, where p_j is **I**-isolated by B_j and $B_j \subset A \cup E_j$, if we need to keep track of the 'witnesses' to the isolation of p_j over $A \cup E_j$.

To avoid double subscripts when quantifying over the elements of an **I**-construction, when $x = e_j$ for some j, we write $Pr(x)$ for the set of e_i with $i < j$ and B_x for the witness to the isolation of $t(x; A \cup Pr(x))$.

The set C is **I**-*constructible* over A if $C = A \cup E$ for some **I**-construction E. Note, however, that a construction is a sequence; thus when we refer to a construction we know both the order of construction and the way in which the p_j were isolated during the construction. We will frequently write C_j for $A \cup E_j$.

2.17 Lemma. *Suppose $\lambda(\mathbf{I})$ is regular. If B is **I**-constructible over $A \subset B$ and $C \subset B$ then there is a C' with $|C'| < |C|^+ + \lambda(\mathbf{I})$ such that $C \subset C' \subseteq B$ and $A \cup C'$ is **I**-constructible over A.*

Proof. For any set $X \subseteq B$ let $X' = \bigcup\{B_x : x \in X\}$; let $X^{n+1} = X \cup (X^n)'$ and let $X^* = \bigcup\{X^n : n < \omega\}$. Note that $|X^*| < \lambda(\mathbf{I}) + |X|^+$. Now C^* is the required subset of B.

As in the previous section, we want the result of any construction to be atomic over the original set. The following axiom specializes to Lemma 1.11 by letting $\mathbf{I} = \mathbf{AT}$ and setting $A = B$.

2.18 Transitivity Axiom. Let $B \subset A$ and let $|B|, |C| < \lambda(\mathbf{I})$. Suppose $t(\bar{d}; A \cup C)$ is **I**-isolated by $B \cup C$ and for each $\bar{c} \in C$, $t(\bar{c}; A)$ is **I**-isolated by B then for each $\bar{c} \in C$, $t(\bar{d}^\frown\bar{c}; A)$ is **I**-isolated by B.

2.19 Exercise. Show that if $\lambda(\mathbf{I})$ is regular then the transitivity axiom is implied by the following simpler formulation. If $t(\bar{d}; A \cup C)$ is **I**-isolated over $A \cup C$ and for each $\bar{c} \in C$, $t(\bar{c}; A)$ is **I**-isolated over B then for each $\bar{c} \in C$, $t(\bar{d}^\frown\bar{c}; A)$ is **I**-isolated over B.

To verify that $\mathbf{SET}_\lambda$ satisfies the transitivity axiom, we need only the following lemma.

2.20 Lemma. *If $B \subseteq A$, $t(\bar{d}; B \cup C) \vdash t(\bar{d}; A \cup C)$ and for each $\bar{c} \in C$, $t(\bar{c}; A)$ is implied by $t(\bar{c}; B)$ then for any $\bar{c}' \in C$, $t(\bar{d}\frown\bar{c}'; B) \vdash t(\bar{d}\frown\bar{c}'; A)$.*

Proof. Fix $\bar{c}' \in C$ and a formula $\phi(\bar{x}, \bar{y}; \bar{a})$ such that $\models \phi(\bar{d}, \bar{c}'; \bar{a})$. Since $t(\bar{d}; B \cup C) \vdash t(\bar{d}; A \cup C)$ there is a formula $\psi(\bar{x}, \bar{c}'; \bar{b}, \bar{c}_1)$ and a $\bar{c}_1$ in C with $\models \psi(\bar{d}, \bar{c}'; \bar{b}, \bar{c}_1)$ and $\models \psi(\bar{x}, \bar{c}'; \bar{b}, \bar{c}_1) \rightarrow \phi(\bar{x}; \bar{c}', \bar{a})$. Now, because $t(\bar{c}\frown\bar{c}_1; B) \vdash t(\bar{c}'\frown\bar{c}_1; A)$, we can choose a formula $\chi(\bar{y}, \bar{z}; \bar{b}_1)$ such that $\models \chi(\bar{c}', \bar{c}_1; \bar{b}_1)$ and

$$\chi(\bar{y}, \bar{z}; \bar{b}_1) \rightarrow (\forall \bar{x})[\psi(\bar{x}, \bar{y}, \bar{b}, \bar{z}) \rightarrow \phi(\bar{x}, \bar{y}; \bar{a})].$$

Now the formula $(\exists \bar{z})(\chi(\bar{y}, \bar{z}, \bar{b}_1) \wedge \psi(\bar{x}, \bar{y}, \bar{b}, \bar{z}))$ is in $t(\bar{d}\frown\bar{c}'; B)$ and implies $\phi(\bar{x}, \bar{y}; \bar{a})$. Choosing such a formula for each $\phi(\bar{x}, \bar{y}; \bar{a}) \in t(\bar{d}\frown\bar{c}'; A)$ shows $t(\bar{d}\frown\bar{c}'; B) \vdash t(\bar{d}\frown\bar{c}'; A)$.

2.21 Exercise. Prove that if λ is regular the transitivity axiom holds for $\mathbf{SET}_\lambda$ and for $\mathbf{AT}_\lambda$.

In Section 3 we discuss in more detail the *transposition axiom* (Axiom V.1 of [Shelah 1978], Axiom IX.3.8 here): If $t(\bar{a}\frown\bar{b}; C)$ is $\mathbf{I}$-isolated by D then $t(\bar{a}; C \cup \bar{b})$ is $\mathbf{I}$-isolated by $D \cup \bar{b}$. The following exercise shows that in the presence of the transposition axiom we obtain another simplification of transitivity.

2.22 Exercise (Weak Transitivity). Show that if $\mathbf{I}$ satisfies the transposition axiom and the weakening of the transitivity axiom obtained by assuming that $t(\bar{d}; A \cup C)$ is isolated by C rather than $B \cup C$ then $\mathbf{I}$ satisfies the transitivity axiom. transitivity of isolation weak transitivity of isolation

In Section 3 we follow the treatment in [Pillay 1983a] and verify the transitivity axiom for $\mathbf{S}_\lambda$ by first verifying the transposition axiom and then the simpler version of transitivity. Illustrating the power of T^{eq}, we give here a direct proof (due to Shelah) of transitivity for $\mathbf{S}_\lambda$.

2.23 Lemma. *For every λ, $\mathbf{S}_\lambda$-isolation satisfies the transitivity axiom.*

Proof. As assumptions we have

i) $stp(\bar{d}; B \cup C) \vdash t(\bar{d}; A \cup C)$ and
ii) $stp(C; B) \vdash t(C; A)$.

We must show that for any $\bar{c} \in C$,

$$stp(\bar{d}\frown\bar{c}; B) \vdash t(\bar{d}\frown\bar{c}; A).$$

Translating to T^{eq} by Theorem VIII.1.10 we have

i)′ $t^{eq}(\bar{d}; \mathrm{cl}(B \cup C)) \vdash t^{eq}(\bar{d}; \mathrm{cl}(A \cup C))$ and
ii)′ $t^{eq}(C; \mathrm{cl}(B)) \vdash t^{eq}(C; \mathrm{cl}(A))$.

Since $\mathrm{cl}(B)$ is algebraically closed, types and strong types over $\mathrm{cl}(B)$ (in L^{eq}) are interchangeable. Applying Lemma IV.3.16 (in T^{eq}) we have

$$t^{\mathrm{eq}}(\mathrm{cl}(B \cup C); B) \vdash t^{\mathrm{eq}}(\mathrm{cl}(B \cup C); A).$$

To use Lemma 2.19, we rewrite and and apply monotonicity to get

i)″ $t^{\mathrm{eq}}(\overline{d}; \mathrm{cl}(B) \cup \mathrm{cl}(B \cup C)) \vdash t^{\mathrm{eq}}(\overline{d}; \mathrm{cl}(A) \cup \mathrm{cl}(B \cup C))$ and
ii)″ $t^{\mathrm{eq}}(\mathrm{cl}(B \cup C); \mathrm{cl}(B)) \vdash t^{\mathrm{eq}}(\mathrm{cl}(B \cup C); \mathrm{cl}(A))$.

By Lemma 2.19 (used in T^{eq}), for any $\overline{e} \in \mathrm{cl}(B \cup C)$,

$$t^{\mathrm{eq}}(\overline{d}^\frown\overline{e}; \mathrm{cl}(B)) \vdash t(\overline{d}^\frown\overline{e}; \mathrm{cl}(A)).$$

Applying Lemma IV.3.16 in the other direction, we have

$$stp(\overline{d}^\frown\overline{e}; B) \vdash t(\overline{d}^\frown\overline{e}; A)$$

and we finish.

We return to the exposition of the general theory of isolation relations.

2.24 Theorem. *Let* **I** *satisfy the transitivity axiom and suppose* $\lambda(\mathbf{I}) = \lambda$ *is regular. If* C *is* **I***-constructible over* A *then* C *is* **I***-atomic over* A.

Proof. We proceed by a series of observations.

i) If $t(\overline{b}; A)$ is **I**-isolated then $A \cup \overline{b}$ is **I**-atomic over A. This is immediate from the transitivity axiom. Specifically, a typical element of $A \cup \overline{b}$ has the form $\overline{a}^\frown\overline{b}$ where $\overline{a} \in A$. Certainly, $t(\overline{a}; A)$ is **I**-isolated over A, say by $\overline{a}$, and by hypothesis, $t(\overline{b}; A)$ is **I**-isolated over A by some B. If $B' = B \cup \overline{a}$ then by the transitivity axiom, $t(\overline{a}^\frown\overline{b}; A)$ is **I**-isolated over A by B', as required.

ii) Suppose $A_1 \subseteq A_2 \subseteq A_3$, A_3 is **I**-atomic over A_2 and A_2 is **I**-atomic over A_1; then A_3 is **I**-atomic over A_1. Suppose $\overline{c} \in A_3$ and let $p = t(\overline{c}; A_2)$. Then there is a $B \subseteq A_2$ with $|B| < \lambda$ and p is **I**-isolated by B. For each $\overline{b} \in B$, let $q_{\overline{b}} = t(\overline{b}; A_1)$ and choose $B_{\overline{b}} \subseteq A_1$ with $|B_{\overline{b}}| < \lambda$ such that $q_{\overline{b}}$ is **I**-isolated by $B_{\overline{b}}$. Let $B_0 = \cup\{B_{\overline{b}} : \overline{b} \in B\}$. Since λ is regular, $|B_0| < \lambda$. Now p is **I**-isolated over $B \cup B_0$ by the monotonicity property and for each $\overline{d} \in B$, $t(\overline{d}; A_1)$ is **I**-isolated over B_0; so by the transitivity axiom, $t(\overline{c}^\frown\overline{d}; A)$ is **I**-isolated over B_0 for each $\overline{d} \in B$. Thus, by i) A_3 is **I**-atomic over A_1.

iii) The theorem now follows easily by induction on the length of the construction.

Note that the regularity of $\lambda(\mathbf{I})$ is needed for the second step.

The following notion tells us when in a construction we have realized 'enough' types.

2.25 Definition. The set C is **I**-*saturated* if for every $B \subseteq C$, whenever $p \in S(B)$ is **I**-isolated then p is realized in C. A model M is **I**-*prime* over A if every elementary embedding of A into an **I**-saturated model, N, can be extended to an elementary embedding of M into N. If M is **I**-constructible over A and **I**-saturated we say M is *strictly* **I**-*prime* over A. If for each $A \subseteq \mathcal{M}$ there is an **I**-prime model over A, we say T *admits I-prime models.* $\lambda_0(\mathbf{I})$ denotes the cardinality of the **I**-prime model over $\emptyset$.

Shelah calls the notion we refer to as strictly **I**-prime, **I**-primary. An **I**-saturated model need not be **I**-prime.

Thus, if T is the theory of algebraically closed fields of characteristic zero, the $\mathbf{SET}_{\aleph_0}$-prime model of T is the algebraically closed field with transcendence degree $\aleph_0$ over the rationals.

2.26 Exercise. Show that any model of REF_ω, indeed of any theory, is **AT**-saturated, but REF_ω has no **AT**-prime model over the empty set.

The usual notion of λ-saturation is just **I**-saturation when **I** is $\mathbf{SET}_\lambda$. For $\mathbf{I} = \mathbf{AT}_\lambda$ we get the notion of a λ-compact model. With this definition we can link the notion of **S**-isolation introduced in this section with the concept of strong saturation considered in Section III.2. The next lemma verifies for $\mathbf{S}_\lambda$ the following observation (which obviously holds for $\mathbf{AT}_\lambda$ and $\mathbf{SET}_\lambda$). The model M is **I**-saturated if and only if every consistent **I**-formula over M is realized in M.

2.27 Lemma. *If $\lambda \geq \kappa(T)$, the model M is strongly λ-saturated iff M is $\mathbf{S}_\lambda$-saturated.*

Proof. Suppose first that M is strongly λ-saturated. We have to show that if $A \subseteq M$, $|A| < \lambda$, and q is almost over A then q is realized in M. Let $\overline{c}$ realize q and let $\overline{d}$ realize an extension of q to a type over $A \cup \overline{c}$ which does not fork over A. Then $t(\overline{d}; A \cup \overline{c})$ is finitely satisfiable almost over A. In particular, $t(\overline{d}; A \cup \overline{c})$ is finitely satisfied in M. By the definition of strong λ-saturation, $t(\overline{d}; A \cup \overline{c})$ and a fortiori q is realized in M.

For the converse, suppose $|B| < \lambda$ and $t(\overline{c}; B)$ is finitely satisfied in M. That is, $\overline{c} \downarrow_M B$. Since $\lambda \geq \overline{\kappa}(T)$, we can choose $A \subseteq M$ with $|A| < \lambda$ such that $B \downarrow_A M$ and $\overline{c} \downarrow_A M$. Then by transitivity $\overline{c} \downarrow_A B$. Now let $\overline{d} \in M$ realize $stp(\overline{c}; A)$. Then $\overline{d} \downarrow_A B$ and $\overline{c} \downarrow_A B$ and they realize the same strong type over A, so $\overline{d}$ realizes $t(\overline{c}; B)$ as required.

Recall that the existence of prime models in Section 1 depended on the density of principal types. The next axiom extends this idea to the general case.

2.28 Existence Axiom. If p is an **I**-formula over $A \subseteq B$ then there is a $q \in S(B)$ such that q is **I**-isolated by an **I**-formula q_0 which implies p.

For $\mathbf{SET}_\lambda$ and $\mathbf{AT}_\lambda$ this axiom can be simplified by saying any **I**-formula extends to an **I**-isolated type. The actual formulation is more clumsy because of the difficulty in describing an 'extension' of a type almost over A.

2.29 Exercise. Show the Existence Axiom implies that for every A and every consistent formula $\phi(\overline{x}; \overline{a})$ there is a $p \in S(A)$ which is **I**-isolated over some subset of A and with $\phi(\overline{x}; \overline{a}) \in p$.

2.30 Exercise. Show that if $p \in S(A)$, $B \subseteq A \subseteq C$, and p is **I**-isolated over B then there is a $q \in S(C)$ such that $p \subseteq q$ and q is **I**-isolated.

2.31 Exercise. Show that if T is ω-stable (and countable), T satisfies the Existence Axiom for **AT**.

It is somewhat more difficult to verify the Existence Axiom for $\mathbf{S}_\lambda$. The key observation is that if $A \subseteq B$ and $p \in S(A)$ has no extension to a type in $S(B)$ which forks over A then for any $\bar{c}$ realizing p, $stp(\bar{c}; A) \vdash stp(\bar{c}; B)$.

2.32 Theorem. *For $\lambda \geq \kappa(T)$, $\mathbf{S}_\lambda$ satisfies the Existence Axiom.*

Proof. Let $A \subseteq B$ and let $p = stp(\bar{c}; A)$ for some $\bar{c}$. Let $p_0 = t(\bar{c}; A)$. If suffices to find a type p' with the domain of p' equal to A' with $A \subseteq A' \subseteq B$ and $|A'| < \lambda$ such that $p_0 \subseteq p'$ and no extension of p' to a complete type over B forks over A'. For, then if d realizes a nonforking extension of p' to a type over B, $stp(d; A')$ is the required q_0. To find p', choose inductively a set of pairs (p_i, A_i) with $A = A_0 \subseteq A_i \subseteq B$ and $p_i \subseteq p_{i+1}$ such that p_{i+1} forks over $A_i = \operatorname{dom} p_i$ and for each i, $|A_i| < \kappa(T) + \lambda$. Using the local character of forking it is easy to construct such a sequence. However, by the definition of $\kappa(T)$ it must cease with some (p_i, A_i) for $i < \kappa(T)$ and p_i is an appropriate p'.

2.33 Exercise. If T is superstable, prove the $\mathbf{S}_{\kappa(T)}$-isolated types are dense in $S(A)$ for each A. (Hint: For each formula $\psi(\bar{x}; \bar{a})$ such that $\phi \to \psi$, minimize $R_C(\phi(\bar{x}; \bar{b}))$.)

2.34 Exercise. If **I** satisfies the Existence Axiom, C is **I**-saturated if every $p \in S(C)$ which is **I**-isolated is realized in C.

2.35 Exercise. Show that if C is **AT**-saturated and T is atomistic then C is a model of T.

We want to construct for any A, an **I**-saturated B containing A and, if possible, of the same cardinality as A. To describe such a construction we require one further invariant. This invariant allows us to discuss the construction of S-saturated models in terms of the realization of ordinary types.

2.36 Definition. $\mu(\mathbf{I})$ is the least cardinal ρ such that any set D which satisfies the following condition on D and ρ is **I**-saturated. If $B \subseteq D$, $|B| < \rho$, and $q \in S(B)$ is **I**-isolated then q is realized in D. If no such ρ exists, $\mu(\mathbf{I}) = \infty$.

$\mu(\mathbf{I})$ is an entirely different notion from the $\mu(T)$ discussed in Section III.4.

2.37 Exercise. Show $\mu(\mathbf{AT}_\kappa) \leq \kappa$ and $\mu(\mathbf{SET}_\kappa) \leq \kappa$.

Thus $\mu(\mathbf{I}) \leq \lambda(\mathbf{I})$ in two of the common examples. However, for $\mathbf{S}_\lambda$, this inequality is no longer always true. The obstruction arises from the fact that a complete type p over a set A may be implied by many different strong types over finite subsets of A.

2.38 Exercise. Show that if T is REF_ω, then $\mu(\mathbf{S}) = \aleph_1$.

2.39 Lemma. $\mu(\mathbf{S}) = \kappa(T) + \aleph_1$.

Proof. Let $\lambda = \kappa(T) + \aleph_1$. Fix a D such that if $B \subseteq D$, $q \in S(B)$, and $|B| < \lambda$ and q is S-isolated then q is realized in D. Let $C \subseteq D$ and suppose $p \in S(C)$ is S-isolated by A. We must show p is realized in D. By Lemma 2.27 choose an extension $\hat{p}$ of p to $S(D)$ which is also S-isolated. Without loss of generality, we may assume $\hat{p}$ is S-isolated over A. Let $\bar{a} \in \mathcal{M}$ realize $\hat{p}$. Now choose $\langle e_i : i \leq \omega\rangle \subseteq D$ such that $t(e_i; A \cup E_i) = t(\bar{a}; A \cup E_i)$. Since for each $i < \omega$, $|A \cup E_i| < \lambda$, we may choose the $\bar{e}_i \in D$. Now, as in Theorem V.2.2, $\bar{e}_\omega$ realizes p.

Now we show that if the Existence Axiom holds, T admits **I**-prime models.

2.40 Theorem. i) *If* **I** *satisfies the Existence Axiom and* $\mu(\mathbf{I}) < \infty$ *then every set* A *is contained in an* **I**-*saturated* B *which is* **I**-*constructible over* A *and with* $|B| \leq |A + 2|^{\mu(\mathbf{I}) + |T|}$.

ii) *If* $\mu(\mathbf{I}) = \aleph_0$ *then* $|B| \leq |A| + \aleph_0$.

Proof. i) For any set C define C' as follows. Enumerate all types p such that for some D with $D \subseteq C$, and $|D| < \mu(\mathbf{I})$, $p \in S(D)$. Note that the number of such types is less than $\max(|C|^{<\mu(\mathbf{I})}, \mu(\mathbf{I})^{|T|})$. Build a sequence $E = \langle e_i : i < \alpha\rangle$ by letting each e_i realize an extension of the ith such type to a complete type over $E_i \cup C$ which is **I**-isolated. Such an extension is guaranteed by the Existence Axiom. Let $C' = C \cup E$. Now define by induction $\langle C_\gamma : \gamma < \mu(\mathbf{I})^+\rangle$ with $C_0 = A$ and $C_{\gamma+1} = C'_\gamma$ and taking unions at limits. By the definition of $\mu(\mathbf{I})$, $C_{\mu(\mathbf{I})^+}$ is **I**-saturated. Note that by induction $|C_\gamma| \leq |\gamma| \times |A|^{\mu(\mathbf{I}) + |T|}$ so $|C_{\mu(\mathbf{I})^+}| \leq |A + 2|^{\mu(\mathbf{I}) + |T|}$.

ii) Under the additional hypothesis, $|C'| \leq \max(|C|, \mu(\mathbf{I})) = |C| + \aleph_0$. Thus, $|B| \leq |A| + \aleph_0$.

Of course, for **AT** if we can construct **I**-saturated models containing A at all, we can require their cardinality to be the same as $\max\{|A|, |T|\}$. But such a strong cardinality restriction is easily seen to fail if we extend even to $\mathbf{AT}_{\aleph_1}$.

2.41 Exercise. Show that if T is countable, $\lambda_0(\mathbf{S}) \leq 2^{\aleph_0}$.

2.42 Corollary. *If* **I** *satisfies the Existence Axiom then for every set* A *there is a strictly* **I**-*prime model* M *over* A *with* $|M| \leq (|A| + 2)^{|T| + \mu(\mathbf{I})}$.

Note that if **I** satisfies the Existence Axiom, every **I**-saturated structure is a model. In particular, if B is **I**-prime over A, then B is a model.

2.43 Exercise. The $\mathbf{AT}_{\aleph_1}$-prime model over the empty set of the theory of infinitely many independent unary predicates has power of the continuum.

2.44 Exercise. The $\mathbf{S}_{\aleph_0}$-prime model of REF_ω has power $2^{\aleph_0}$.

2.45 Exercise. Show if T is an uncountable theory such that for every $A \subseteq \mathcal{M}$, $F(A)$ is atomistic then T admits **AT**-prime models over all sets.

It is easy to prove the following result by induction.

2.46 Theorem. i) *If B is* **I***-constructible over A and $C \supseteq A$ is* **I***-saturated then there is an elementary embedding of B into C.*

ii) *If B is* **I***-constructible over A, and B is* **I***-saturated then B is* **I***-prime over A.*

We can also show that every **I**-prime stucture is **I**-atomic.

2.47 Theorem. *If B is* **I***-prime over A, then B is* **I***-atomic over A.*

Proof. Choose an **I**-saturated set C containing A which is **I**-constructible as in Theorem 2.40. Then B can be embedded into C and C is **I**-atomic over A by Theorem 2.24 so B is **I**-atomic over A.

It is not nearly so easy to show that if B is **I**-prime over A then B is **I**-constructible over A. This has been shown only for countable stable theories which admit prime models over all sets. Indeed, the following question remains open. Let T be a countable stable theory and let M be prime over A (in the usual, **AT**, sense). Must M be **AT**-constructible over A?

2.48 Exercise (Sacks). The countable first order theory T is *quasi-totally transcendental* (q.t.t.) if for each A the subset of $S(A)$ consisting of points whose Morley rank is less than ∞ is dense in $S(A)$. Show that if T is q.t.t. **AT** still satisfies all the properties it was shown to satisfy for an ω-stable theory T in this section. (Only the Existence Axiom is nontrivial.) Conclude that T admits prime models over sets.

2.49 Exercise. Suppose $p \in S(A)$ is **AT**-isolated and the multiplicity of p is finite. Show that for any $\overline{c}$ realizing p, $stp(\overline{c}; A)$ is realized in every model containing A.

2.50 Exercise. Show that an increasing union of sets which are **I**-atomic over A is **I**-atomic over A. Conclude that if $A \subseteq N$ there is a maximal subset of N which is atomic over A. Show that the analogous result, replacing atomic by constructible fails.

2.51 Historical Notes. The importance of discussing several notions of isolation was discovered by Shelah. He presents two such families is [Shelah 1970] and [Shelah 1978]. This section is primarily a translation of Sections IV.1 and IV.2 of [Shelah 1978]. Shelah's Chapter IV contains not only a larger collection of isolation relations but a valuable chart summing up the properties of the various notions.

3. Bookkeeping Axioms for Isolation Relations

This brief section introduces some 'bookkeeping' axioms for isolation notions. We separate them from the axioms in Section 2 because they are somewhat more technical and are required only for the uniqueness as opposed to the existence of prime models. In the main these axioms deal with the continuing problem of how our various properties behave with respect to the concatenation of two sequences. That is they attempt to capture the relation between the isolation of the types $t(\overline{c}; A)$ and $t(\overline{d}; A)$ separately and the isolation of $t(\overline{c}\frown\overline{d}; A)$. We began this discussion for isolation with the transitivity axiom in Section 2. The necessity to justify these axioms could be regarded as a defect in our notation.

We show that under certain conditions if N is prime over A and $\overline{b} \in N$ then N is prime over $A \cup \overline{b}$. Note that the obvious statement of a symmetry principle for isolation, 'If $t(\overline{a}; A \cup \overline{b})$ is **I**-isolated then $t(\overline{b}; A \cup \overline{a})$ is **I**-isolated' is false. This principal obviously fails if $\overline{a} \in A$ and $\overline{b}$ realizes a nonisolated type over $A \cup \overline{a}$. The following example is a little more complex.

3.1 Exercise. Consider Ehrenfeucht's example [Vaught 1961] of a theory with 3 countable models: $\mathrm{Th}(Q, <, 0, 1, 2, ...)$. Let M be the model of T with a the least upper bound of $\{0, 1, 2, ...\}$ and $b > a$. Show $t(b; a)$ is **AT**-isolated but $t(a; b)$ is not.

3.2 Exercise. Find an example of two sequences $\overline{a}$ and $\overline{b}$ and a set A such that both $t(\overline{a}; A)$ and $t(\overline{b}; A)$ are **AT**-isolated but $t(\overline{a}\frown\overline{b}; A)$ is not.

We can, however, weaken this version of the symmetry property and obtain a principle both useful and true.

3.3 Symmetry Axiom for Isolation. If $t(\overline{a}; B)$ is **I**-isolated by $C \subseteq B$ and $t(\overline{b}; B)$ is **I**-isolated by $D \subseteq B$ then $t(\overline{a}; B \cup \overline{b})$ is **I**-isolated by C iff $t(\overline{b}; B \cup \overline{a})$ is **I**-isolated by D.

Remember that we formulated **I**-isolation in terms of '$\vdash$' and a collection of **I**-formulas. The difficulty with symmetry comes not from '$\vdash$' but from the difficulty in making the set of formulas in the hypothesis an **I**-formula. The key to seeing that the symmetry principle holds is to realize the notation $t(A; C) \vdash t(A; C \cup D)$ disguises the role of D. Remember that $t(A; C \cup D)$ may be thought of as a collection of sentences with names for the elements of A, C, and D. The side of the semicolon on which a particular name appears changes our view of the name but not its role in the set of sentences. Thus, $\mathrm{Diag}(A \cup B)$ and $t(A; B)$ contain the same formulas. In more detail, consider the meaning of the statements:

$$t(A; B) \vdash t(A; B \cup C) \text{ and } t(C; B) \vdash t(C; B \cup A).$$

The first means:

$$T \cup \mathrm{Diag}(B \cup C) \cup t(A; B) \vdash t(A; B \cup C).$$

The second means:

$$T \cup \mathrm{Diag}(B \cup A) \cup t(C; B) \vdash t(C; B \cup A).$$

Consideration of the latter two statements should easily show you that the first two are equivalent. In our earlier notation all of these statements mean $t(A; B) \perp^w t(C; B)$.

3.4 Exercise. Verify that **AT** satisfies this axiom.

3.5 Lemma. $\mathbf{AT}_\lambda$ *satisfies the symmetry axiom.*

Proof. From the hypotheses of the symmetry axiom, we have $p \vdash t(\overline{a}; B)$ and $q \vdash t(\overline{b}; B \cup \overline{a})$ where $p \subseteq t(\overline{a}; C)$, $q \subseteq t(\overline{b}; D)$ for some C, D contained in B and $|p|, |q| < \lambda$. That is,

$$p \cup \text{Diag}(B) \vdash t(\overline{a}; B)$$

$$q \cup \text{Diag}(B \cup \overline{a}) \vdash t(\overline{b}; B \cup \overline{a}).$$

But then,

$$p \cup \text{Diag}(B \cup \overline{b}) \vdash t(\overline{a}; B \cup \overline{b})$$

since $\text{Diag}(B \cup \overline{b}) \vdash q$ and $p \cup \text{Diag}(B) \vdash \text{Diag}(B \cup \overline{a})$. Thus, $t(\overline{a}; B \cup \overline{b})$ is $\mathbf{AT}_\lambda$-isolated by C.

Note that in the proof we did not use the hypothesis that $t(\overline{b}; B)$ is $\mathbf{I}$-isolated; it is required for the converse.

Invoking T^{eq} we proved the strong form of the transitivity axiom for $\mathbf{S}_\lambda$ in Lemma 2.23. Using a rudimentary form of symmetry for strong types, we will verify the weak version of transitivity for $\mathbf{S}_\lambda$. Recall from Exercise 2.22 that in the presence of transposition the two forms of transitivity are equivalent. Now we formally introduce the axiom of transposition and verify that it holds for $\mathbf{AT}_\lambda$, $\mathbf{SET}_\lambda$, and $\mathbf{S}_\lambda$. With the aid of transposition we will show that symmetry holds for $\mathbf{S}_\lambda$. We require two preliminary lemmas.

3.6 Lemma. $stp(\overline{a}; B) \vdash stp(\overline{a}; B \cup \overline{c})$ *implies* $stp(\overline{c}; B) \vdash stp(\overline{c}; B \cup \overline{a})$.

Proof. The hypothesis yields that for any $\overline{c}'$ realizing $stp(\overline{c}; B)$, $\overline{c}' \downarrow_B \overline{a}$. The lemma follows since strong types are stationary.

3.7 Lemma. *If* $t(\overline{d}; A \cup C)$ *is* $\mathbf{S}_\lambda$*-isolated by* C *and for each* $\overline{c} \in C$, $t(\overline{c}; A)$ *is* $\mathbf{S}_\lambda$*-isolated by* B *then for each* $\overline{c} \in C$, $t(\overline{d}^\frown\overline{c}; A)$ *is* $\mathbf{S}_\lambda$*-isolated by* B.

Proof. We want to show that $stp(\overline{d}^\frown\overline{c}; B) \vdash t(\overline{d}^\frown\overline{c}; A)$ for each $\overline{c} \in C$. By Lemma 3.6, it suffices to show that for each $\overline{a} \in A$,

$$stp(\overline{a}; B) \vdash stp(\overline{a}; B \cup C \cup \overline{d}).$$

To see this, fix $\overline{a} \in A$. By hypothesis, $stp(\overline{c}; B) \vdash stp(\overline{c}; A)$ so by Lemma 3.6, $stp(\overline{a}; B) \vdash stp(\overline{a}; B \cup \overline{c})$. But, $stp(\overline{d}; C) \vdash stp(\overline{d}; A \cup C)$ so clearly $stp(\overline{d}; B \cup C) \vdash stp(\overline{d}; A \cup C)$. For any $\overline{a} \in A$, by Lemma 3.6 again, $stp(\overline{a}; B \cup C) \vdash stp(\overline{a}; B \cup C \cup \overline{d})$. But then $stp(\overline{a}; B) \vdash stp(\overline{a}; B \cup C \cup \overline{d})$.

We often make use of the observation that $t(\overline{a}^\frown\overline{b})$ and $t(\overline{a}; \overline{b})$ carry the same information. When dealing with isolation we must be a little fussy. We encode this fussiness in the transposition axiom.

3.8 Transposition Axiom. If $t(\overline{a}^\frown\overline{b}; C)$ is $\mathbf{I}$-isolated by D then $t(\overline{a}; C \cup \overline{b})$ is $\mathbf{I}$-isolated by $D \cup \overline{b}$.

3.9 Lemma. *The transposition axiom holds for* $\mathbf{SET}_\lambda$.

Proof. Let $p = t(\overline{a}^\frown\overline{b}; C)$ and fix $q \in S(D)$ such that $q \vdash p$. Let q' be $\{\phi(\overline{x}; \overline{b}) : \phi(\overline{x}; \overline{y}) \in q\}$. Clearly $q' \vdash t(\overline{a}; C \cup \overline{b})$.

3.10 Exercise. Verify that $\mathbf{AT}_\kappa$ satisfies the transposition axiom.

The verification of this axiom for $\mathbf{S}_\lambda$ is only slightly more difficult.

3.11 Lemma. *$\mathbf{S}_\lambda$ obeys the transposition axiom.*

Proof. Note that $stp(\overline{a}; D \cup \overline{b}) = stp(\overline{a}'; D \cup \overline{b})$ implies (Lemma IV.3.19) $stp(\overline{a}\frown\overline{b}; D) = stp(\overline{a}'\frown\overline{b}; D)$. This, in turn, implies $t(\overline{a}\frown\overline{b}; C) = t(\overline{a}'\frown\overline{b}; C)$ which implies $t(\overline{a}; C \cup \overline{b}) = t(\overline{a}'; D \cup \overline{b})$ and we finish.

3.12 Lemma. *The symmetry axiom holds for $\mathbf{S}_\lambda$.*

Proof. We have $stp(\overline{a}; C) \vdash stp(\overline{a}; B \cup \overline{b})$ and $stp(\overline{b}; D) \vdash stp(\overline{b}; B)$. We want to show $stp(\overline{b}; D) \vdash stp(\overline{b}; B \cup \overline{a})$. Clearly, $stp(\overline{a}; C) \vdash stp(\overline{a}; B \cup \overline{b})$ implies $stp(\overline{a}; B) \vdash stp(\overline{a}; B \cup \overline{b})$. But then, Lemma 3.6 yields $stp(\overline{b}; B) \vdash stp(\overline{b}; B \cup \overline{a})$. As, $stp(\overline{b}; D) \vdash stp(\overline{b}; B)$ we have the result.

If $\lambda(\mathbf{I})$ is regular, we can extend the transposition axiom.

3.13 Lemma (Generalized Transposition). *Suppose $\mathbf{I}$ satisfies the transposition axiom, $|C| < \lambda(I)$, $\lambda(\mathbf{I})$ is regular, and $\overline{a} \cup C$ is $\mathbf{I}$-atomic over B then $t(\overline{a}; B \cup C)$ is $\mathbf{I}$-isolated.*

Proof. By the transposition axiom, for each $\overline{c} \in C$ $t(\overline{a}; B \cup \overline{c})$ is $\mathbf{I}$-isolated. By the first continuity property (Exercise 2.10) we have the result.

With the aid of the transitivity axiom, we can prove a converse to the transposition axiom which allows us to decompose sequences for isolation just as we did for independence in Chapter II.

3.14 Lemma. *Suppose $\mathbf{I}$ satisfies the transposition and transitivity axioms. Then the following are equivalent.*

i) *$t(\overline{a}\frown\overline{b}; A)$ is $\mathbf{I}$-isolated over B.*
ii) *$t(\overline{a}; A \cup \overline{b})$ is $\mathbf{I}$-isolated over $B \cup \overline{b}$ and $t(\overline{b}; A)$ is $\mathbf{I}$-isolated over B.*

Proof. The result is immediate from the axioms cited.

3.15 Historical Notes. Most of this material is from Chapter IV of [Shelah 1978]. The proof of transitivity for $\mathbf{S}$ without recourse to T^{eq} is taken from Chapter 8 of [Pillay 1983a]. The proof here of symmetry for $\mathbf{AT}_\lambda$ seems to be the natural attack. The more complicated argument used here for $\mathbf{S}_\lambda$ was used by Shelah in both cases.

4. Uniqueness Of Strictly Prime Models

In this section we prove, assuming that $\lambda(\mathbf{I})$ is regular, that if M and N are both $\mathbf{I}$-constructible over A and $\mathbf{I}$-saturated, that is, if M and N are strictly $\mathbf{I}$-prime over A, then M and N are isomorphic over A. In Chapter X, we will invoke properties of the freeness relation to show that this result extends to prime models over A if T is a countable stable theory which admits prime models. In particular, these arguments yield that prime models for two of the categories we have discussed in detail, **AT**-prime and **S**-prime, are unique, at least when the theory is countable and stable.

Since strictly prime models are prime it is immediate that two strictly prime models over a set A can each be embedded in the other. Thus, by the Schröder-Bernstein theorem they have the same cardinality. We want to extend this Schröder-Bernstein phenomena from the category of set mappings to the category of elementary embeddings. We will construct a back-and-forth to create the isomorphism.

There are three steps to the argument showing that there is a unique (up to isomorphism over A) **I**-saturated, **I**-constructible model over A. We first note that this result holds outright if the construction is not too long. Then we show how to break an arbitrary construction into pieces to which the first argument applies. Finally, we stitch the pieces together by the general methods of back-and-forths. The next two lemmas combine to prove, in Theorem 4.3, the 'short' case of the theorem. They follow easily from the generalized transposition property (Lemma 3.13) and we leave them as an exercise.

4.1 Lemma. *If $E = \langle e_i : i < \alpha\rangle$ is an **I**-construction over A, $C \subseteq E$, $|C| < \lambda(\mathbf{I})$, and $\lambda(\mathbf{I})$ is regular then $\langle e_i : i < \alpha\rangle$ is an **I**-construction over $A \cup C$.*

The following somewhat weaker conclusion also holds: $D = E - C = \langle d_j : j < \delta\rangle$ is constructible over $A \cup C$.

4.2 Lemma. *If $E = \langle e_i : i < \alpha\rangle$ is an **I**-construction over A and $B \subseteq E$ with $|B| < \lambda(\mathbf{I}))$ and $\lambda(\mathbf{I})$ is regular then B is **I**-constructible over A.*

4.3 Theorem. *If M and N are strictly **I**-prime over A and $|M| \leq \lambda(\mathbf{I})$ then $M \approx N$.*

4.4 Exercise. Prove Lemma 4.1, Lemma 4.2, and Theorem 4.3.

4.5 Exercise. If $|E| < \lambda(\mathbf{I})$, $\lambda(\mathbf{I})$ is regular, and E is constructible over A then E is constructible over A under any wellordering of E with ordinal less than $\lambda(\mathbf{I})$.

4.6 Exercise. Show that although the rationals may be constructed as an **AT**-prime model of the theory of dense linear order, there are wellorderings of type greater than ω under which they are not constructible.

4.7 Exercise. Suppose $|A|$, $|B| < \lambda(\mathbf{I})$ and $\lambda(\mathbf{I})$ is regular. Show that if $B \subseteq N$ for some N which is **I**-prime over $M \cup A$, then if N' is a strictly **I**-prime model over $M \cup B$ and $A \subseteq N'$, N' is strictly **I**-prime over $M \cup A$.

We are able to show the isomorphism exists in Theorem 4.3 because each initial segment of a short construction is sufficiently nice. For the general case, it suffices by [Kueker 1970] to find a set, J, of partial isomorphisms between M and N each with cardinality less than M such that J is closed under increasing unions with cardinality less than $\mathrm{cf}(|M|)$ and J satisfies the following back and forth conditions.

$$(\forall g \in J)(\forall m \notin \operatorname{dom} g)(\exists g_1 \in J)(g_1 \supseteq g \wedge m \in \operatorname{dom} g_1)$$
$$(\forall g \in J)(\forall n \notin \operatorname{rng} g)(\exists g_1 \in J)(g_1 \supseteq g \wedge n \in \operatorname{rng} g_1).$$

The actual construction of J will require further restrictions on the domains of the partial isomorphisms. That is, we must find some nice subsets of a construction such that if the domain and range of a partial isomorphism are nice then it can be extended. The crucial step in establishing uniqueness is to see that any subset of a construction is contained in a nice set which is not much larger than itself. The criteria for B being a nice subset of the construction of E over A is that both B should be constructible over A and $E - B$ should be constructible over $A \cup B$. The following definition provides a more concrete way to identify such subsets.

4.8 Definition. If $\{\langle e_i, B_i\rangle : i < \alpha\}$ is a construction over A, then a subset $C \subseteq E$ is *closed* (relative to the construction E) if for each $e_i \in C$, $B_i \subseteq C \cup A$.

If C is a closed subset of the construction E then we can reorder the construction and put C first. More formally:

4.9 Notation. i) Let $E = \langle e_i : i < \alpha\rangle$ and suppose $C \subseteq E$. Then $<_C$ is the following linear order on α.

a) If $e_i \in C$ and $e_j \in E - C$ then $i <_C j$.
b) Restrict the ordering $<$ to C and to $E - C$.

ii) Let γ be the ordinal of $<_C$ restricted to C and write $C = \{c_i : i < \gamma\}$. Let D denote $E - C = \{d_i : i < \delta\}$. Define maps $f : \delta \mapsto \alpha$ and $g : \gamma \mapsto \alpha$ by $f(i) = j$ if $e_j = d_i$ and $g(i) = j$ if $e_j = c_i$.

It is easy to see that if C is a closed subset of the construction E over A then $C \cup A$ is constructible over A by $\{\langle c_i, B_{g(i)}\rangle : i <_C \gamma\}$. We proceed to the somewhat harder fact that in this situation $E \cup A$ is constructible over $A \cup C$.

4.10 Lemma. *If C is a closed subset of the construction E over A then $E - C = D$ is constructible over $A \cup C$ by $\{\langle d_j, B_{f(j)}\rangle : j < \delta\}$.*

Proof. We have to show that for each $j < \delta$, $p_j = t(d_j; A \cup C \cup D_j)$ is **I**-isolated by $B_{f(j)}$. For this, we fix j and show by induction on $k < \gamma$ that $t(d_j; D_j \cup A \cup C_k \cup B_{f(j)})$ is **I**-isolated by $B_{f(j)}$. Since $C_0 = \emptyset$, the ground case of the induction is trivial. The second continuity property described in Exercise 2.12 guarantees that the condition is preserved at limit ordinals. So suppose $k = l + 1$. If c_k preceded d_j in the original ordering the result is clear by monotonicity. So suppose $f(j) < g(k)$ and thus that $B_{f(j)} \subseteq C_k \cup D_j \cup A$ with $t(c_k; A \cup C_l \cup D_{j+1})$ **I**-isolated. By induction, $t(d_j; D_j \cup C_l \cup A)$ is **I**-isolated by $B_{f(j)}$. Applying 3.3 (the symmetry axiom for isolation) $t(d_j; D_j \cup C_k \cup A)$ is **I**-isolated by $B_{f(j)}$ as required.

4.11 Corollary. *If E is constructible over A and $C \subseteq C' \subseteq E$ with C and C' closed, then C' is closed in the construction $\langle d_j : j < \delta \rangle$ of $A \cup D$ over $A \cup C$. Thus $A \cup C'$ is constructible over $A \cup C$.*

Proof. By Lemma 4.10, $t(d_j; A \cup C \cup D_j)$ is $\mathbf{I}$-isolated by $B_{f(j)}$. Since $<_C$ agrees with $<$ on D, $B_{f(j)} \subseteq D_j \cup C \cup A$.

It is even easier to see that if E is constructible over A, C is closed in E, and $D \subseteq E - C$ is closed in the $<_C$-ordered construction of $E - C$ over $A \cup C$ then $C \cup D$ is closed in E.

We now state the main theorem of this section. We then sketch the proof, indicating two lemmas which we still must prove. Then we will prove those two lemmas.

4.12 Theorem. *Let E and F be $\mathbf{I}$-prime and $\mathbf{I}$-constructible over A. If $\lambda(\mathbf{I})$ is regular then E is isomorphic to F over A.*

Proof. We noticed in Theorem 4.3 that it suffices to consider the case $|E| > \lambda(\mathbf{I})$. Let J be the collection of partial A-isomorphisms between E and F such that each $g \in J$ satisfies the following conditions.

i) $\operatorname{dom} g$ is a closed subset of E.
ii) $\operatorname{rng} g$ is a closed subset of F.
iii) $|\operatorname{dom} g| < |E|$

We must show that J is closed under the union of increasing chains with cardinality less than $\operatorname{cf}(|E|)$ and that J satisfies the following back and forth conditions.

$$(\forall g \in J)(\forall e \notin \operatorname{dom} g)(\exists g_1 \in J)(g_1 \supseteq g \wedge e \in \operatorname{dom} g_1)$$

$$(\forall g \in J)(\forall f \notin \operatorname{rng} g)(\exists g_1 \in J)(g_1 \supseteq g \wedge f \in \operatorname{rng} g_1).$$

The closure under increasing chains is immediate. In order to treat the successor stages we have to show each point can be embedded in a small closed set (Lemma 4.12) and then use the results on short sequences to extend the map to the closed sets. The following lemma is a slightly refined version of Lemma 2.17.

4.13 Lemma. *Suppose $\lambda(\mathbf{I})$ is regular. Let $E = \{\langle e_i, B_i \rangle : i < \alpha\}$ be a construction over A and $D \subseteq E$ with $|D| < \lambda(\mathbf{I})$. There exists a closed subset C of E with $D \subseteq C$ and $|C| < \lambda(\mathbf{I})$.*

Proof. Define $\langle C_i : i < \omega \rangle$ by induction: $C_0 = D$, $C_{i+1} = \cup\{B_j : e_j \in C_i\}$ and let $C = \bigcup_{i<\omega} C_i$. Clearly, if $\lambda(\mathbf{I}) > \aleph_0$ then $|C| < \lambda(\mathbf{I})$. If $\lambda(\mathbf{I}) = \aleph_0$, define a partial order $<$ on C by: e_i is immediately below e_j if and only if $e_i \in B_j$. It is easy to see that $(C, <)$ is a finite branching tree with no infinite branch, so, by König's lemma, the tree has only finitely many nodes and so $|C| < \aleph_0$.

There is one more crucial lemma (Lemma 4.14) which enables us to conclude Theorem 4.12. To apply Lemma 4.14 to the proof of Theorem

4.12, let B from Lemma 4.14 be $A \cup \operatorname{rng} g = A \cup \operatorname{dom} g$, where we identify the domain and range of g via g.

4.14 Lemma. *Suppose $\lambda(\mathbf{I})$ is regular. Let E and F be* **I**-*constructible and* **I**-*saturated over B and $e \in E$. There is a partial B-isomorphism g between E and F with $e \in \operatorname{dom} g$ and such that both* $\operatorname{dom} g$ *and* $\operatorname{rng} g$ *are closed, and* $|\operatorname{dom} g| < \lambda(\mathbf{I})$.

Proof. We will define an increasing sequence of maps $\langle g_i : i < \omega \rangle$ such that, denoting $\operatorname{dom} g_i$ by C_i and $\operatorname{rng} g_i$ by D_i, we have $e \in C_0$, $|C_i|$, $|D_i| < \lambda(\mathbf{I})$, and C_{2i} (D_{2i+1}) is closed in E (F). Choose by Lemma 4.13 a closed subset C_0 containing e with $|C_0| < \lambda(\mathbf{I})$. Let g_0 map C_0 into F and let D_{-1} be empty. Suppose g_k, C_k and D_l for $k \leq 2i$, $l < 2i$ have been chosen as required. We will construct D_{2i}, D_{2i+1}, C_{2i+1}, and C_{2i+2}. Let D_{2i} be the range of g_{2i}. Choose by Lemma 4.13 a set D_{2i+1} containing D_{2i} which is closed and with $|D_{2i+1}| < \lambda(\mathbf{I})$. Since F is constructible over B and $|D_{2i}| < \lambda(\mathbf{I})$, by Lemma 4.1 F is constructible over $B \cup D_{2i}$. Since $|D_{2i+1} - D_{2i}| < \lambda(\mathbf{I})$ by Lemma 4.2, $D_{2i+1} - D_{2i}$ is constructible over $B \cup D_{2i}$. So by Theorem 2.46 there is an $(B \cup D_{2i})$-embedding of $B \cup D_{2i+1}$ into E. Call this map, which extends g_{2i}, g_{2i+1}^{-1}. Let C_{2i+1} be the image of this map and construct C_{2i+2} by Lemma 4.13. Now, $g = \bigcup g_i$ is the required map.

4.15 Historical Notes. A full discussion of the history of the uniqueness theorem for prime models is given in Section X.3 We only mention here that the straightforward proof for strictly prime models is due to Ressayre (unpublished).

5. Locally Atomic Models

One of the principal difficulties in the study of general stable theories is the lack of prime models over arbitrary sets. This section is devoted to a substitute notion, locally atomic models. We develop a notion of **L**-atomic such that every subset of $\mathcal{M}$ for a countable stable theory can be embedded in an **L**-saturated model. Unfortunately, there is no natural algebraic version of this notion of atomicity. That is, there does not seem to be a reasonable meaning for the phrase locally prime. However, as we will see in this section, locally atomic models are very useful for proving two cardinal theorems. They work well for this purpose because locally atomic models allow one to omit nonprincipal Δ-types for finite sets of formulas Δ and this is exactly what is needed for two cardinal theorems. Moreover, we will be able to use the existence theorem proved in this section to partially extend results about ω-stable theories to small superstable theories. Throughout this section T is stable.

The development here can not be given as a straightforward example of the type of isolation relation considered in Section IX.2. Shelah [Shelah

1978] developed a rubric which included both examples. However, to simplify the earlier exposition we used a scheme into which this notion does not fit very well. In this context we must modify our scheme somewhat. We define the class of **L**-*formulas* to be exactly the set **AT** of first order formulas. We vary the notion of isolation by substituting a weakened form of implication.

5.1 Definition. Let T be a countable stable theory. The type $p \in S(A)$ is **L**-*implied* (written $\vdash_L$) by $p|B$ for some countable $B \subseteq A$ if for every formula $\phi(\overline{x};\overline{y})$ there is a formula $\psi_\phi(\overline{x};\overline{b}) \in p|B$ such that $\psi_\phi(\overline{x};\overline{b}) \vdash p_\phi$. We also say $p|B$ *locally isolates* (**L**-isolates) p. Let $\lambda(\mathbf{L}) = \aleph_1$. As in Section IX.2, we have a corresponding notion of **L**-*atomic*. The structure A is **L**-saturated if every consistent **L**-formula over A is realized in A.

We noticed in Section IX.2 the equivalence (for the **I** discussed there) of two notions of **I**-saturation: i) Each type implied by a consistent **I**-formula is realized; ii) Each **I**-formula is realized. Since an **L**-isolated type need not be implied by a single **L**-formula this equivalence does not hold here. This divergence accounts for the nonuniformity of the proofs of Theorem X.1.18, Corollary X.1.19, etc. in the next chapter. Under the notion of **L**-saturation which we have chosen every model of a stable theory is **L**-saturated.

The choice of $\lambda(\mathbf{L})$ as $\aleph_1$ makes $\lambda(\mathbf{L})$ the bound on the size of an isolating set so it shares the crucial property of the $\lambda(\mathbf{I})$ for the other **I**. This definition of local isolation differs slightly from the definition of $F^l_{\aleph_1}$ in [Shelah 1978], IV.2. Pillay pointed out that the two definitions are equivalent. The following exercise verifies this assertion.

5.2 Exercise. Show that the type $p \in S(A)$ is **L**-implied by $p|B$ for some countable $B \subseteq A$ if for every $\phi(\overline{x};\overline{y})$ there is a formula $\psi_\phi(\overline{x};\overline{b}) \in p|B$ such that $\psi_\phi(\overline{x};\overline{b}) \vdash p^+_\phi$ where $p^+_\phi = \{\phi(\overline{x};\overline{a}) : \phi(\overline{x};\overline{a}) \in p\}$. (Hint: Show that if $\psi_\phi \vdash p^+_\phi$ and $\psi'_\phi \vdash p^+_{\neg\phi}$ then $\psi_\phi \wedge \psi'_\phi \vdash p_\phi$.)

Note that p^+_ϕ differs from p_ϕ since it does not contain instances of the negation of ϕ.

Although the isolating formulas for **L**-isolation involve only finitely many parameters we can in general isolate the type of an element only by using countably many elements.

5.3 Exercise. Show that if p is **L**-isolated over B, then p is $\mathbf{AT}_{\aleph_1}$ isolated over B.

Just as in Definition 2.7 we can define the notions of p being locally isolated by B or over B.

5.4 Exercise. Verify that **L**-isolation satisfies the continuity and monotonicity properties elaborated in Exercises 2.10 through 2.14.

To show that every subset of a model of a countable stable theory can be embedded in an **L**-saturated model we show that **L** satisfies the existence and transitivity axioms. For this task we use the local rank defined in Definition III.1.10.

5.5 Lemma. *For any set A and any formula $\chi(\overline{x};\overline{a})$ over A, there is a type $p \in S(A)$ with $\chi(\overline{x};\overline{a}) \in p$ and with p locally isolated by a countable subset of A.*

Proof. Fix $\chi(\overline{x};\overline{a}) \in F(A)$ and an enumeration $\langle \phi_i(\overline{x};\overline{y}) : i < \omega \rangle$ of $F(T)$. We define by induction a sequence of types p_n such that $p = \bigcup_{n<\omega} p_n$ locally implies a complete type over A and $\chi(\overline{x};\overline{a}) \in p$. First, let $p_0 = \{\chi(\overline{x};\overline{a})\}$. If p_n has been defined, let $p_{n+1} = p_n \cup \{\psi_n(\overline{x};\overline{a}_n)\}$ where ψ_n is chosen so that $R(p_{n+1}, \phi_n)$ is minimal among all consistent extensions of p_n over A. It suffices to show that for any $\psi(\overline{x};\overline{c}) \in F(A)$, if $p \cup \psi(\overline{x};\overline{c})$ is consistent then for some $n < \omega$, $p_n \vdash \psi(\overline{x};\overline{c})$. For, then p has a unique consistent extension $q \in S(A)$ and $p \vdash_L q$. To see this, fix such a $\psi(\overline{x};\overline{c})$ and let $\psi(\overline{x};\overline{y})$ be ϕ_m. If $p_{m+1} \cup \neg\psi(\overline{x};\overline{c})$ is also consistent then both that type and $p_{m+1} \cup \psi(\overline{x};\overline{c})$ have ranks at least as great as $R(p_{m+1}, \phi_m)$ by the choice of p_{m+1}. But this contradicts the definition of $R(p_{m+1}, \phi_m)$.

5.6 Exercise. Show that for uncountable T, we can define $\mathbf{L}$-isolation as we did for countable languages but that we must take $|T|^+$ for $\lambda(\mathbf{L})$.

5.7 Lemma. $\mathbf{L}$-*isolation satisfies the transitivity axiom.*

Proof. Assume we have $B \subseteq A$ such that $p = t(\overline{d}; A \cup C)$ is locally isolated over $B \cup C$ and for each $\overline{c} \in C$, $t(\overline{c}; A)$ is locally isolated over B. We must show that for any $\overline{c}' \in C$, $t(\overline{d}^\frown\overline{c}'; A)$ is locally isolated over B. That is, for any formula $\phi(\overline{x},\overline{y};\overline{z})$ we must find a formula $\psi(\overline{x},\overline{y};\overline{b})$ over B such that if $\models \phi(\overline{d},\overline{c}';\overline{a})$ with $\overline{a} \in A$ then $\models \psi(\overline{x},\overline{y};\overline{b}) \to \phi(\overline{x},\overline{y};\overline{a})$. We have a formula $\chi(\overline{x};\overline{b}_1,\overline{c}_1)$ such that $\chi(\overline{x};\overline{b}_1,\overline{c}_1) \to \phi(\overline{x};\overline{c},\overline{a})$ where $\overline{b}_1 \in B$ and $\overline{c}_1 \in C$ since $t(\overline{d}; A \cup C)$ is locally isolated over $B \cup C$. Moreover, there is a $\phi_1(\overline{y},\overline{z};\overline{b}_2)$ such that $\models \phi_1(\overline{c},\overline{c}_1;\overline{b}_2)$ and

$$\models [\phi_1(\overline{y},\overline{z};\overline{b}_2) \to ((\forall\overline{x})\chi(\overline{x},\overline{b}_1,\overline{z}) \to \phi(\overline{x},\overline{y},\overline{a}_1))]$$

since $t(\overline{c}^\frown\overline{c}_1; A)$ is locally isolated over B. Now, $(\exists\overline{z})\phi_1(\overline{y},\overline{z},\overline{b}_2) \wedge \chi(\overline{x},\overline{b}_1,\overline{z})$ is the required formula $\psi(\overline{x},\overline{y};\overline{b})$ where $\overline{b} = \overline{b}_1{}^\frown\overline{b}_2$.

Recall Definition 2.36 of $\mu(\mathbf{I})$. The following exercise depends crucially on our definition (Definition 5.1) of $\mathbf{L}$-saturation.

5.8 Exercise. Show that $\mu(\mathbf{L}) = \aleph_0$.

Now by a proof like that of Theorem 2.40 we can deduce

5.9 Theorem. *If T is a countable stable theory then every set A is contained in a model M with $|M| \leq |A| + \aleph_0$ which is* $\mathbf{L}$*-saturated and* $\mathbf{L}$*-atomic over A.*

$\mathbf{L}$-isolation does not satisfy the transposition property (Axiom 3.8). Thus, the proof of the uniqueness of strictly prime models given in Section 4 does not work for this notion.

5.10 Exercise. Verify the failure of the transposition property by considering the theory, T, of an infinite set. If $M \models T$ and $m \in M$, then $t(M;\emptyset)$ is $\mathbf{L}$-atomic but $t(m; M - \{m\})$ is not $\mathbf{L}$-isolated.

5.11 Exercise. Show that if T is the theory of infinitely many independent unary predicates, T has $2^{\aleph_0}$ non-isomorphic models which are all **L**-constructible over the empty set.

Now we use this construction to prove the two cardinal theorem for stable theories. This will complete the proof of Morley's categoricity theorem from Section I.3.

5.12 Definition. The theory T *admits the pair of cardinals* (κ, λ) if for some formula $\phi(\overline{x}; \overline{m})$ and some model M of T, $|M| = \kappa$ and $|\phi(M, \overline{m})| = \lambda$. We also say, T has a (κ, λ) model.

5.13 Theorem. *If a countable stable theory admits some pair* (κ, λ) *with* $\kappa > \lambda$ *then it admits any pair* (κ', λ') *with* $\kappa' \geq \lambda'$.

Proof. Assuming the hypothesis, it is easy to show by the Löwenheim-Skolem theorem that there is a pair of models M, N of T with $|M| = |N| = \lambda'$ such that N is a proper submodel of M but $\phi(N, \overline{n}) = \phi(M, \overline{n})$ for some $\overline{n} \in N$. We will show how to construct from any such pair of models a third model M' which is a proper elementary extension of M but with $\phi(M', \overline{n}) = \phi(N, \overline{n})$. With this in hand it is easy to construct by induction on $\alpha < \kappa'$ a chain of models M_α such that for all α, $\phi(M_\alpha, \overline{n}) = \phi(N, \overline{n})$. The union of this chain demonstrates that T admits (κ', λ').

To construct M', let a be an arbitrary element of $M - N$ and let c realize the nonforking extension over M of $t(a; N)$. We claim that for any formula $\psi(x, \overline{m}, y)$ over M, if $\models (\exists y)(\psi(c, \overline{m}, y) \wedge \phi(y, \overline{n}))$ then for some c' in $\phi(N, \overline{n})$, $\models \psi(c, \overline{m}, c')$. To see this, let d define $t(a; N)$. For every $\overline{n}' \in N$, if $\models (\exists y)(\psi(a, \overline{n}', y) \wedge \phi(y, \overline{n}))$ then for some $c_0 \in N$, $\models \psi(a, \overline{n}', c_0) \wedge \phi(c_0, \overline{n})$. Thus

$$(\forall \overline{z})\,[d\,((\exists y)\psi(x, \overline{z}, y) \wedge \phi(y, \overline{n}))\,[\overline{z}, y] \rightarrow (\exists y)(\phi(y, \overline{n}) \wedge d(\psi(x, \overline{z}, y))[\overline{z}, y])]$$

is true. Since this formula also holds of c we establish the claim.

Now let M' be locally atomic over $M \cup c$. If some element d of $M' - N$ satisfies $\phi(d, \overline{n})$, then by local atomicity there is a formula $\psi(c, \overline{m}, x)$ such that $\models \psi(c, \overline{m}, x) \rightarrow \phi(x, \overline{m})$ and $\models \psi(c, \overline{m}, x) \rightarrow x \neq n$ for each n in $\phi(N, \overline{n})$. But this contradicts the choice of c and we finish the theorem.

5.15 Exercise. Verify the first sentence of the proof of Theorem 5.14. (Hint: Let M witness that T admits (κ, λ). Choose an elementary submodel of M which contains $\phi(M, \overline{m})$ but has cardinality λ. Add to the language a unary predicate to pick out this submodel and then apply the Löwenheim-Skolem theorem.)

5.16 Exercise. Prove that any countable stable theory which admits a two cardinal model admits (ω_1, ω). Apply the technique from Theorem 5.14 to build a chain of models but replace the use of locally atomic models by an application of the omitting types theorem.

We can extend the two cardinal theorem to uncountable languages as follows. Recall that Vaught's two cardinal theorem for cardinals far apart,

[Vaught 1965], asserts that if a theory T has a (κ, λ) model with $\kappa \geq \beth_\omega(\lambda)$ then T has a (κ', λ') model for any pair $\kappa' \geq \lambda' \geq |T|$. The key step in Vaught's proof is to show that in the Skolem theory of T the set of sentences which express the following properties is consistent. First, $\{\langle c_n : n < \omega \rangle\}$ is a set of order indiscernibles in $L(T)$. Second, if U denotes the distinguished predicate and $\overline{c}, \overline{c}'$ are increasing sequences of the indiscernibles which satisfy U, then for any Skolem function τ, $\tau(\overline{c}) = \tau(\overline{c}')$. Since every reduct of a stable theory T to a countable language is stable, we can easily establish the consistency of this set of sentences by appealing to the two cardinal theorem for countable stable theories. We have outlined the proof of the following theorem.

5.17 Theorem. *If a stable theory admits some pair (κ, λ) with $\kappa > \lambda$ then it admits any pair (κ', λ') with $\kappa' \geq \lambda' \geq |T|$.*

We pointed out in Exercise 5.6 that the notion of locally atomic models extends to uncountable languages. Using this observation there is another approach to the two cardinal theorem for uncountable theories.

5.18 Exercise. Derive Theorem 5.17 by a proof in the style of the proof given for Theorem 5.14.

5.19 Historical Notes. The study of two cardinal theorems began with Vaught's proof [Vaught 1961] that an $\aleph_1$-categorical theory could not have a two cardinal model. He deduced this result from the transfer principle $(\kappa, \lambda) \mapsto (\aleph_1, \aleph_0)$ which he showed holds for an arbitrary countable first order theory in that same paper. Vaught [Vaught 1965] and Morley [Morley 1965a] then independently showed that for a countable theory T, $(\beth_\omega, \omega) \mapsto (\kappa, \omega)$ for any κ. Attempts to extend these results to arbitrary first order theories led via the Chang two cardinal theorem [Chang & Keisler 1973] to problems which are more set theoretic than model theoretic in character. Shelah [Shelah 1969] and Lachlan [Lachlan 1972] restored the subject to the realm of model theory by discovering the relevance of assuming that the theory is stable. The general outline of the proof of Theorem 5.14 comes from [Lachlan 1972]. The exact construction here is from [Lascar 1973]; a similar version of the proof using conservative extensions and compactness rather than definability occurs in [Baldwin 1975]. The concept of a locally atomic model, though not the name, occurs in [Lachlan 1972] and [Shelah 1971]. Lachlan dealt only with countable languages while Shelah included the uncountable case. There are various extensions of this notion in [Shelah 1978]. The two cardinal theorem for uncountable stable theories was originally proved by Shelah in his thesis invoking an absoluteness argument. Harnik outlines five different proofs of the theorem in [Harnik 1975]. Alan Mekler brought this fairly direct proof to my attention when he discovered it; Harnik had earlier noted its existence in the cited paper.

6. The Number of Models of Strictly Stable Theories

This section is devoted to a proof that a strictly stable (i.e. a stable but not superstable theory) has 2^λ models of power λ if $\lambda > |T|$ and λ is regular. In Chapter VIII of [Shelah 1978], the result is proved by much more complicated extensions of the methods here for arbitrary $\lambda > |T|$. The idea of Shelah's proof was discussed in Section I.5.

Our proof will require three ingredients. The first is the existence of a certain indiscernible tree if T is stable but not superstable. This was proved in Theorem III.4.25. The second is the development of the properties of a further notion of isolation. The third is certain combinatorial facts about stationary sets.

We begin by describing a notion of isolation which does not fit particularly well with the scheme we have described in Section 2; however, we are already very familiar its properties. Throughout this section we assume that T is stable.

6.1 Definition. The type $p \in S(A)$ is $\mathbf{F}_\lambda$ isolated if there is a subset B contained in A with $|A| < \lambda$ such that p does not fork over B. We will write $\mathbf{F}$ for $\mathbf{F}_{\aleph_0}$.

Shelah calls this notion $\mathbf{F}_\lambda$-isolation.

Note that we can squeeze this notion into our general scheme by taking as the set of isolating formulas the collection $\mathbf{AT}_\lambda$ but replacing the relation of provability by the notion q 'pseudoproves' p if p is a nonforking extension of q. With this in mind we can continue to use the terminology defined in Sections 2 and 3.

6.2 Exercise. Show that any p is $\mathbf{F}_{\kappa(T)}$-isolated.

Thus, when discussing countable models the only interesting case is that of $\mathbf{F}_{\aleph_0}$ and then only when T is not superstable. It is easy to see that the basic properties of an isolation relation discussed in section 2 hold for $\mathbf{F}_\lambda$-isolation.

6.3 Exercise. Show that for any λ, $\mathbf{F}_\lambda$-isolation satisfies the monotonicity and continuity properties discussed in Exercises 2.9 through 2.14.

6.4 Exercise. Deduce the symmetry, transitivity, and transposition properties of $\mathbf{F}_\lambda$-isolation from the analogous statements in Chapter III.

The anomaly of suddenly treating as a dependence notion the concept we have regarded as a notion of independence for much of the book has no clear explanation. There is one property of the other isolation relations which is not shared by $\mathbf{F}_\lambda$ and this provides some insight. For any of the other isolation relations, $\mathbf{I}$, there is a cardinal $\mu(\mathbf{I}) < \infty$ which enables us (Theorem 2.40), in the presence of the Existence Axiom, to embed each set in an **I**-saturated model. In this case no such cardinal exists and, in fact, there are no $\mathbf{F}_\lambda$-saturated models.

However, the Existence Axiom, 2.28, is easily seen to hold for $\mathbf{F}_\lambda$. Thus, a variant of the proof of Theorem 2.40 shows the existence of $\mathbf{F}_\lambda$-constructible models over arbitrary sets.

6.5 Lemma. *If T is stable then over A there is an $\mathbf{F}_\lambda$-constructible model.*

Proof. Enumerate the types over A and realize them in order by an element which does not fork over its predecessors in the sequence. Iterating this procedure ω times produces a model of T.

Note that this model is by no means prime in the normal sense.

6.6 Exercise. Show that there are two models of the theory of infinitely many independent unary predicates which are each **F**-constructible over the empty set such that neither can be embedded in the other.

Before proving the main theorem, we require the following combinatorial notions.

6.7 Definition. A subset C of the cardinal λ is *closed and unbounded (cub)* if

i) for every $\alpha < \lambda$ there is a $\beta \in C$ with $\alpha < \beta$, and
ii) if $X \subseteq C$ and $|X| < \lambda$ then $\sup(X) \in C$.

A subset S of λ is *stationary* if S intersects every *cub*.

The key facts needed about these sets are contained in the following lemma. The proof of this lemma can be found in a number of works on set theory, e.g. [Kunen 1980].

6.8 Lemma. *Let $\lambda > \aleph_0$ be regular.*

i) *The set, Λ_0, of $\alpha < \lambda$ with cofinality ω is stationary.*
ii) *For any strictly increasing continuous function (in the order topology on λ) g from λ to itself, the set*

$$C = \{i < \lambda : \textit{for all } j < i,\ g(j + \omega) < i\}$$

is a cub.
iii) *There exist a family of λ pairwise disjoint stationary subsets of λ.*
iv) *There exist a family $\langle W_i : i < 2^\lambda \rangle$ of stationary subsets of λ such that if $i \neq j$ then $W_i - W_j$ is stationary. The same result holds replacing λ by Λ_0.*

Statements i) and ii) of the lemma are fairly straightforward from the definitions. Statement iii) is a well known theorem of Ulam (See Theorem 1.3.2 in the Appendix of [Shelah 1978] or Corollary 6.12 of [Kunen 1980].) The following exercises indicate how iv) can be fairly easily deduced from iii).

6.9 Exercise. Show there exist a family of 2^λ pairwise incomparable subsets of λ. (That is, subsets satisfying iv) but with $W_i - W_j \neq \emptyset$ replacing the requirement that $W_i - W_j$ is stationary.) (Hint: Almost disjoint sets suffice. Two sets are almost disjoint if their intersection is finite.)

6.10 Exercise. Combine the previous exercise with iii) to obtain iv).

To aid the comprehension of the following complicated argument we fix a number of notational conventions for this section.

6.11 Notation. Elements of $\lambda^{<\omega}$ (finite sequences of ordinals $< \lambda$) are denoted by s or t; elements of λ^ω (countable sequences of ordinals $< \lambda$) are denoted by σ or τ; elements of $\lambda^{\leq\omega} = \lambda^{<\omega} \cup \lambda^\omega$ are denoted by ν or η.

We denote by I the structure $\langle \lambda^{\leq\omega}, \subseteq, < \rangle$ with the indicated universe, with $\subseteq$ interpreted as initial segment relation between functions, and with $<$ interpreted as lexicographic order.

The following assertion can be easily derived from Theorem III.4.25.

6.12 Lemma. *If T is stable but not superstable then for every infinite cardinal λ there is a model M, which contains a set $A = \{\bar{a}_\eta : \eta \in \lambda^{\leq\omega}\}$ with the following property. For each $B \subseteq A$ and each $\eta \in \lambda^\omega$, if B contains $\{\bar{a}_{\eta|n} : n < \omega\}$ then $t(\bar{a}_\eta; B)$ is not* **F***-isolated. Moreover, A is an independent tree in the sense of Definition II.2.25.*

We fix the tree A from Lemma 6.12 for the remainder of this section.

6.13 Theorem. *If T is stable but not superstable then for every regular $\lambda > |T|$, $I(\lambda, T) = 2^\lambda$.*

Proof. Recall that Λ_0 denotes the set of ordinals less than λ which have cofinality ω. Let W be the set of 2^λ pairwise incomparable subsets of Λ_0 guaranteed by Lemma 6.8. For each set of ordinals $w \in W$ we define a model M^w of T as follows. For each $\delta \in \Lambda_0$ fix one strictly increasing function $\sigma_\delta \in \lambda^\omega$ which witnesses that $\mathrm{cf}(\delta) = \omega$. For $w \in W$, let

$$I^w = \lambda^{<\omega} \cup \{\sigma_\delta : \delta \in w\}.$$

Let M^w be **F**-constructible over $A^w = \{\bar{a}_\nu : \nu \in I^w\}$. Now if w, u are distinct elements of W, we will deduce a contradiction from the assumption that there is an isomorphism between M^w and M^u.

In order to effect this contradiction we will approximate the tree I^w by the set of functions with bounded range. For any $i < \lambda$, $I_i^w = \bigcup_{\beta < i} \beta^{\leq\omega} \cap I^w$. Similarly, $A_i^w = \{\bar{a}_\eta : \eta \in I_i^w\}$.

Now the crucial fact is contained in the following assertion.

6.14 Claim. *For any $w \in W$, A^w is* **F***-atomic over A_δ^w if and only if $\delta \notin w$.*

Proof. This follows routinely from the fact that A is an independent tree. For, if $B \subseteq A^w$, $t(B; A_\delta^w)$ is **F**-isolated if and only if the downward closure of B, i.e. $\{\bar{a}_\sigma : \sigma \subseteq \tau \wedge \bar{a}_\tau \in B\}$, intersects A_δ^w in a finite set. But, if $\delta \notin w$, it easy to check that this condition holds. Moreover, the condition clearly fails if $\delta \in w$.

Let M^w be constructed by the sequence $\{c^w_\alpha : \alpha < \lambda\}$. Suppose that $t(c^w_\alpha; A^w \cup C^w_\alpha)$ is **F**-isolated by $\overline{b}^w_\alpha$. (Remember C^w_α is $\{c^w_\beta : \beta < \alpha\}$.) Now we construct the closed unbounded subsets of Λ_0 whose intersection will lead to the contradiction. Let G^w denote the set of α such that for each $\beta < \alpha$, $\overline{b}^w_\beta \subseteq A^w_\alpha \cup C^w_\alpha$.

6.15 Exercise. Show G^w is the set of α such that $A^w_\alpha \cup C^w_\alpha$ is the universe of an **F**-construction.

Now, it is easy to see that G^w, G^u and the set S of those α such that g maps $A^w_\alpha \cup C^w_\alpha$ onto $A^u_\alpha \cup C^u_\alpha$ are closed unbounded sets. Thus there is an ordinal δ in $G^w \cap G^u \cap S \cap w$ and so not in u. This δ will yield the contradiction.

We first argue that M^u is **F**-atomic over $A^u_\delta \cup C^u_\delta$. Claim 6.14 implies, that A^u is **F**-atomic over A^u_δ. Since $\delta \in G^u$, C^u_δ is **F**-atomic over A^u_δ. We can prove from the definition of G^u by induction on the length of C^u_δ that $C^u_\delta \downarrow_{A^u_\delta} A^u$. Now by transitivity of the nonforking relation, A^u is **F**-atomic over $A^u_\delta \cup C^u_\delta$. By transitivity of isolation we conclude M^u is **F**-atomic over $A^u_\delta \cup C^u_\delta$. In particular, $t(g(\overline{a}^w_{\sigma_\delta}); A^u_\delta \cup C^u_\delta)$ is **F**-isolated. But this contradicts the assumption that g is an isomorphism. For, since $\delta \in S$, we can assume that g maps $A^u_\delta \cup C^u_\delta$ to $A^w_\delta \cup C^w_\delta$ and $t(\overline{a}_{\sigma_\delta}; A^w_\delta \cup C^w_\delta)$ is not **F**-isolated. As, if it were, we could apply the fact that $\delta \in G^w$ to see that C^w_δ is **F**-atomic over A^w_δ which implies by Theorem II.2.5 that $C^w_\delta \cup a^w_\delta$ is **F**-atomic over A^w_δ. But this contradicts the fundamental fact about the construction, Claim 6.12, that $t(\overline{a}_\delta; A^w_\delta)$ is not **F**-isolated since $\delta \in w$.

6.16 Corollary. *If T is a stable but not superstable theory the function $I(\kappa, AT)$ which gives the number of models of T of power κ is increasing.*

Proof. Observe that the previous proof actually shows that if λ is a regular cardinal less than or equal to κ then $I(\kappa, AT) \geq 2^\lambda$. Thus, for regular cardinals we have the maximum number of models and the function does not decrease at a singular cardinal.

6.17 Historical Notes. This proof given here is the proof of Theorem VIII.2.7 of [Shelah 1978]. The contrast with proofs using Ehrenfeucht-Mostowski models which dominate Chapter VIII of [Shelah 1978] is discussed in Section I.5. Pillay [Pillay 1981] provides a very nice exposition of the Ehrenfeucht-Mostowski version of the proof when λ is regular. Hodges [Hodges 198?] is writing an exposition of Shelah's approach to the nonstructure results. Conversations with Jürgen Saffe greatly aided my writing of this section.

Chapter X

Freeness and Isolation

In this chapter we investigate the relation between the notion of isolation (analogous to dependence) and the notion of freeness (analogous to independence). Since in fact we study only one notion of freeness (nonforking) but several notions of isolation, we could consider this study as the further investigation of properties of isolation relations. In fact, one of the notions of isolation, namely **S**-isolation, will be seen to play a rather privileged role. Of the dependence relations we consider, it is the most natural counterpart to the independence relation of nonforking. Unfortunately, study of **S**-isolation gives direct information about only the **S**-saturated structures. By making more restrictive assumptions on the theory we are able to widen the class of models treated. In particular, for ω-stable theories $\mathbf{AT}_{\aleph_0}$ will be covered by our discussion. In Section 1 we develop a set of axioms relating the notions of isolation and independence. Section 2 explores the notion of a 'powerful' isolation relation. In Section 3 we prove the uniqueness of prime models. Section 4 discusses the posssible sizes of indiscernible sequences over a set A which lie in prime models over A.

WARNING: While we have attempted to isolate in Chapter II all relevant features of nonforking, in Chapter IX the relevant features of isolation relations and here the properties which govern their interaction, attempts to apply these results to other isolation relations should be undertaken with caution. In particular, we have not carefully surveyed the situation for **F**-isolation. Our major purpose in this axiomatization has been to systematize the exposition of the properties of $\mathbf{AT}_\lambda$, $\mathbf{SET}_\lambda$, and $\mathbf{S}_\lambda$ for regular λ. We also include some material on **L**-isolation for countable languages.

1. *Axioms Relating Freeness and Isolation*

In this section we discuss four principles from which we derive all the information relating forking and the formation of prime models (subject to the caveat in the introduction to this chapter). The first of these makes the rather obvious assertion that a type which is isolated over a set A does not fork over A. The second is a transfer principle which asserts that if an element is independent from a set then it is independent from all points whose types are isolated over that set. This property has immediate algebraic consequences. The third principle is the topologically oriented open mapping theorem while the fourth is a rather technical consequence of the finite character of forking.

We also introduce in this section the notion of a relatively **I**-saturated substructure. This concept clarifies some of the arguments here and extends to prove similar results for independent trees in Chapter XVII. The following notations will simplify much of the discussion.

1.1 Notation. Let M be **I**-saturated and A a set; $M[A]$ denotes a strictly **I**-prime model over $M \cup A$. We also frequently write $M[\overline{a}]$, ignoring the distinction between the sequence $\overline{a}$ and its range. If $p \in S(M)$ then $M[p]$ denotes $M[\overline{a}]$ for some $\overline{a}$ realizing p.

There are several potential ambiguities in this notation. A minor ambiguity is that we suppress mention of **I**; the choice of **I** will always be clear from context. A more serious ambiguity is the abuse of functional notation to denote an operation which is unique only up to isomorphism. We use the notation as a shorthand since in most cases it causes no problems. However, sometimes the *specific* structure which is strictly prime over $M \cup A$ is important. In such situations one must be very careful not to be misled by the notation. This difficulty is discussed in detail in connection with Theorem 1.28.

This problem becomes more serious if **I** is taken as **L**. For, as we observed in Section IX.4, there is no uniqueness theorem for **L**-constructible models. In general, we try to avoid the $M[a]$ notation in this context. It occasionally appears because theorems which are proved with the other **I** in mind apply to **L** if $M[\overline{a}]$ is taken as a meaning, 'an arbitrary **L**-constructible model containing $M \cup \overline{a}$.'

A lesser problem is the distinction between $M[p]$ and $M[\overline{a}]$ when $\overline{a}$ is a realization of p. The next two exercises illustrate this distinction.

1.2 Exercise. Deduce from the uniqueness of strictly **I**-prime models that i) $M[\overline{a}]$ is unique up to isomorphism over $M \cup \overline{a}$, while ii) $M[p]$ is unique only up to isomorphism over M.

1.3 Exercise. Let T be $\mathrm{Th}(Z_4^{\aleph_0})$. Let M be a countable model of T and let p, q be the nonforking extensions to M of the types of an element of order four and an element of order two respectively. Let a realize p and b realize q. Show that there are many copies of $M[b]$ (all isomorphic over $M \cup b$) but that within $\mathcal{M}$, $M[a]$ is unique, period, not just up to isomorphism.

Recall that if p is a type over A then p does not fork over A. We strengthen this requirement by demanding that if p is isolated over A then p does not fork over A.

1.4 Axiom FI$_0$. If $A \subseteq B$ and $p \in S(B)$ is isolated by A then p does not fork over A.

For $\mathbf{AT}_\lambda$, L, and $\mathbf{SET}_\lambda$ the verification of this axiom is immediate since we have a type $q \subseteq p$ such that $q \vdash p$ and q is over A. But for $\mathbf{S}_\lambda$, the result is equally clear, noting only that if q is almost over A then q does not fork over A.

Consider the following situation: $(\overline{a}\downarrow\overline{b}; M)$ and $t(\overline{c}; M \cup \overline{b})$ is **I**-isolated by $D \cup \overline{b}$. What can be said about the relation of $\overline{a}$ and $\overline{c}$? Perhaps the most intuitive idea is that if $\overline{a}$ is independent from $\overline{b}$ then $\overline{a}$ is independent from any element whose type over $\overline{b}$ is **I**-isolated. Before stating the axiom we introduce two further concepts which will simplify its verification. The crucial relation between $M \cup \overline{a}$ and $M \cup \{\overline{a}, \overline{b}\}$ is crystalized in the following definition.

1.5 Definition. The set A is *relatively* **I**-*saturated* in B if each **I**-formula q over A which is satisfied in B is satisfied in A. We write $A \leq_{\mathbf{I}} B$.

Shelah has defined several special cases of this notion. When **I** is **AT** we have Shelah's notion of the *Tarski-Vaught property.* This concept also encompasses Shelah's *strong elementary submodel* [Shelah 198?]. Thus, his notions of $\subseteq_g$ and $\subseteq_h$ correspond to relative $\mathbf{SET}_{\aleph_0}$ and $\mathbf{S}_{\aleph_0}$ saturation respectively.

1.6 Exercise. Show that $\leq_{\mathbf{I}}$ is transitive.

1.7 Exercise. If M is an **I**-saturated model of T then for any A with $M \subseteq A$, $M \leq_{\mathbf{I}} A$.

The next two exercises were suggested by Leo Harrington.

1.8 Exercise. Show that if $A \leq_{AT} B$ then any formula almost over A which is satisfied in B is satisfied in A.

1.9 Exercise. Prove that if $A \leq_{\mathbf{AT}} B$ and $c \downarrow_A B$ then $t(B; A \cup c)$ is finitely satisfiable in A. Find an example where $t(c; B)$ is not finitely satisfied in A. (Hint: For the proof use the fact that $t(c; B)$ is definable almost over A and the previous exercise.)

The last exercise shows that although forking is symmetric when T is stable, the motivating notion of finite satisfiability may not be.

We now extend the concept of strong saturation discussed in Section III.2 to a more general context. We will verify that **I**-saturation implies strong **I**-saturation and derive several extremely useful corollaries from this result.

1.10 Definition. The structure M is *strongly* **I**-*saturated* if for any set A with $|A| < \lambda(\mathbf{I})$ and any **I**-formula, q, over $M \cup A$, if there is a $\overline{b}$ with $\lg(\overline{b}) < \lambda(\mathbf{I})$, realizing q such that $\overline{b} \downarrow_M A$, then there is a $\overline{b}' \in M$ which realizes q.

1.11 Theorem. *Let T be a stable theory. If* **I** *is* $\mathbf{AT}_\lambda$, $\mathbf{S}_\lambda$ *or* **L** *every* **I**-*saturated model is strongly* **I**-*saturated. If $\lambda > |T|$, any* $\mathbf{SET}_\lambda$-*saturated model is strongly* $\mathbf{SET}_\lambda$-*saturated.*

Proof. For the **AT** and **L**, the result follows immediately from the definition of coheir. For **S** the result is obvious. For the other isolation notions slight variations of the proof of Lemma III.2.29 yield the result.

We can rephrase strong **I**-saturation in terms of relative **I**-saturation.

1.12 Lemma. *If the strongly* **I***-saturated model* M *is contained in* $A \cap B$ *then* $A \downarrow_M B$ *if and only if* $A \leq_{\mathbf{I}} A \cup B$.

Proof. Let q be an **I**-formula over A which is realized in B. Since the conjunction of any finite subset of q is realized in M, strong **I**-saturation guarantees that q is realized in M.

The following stronger version will be applied both here and in Chapter XII (to aid the discussion of regular types). For **L** and **AT** the Corollary does not really add anything to the theorem.

1.13 Corollary. *Suppose the strongly* **I***-saturated model* M *is contained in* $A \cap B$ *and* $A \downarrow_M B$. *Let the* **I***-formula* $q(\overline{x}, \overline{a}, \overline{b})$ *be consistent (where* $\lg(\overline{a}), \lg(\overline{b}) < \lambda(\mathbf{I})$*). Then, there is a* $\overline{b}' \in M$ *such that* $q(\overline{x}, \overline{a}, \overline{b}')$ *is consistent.*

Proof. The type $\{(\exists \overline{x})[\phi(\overline{x}, \overline{a}, \overline{y})] : \phi(\overline{x}, \overline{a}, \overline{x}) \in q\}$ is an **I**-formula. By the definition of strong **I**-saturation it must be realized in M by some $\overline{b}'$. Now the required type over $\overline{a} \cup \overline{b}'$ is easily seen to be consistent since we may assume q is closed under finite conjunction.

1.14 Exercise. Show that if **I** contains $\mathbf{AT}_{\aleph_0}$ then any structure which is algebraically closed and strongly **I**-saturated must be a model.

The following axiom establishes the relation between $M[\overline{a}]$ and $\overline{b}$ when $\overline{a} \downarrow_M \overline{b}$.

1.15 Axiom $\mathbf{FI_1}$. Let $M \models T$ be strongly **I**-saturated. If $\overline{a} \downarrow_M \overline{b}$ and $t(\overline{c}; M \cup \overline{b})$ is **I**-isolated then $t(\overline{c}; M \cup \overline{b}) \vdash t(\overline{c}; M \cup \{\overline{a}, \overline{b}\})$.

This axiom clearly implies the following more precise formulation. Let $M \models T$ be **I**-saturated. If $\overline{a} \downarrow_M \overline{b}$ and $t(\overline{c}; M \cup \overline{b})$ is **I**-isolated by $D \cup \overline{b}$ with $D \subseteq M$ then $t(\overline{c}; M \cup \{\overline{a}, \overline{b}\})$ is **I**-isolated by $D \cup \overline{b}$.

It is easy to see that this axiom requires some hypothesis on the set M.

1.16 Exercise. Let T be the theory of an infinite set and let $a, b, c,$ be elements of a model of T. Show $t(c; a)$ is isolated by a and $a \downarrow_\emptyset b$, but $t(c; \{a, b\})$ is not isolated over a.

We derive Axiom FI_1 from a stronger result which will be applied in Chapter XVII. The stronger result arises by generalizing the context, replacing $M \cup \overline{a}$ and $M \cup \{\overline{a}, \overline{b}\}$ by any A, B with $A \leq_{\mathbf{I}} B$. For our standard isolation relations we can deduce the theorem from Corollary 1.13. Because of the weakness of **L**-saturation, this proof does not suffice in the **L** case but a minor variant does.

1.17 Theorem. *Suppose* $A \leq_{\mathbf{I}} B$ *and* $t(\overline{c}; A)$ *is* **I***-isolated over* A. *Then* $t(\overline{c}; A) \vdash t(\overline{c}; B)$.

Proof. First, let **I** be any of the isolation relations except **L** and choose an **I** formula $q(\overline{x};\overline{a})$ such that $q(\overline{x};\overline{a}) \vdash t(\overline{c};A)$. Suppose there is a formula $\phi(\overline{x};\overline{b}) \in F(B)$ with both $q(\overline{x};\overline{a}) \cup \{\phi(\overline{x};\overline{b})\}$ and $q(\overline{x};\overline{a}) \cup \{\neg\phi(\overline{x};\overline{b})\}$ consistent. Then, by Corollary 1.13 there is a $\overline{b}' \in A$ such that both $q(\overline{x};\overline{a}) \cup \{\phi(\overline{x};\overline{b}')\}$ and $q(\overline{x};\overline{a}) \cup \{\neg\phi(\overline{x};\overline{b}')\}$ are consistent. But this contradicts the choice of q.

Now suppose **I** is **L** and let p denote $t(\overline{c};A)$. Fix $\phi(\overline{x};\overline{b}) \in F(B)$ so that both $p \cup \{\phi(\overline{x};\overline{b})\}$ and $p \cup \{\neg\phi(\overline{x};\overline{b})\}$ are consistent. Now for $\phi(\overline{x};\overline{y})$, choose $\psi_\phi(\overline{x};\overline{a})$ such that $\psi_\phi(\overline{x};\overline{a}) \vdash p_\phi$. Then both $\psi_\phi(\overline{x};\overline{a}) \cup \{\phi(\overline{x};\overline{b})\}$ and $\psi_\phi(\overline{x};\overline{a}) \cup \{\neg\phi(\overline{x};\overline{b})\}$ are consistent. Since $A \leq_L B$ there is a $\overline{b}' \in A$ with both $\psi_\phi(\overline{x};\overline{a}) \cup \{\phi(\overline{x};\overline{b}')\}$ and $\psi_\phi(\overline{x};\overline{a}) \cup \{\neg\phi(\overline{x};\overline{b}')\}$ consistent. This contradicts the choice of ψ_ϕ.

From Lemma 1.12 and Theorem 1.17 we deduce that FI_1 holds for each **I**. The following version of Theorem 1.17 is easier to apply. As in Theorem 1.17, for ii) the case **I** is **L** requires a slightly different proof which we omit this time.

1.18 Corollary. *If $A \leq_{\mathbf{I}} B$ and C is* **I***-atomic over A then*

i) $C \downarrow_A B$.
ii) $A \cup C \leq_{\mathbf{I}} B \cup C$.

Proof. Statement i) is immediate from Theorem 1.17. For ii), suppose that $\phi(\overline{a};\overline{b};\overline{c})$ holds and that the **I**-formula $q(\overline{a}';\overline{x})$ implies $t(\overline{c};A)$. By Lemma 1.14, and the completeness theorem there is a formula $\theta(\overline{x};\overline{a}_1)$ (almost) over A such that $\models (\forall\overline{x})\theta(\overline{x};\overline{a}_1) \rightarrow \phi(\overline{a},\overline{b};\overline{x})$. Since $A \leq_{\mathbf{I}} B$, there exists $\overline{b}' \in A$ such that $\models (\forall\overline{x})\theta(\overline{x};\overline{a}_1) \rightarrow \phi(\overline{a},\overline{b}';\overline{x})$. Since $\models \theta(\overline{c};\overline{a}_1)$, we can deduce $\models \phi(\overline{a};\overline{b}';\overline{c})$ and finish.

1.19 Exercise. Prove Corollary 1.18 when **I** is **L**.

The following result is now immediate.

1.20 Corollary. *Suppose* FI_0 *and* FI_1 *hold, M is strongly* **I***-saturated and* $\lambda(\mathbf{I})$ *is regular. Then,* $\overline{b} \rhd_M M[\overline{b}]$.

1.21 Exercise. Suppose FI_0 and FI_1 hold, M is strongly **I**-saturated and $\lambda(\mathbf{I})$ is regular. Prove that for any $\overline{b}$ if $\overline{c} \in M[\overline{b}] - M$ then $\overline{c} \not\downarrow_M \overline{b}$.

The following theorem is proved for the isolation relations other than **L**. The proof for **L** is a minor variant but we don't rely on that case. The full significance of this result will only become apparent with the discussion of strongly regular types in Section XII.1.

1.22 Corollary. *If M is strongly* **I***-saturated, $\overline{a} \downarrow_M D$ but $(\overline{b} \not\downarrow \overline{a}; D \cup M)$, and N is any* **I***-saturated model containing $M \cup \overline{a}$. Then for any* **I***-formula q with* $\mathrm{dom}^*\, q$ *contained in M such that $\models q(\overline{b})$, there exists $\overline{b}'$ such that the following hold.*

i) $\overline{a} \not\downarrow_M \overline{b}'$

ii) $\overline{b}' \in q(N)$
iii) $\overline{b}' \notin M$.

Proof. (Fig. 1). Let A denote dom* q. Without loss of generality (since we assume $\lambda(\mathbf{I}) \geq \kappa(T)$), $\overline{a} \downarrow_A M$. Thus $(\overline{a} \not\downarrow \overline{d}\frown\overline{b}; A)$. By the local character of forking there is a formula $\theta(\overline{x}; \overline{y}\frown\overline{z})$ such that $\models \theta(\overline{a}; \overline{b}\frown\overline{d})$ for some $\overline{d} \in D$ and for any $\overline{b}', \overline{d}'$, if $\models \theta(\overline{a}; \overline{b}', \overline{d}')$ then $(\overline{a} \not\downarrow \overline{b}'\frown\overline{d}'; A)$. Now by Corollary 1.13, there exist $\overline{b}', \overline{d}'$, with $\overline{d}' \in M$ such that $\models \theta(\overline{a}; \overline{b}', \overline{d}')$. Since $\overline{a} \downarrow_A M$, this implies $\overline{a} \not\downarrow_M \overline{b}'$ so $\overline{b}' \notin M$. Since the requirements on $\overline{b}'$ were expressed by an $\mathbf{I}$-formula over $A \cup \overline{a}$, without loss of generality $\overline{b}' \in M[\overline{a}]$.

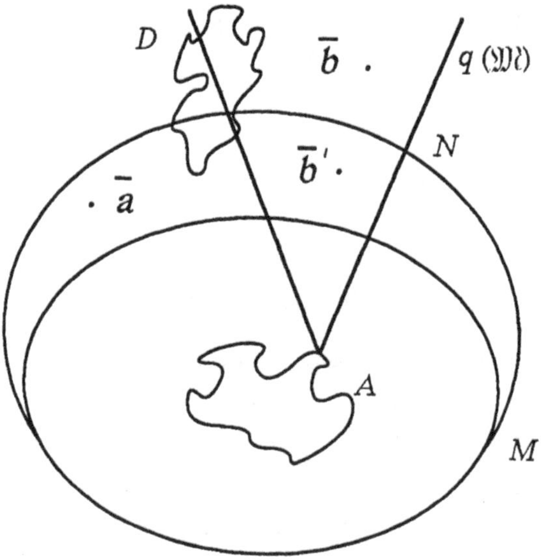

Fig. 1. Corollary X.1.22

The following exercises provide further rephrasings of axiom FI_1 and Corollary 1.20. In all cases M is to be assumed strongly $\mathbf{I}$-saturated.

1.23 Exercise. Show that if p is realized in $M[q]$ then $q \triangleright_M p$.

1.24 Exercise. Show that if X is an independent set over M then X is a maximal independent (over M) subset of $M[X]$.

The following exercise emphasizes that dependence is not a simple negation of nonforking by showing $\overline{b}$ can depend on $\overline{a}$ over M without being in $M[\overline{a}]$.

1.25 Exercise. Using the theory of a single equivalence relation with infinitely many infinite classes show the following assertion is false for $\mathbf{AT}_{\aleph_0}$. If $b \notin M[a]$ then $b \downarrow_M M[a]$.

This phenomenon can be made even sharper.

1.26 Exercise. Show using the theory of two refining equivalence relations that there exist M, a, and b such that $b \notin M[a]$ but $b \not\downarrow_{M \cup a} M[a]$.

1.27 Exercise. Prove that $p \perp q$ implies $M[p] \downarrow_M M[q]$. Indeed, $p \perp q$ implies $t(M[p]; M) \perp t(M[q]; M)$.

What is the relation between $M[\overline{a}, \overline{b}]$ and $M[\overline{a}][\overline{b}]$? It is easy to see that there are embeddings $M[\overline{a}] \mapsto M[\overline{a}, \overline{b}] \mapsto M[\overline{a}][\overline{b}]$. There are several variations on the question of whether this last map has an inverse. These variations depend on such subtleties as whether we are working in the category of constructions or the category of models and embeddings. The next theorem gives the strongest possible positive answer when $\overline{a} \downarrow_M \overline{b}$.

1.28 Theorem. i) *Let M be strongly* **I**-*saturated and $A = \{\overline{a}_0, \ldots, \overline{a}_n\}$ be an independent set over M. Suppose $M_{i+1} = M_i[\overline{a}_i]$. Then M_n is* **I**-*constructible over $M \cup A$.*

ii) *More precisely, if E_1 is an* **I**-*construction of N_1 over $M \cup \overline{a}$, E is an* **I**-*construction of N over $N_1 \cup \overline{b}$, and $\overline{a} \downarrow_M \overline{b}$ then N is* **I**-*constructible over $M \cup \{\overline{a}, \overline{b}\}$.*

Proof. The proof of ii) is included in the proof of i). We show i) by induction on $|A|$. Suppose $A = B \cup \overline{a}$, $|B| = m$ and, by induction, that M_m is **I**-constructible over $M \cup B$. Note that by FI_1 the construction of M_m over $M \cup B$ is, in fact, a construction of M_m over $M \cup A$. Now, appending the construction of M_{m+1} over $M_m \cup \overline{a}$ yields a construction of M_{m+1} over $M \cup A$ as required.

1.29 Exercise. Extend Theorem 1.28 by induction to infinite indiscernible sequences.

1.30 Exercise. Show that if $\lambda(\mathbf{I})$ is regular, FI_0 and FI_1 hold and $\overline{a} \downarrow_M \overline{b}$ then $t(M[\overline{b}]; M \cup \overline{b}) \vdash t(M[\overline{b}]; M \cup \overline{a} \cup \overline{b})$.

One important fact about **AT** is the following variant of the open mapping theorem: If $t(\overline{a}; A \cup \overline{b})$ is **AT**-isolated and $\overline{a} \downarrow_A \overline{b}$ then $t(\overline{a}; A)$ is **AT**-isolated. For the topological explanation of this nomenclature see Section III.3. We designate this property as the third basic relation between freedom and isolation. Both this property and FI_3 which follows are closely related to the finite character of forking, so there may be some more elegant way to unify these two notions.

1.31 Axiom FI_2 (The Open Mapping Theorem). If both $(\overline{a}\downarrow\overline{b}$;A) and $t(\overline{a}; A \cup \overline{b})$ is **I**-isolated over $A \cup \overline{b}$ then $t(\overline{a}; A)$ is **I**-isolated.

We have already seen (III.3.24) that FI_2 holds for $\mathbf{AT}_{\aleph_0}$. We now verify it for **S**-isolation.

1.32 Lemma. *If $\overline{a} \downarrow_A B$ and $t(\overline{a}; B)$ is* **S**-*isolated then $t(\overline{a}; A)$ is* **S**-*isolated.*

Proof. Choose $C \subseteq B$ such that $stp(\overline{a}; C) \vdash t(\overline{a}; B)$, and $D \subseteq A$ such that $\overline{a} \downarrow_D A$, $|D| < \kappa(T)$ and D contains $C \cap A$. We claim that $stp(\overline{a}; D) \vdash t(\overline{a}; A)$. This is a property of A, D, and an arbitrary realization $\overline{a}'$ of $stp(\overline{a}; D)$. Thus, we are completely free to choose the type of B over $A \cup \overline{a}'$. So without loss of generality we may assume that $A^\frown\overline{a}' \downarrow_D B$. Then since $stp(\overline{a}; D)$ is stationary, $stp(\overline{a}'; C) = stp(\overline{a}; C)$. But $stp(\overline{a}; C) \vdash t(\overline{a}; B)$ so $t(\overline{a}'; B) = t(\overline{a}; B)$ which is more than is required.

An alternative exposition of this proof would involve a new copy B' of B rather than relying on 'without loss of generality'. Although we concluded that $t(\overline{a}; B) = t(\overline{a}'; B)$, easy examples show that this phenomenon depended on our auspicious choice of B.

1.33 Exercise. Verify that FI_2 holds for **L**-isolation. (Hint: This is easily deduced from the original Open Mapping Theorem.)

There is a final axiom connecting isolation and nonforking which holds in all the classes we consider. In order to describe this principle we need some additional notation.

1.34 Notation. Let $A \subseteq \mathcal{M}$ and $p \in S(A)$. Suppose p is an n-type. We let $p^*(\mathcal{M})$ denote $\{c \in \mathcal{M} : \text{for some } \overline{c} \text{ realizing } p, c \in \mathrm{rng}(\overline{c})\}$.

Thus $p(\mathcal{M})$ denotes a set of n-tuples while $p^*(\mathcal{M})$ denotes a subset of $\mathcal{M}$.

1.35 Axiom $\mathbf{FI_3}$ (The Local Nature of Implication). Suppose that M is **I**-saturated and p is an **I**-formula with $\mathrm{dom}^*\, p = B \subseteq M$. If q is any extension of p to $S(M)$ then $q|(B \cup p^*(M)) \vdash q$.

1.36 Exercise. Deduce from FI_3 that if $p \in S(A)$, $(\overline{c} \not\downarrow M; A)$ and $\overline{c}$ realizes p then $(\overline{c} \not\downarrow p^*(M); A)$.

If $p \in S(A)$, keeping in mind that if $\overline{a}^\frown\overline{b} \in p(\mathcal{M})$ and $q = t(\overline{a}^\frown\overline{b}; A)$ then $q^*(\mathcal{M}) \subseteq p^*(\mathcal{M})$, we easily see

1.37 Proposition. *Suppose* FI_3 *holds. Let* M *be* **I**-*saturated. For* $\overline{a}, \overline{b}$ *realizing* p, *if* $q = t(\overline{a}^\frown\overline{b}; M)$ *then, letting* $B = \mathrm{dom}^*\, p$, $q|(B \cup p^*(M)) \vdash q$.

The following result, which easily yields that the set of models of an ω-stable (indeed an arbitrary stable theory) satisfies FI_3 for both **AT** and **L**-isolation, was first enunciated in this form by Lascar and Bouscaren in [Bouscaren & Lascar 1983].

1.38 Theorem. *Let* M *be a model of a stable theory,* $p \in S(M)$, $\phi(\overline{x}, \overline{a}) \in p$ *and set* $B = \phi(M, \overline{a}) \cup \overline{a}$. *Then,* $p|B \vdash p$.

Proof. Assume for contradiction that there are $\psi(\overline{x}; \overline{m}) \in F(M)$ and $\overline{c}$, $\overline{c}'$ in $\mathcal{M}$ such that $\overline{c}$ realizes p and $t(\overline{c}; B) = t(\overline{c}'; B)$ but $\models \psi(\overline{c}; \overline{m}) \wedge \neg\psi(\overline{c}'; \overline{m})$. Let d define $t(\overline{m}; B)$. In particular, $d\psi(\overline{y}) \in F(B)$ and for any $\overline{b} \in B$, $\models d\psi(\overline{b})$ iff $\models \psi(\overline{b}; \overline{m})$. Since $t(\overline{c}; B) = t(\overline{c}'; B)$ and $\models \phi(\overline{c})$, we have $\models \phi(\overline{c}')$. Suppose $\models d\psi(\overline{c})$ and thus $\models d\psi(\overline{c}')$. (If $\models \neg d\psi(\overline{c})$ we obtain a similar contradiction.) Since M is an elementary submodel of $\mathcal{M}$, there is a $\overline{c}'' \in M$ such that

$\models \phi(\overline{c}'') \wedge d\psi(\overline{c}'')$ and $\models \neg\psi(\overline{c}''; \overline{m})$. But then $\overline{c}'' \in B$ and this contradicts the choice of d.

With a similar diagonal argument we extend this result to saturated models. This imposes a stricter requirement on the model M but makes the subset on which the type depends smaller. We can easily conclude that FI_3 holds for $\mathbf{SET}_\lambda$ and $\mathbf{S}_\lambda$.

1.39 Theorem. *If T is stable, M is a λ-saturated model of T, $B \subseteq M$ with $|B| < \lambda$, and $p \in S(M)$ with $p|B = p_0$ then $p|(B \cup p_0^*(M)) \vdash p$.*

Proof. Let $A = B \cup p_0^*(M)$. For each $\overline{n} \in M$ and each formula $\phi(\overline{x}; \overline{n})$, the definition d of $t(\overline{n}; A)$ yields a $d\phi$ such that:

$$\text{For all } \overline{a} \in A, M \models \phi(\overline{a}; \overline{n}) \text{ iff } M \models d\phi(\overline{a}). \qquad (*)$$

But if $p|B \cup d\phi(\overline{x}) \cup \neg\phi(\overline{x}; \overline{n})$ is consistent then it is realized by some $\overline{a}' \in M$ contradicting $(*)$.

1.40 Exercise. Verify that for each $i < 4$ and each regular $\lambda \geq \kappa(T)$ the axiom FI_i holds for $\mathbf{SET}_\lambda$ and $\mathbf{S}_\lambda$.

1.41 Exercise. Suppose FI_3 holds and M is $\mathbf{I}$-saturated. For $\overline{a}$, $\overline{b}$ realizing p, if $\overline{a}^\frown\overline{b} \not\downarrow_A M$ then there exists an $X \subseteq p^*(M)$ such that $\overline{a}^\frown\overline{b} \not\downarrow_A X$

1.42 Historical Notes. This section unifies various arguments scattered in [Shelah 1978], [Lascar & Poizat 1979], [Bouscaren & Lascar 1983] and [Makkai 1984]. The fact that Corollary 1.20 holds for local isolation is new here. The weaker Exercise 1.21 was proved (for $\mathbf{L}$) earlier by Pillay and Steinhorn [Pillay & Steinhorn 1985]. The origins of Theorem 1.33 are a little murky. Although, first given explicitly in this form in [Bouscaren & Lascar 1983],there are variations in both [Lascar & Poizat 1979] and [Shelah 1978] (V.1.11).

2. Powerful Isolation Relations

We introduce here an important strengthening of the isolation concept. Most of the notions we have considered, but not $\mathbf{AT}$, satisfy this additional condition. These results show that under certain conditions the concepts of isolation and domination are very closely related. Corollary 2.5 records another crucial property of $\mathbf{S}$-isolation.

2.1 Definition. The isolation relation $\mathbf{I}$ is *powerful* if for any finite sequence $\overline{a}$ and any set B with $|B| < \lambda(\mathbf{I})$, $t(\overline{a}; B)$ is $\mathbf{I}$-isolated.

Of course, if $t(\overline{a}; B)$ is an $\mathbf{I}$-formula, it is $\mathbf{I}$-isolated, so for any λ, $\mathbf{SET}_\lambda$ and $\mathbf{S}_\lambda$ are powerful isolation relations. It is equally easy to see that even for ω-stable T, $\mathbf{AT}_{\aleph_0}$ is not always a powerful isolation relation (see Exercise 2.4).

2.2 Exercise. Show that if T is a countable, ω-stable, $\aleph_0$-categorical theory then $\mathbf{AT}_{\aleph_0}$ is a powerful isolation relation.

The following theorem is one of the main reasons for introducing this notion.

2.3 Theorem. *Let M be $\mathbf{I}$-saturated for a powerful isolation relation $\mathbf{I}$. Suppose also that if M is $\mathbf{I}$-saturated, then for any $\bar{a}$, $t(\bar{a}; M)$ is strongly based on a subset of M with fewer than $\lambda(\mathbf{I})$ elements. Then for any $\bar{a}$ and $\bar{b}$, $\bar{a} \rhd_M \bar{a}^\frown\bar{b}$ if and only if there exists a structure N which is $\mathbf{I}$-strictly prime over $M \cup \bar{a}$ and contains $\bar{b}$.*

Proof. Suppose $\bar{a} \rhd_M \bar{a}^\frown\bar{b}$; let $p = t(\bar{a}; M)$ and $r = t(\bar{b}; M \cup \bar{a})$. Choose $B \subseteq M$ such that $|B| < \lambda(\mathbf{I})$, $\bar{a}^\frown\bar{b} \downarrow_B M$, and $t(\bar{a}^\frown\bar{b}; M)$ is strongly based on B. Then, by Theorem VI.3.12, $t(\bar{a}; B) \rhd_B t(\bar{a}^\frown\bar{b}'; B)$ for any $\bar{b}'$ realizing $r_0 = r|(B \cup \bar{a})$. Since $\mathbf{I}$ is powerful, r_0 is $\mathbf{I}$-isolated. We now show r_0 $\mathbf{I}$-isolates $t(b; M \cup \bar{a})$. If $\bar{b}'$ realizes r_0, $\bar{a} \downarrow_B M$ implies $\bar{a}^\frown\bar{b}' \downarrow_B M$. Thus $t(\bar{a}^\frown\bar{b}'; M) = t(\bar{a}^\frown\bar{b}; M)$ and, in particular, $t(\bar{b}; M \cup \bar{a}) = t(\bar{b}'; M \cup \bar{a})$ as required. Thus, a strictly $\mathbf{I}$-prime model can be constructed over $M \cup \bar{a}$ beginning with $\bar{b}$. This is more than is required. The converse is an easy consequence of Corollary 1.20.

2.4 Exercise. Show that Theorem 2.3 fails for $\mathbf{AT}_{\aleph_0}$. (Hint: Consider the theory of an equivalence relation with infinitely many infinite classes and a model of $\mathrm{Th}(Z, S)$ in each class.)

Note that as phrased the result applies to $\mathbf{SET}_\lambda$ for $\lambda \geq |T|^+$. If T is a countable ω-stable theory, Theorem 2.3 applies to $\mathbf{SET}_{\aleph_0}$ but not if T is strictly superstable.

2.5 Corollary. *Let M be $\mathbf{I}$-saturated and good, $\mathbf{I}$ a powerful isolation relation, and p, $q \in S(M)$. Then $p \rhd q$ if and only if q is realized in $M[p]$.*

Proof. Let $\bar{a}$, $\bar{b}$ realize p and q and witness $p \rhd q$. Choose $B \subseteq M$ with $|B| < \lambda(\mathbf{I})$, p, q strongly based on B, and $\bar{a}^\frown\bar{b} \downarrow_B M$. Now, if $\bar{b}'$ realizes $t(\bar{b}; B \cup \bar{a})$, Lemma VI.3.12 implies $\bar{b}' \downarrow_B M$ and thus $\bar{b}'$ realizes q. But, certainly $t(\bar{b}; (B \cup \bar{a})$ is realized in $M[\bar{a}]$. The other implication is immediate from Theorem 2.3.

2.6 Exercise. Show, using the same example as before, that Corollary 2.5 does not hold for $\mathbf{AT}$.

2.7 Historical Notes. The notion of a powerful isolation relation arose from a desire to unify some proofs from [Buechler 1984b] (only in the preprint) and [Prest 1985]. Although Prest proves Theorem 2.3 for a notion of isolation in modules, I have not succeeded in reducing his argument to this rubric. Nevertheless, the notion seems to isolate one of the essential distinctions between $\mathbf{AT}$ and $\mathbf{S}$. The example in Exercise 2.4 was first suggested to me by Shelah; it was also discovered by Pillay and Steinhorn. Pillay corrected an error in the original proof of Theorem 2.3.

3. Uniqueness of Prime Models

In this section we prove that if T is a countable stable theory which has prime models over every set A then these models are unique up to A-isomorphism. More generally, we show that for any reasonable notion of isolation **I**, any subset of an **I**-constructible model is **I**-constructible. Intuitively, one could justify such a claim by asserting that if the order of a construction was really essential it would be reflected by some ordering in the theory; thus, the theory would be unstable. This intuition is made precise in [Shelah 1975] where Shelah proves the uniqueness of prime models over a set of power $\aleph_1$ using the techniques of stationary sets and relying on stability theory only for the fact that no formula in a stable theory can have the order property. Here, we adopt a more model theoretic course and invoke the machinery of nonforking we have developed earlier.

3.1 Theorem. *Let* **I** *be an isolation relation satisfying the axioms listed in Sections IX.2, IX.3, and X.1. Suppose* $\lambda(\mathbf{I})$ *is regular and* $\kappa(T) \leq \lambda(\mathbf{I})^+$. *If* E *is strictly* **I**-*prime over* A *and* $A \subseteq M \subseteq E$ *for some* $M \models T$ *then* M *is strictly* **I**-*prime over* A.

Proof. Fix A, B which is **I**-prime over A, and E which is strictly **I**-prime over A. Without loss of generality $A \subseteq B \subseteq E$. By Theorem IX.4.12, it suffices to show that B is **I**-constructible. The proof proceeds by induction on $|E - A|$. If $|E - A| \leq \lambda(\mathbf{I})$ the result follows from Lemma IX.4.2. Suppose $|E - A| = \mu$ and E is given by the construction $\langle e_i : i < \mu \rangle$. The idea of the induction is to partition the construction of E so that the conclusion of the theorem can be applied to a sequence of smaller models C_{i+1} for $i < |E|$. The main theorem is illustrated by the first diagram (Fig. 2) which concerns E, B, and A. To perform the induction we partition E as in

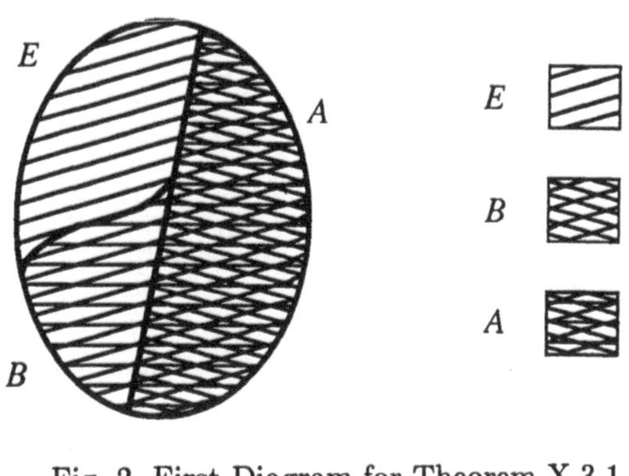

Fig. 2. First Diagram for Theorem X.3.1

the second diagram and apply the hypothesis of induction to $C_{i+1} \cup A$, $C_i \cup A \cup (C_{i+1} \cap B))$, $C_i \cup A$. (Fig. 3). Formally we build $\langle C_i : i < \mu \rangle$ to satisfy the following five conditions.

i) $\langle C_i : i < \mu \rangle$ is an increasing continuous sequence of subsets of $A \cup E$ and $\bigcup C_i = E \cup A$.
ii) $|C_i| \leq |i| + \lambda(\mathbf{I})$, so $|C_{i+1} - C_i| < |E - A|$.
iii) For each $\overline{c} \in C_i$, $\overline{c} \downarrow_{B \cap C_i} B$ and $\overline{c} \downarrow_{A \cap C_i} A$.
iv) C_i is closed. (Recall Definition IX.4.8)
v) C_{i+1} is $\mathbf{I}$-constructible over $A \cup C_i$.

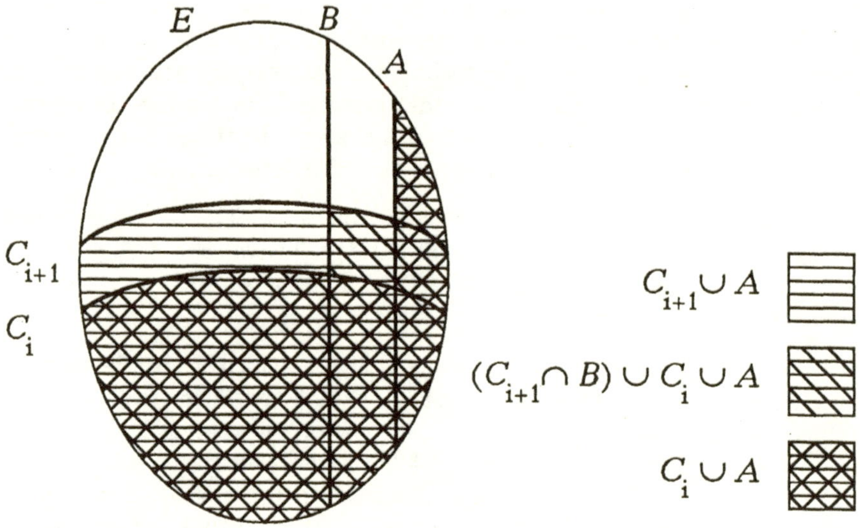

Fig. 3. Second diagram for Theorem X.3.1

The C_i are constructed by induction. Let C_0 be empty. If we have C_i, we define C_i^n for $n < \omega$ by induction and let $C_{i+1} = \bigcup_n C_i^n$. Let C_i^0 contain C_i and a point from $E - C_i$. Suppose we have constructed C_i^{2n}. For each $\overline{c}$ in C_i^{2n}, there exists a $D_{\overline{c}} \subseteq B$ with $|D_{\overline{c}}| < \kappa(T)$ such that $\overline{c} \downarrow_{D_{\overline{c}}} B$. Similarly, there exists an $E_{\overline{c}} \subseteq A$ with $\overline{c} \downarrow_{E_{\overline{c}}} A$. Let $C_i^{2n+1} = \bigcup_{\overline{c} \in C_i^{2n}} D_{\overline{c}} \cup \bigcup_{\overline{c} \in C_i^{2n}} E_{\overline{c}}$. Now for each $e_j \in C_i^{2n+1}$, there is an $F_j \subseteq E_j \cup A$ with $|F_j| < \lambda(\mathbf{I})$ such that $t(e_j; A \cup E_j)$ is isolated by F_j. Let $C_i^{2n+2} = \bigcup_{e_j \in C_i^{2n+1}} F_j$. Now C_{i+1} is a closed subset of E when a) E is regarded as constructed over A and when b) E is regarded as constructed over $A \cup C_i$ (cf. Theorem IX.4.11). Then a) guarantees iv) and b) guarantees v). The choice of C_{i+1} guarantees iii) and i); it is easy to verify ii).

Applying iv) and the induction hypothesis to $A \cup C_i$, $A \cup C_i \cup (C_{i+1} \cap B)$, and $A \cup C_{i+1}$, we conclude $A \cup C_i \cup (C_{i+1} \cap B)$ is $\mathbf{I}$-constructible over $A \cup C_i$. Thus for some α, $(B \cap C_{i+1}) - (A \cup C_i) = \langle b_j : j < \alpha \rangle$. Let $B_j = \{b_k : k < j\}$. For each j, $q_j = t(b_j; A \cup C_i \cup B_j)$ is $\mathbf{I}$-isolated. If we can show q_j does not fork over $(A \cup (B \cap C_i) \cup B_j)$ for each j, we can conclude from the Open Mapping Theorem that $t(b_j; A \cup (B \cap C_i) \cup B_j)$ is $\mathbf{I}$-isolated. This yields that $B \cap C_{i+1}$ is constructible over $A \cup (B \cap C_i)$. But if this holds for each $i < \mu$, B is constructible over A and we finish.

Consider any $\bar{c} \in C_i$. By iii) $\bar{c} \downarrow_{B \cap C_i} B$ so by the monotonicity of nonforking

$$\bar{c} \underset{B \cap C_i}{\downarrow} A \cup (B \cap C_{i+1}).$$

By monotonicity again,

$$\bar{c} \underset{A \cup (B \cap C_i) \cup B_j}{\downarrow} A \cup (B \cap C_i) \cup B_{j+1}.$$

Then by symmetry

$$b_j \underset{A \cup (B \cap C_i) \cup B_j}{\downarrow} A \cup (B \cap C_i) \cup B_j \cup \bar{c}.$$

Since this holds for each $\bar{c} \in C_i$, we have by the finite character of forking that

$$b_j \underset{A \cup (B \cap C_i) \cup B_j}{\downarrow} A \cup C_i \cup B_j$$

as required.

Note that this proof does not require the Existence Axiom IX.2.7. We can state a less technical result if we assume it as well.

3.2 Corollary. *Let* $\mathbf{I}$ *be one of the isolation relations defined in IX.2.1. Suppose* $\lambda(\mathbf{I})$ *is regular,* $\kappa(T) \leq \lambda(\mathbf{I})^+$, *and that for every set* $A \subseteq \mathcal{M}$ *there is a strictly* $\mathbf{I}$*-prime model over* A. *For any* $A \subseteq \mathcal{M}$, *any two* $\mathbf{I}$*-prime models over* A *are isomorphic.*

3.3 Historical Notes. Shelah first proved the uniqueness of prime models for a countable ω-stable theory by induction on rank in [Shelah 1972]. He gave an alternative proof using the machinery of stationary sets rather than the technology of stability theory in [Shelah 1975]. The proof given here is taken from [Shelah 1979]. That proof was inspired by the (unpublished) proof by Ressayre, which we discussed in Section IX.4, that strictly prime models are unique. In [Shelah 1979] Shelah showed that both the hypotheses of countability and stability are necessary for the theorem. First he shows that the theory $\mathrm{REF}_{\aleph_1}$ of $\aleph_1$ refining equivalence relations has prime models over all sets but does not have unique prime models. Then he notes that replacing the equivalence relations by ternary relations such that a linear order on one set indexes a collection of refining equivalence relations on another produces a countable unstable example of the same phenomenon.

This line of argument produced the first proof of the uniqueness of "differential closure". Blum [Blum 1968] showed the theory of differentially closed fields of characteristic zero is ω-stable and deduced the uniqueness from [Shelah 1972]. This result is extended to differentially closed fields of characteristic p by the proof that the theory is stable ([Wood 1973], [Wood 1976], [Wood 1979]) and the application of Theorem X.3.1. These proofs were later given more algebraic form in [Kolchin 1973].

4. Indiscernible Sets in Prime Models

Axiom FI_1 clearly implies that if X is an independent set over M, then X is a maximal independent set in $M[X]$. We would like to improve this result to: if X is an independent set of indiscernibles then $\dim(X, M, M[X]) = |X|$. (Recall Definition V.2.3.) In general, however, this assertion is false. In this section we show that it is true if the cardinality of X is sufficiently large. In Chapter XII we show that for regular types it is always true.

Let A be a set and suppose that M is $\mathbf{I}$-constructible over A. We will show that $\lambda(\mathbf{I})$ provides an upper bound for the length of independent sets over A in M. This upper bound will be a key to the construction of many non-isomorphic models of T in Part D. It is easy to see from Exercise X.1.21 that this property holds if we start with a model.

For simplicity we assume in this section that $\lambda(\mathbf{I})$ is regular. We could strengthen the results to deal with singular $\lambda(\mathbf{I})$ by replacing such hypotheses as $\lambda(\mathbf{I}) \geq \kappa(T)$ by $\mathrm{cf}(\lambda(\mathbf{I})) \geq \kappa(T)$.

4.1 Exercise. Show that if $\lambda(\mathbf{I}) \geq \kappa(T)$, $|A| < \kappa(T)$ and M is $\mathbf{I}$-saturated then every subset of $M[A]$ which is indiscernible over $M \cup A$ has power at most $\kappa(T)$.

It is not so easy to see this when we do not start over a model. We begin by describing a further property of the isolation relations introduced in Chapter IX. It is unclear whether this property follows from the axioms in Chapter IX but it is trivial to verify it for the isolation relations defined in Section IX.2.

4.2 Lemma. *If q is an $\mathbf{I}$-formula with $\mathrm{dom}^* q \subseteq A$, E is a set of indiscernible over A realizations of q and $\overline{e}$ realizes $\mathrm{Av}(E, A)$ then $\models q(\overline{e})$.*

Proof. If $\mathbf{I}$ is $\mathbf{SET}_\lambda$ or $\mathbf{AT}_\lambda$, the result follows immediately from the definition of average type. For $\mathbf{S}_\lambda$, apply Lemma V.1.7.

4.3 Lemma. *Suppose $E \subseteq A$ is a set of indiscernibles over $A_0 \subseteq A$ with $|E| \geq \kappa(T)$ and $\overline{e}$ realizes $\mathrm{Av}(E; A)$. Suppose also that for some $\mathbf{I}$ with $\lambda(\mathbf{I}) \geq \kappa(T)$ and some $\mathbf{I}$-formula q with $\mathrm{dom}^* q = A_0$, $q \vdash t(\overline{c}; A)$. Then $q \vdash t(\overline{c}; A \cup \overline{e})$.*

Proof. For any formula $\psi(\overline{x}, \overline{y}; \overline{a})$ such that $\models \psi(\overline{c}, \overline{e}; \overline{a})$ there is a set $E_{\overline{a}} \subset E$ with $|E_{\overline{a}}| < \kappa(T)$ such that $E - E_{\overline{a}}$ is a set of indiscernibles over $E_{\overline{a}} \cup \overline{a} \cup \overline{c} \cup A_0$. Now $\models \psi(\overline{c}, \overline{e}; \overline{a})$ if and only if for one (and thus every) $\overline{e}' \in E - E_{\overline{a}}$, $\models \psi(\overline{c}, \overline{e}'; \overline{a})$. Since $q \vdash t(\overline{c}; A)$ there is a formula $\theta \in q$, with $\models \theta(\overline{x}) \rightarrow \psi(\overline{x}, \overline{e}'; \overline{a})$. Then the $\mathbf{I}$-formula $(\forall \overline{x})\theta(\overline{x}) \rightarrow \psi(\overline{x}, \overline{e}'; \overline{a})$ holds for all $\overline{e}'$. By Lemma 4.2, we conclude $(\forall \overline{x})\theta(\overline{x}) \rightarrow \psi(\overline{x}, \overline{e}; \overline{a})$. Since this holds for any formula ψ we have the lemma.

It is now fairly easy to show

4.4 Corollary. *If M is* **I***-constructible over A and $E \subset A$ is a set of indiscernibles with $|E| \geq \lambda(\mathbf{I})$ then* $\mathrm{Av}(E; A) \vdash \mathrm{Av}(E; M)$.

Proof. The proof is a straightforward induction on the length of the construction. For the crucial successor stage we must show that if $E \subset A'$, $t(c; A')$ is **I**-isolated by some q with $\mathrm{dom}^* q \subseteq A'$, $\bar{e}$ realizes $\mathrm{Av}(E; A)$ and $t(\bar{e}; A) \vdash t(\bar{e}; A')$ then $t(\bar{e}; A) \vdash t(\bar{e}; A' \cup \bar{c})$. Since $|E| \geq \lambda(\mathbf{I})$ we may assume E is indiscernible over $\mathrm{dom}^* q$. By Lemma 4.3 we know $q \vdash t(\bar{c}; A' \cup \bar{e})$. By the symmetry axiom for **I**-isolation (Axiom IX.3.3) we deduce $t(\bar{e}; A' \cup \bar{c})$ is **I**-isolated over A. If **I** is $\mathbf{SET}_\lambda$ or $\mathbf{AT}_\lambda$ the result is immediate. For $\mathbf{S}_\lambda$ it follows from Lemma IV.3.12.

Now we can show one of the two components of the characterization of **I**-prime models. We show here that if M is **I**-prime over A then M is **I**-atomic over A and every set of indiscernibles in M has cardinality at most $\lambda(\mathbf{I})$. The necessity of the first condition is almost obvious and now we deduce the necessity of the second. The converse to this theorem is proved by induction on rank in Chapter IV of [Shelah 1978].

4.5 Theorem. *Suppose $\lambda(\mathbf{I}) \geq \kappa(T)$ is regular. Assume T admits* **I***-prime models. If M is strictly* **I***-prime over A and $E \subseteq M$ is a set of indiscernibles over A then $|E| \leq \lambda(\mathbf{I})$.*

Proof. Suppose $|E| > \lambda(\mathbf{I}) \geq \kappa(T)$. By Theorem V.1.23 choose $E_0 \subseteq E$ with $|E_0| < \kappa(T)$ such that $\mathrm{Av}(E; M)$ is strongly based on $A \cup E_0$. Let p denote the restriction of $\mathrm{Av}(E; M)$ to $A \cup E_0$. Let F be an independent sequence of $\lambda(\mathbf{I})$ realizations of p and let M' be **I**-prime over $B = A \cup E_0 \cup F$. Corollary 4.4 implies $\mathrm{Av}(F; B) \vdash \mathrm{Av}(F; M)$. Thus, $\dim(F, A \cup E_0, M')$ is $\lambda(\mathbf{I})$. By Lemma IX.4.1, M is **I**-prime over $A \cup E_0$. Let f imbed M into M' fixing $A \cup E_0$. Now $f(E)$ and F are equivalent indiscernible sets with at least $\kappa(T)$ elements. Thus, by Theorem V.2.4 they should have the same dimension in M'. But, clearly $\dim(f(E), A \cup E_0, M') > \lambda(\mathbf{I})$ and we finish.

The next result will play an important role in the construction of many non-isomorphic models. It will be used in conjunction with Theorem 4.5 but is proved solely on the basis of the methods of Section 1.

4.6 Theorem. *Let M be* **I***-saturated and $p, q \in S(M)$ with $p \perp q$. If E is an independent set of realizations of q and N is* **I***-prime over $M \cup E$ then* $\dim(p, N) = 0$.

Proof. Clearly, if $\bar{a} \in N - M$ realizes p, $\bar{a} \not\downarrow_M E$. But then $p^\omega \not\downarrow^w q^\omega$, whence by Corollary VI.2.18 $p \not\downarrow q$.

4.7 Definition. The model M is **I**-minimal over A if M is **I**-saturated and there is no **I**-saturated model N containing A and properly contained in M.

The following lemma allows us to characterize minimal models by the cardinalities of maximal sets of indiscernibles. To apply it we must know every **I**-saturated model is good. We showed in Corollary IV.2.9 and Theorem IV.3.22 that **AT** has that property if T is a countable ω-stable theory and **S** has that property for every superstable theory.

4.8 Lemma. *Fix* **I** *with* $\lambda(\mathbf{I}) \geq \kappa(T)$ *and such that every* **I***-saturated model is good. Let* M *be* **I***-atomic over* A *and suppose* $N \subseteq M$ *is* **I***-saturated. Then for any* $\overline{a} \in M - N$*, there is an infinite set* $E \subseteq N$ *of indiscernibles over* A *such that* $t(\overline{a}; N) = \text{Av}(E, N)$.

Proof. Suppose there exists $c \in M - N$. Choose $C \subseteq N$ such that $t(c; N)$ is strongly based on $A \cup C$. Note that $stp(c; A \cup C)$ is **I**-isolated. Now choose $c_i \in N$ for $i < \omega$ so that c_n realizes $stp(c; A \cup C_n)$. The c_i are a strongly independent sequence and therefore indiscernible by Lemma V.1.8 so we have the lemma.

The next result characterizes minimal models.

4.9 Theorem. *Let* T *be superstable and countable.* M *is* **I***-minimal over* A *if and only if every set* E *of indiscernibles over* A *in* M *is finite.*

Proof. Suppose E is an infinite set of indiscernibles in M. Choose $N \subseteq M$ to be **I**-prime over $A \cup (E - \{e\}$, for some $e \in E$. By Corollary 4.4, $e \notin N$ so M is not minimal.

The converse is immediate from the lemma.

4.8 Historical Notes. These results are scattered in [Shelah 1978] but Section IV.4 especially Theorem IV.4.9 is most relevant.

Part C

Local Dimension Theory

In Part A, we began the study of an independence relation for elements of models of stable theories. In Chapter II we noticed that, taking dependence to be the negation of independence, our notion does not in general satisfy the transitivity axiom for dependence. In Part B, we sought a global remedy for this problem. That is, we considered several notions of dependence which are transitive and thus provide a notion of the 'closure' of a set. Since none of these are the negation of the independence relation, they do not immediately yield a notion of dimension. In this part, we return to the study of nonforking and study those types, p, such that forking is transitive when restricted to realizations of p. We will be able to assign a well defined dimension, even in the finite case, to the set of realizations of such a 'regular' type.

There are, however, several complications. The most important is that we can not single out the 'regular' types with one definition which is appropriate for any theory T. Rather, we must define a notion of K-regularity where K is the set of $\mathbf{I}$-saturated models for an isolation relation $\mathbf{I}$ and vary the choice of $\mathbf{I}$ according to the stability class of T.

Our goal is to provide a structure theory for the models of a theory T. As a test problem we try to calculate the spectrum function of T. In general, this problem is too difficult to be attacked directly. We can directly attack the problem of calculating the number of $\mathbf{S}$-models of a superstable theory and the number of models of an ω-stable theory. The methods of attack on these two problems in [Shelah 1978], [Shelah 1982], and [Harrington & Makkai 1985] are parallel. Thus, in this book we have attempted to develop a single theory which specializes to the two cases. For this, we define in Chapter XI the notion of an acceptable class K. In so far as is possible, we provide a common development of the theory for any such class. In fact, the class of $\mathbf{S}$-saturated models plays a distinguished role which prevents a completely uniform treatment. The formulation of the notion of acceptable class K given here and the general treatment of the concepts of regularity are only a first attempt at a theory which we hope can be extended eventually, e.g. to the infinitary case.

The bulk of Part C is devoted to the development of what is sometimes referred to as 'intermediate' stability theory. In Part D, we begin the study of 'advanced' stability theory, the use of the tools developed here to analyze models. The building blocks for models in this analysis will be the regular and weight one types defined in Chapters XII and XIII. The mortar between these blocks is the compulsion notion which is introduced in Chapter XI and studied throughout this part of the book.

Chapter XI

Acceptable Classes

Section 1 of this chapter describes the framework for the rest of this book. Our primary concerns are with the class of all models of an ω-stable theory and the class of $\mathbf{S}_{\aleph_0}$-saturated models of a superstable theory. However, the notion of acceptable class will encompass a number of other classes such as the class of $\mathbf{AT}_\lambda$-saturated models of T if $\lambda \geq \kappa(T)$. Section 2 contains some more detailed information about the class of $\mathbf{S}$-models.

1. Basic Properties of an Acceptable Class K

Two classes K are the primary focus of our analysis:

i) the strongly $\kappa(T)$-saturated models of a superstable theory,
ii) all models of an ω-stable theory.

The notion of an acceptable class is somewhat broader. It includes, for example, the class of $\mathbf{AT}_\lambda$-saturated models of T if $\lambda \geq \kappa(T)$ and the class of strongly $\kappa(T)$-saturated models of a stable theory. We try to isolate some of the properties of an acceptable class which would have to be preserved to extend the theory to include, for example, the infinitary case or the class of pure-injective submodules of models of a theory of modules.

The class of $\mathbf{S}$-models of a superstable theory is the easiest to analyze. In some sense it was defined exactly to make the analysis work. The difficulty is to see what variations in the arguments used in that situation will allow the results to carry over to the other cases. For strongly $\kappa(T)$-saturated models the superstable case is distinguished from the stable case by the abundant supply of regular types (cf. Section XII.2).

1.1 Definition. K is an *acceptable class* of structures for the theory T if

i) K is the class of $\mathbf{I}$-saturated models for an isolation relation $\mathbf{I}$ described in Definition IX.2.2 for which $\lambda(I) \geq \kappa(T)$.
ii) T admits K-prime models over all sets.

We denote by $\lambda_0(K)$ or $\lambda_0(\mathbf{I})$ the least cardinality of a member of K.

We will define a number of notions relative to an acceptable class K. These may be prefixed either by **I** as in **I**-prime or K as in K-strongly regular. We sometimes refer to members of K as K-models.

The following easy exercise lists the most interesting acceptable classes.

1.2 Exercise. Verify that if $\lambda \geq \aleph_1$ and T is stable then the classes of $\mathbf{AT}_\lambda$ and $\mathbf{S}_\lambda$-saturated models are acceptable. Moreover, if T is ω-stable the class of all models is acceptable. Finally, for any superstable T, the class of **S**-saturated models is acceptable.

The properties of the class K described in the following exercise suffice to prove all the results in Section XII.1.

1.3 Exercise. Show that if K is acceptable class it has the following property.

For each $M \in K$ and each set A with $|A| < \lambda(I)$ there is a model $M[A] \in K$ containing $M \cup A$ which is **I**-saturated and satisfies the following conditions.

i) $A \rhd_M M[A]$.
ii) If $M, N \in K$ and $M \cup \bar{a} \subseteq N$ then there is an **I**-constructible $M' \in K$ with $M \cup \bar{a} \subseteq M' \subseteq N$.

Of course, since T admits K-prime models $M[A]$ is just a K-prime model over $M \cup A$. We have explicitly excluded L as an acceptable class. The following exercise explains why.

1.4 Exercise. Show that for any stable theory T, the class of **L**-saturated models satisfies condition i) of the last exercise. Find an example to show ii) may fail.

To simplify the statement of the results on the spectra of the classes K, we introduce the following notation.

1.5 Notation. For any cardinal $\aleph_\alpha$ we denote by K_α the collection of members of K with cardinality $\aleph_\alpha$ and by $K_{\leq\alpha}$ the collection of members of K with cardinality at most $\aleph_\alpha$. We write $I(\aleph_\alpha, K)$ for $|K_\alpha|$ and $I^*(\aleph_\alpha, K)$ for $|K_{\leq\alpha}|$.

Occasionally, we will depart from the $\aleph_\alpha$ notation and write $I(\kappa, K)$ for $|\{M \in K : |M| = \kappa\}|$. The basic notation is defined in terms of the $\aleph_\alpha$ notation because the spectrum is often most easily computed in terms of the index. Similarly, it turns out that the formulas for the cumulative hierarchy $I^*(\aleph_\alpha, K)$ are often easier to state than those for $I(\aleph_\alpha, K)$. Clearly, the two functions are interdefinable.

We now define an extremely powerful and natural partial ordering on types. Intuitively, p compels q if q is realized in every model which realizes p. One of the great strengths of stability theory is the ability to translate this intuitively appealing idea into a practical technical tool.

1.6 Definition. Let $M \in K$ and $p, q \in S(M)$. Then we say p *compels* q over K and write $p \mapsto_K q$ if for every $\overline{a}$ realizing p and every $N \in K$ which contains $M \cup \overline{a}$, q is realized in N. If $p \mapsto_K q$ and $q \mapsto_K p$ we write $p \underset{\sim}{\mapsto}_K q$.

We omit the subscript K when it is clear from context.

1.7 Exercise. Show that for any class K, $\mapsto_K$ is a transitive reflexive relation.

1.8 Exercise. Show that when T admits K-prime models the reference in Definition 1.6 to 'all $N \in K$ which contain $M \cup \overline{a}$' can be replaced by '$M[\overline{a}]$'.

1.9 Definition. Let $p \in S(M)$. If for every $q \in S(M)$ which is not realized in M, $p \mapsto q$ implies $q \mapsto p$ then p is a *K-minimal type.*

1.10 Exercise. Show $\underset{\sim}{\mapsto}$ is an equivalence relation on K-minimal types over M.

The following analogy will make it easier to visualize the properties of $\mapsto$. We will prove a precise form of it in Theorem XIII.3.6. Think of the equivalence classes of types over M mod $\underset{\sim}{\mapsto}$ as natural numbers. The relation $\mapsto$ corresponds to divisibility; the K-minimal types are primes.

1.11 Exercise. Let $M \in K$, $p, q \in S(M)$. Show that if $p \mapsto q$ then $p \not\perp^a q$.

Thus far, $\mapsto$ is only defined relative to a model M. In the next section we will see that if K is the class of strongly $\kappa(T)$-saturated models then this restriction can be removed. For now, we show that for any acceptable class K, $\mapsto_K$ is preserved under nonforking extension. It is unclear whether this relation is preserved downwards (for arbitrary acceptable K).

1.12 Lemma. *If $M, N \in K$, p, $q \in S(M)$ and $p \mapsto_K q$ then if p' and q' are nonforking extensions to $S(N)$, $p' \mapsto_K q'$.*

Proof. Let $\overline{b}$ realize p'. Choose $M[\overline{b}] \prec N[\overline{b}]$ and $\overline{c} \in M[\overline{b}]$ to realize q. By FI_1 (letting $N - M$ play the role of $\overline{a}$) $\overline{c} \downarrow_M N$ and we finish.

1.13 Historical Notes. This notion of compulsion, though not the name, was introduced by Shelah in Definition V.2.1 of [Shelah 1978]. However, his emphasis was not on an ordering between types but on the order between the associated indiscernible sets (built by taking a nonforking sequence of realizations of a stationary type). Lascar (independently) introduced the notion as a relation on types [Lascar 1976]. Since for models of arithmetic the relation can be interpreted as the Rudin-Keisler order on ultrafilters, Lascar dubbed it the RK-order. As this connection is somewhat strained we don't use that term.

If a suitable notion **I** of isolation were defined such that all models of a small superstable theory were I-saturated and Exercise 1.3 held for **I**, this would at least clarify matters and perhaps lead to some new results for those theories. However, as Saffe so aptly put it, "I know (?) a lot of things in this direction which are not true."

2. S-Models

This section is devoted to describing a few further properties of the most useful acceptable class, **S**. Precisely because of its usefulness the models in this class have received a variety of names in the past. At various times this concept has been known as an $F^a_{\kappa(T)}$-saturated model, an $\aleph_\epsilon$-saturated model, an ϵ-saturated model, an a-model, and an **S**-model. We showed in Lemma IX.2.13 that **S**-models are the $\mathbf{S}_{\kappa(T)}$-saturated models. We note now that for any stable theory T, the **S**-models of T form an acceptable class. We will primarily be interested in superstable T so the **S**-models are the strongly $\aleph_0$-saturated models.

2.1 Theorem. *For any stable theory T, the class of $\mathbf{S}_{\kappa(T)}$-models of T is an acceptable class.*

Proof. We showed in Chapter X that $\mathbf{S}_{\kappa(T)}$ is a powerful isolation relation which satifies FI_1, FI_2, and FI_3.

The following lemma spells out the meaning of Theorem IV.3.22 which asserts that every strongly $\kappa(T)$-saturated model is good.

2.2 Lemma. *If T is stable and M is strongly $\kappa(T)$-saturated then for each $p \in S(M)$ there is a subset A of M with $|A| < \kappa(T)$ such that p does not fork over A and $p|A$ is stationary.*

There are two other properties of strongly saturated models which will prove useful later. Theorem X.2.27 demonstrated that our **S**-models are the same as the a-models of [Makkai 1984]. Exercise 2.3 i) is the special case of Theorem X.1.10 which shows 'strong' **S**-saturation is the same as **S**-saturation. Note that if $\bar{b}$ is omitted from condition i) of Theorem 2.3, we have the hypothesis of condition ii).

2.3 Exercise. i) If $\bar{a} \downarrow_M \bar{b}$ and M is strongly λ-saturated with $\lambda \geq \kappa(T)$ then for any $A \subseteq M$ with $|A| < \kappa(T)$, $stp(\bar{a}; A \cup \bar{b})$ is realized in M.

ii) If for every $B \subseteq M$ with $|B| < \lambda$ and every $\bar{a}$, $stp(\bar{a}; B)$ is realized in M then M is strongly λ-saturated.

2.4 Theorem. *The model M is **S**-saturated iff for all $p \in S(M)$ there is a set of indiscernibles $I \subseteq M$, $|I| \geq \kappa(T)$ with $\mathrm{Av}(I, M) = p$.*

Proof. If M is **S**-saturated choose $A \subseteq M$ such that p is based on A and $|A| < \kappa(T)$. It is then routine to choose an independent coherent sequence of elements $\bar{a}_i$ of M such that $\bar{a}_i$ realizes $p|A_i$ and the conclusion is clear.

For the converse, let $A \subseteq M$, $|A| < \kappa(T)$, and $c \in C$. We must show $stp(\bar{c}; A)$ is realized in M. Choose $\bar{c}_0$ such that $stp(\bar{c}_0; A) = stp(\bar{c}; A)$ and $t(\bar{c}_0; M) = p$ does not fork over A. Now, if I is a set of indiscernibles contained in M with $|I| \geq \kappa(T)$ and $\mathrm{Av}(I, M) = p$, by Corollary V.1.19 there is an $I_0 \subseteq I$ with $|I_0| < \kappa(T)$ such that $I - I_0$ is a set of indiscernibles realizing $stp(\bar{c}; A)$. Any element of $I - I_0$ verifies that M is **S**-saturated.

2.5 Theorem. *If T is a countable ω-stable theory then every $\aleph_0$-saturated model is* **S***-saturated.*

Proof. By Corollary IV.2.15 for each $p \in S(M)$ there is an $A \subseteq M$ with $|A| < \kappa(T)$ and p strongly based on A. Choose $E = \langle e_i : i < \omega \rangle$ so that e_i realizes $p|A \cup E_i$. Clearly $p = \mathrm{Av}(E, M)$ and the theorem follows from Theorem 2.4.

Just as for domination we can extend the notion of compulsion to a property of global types. But as before there is no general guarantee that if $p \mapsto^e_k q$ then $p \mapsto_K q$. One of the major advantages of **S**-models is that when K is the class of **S**-models then if $p \mapsto^e_S q$ holds it holds for any restriction of the global types to **S**-models. We show this in Lemma 2.9.

2.6 Definition. Let p and q be stationary types. Then $p \mapsto^e_K q$ if for some K-saturated M containing $\operatorname{dom} q \cup \operatorname{dom} p$ and some nonforking extensions p', q' of p and q to M, we have $p' \mapsto_K q'$.

The following lemma shows that this definition depends only on the parallelism class of p and q.

2.7 Lemma. *If p, p', q, q' are stationary, $p \parallel p'$, and $q \parallel q'$ then $p \mapsto_K q$ if and only if $p' \mapsto_K q'$.*

Proof. It suffices to show that $p \mapsto_K q$ implies $p' \mapsto_K q'$. Choose nonforking extensions, p_1, q_1 of p and q respectively to witness $p \mapsto_S q$ according to Definition 2.6. Then $p_1 \rhd q_1$. Let p'', q'' be nonforking extensions to a K-model of, respectively, p_1, p' and q_1, q'. Then if $p'' \not\mapsto_K q''$, by Corollary X.2.5, $p'' \not\rhd q''$. But, by applying Corollary VI.3.13 to p_1 and q_1, we have $p'' \rhd q''$.

Deduce the following exercise from Corollary X.2.5. It shows that $\mapsto_{\mathbf{S}}$ can be defined without reference to the acceptable class **S**.

2.8 Exercise. Show that $\mapsto^e_S$ is equivalent to eventual dominance (Definition VI.3.14).

In the light of Exercise 2.8 and Definition VI.3.14 we usually write $p \sqsupseteq^e q$ rather than $p \mapsto^e_{\mathbf{S}} q$.

2.9 Lemma. *Let T be superstable, M an* **S***-model and $p, q \in \mathbf{S}(M)$. Then, $p \mapsto^e_{\mathbf{S}} q$ if and only if $p \mapsto_{\mathbf{S}} q$*

Proof. Let M' be an **S**-model and p', q' nonforking extensions of p, q to $S(M')$ realized by $\overline{a}'$ and $\overline{b}'$ respectively. Choose successively $A \subseteq M$ and $C \subseteq M'$ such that $|A|, |C| < \kappa(T)$, both p and q (hence p' and q') are strongly based on A, $t(\overline{a}'{}^\frown\overline{b}'; M')$ strongly based on C and $A \subseteq C$. By the $\kappa(T)$ saturation of M, choose an isomorphism α fixing A and mapping C into M. Choose $\overline{a}$ and $\overline{b}$ independent from M over $\alpha(C)$ and realizing $\alpha p'|C$, $\alpha q'|C$ respectively. Now the choice of A guarantees that $\overline{a}$ and $\overline{b}$ realize p and q. But the choice of C guarantees (by Lemma VI.3.12) that

$\overline{a}' \rhd_C \overline{b}'$; thus, $\overline{a} \rhd_{\alpha(C)} \overline{b}$ and by Corollary VI.3.13 $\overline{a} \rhd_M \overline{b}$. Thus $p \rhd_M q$ and by Corollary X.2.5 $p \mapsto_{\mathbf{S}} q$.

It is becoming customary to think of stationary types as representatives of their parallelism equivalence class. Thus in recent papers authors often write $p \sqsupseteq q$ where p and q are stationary (but with no mention of the exact domain) where we write $p \sqsupseteq^e q$. This practice is somewhat risky. For, if specific representatives p', q' of the global types are chosen whether $p' \rhd q'$ depends on the relation between p' and q'. Lemma 2.9 shows that for safety it suffices to choose the representatives over **S**-models. The key fact to remember is that $\mapsto_{\mathbf{S}}$ and $\rhd$ are preserved upwards by nonforking extension of the individual types but downward only if the type of the pair does not fork.

2.10 Historical Notes. Shelah [Shelah 1978] first discovered the importance of **S**-models. He called them $F^a_{\kappa(T)}$-saturated models and defined them by the first condition of Exercise 2.3. The properties of this notion discussed in this section are scattered in [Shelah 1978]. The equivalence of his notion with strong saturation is from [Baldwin 1984]. The proof of Lemma 2.9 is adapted from [Lascar 1984].

Chapter XII

Regular Types

In this chapter we introduce the most important concepts involved in assigning dimension to models. We saw in Section 1.3 that the simplest sort of $\aleph_1$-categorical theory is one in which the universe of each model is strongly minimal. A strongly minimal set D has two crucial features: i) its dimension is well-defined, ii) if $M \subseteq N$, every element of $D(N) - D(M)$ is independent from M over the empty set. The possession of both of these features is no accident; we will show that properly formulated versions of the two properties are equivalent. For any acceptable class K, we define the concept of a K-strongly regular type. The definition given in Section 1 is less intuitive but technically more useful than either of the properties just described. We then recast this definition in terms of the second of these properties. This recasting simplifies the construction of regular types in Section 2. In Section 3 we analyze invariance of dimension in terms of the transitivity properties of the forking relation. This approach makes it clear that for any K-strongly regular (stationary) type based on $A \subseteq M$, $\dim(p, M)$ is well defined. Finally, in Section 4 we show that the two approaches to defining regular types yield the same class of types. With this we can describe the relation between $\not\perp$ and $\triangleright^e$. Two types are orthogonal if and only if they are disjoint in the partial ordering imposed by domination. We further conclude that $\not\perp$ is an equivalence relation on the K-strongly regular types. Moreover, if $M \prec N$ and q is the nonforking extension of p to $S(N)$ then $\dim(p, N) = \dim(p, M) + \dim(q, N)$.

Throughout this chapter we assume that K is an acceptable class of models.

1. Weak Isolation and Regular Types

The usual notion of regular type is primarily associated with **S**-models and superstable theories; strongly regular types are similarly associated with arbitrary models of an ω-stable theory. We develop in this chapter a common framework for the two notions.

The first three paragraphs of this section define the notions of regularity, strong regularity and weak isolation. The next three pages expound technical properties of these notions and establish their parallelism invariance. These technicalities arise again, for example in Theorem XIII.3.3. The theories described in Example 1.11 illustrate the fine distinctions made earlier in the discussion. The last three results concern the structure of $M[p]$ where p is regular.

We will describe a collection of types with the following property. On realizations of a type in the class the dependence relation arising from forking has a good notion of dimension. That is, we will be able to find a 'basis' for the set of realizations of such a type. Intuitively, a basis is both a maximal independent set and a minimal generating set. Suppose that a, b, c are the realizations of such a type in M and that a is independent from b. If both c depends on a and b depends on c over a, we have that $\{a, b\}$ is a maximal independent set but $\{a\}$ is a smaller generating set. Thus, we need to guarantee that if a and b are independent and c depends on a then b does not depend on c over a or, by symmetry, that c does not depend on b over a. The following definition is, a priori, somewhat stronger than is required. We will see in Section 4 that the weaker intuition suffices but the chosen definition is more useful in practice.

1.1 Definition. The non-algebraic type $p \in S(A)$ is *regular* if and only if every extension of p is orthogonal to p or parallel to p.

If $q = stp(\bar{b}; B)$ then q is regular if the unique nonforking extension of q to a complete type over $B \cup \bar{b}$ is regular.

There are two difficulties with this definition. First, it makes the verification of the existence of regular types inconvenient. More important, a surprising but crucial consequence of this definition—that regular types are **S**-minimal—does not extend to the ω-stable case. For this, we must sharpen the definition of regularity. The sharpened version deals with a pair (p, q) where p is a complete type and q is an **I**-formula.

1.2 Definition. i) For any acceptable class K, any non-algebraic type $p \in S(A)$, and any **I**-formula q over A with $q \subseteq p$, (p, q) is *K-strongly regular* if for every $B \supseteq A$ and for every $r \in S(B)$ which extends q, either r is a nonforking extension of p or $r \perp p$.

ii) The pair (p, q) is called *stationary and strongly regular* if, in addition, p is stationary.

iii) We may write $p \in S(A)$ is *K-strongly regular* to mean that for some **I**-formula over A, (p, q) is K-strongly regular.

Unless explicitly declared otherwise, when we refer to a K-strongly regular pair (p, q), we denote $\operatorname{dom} p$ by A.

Following [Shelah 1978] and in contrast to [Makkai 1984], we do not require that a K-strongly regular type be stationary. Note that in Definition 1.2 i) the requirement is that r does not fork over A, not that r does not fork over $\operatorname{dom} q$. Neither p nor r is required to be a nonforking extension of q.

This distinction is illustrated in Example 1.11 iv). In fact, this requirement could be added without invalidating the existence proof. We omitted it to maintain uniformity with [Shelah 1978].

The role of q in the definition of strong regularity is suggested by the relation between a strongly minimal formula, $\phi(x)$, and the unique non-algebraic type p containing ϕ. That is, ϕ 'weakly isolates' p in the sense that if the parameters of ϕ are from a model M and N is a proper elementary extension of M, then every realization of ϕ in $N - M$ actually realizes p. The following definition of one type weakly isolating another extends this concept to an arbitrary acceptable class K. After showing this relation is a property of the parallelism class of p, we will prove that (p, q) is K-strongly regular if and only if p is weakly isolated by q.

1.3 Definition. (Fig. 1). Let $M \subseteq N \models T$. Suppose M is **I**-saturated, q is an **I**-formula over $A \subseteq M$, and p is a nonforking extension of q to $S(M)$.

i) *q weakly isolates p* in N if $q(N) - M \neq \emptyset$ and every $\bar{b} \in q(N) - M$ both satisfies $p|A$ and is independent from M over A.
ii) q weakly isolates the type $p \in S(M)$ if for some N containing M with $q(N) - M \neq \emptyset$, q weakly isolates p in N.

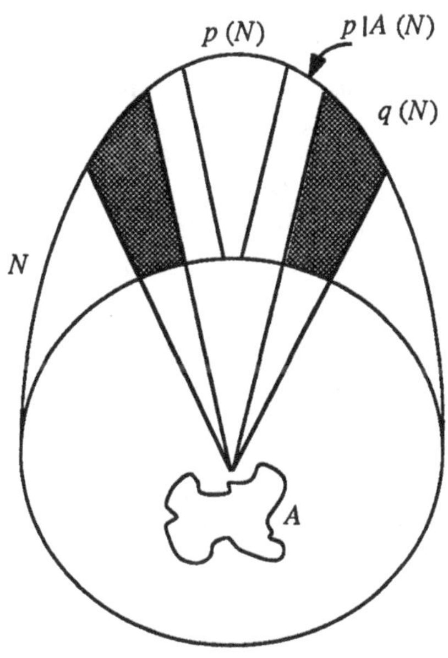

Fig. 1. q weakly isolates p in N. If $p|A$ is stationary then $p|A\,(N) - p(N)$ is empty.

We will need in Section XIII.3 the following more complicated version of weak isolation.

iii) *q weakly isolates p* in (N_1, N), where $M \subseteq N_1 \subseteq N$, if $q(N) - N_1 \neq \emptyset$ and for every $\bar{b} \in q(N_1) - N$, $\bar{b} \downarrow_A M$ and $\bar{b}$ realizes $p|A$.

Note that i) is the special case of iii) obtained by identifying N_1 and N.

The definition of weak isolation is actually relative to the class K. Since this class is invariably fixed by the context we have not made it explicit in the definition.

In considering this definition recall that the notation $\bar{b} \in q(N) - M$ means $\bar{b} \in N$ realizes q and some member of the sequence $\bar{b}$ is not in M. The conclusion $\bar{b} \downarrow_A M$ implies $\bar{b} \cap M \subseteq cl(A)$. Thus, a minor consequence of q weakly isolating p in N is that no realization $\bar{b} \in N$ of q can contain points from M which are not algebraic over A.

Suppose that a nonalgebraic type $p \in S^2(A)$ is realized by $\langle a, b\rangle$ with $a \downarrow_A b$. Then we can choose $\langle c, d\rangle$ such that $a^\frown b \downarrow_A c^\frown d$ and each of $\langle a, b\rangle$, $\langle c, b\rangle$, and $\langle c, d\rangle$ realize p. This choice quickly yields that p is not regular. Thus, although we consider regular n-types for $n > 1$, any sequence realizing such a regular type must be dependent.

When K is the class of **AT**-saturated (i.e. all models) we have the notion called strongly regular in Shelah [Shelah 1978]. In discussing ω-stable theories, we will frequently write strongly regular for **AT**-strongly regular.

1.4 Exercise. Show that if q weakly isolates $p \in S(M)$, where q is an **I**-formula and $M \in K$, then q weakly isolates p in $M[\bar{a}]$ for any $\bar{a}$ realizing p.

The next lemma is a technical point to show that weak isolation is a property of the parallelism class of p where $p \in S(M)$. Note that if q weakly isolates p in N then q weakly isolates p in the image of N under any isomorphism which fixes M. We will show that if q weakly isolates p then q weakly isolates every p' which extends q and is parallel to p. We first show this if p' is a nonforking extension of p (or vice versa). The result then extends to arbitrary parallel p and p' by passing through their common nonforking extension.

1.5 Theorem. *Let q be an **I**-formula and M be an **I**-saturated model. Suppose $p \in S(M)$ and $p' \in S(M')$ is a nonforking extension of p. Then q weakly isolates p if and only if q weakly isolates p'.*

Proof. Suppose first that q weakly isolates p' and let $\bar{a}$ realize p'. Suppose that $N' \supseteq M' \cup \bar{a}$ witnesses the weak isolation of p' by q. Let $N = M[\bar{a}]$. It suffices to show that q weakly isolates p in N. Let $\bar{b} \in q(N) - M$. Let $\bar{b}_1$ denote $\bar{b} \cap (N - M)$ and $\bar{b}_2$ denote $\bar{b} \cap M$. Then $\bar{a} \rhd_M \bar{b}_1$. Since $\bar{a} \downarrow_M M'$ it follows that $\bar{b}_1 \downarrow_M M'$. By monotonicity and since $\bar{b}_2 \in M$, $\bar{b}_1 \downarrow_M M' \cup \bar{b}_2$ and thus $\bar{b} \downarrow_M M'$. So, $\bar{b} \in q(M'[\bar{a}]) - M'$. Now, as q weakly isolates p' we have $\bar{b}$ realizes $p'|A = p|A$ and $\bar{b} \downarrow_A M'$. By monotonicity $\bar{b} \downarrow_A M$, so q weakly isolates p in N

Now suppose that q weakly isolates p and let $\bar{a}$ realize p'. Let $N' = M'[\bar{a}]$. Assume for contradiction that $\bar{b} \in N'$ realizes q but either $\bar{b}$ does not realize $p|A$ or $\bar{b} \not\downarrow_A M'$. We prove the following strong result which not only yields the theorem at hand but will be applied again to prove Theorem 1.7 and Corollary 1.8.

1.6 Lemma. *Suppose the type q over $A \subseteq M$ weakly isolates the type $p \in S(M)$ where q is an* **I**-*formula and M is* **I**-*saturated. Let $\overline{a}$ realize p and suppose $\overline{a} \downarrow_A M$ and $\overline{a} \downarrow_M D$. For any $\overline{b}$ realizing q, if $\overline{b} \not\downarrow_{M \cup D} \overline{a}$ then $t(\overline{b}; M \cup D)$ is a nonforking extension of $p|A$.*

Proof. By monotonicity, $\overline{a} \not\downarrow_A M \cup D \cup \overline{b}$ so applying the local character of forking, there is a formula $\chi_0(\overline{x}, \overline{y}, \overline{z}) \in F(A)$ such that $\chi_0(\overline{a}, \overline{b}, \overline{m})$ for some $\overline{m} \in M \cup D$ and for any $\overline{b}'$ and $\overline{m}'$, if $\models \chi_0(\overline{a}, \overline{b}', \overline{m}')$ then $\overline{a} \not\downarrow_A \overline{b}' \frown \overline{m}'$. We argue by contradiction. If $t(\overline{b}; A) \neq p|A$ let $\chi_1(\overline{x})$ be a formula over A which is satisfied by $\overline{b}$ but is not in $p|A$. Otherwise, choose $\chi_1(\overline{x}, \overline{y}) \in F(A)$ and $\overline{d} \in M \cup D$ so that $\models \chi_1(\overline{b}, \overline{d})$ and for every $\overline{b}'$ realizing $t(\overline{b}; A)$, if $\models \chi_1(\overline{b}', \overline{d}')$ then $\overline{b}' \not\downarrow_A \overline{d}'$. Now apply Corollary X.1.13 to the **I**-formula $\{\chi_0(\overline{a}, \overline{y}, \overline{z}) \wedge \chi_1(\overline{y}, \overline{w})\} \cup q(\overline{y})$ to obtain $\overline{m}', \overline{d}' \in M$ and $\overline{b}'$ such that $\chi_0(\overline{a}, \overline{b}', \overline{m}') \wedge \chi_1(\overline{b}', \overline{d}') \wedge q(\overline{b}')$. Let N be any model in K containing $M \cup \overline{a} \cup \overline{b}'$. As $\overline{a} \not\downarrow_A \overline{b}' \frown \overline{m}'$, $\overline{b}' \notin M$. Since q weakly isolates p, $t(\overline{b}'; A) = p|A$. This contradicts the choice of χ_1 in the case $t(\overline{b}; A) \neq p|A$. In the second case, we have $\overline{b}' \not\downarrow_A \overline{d}'$ which also contradicts the weak isolation of p by q. This concludes the proof of the lemma.

To finish the proof of Theorem 1.5, let $M' = D$. Then by FI_2, $\overline{b} \not\downarrow_{M'} M' \cup \overline{a}$ so $t(\overline{b}; M) \not\perp p$. Hence, by Lemma 1.6, $t(\overline{b}; M')$ is a nonforking extension of p.

With these technical preliminaries disposed of we can recast the K-strong regularity in terms of weak isolation. We will profit from this reformulation in showing that strong regularity is a parallelism invariant and when constructing K-strongly regular types in Section 2.

1.7 Theorem. *Let $p \in S(A)$ and $q \subseteq p$ be an* **I**-*formula over A. The pair (p, q) is K-strongly regular if and only if there is an $M \in K$ which contains A and an $\overline{a}$ realizing p with $\overline{a} \downarrow_A M$ such that q weakly isolates $t(\overline{a}; M)$.*

Proof. Suppose that p is K-strongly regular. Let $A \subseteq M \in K$ and let p' be a nonforking extension of p to $S(M)$. Let $\overline{a}'$ realize p'.

We claim q weakly isolates p' in $M[\overline{a}']$. To see this, let $\overline{b} \in M[\overline{a}']$ realize q and let r denote $t(\overline{b}; M)$. By the definition of K-strong regularity, $r \perp p$ or r is a nonforking extension of p. The first is impossible. For, $r \not\perp p'$ because r is realized in $M[\overline{a}']$; thus, r is a nonforking extension of p as required.

Note that we are not required to conclude that $\overline{b}$ realizes p'. Without assuming that p is stationary we could not reach that conclusion.

Suppose the second condition holds. We must show that if $B \supseteq A$, $r \in S(B)$, and $r \supseteq q$ then $r \perp p$ or r is a nonforking extension of p. By taking a nonforking extension of r we can assume without loss of generality that B is the universe of a model $M \in \mathbf{S}$. Then, if $r \not\perp p$ there exist a $\overline{b}$ realizing r, an $\overline{a}$ realizing p such that $\overline{a} \downarrow_A M$ and a set D such that $\overline{a} \downarrow_M D$, $\overline{b} \downarrow_M D$ but $(\overline{a} \not\downarrow \overline{b}; M \cup D)$. By Lemma 1.6, r is a nonforking extension of p.

An easy application of Theorem 1.7 yields that if the **I**-formula q weakly isolates $p \in S(M)$ then (p, q) is K-strongly regular. We now give sufficient conditions for the preservation of regularity and strong regularity 'up' and 'down' under nonforking extensions. We have so far avoided requiring regular types to be stationary. That requirement is essential for most of the preservation results. The proof of Theorem 1.8 ii) is a natural extension of the second paragraph of the proof of Theorem 1.7.

1.8 Corollary. *Let $p' \in S(A')$ be a nonforking extension of $p \in S(A)$.*

i) *If p is regular then p' is regular. If p is stationary and p' is regular then p is regular.*

ii) *Suppose the stationary type p contains the* **I**-*formulas q and for some $M \in K$, q weakly isolates p^M. Then (p', q) is K-strongly regular.*

iii) *Suppose p is stationary. Then for any* **I**-*formula $q \subseteq p$, (p', q) is K-strongly regular if and only if (p, q) is K-strongly regular.*

Proof. i) Preservation 'up' is immediate from the definitions. Suppose for the converse that $p' \in S(A')$ is a nonforking extension of the stationary type p and p' is regular. To show p is regular, let $A \subseteq C$ and suppose $\bar{b}$ realizes a forking extension r of p to $S(C)$. Without loss of generality, we may choose $C^\frown\bar{b} \downarrow_A A'$. Thus, $t(\bar{b}; C \cup A')$ is a forking extension of p' (since p is stationary and $\bar{b} \downarrow_A A'$). So $t(\bar{b}; C \cup A')$ is orthogonal to p' and thus to p. But, by monotonicity, $\bar{b} \downarrow_C A' \cup C$ so $r \perp p$ as required.

ii) Let $r \in S(B)$ be an extension of p'. Suppose N contains $M \cup B$ and r' is a nonforking extension of r to $S(N)$. Now if $r \not\perp p$, the preservation of nonorthogonality by parallelism guarantees $r' \not\perp p^N$. Choose $\bar{b}'$ realizing r', $\bar{a}$ realizing p^N and D so that $\bar{b}' \downarrow_N D$ and $\bar{a} \downarrow_N D$ but $\bar{b}' \not\downarrow_{D \cup N} \bar{a}$. By Lemma 1.5, q weakly isolates p^N. So, by Lemma 1.6 r' is a nonforking extension of p. Since p is stationary, r' is a nonforking extension of p as required.

iii) Suppose (p, q) is K-strongly regular. By Theorem 1.7, for some M in K, q weakly isolates p^M. By ii) we conclude (p', q) is K-strongly regular. Conversely, if (p', q) is K-strongly regular then for some $M \in K$, q weakly isolates $(p')^M$. But, $(p')^M = p^M$ so the result follows by ii).

1.9 Exercise. Prove the 'up' direction of Corollary 1.8 i). Deduce from Corollary 1.8 that if $p \in S(A)$ is regular and $\bar{a}$ realizes p then $stp(\bar{a}; A)$ is regular.

1.10 Exercise. Show that the hypothesis that p is stationary is necessary to prove that regularity is preserved downwards by nonforking extensions. (Hint: Consider the following theory with three equivalence relations E_1, E_2, E_3. Restricted to E_1 and E_2, T is the theory CEF_2. E_3 has two classes, one coincides with a single E_1 class. In this theory there are two types over the empty set: p_1 contains $(\exists y)[E_3(x, y) \wedge \neg E_1(x, y)]$ while p_2 does not. However, it is easy to see that if there were a unique type over the empty set it would have both a nonforking extension which is regular and one which is not. David Marker has suggested the following device to modify the theory and obtain a single type over the empty set. Add to the language three unary predicates A, B, and C. Let the extension of C in a model of T

be the structure described above. Now consider a ternary relation R with the following properties.

i) $R(x, y, z) \to A(x) \wedge B(y) \wedge C(z)$.
ii) For each $a \in A$, there are infinitely many $b \in B$ (this subset of B is called U_a^1) such that $R(a, b, c)$ if and only if b realizes p_1.
iii) For each $a \in A$, there are infinitely many $b \in B$ (this subset of B is called U_a^2) such that $R(a, b, c)$ if and only if b realizes p_2.
iv) For each $a \in A$, there are infinitely many $b \in B$ (this subset of B is called U_a^3) such that there is no $c \in C$ satisfying $R(a, b, c)$.
v) If $a_0 \neq a_1$, $U_{a_0}^1 \cup U_{a_1}^2 \subseteq U_{a_1}^3$.

The first four conditions tranform the pair of types into a single type. The fifth condition makes the theory complete and ω-stable.)

The following examples illustrate a number of the subtleties of the notion of K-strong regularity.

1.11 Examples. i) If T is ω-stable and K is the class of **AT**-saturated models then for any strongly minimal set D, if p is the type of an unrealized element of D, the pair (p, D) is K-strongly regular.

ii) Let T be the theory of an equivalence relation with infinitely many infinite classes each of which is a model of $\mathrm{Th}(Z, S)$. Consider the 2-type, p, of a pair of elements, $\langle a, b\rangle$, which are equivalent but not on the same Z-segment. Whether p is K-strongly regular depends on our choice of K. If we fix K_1 to be **AT** and thus **I**-formulas are single formulas then p is not K_1-strongly regular. For, if $\phi(x, y)$ is any formula in p, there is a nonforking extension of ϕ realized by two elements which are only finitely distant as well as one which is realized by two points which are infinitely far apart. Thus, there is no choice of an **I**-formula ϕ to make (p, ϕ) K_1-strongly regular.
On the other hand, letting K_2 be the class of (strongly)-$\aleph_0$-saturated models of T (thus the **I**-formulas are types which are (almost) over finite sets) it is easy to see that $(p, p|\emptyset)$ is K_2-strongly regular.

Both example i) and ii) (when $K = \mathbf{S}$) have the property that if we pass to $\mathcal{M}^{eq}$ there is an obvious 'quotient' type of the K-strongly regular type which contains a strongly minimal formula. The next example does not have that property.

iii) Let T be the theory REI_ω of countably many refining equivalence relations with infinite splitting. If M is a model of T, $p \in S(M)$, and p specifies all the equivalence classes, then (p', p) is $\mathbf{AT}_{\aleph_1}$-strongly regular for any nonforking extension p' of p. But no strongly minimal set is associated with p (even in $\mathcal{M}^{eq}$).

The next two examples illustrate some difficulties in trying to strengthen the properties of the weakly isolating formula in a strongly regular pair.

iv) Let T be the theory of two crosscutting equivalence relations, E_1, E_2, and let p be the complete type over a containing the formula

$\{E_1(x;a) \wedge E_2(x;a)\}$. Then $(p, x = x)$ is **AT**-strongly regular but p forks over the empty set. (This example was suggested by Steve Buechler.)

It is essential for Theorem 1.7 that p be a complete type over A. Moreover, it is impossible to revise the definition of strongly regular type to make it a property of one type rather than a pair of types. The next example illustrates the dependence on p.

v) Let T be the theory of infinitely many disjoint infinite unary predicates. If M is the prime model of T and p is any non-algebraic type over M, $(p, x = x)$ is **AT**-strongly regular. But there is no stationary formula q so that (p, q) is **AT**-strongly regular. Here, we say a formula, ϕ, is stationary if the type $\{\phi\}$ is stationary.

The following result emphasizes the importance of the **S**-saturated models.

1.12 Theorem. *The following are equivalent.*

i) *There is an* **S**-*formula* q *such that* (p, q) *is* **S**-*strongly regular.*
ii) *For some acceptable class* K, *there is an* **I**-*formula* q *such that* (p, q) *is* K-*strongly regular.*
iii) *The type* p *is regular.*

Proof. Clearly, i) implies ii). Definition 1.2 yields immediately that ii) implies iii). To see that iii) implies i), let $p \in S(B)$ and choose $A \subseteq B$ with $|A| < \kappa(T)$ such that p does not fork over A. For some $\overline{a}$ realizing p, let $q = stp(\overline{a}; A)$. By Theorem 1.8 i) q is regular. Now, let $M \supseteq B$ be **S**-saturated and $p' \in S(M)$ be a nonforking extension of p. If $\overline{a}'$ realizes p' then for any $\overline{b}'$ in $q(M[\overline{a}']) - M$, $t(\overline{b}'; M) \not\perp p'$ by FI_2. Thus by the definition of regular, $t(\overline{b}'; M)$ is a nonforking extension of q and thus of p as required.

K-strongly regular types should be viewed as a particuarly good sort of K-minimal type. The next theorem verifies that they are, in fact, K-minimal. Sections 3 and 4 will reveal their sterling qualities.

1.13 Theorem. *If* $M \in K$, $p \in S(M)$ *and* (p, q) *is* K-*strongly regular then* p *is* K-*minimal.*

Proof. Suppose $r \in S(M)$, r is not realized in M, and $p \mapsto_K r$. We must show $r \mapsto_K p$. To this end, let $\overline{b}$ realize p and $\overline{a} \in M[\overline{b}]$ realize r. By Corollary X.1.22 (letting $D = \emptyset$) there exists a $\overline{b}' \in M[\overline{a}] - M$ which realizes q. By the definition of a strongly regular pair, $\overline{b}'$ realizes p and we finish.

The following theorem of Pillay shows a good deal about the structure of the K-models prime over K-strongly regular types.

1.14 Theorem. *If* $M \in K$, $p \in S(M)$ *and* (p, p_0) *is* K-*strongly regular then for any* $\overline{a}$ *realizing* p *and any* $\overline{b} \in M[\overline{a}] - M$, $t(\overline{a}; M \cup \overline{b})$ *is* **I**-*isolated.*

Proof. Since $\bar{a} \not\downarrow_M \bar{b}$, there is a formula $\chi(\bar{x};\bar{b})$ such that $\models \chi(\bar{a};\bar{b})$ and $\chi(\mathcal{M};\bar{b}) \cap M = \emptyset$. Let the **I**-formula $q_0(\bar{a};\bar{y})$ isolate $t(\bar{b}; M \cup \bar{a})$. Then the type $p(\bar{x}) \cup q_0(\bar{x};\bar{y})$ implies a complete type in $\bar{x}^\frown\bar{y}$ over M. Let r_0 be the **I**-formula, $p_0(\bar{x}) \cup q_0(\bar{x};\bar{b}) \cup \chi(\bar{x};\bar{b})$. Now, any point in $M[\bar{a}]$ which realizes r_0 (in particular, which realizes $p_0 \cup \chi(\bar{x};\bar{b})$) is in $M[\bar{a}] - M$ and thus realizes p. Since $M[\bar{a}]$ is **I**-saturated, this implies that $r_0 \vdash p$. We deduce $r_0(\bar{x}) \vdash p(\bar{x}) \cup q_0(\bar{x};\bar{b}) \vdash t(\bar{a}; M \cup \bar{b})$ and finish.

We can immediately conclude

1.15 Corollary. *If $M \in K$, $p, q \in S(M)$ are K-strongly regular, and $p \vDash\!\!\!\!\sim_K q$ then $M[p] \approx M[q]$.*

Proof. Let $\bar{a}$ realize p, $\bar{b}$ realize q. Applying Theorem 1.14 it is easy to see that $M[\bar{a}]$ is **I**-constructible over $M \cup \bar{b}$. By Theorem IX.4.12, $M[\bar{a}]$ is isomorphic to $M[\bar{b}]$.

1.16 Historical Notes. Shelah introduces the notion of regular type in Section V.1 and that of strongly regular type in Section V.3 of [Shelah 1978]. Our treatment has been greatly influenced by [Makkai 1984] which in turn was influenced by [Lascar 1984]. The direct argument for Theorem 1.13 is due to Lascar [Lascar 1984]. Theorem 1.14 is due to Pillay, [Pillay 1982] (also quoted in [Lascar 1984]). We also learned a good deal in conversations with Buechler and Pillay. Another approach is taken in [Pillay 1984]. Exercise 1.10 provides a negative answer to question 1.1 on page 224 of [Shelah 1978].

2. *Existence of Strongly Regular Types*

This section is devoted to finding strongly regular types for various acceptable classes K. We will see that for superstable theories regular types are plentiful. One of the difficulties in investigating theories which are stable but not superstable is that analogs of the next several results do not hold.

For the existence theorems beginning this section we do not assume that T admits K-prime models. In particular, one of the exercises shows that a small superstable theory admits **AT**-strongly regular types (which may not be stationary). We will reap some benefits from this added generality in Chapter XIII. However, the main cases are **S**-models of superstable theories and arbitrary models of ω-stable theories.

The exercise after the following definition describes the 'useful' version of that definition. That is, it provides for the existence of all regular types needed for most of the arguments in this chapter. For Theorem XIII.3.3 a more refined existence theorem is necessary. The definition itself, which relies on part iii) of Definition 1.3, formulates this condition. Since the definition only yields a $b \in N - N_1$ which is weakly isolated over M, not N_1, the definition is not a very significant strengthening of the exercise.

The extremely difficult Theorem XIII.3.3 is necessary to get any benefit from the extension.

2.1 Definition. The acceptable class K *admits* (K*-strongly*) *regular types* if for every triple of K-models $M \subseteq N_1 \subseteq N$ with M properly contained in N there is an **I**-formula, p_0, over $A \subseteq M$ with $|A| < \lambda(I)$ and a $b \in N - N_1$ such that p_0 weakly isolates $t(b; M)$ in (N_1, N).

We say K *admits stationary regular types* if we can demand in addition that $t(b; A)$ is stationary. The theory T admits K-strongly regular types if the class of K-models of T does so.

Although we spoke in Section 1 of regular type of arbitrary finite length, this definition specifically provides for the existence of regular 1-types.

2.2 Exercise. Show that by taking $N_1 = N$ we can deduce from the assertion that K admits regular types the following simpler assertion. For every M, $N \in K$ with N a proper extension of M, there is a $b \in N$ such that for some $A \subseteq M$ and some **I**-formula, q, over A, $(t(b; M), q)$ is K-strongly regular.

2.3 Theorem. *If T is a superstable theory then* **S** *admits stationary regular types.*

Proof. (Fig. 2). Let $M \subseteq N_1 \subseteq N$ with N_1 properly contained in N. For as

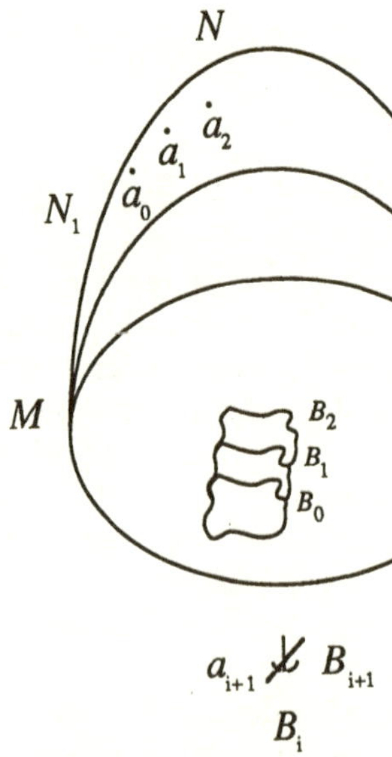

Fig. 2. Theorem XII.2.3.

long as possible, choose a coherent sequence $\langle \overline{a}_i : i < \alpha \rangle$ and $\langle B_i : i < \alpha \rangle$ with $B_i \subseteq B_{i+1} \subseteq M$ such that $\overline{a}_i \in N - N_1$, $t(a_i; B_i)$ is stationary, $|B_i| < \kappa(T)$, $|B_{i+1} - B_i| < \omega$, and $\overline{a}_{i+1} \not\downarrow_{B_i} B_{i+1}$. If continued infinitely this sequence contradicts the assumption that $\kappa(T) = \omega$; when it stops $stp(\overline{a}_i; B_i)$ weakly isolates $t(a_i; M)$ in (N_1, N).

This argument does not continue past a limit ordinal. Thus, we are unable to prove that a countable stable theory admits regular types for $K = \mathbf{AT}_{\aleph_1}$, which is the natural candidate when T is merely stable. It is not easy to find an example to illustrate this problem. There may be one in the second edition of [Shelah 1978].

2.4 Exercise. Modify the argument for Theorem 2.3 to show that if M is $\mathbf{S}$-saturated, T is superstable and M is a proper submodel of the $\mathbf{S}$-model N then there is a $b \in N - M$ with $t(b; M)$ regular.

An alternative proof of Theorem 2.3 invokes rank to find the element realizing the regular type.

2.5 Exercise. Show that if T is superstable and $R_C(t(a; M))$ is minimal among all types over M realized in N then $t(a; M)$ is regular.

In the remaining arguments in this chapter we shall use several of the notions of rank discussed in Chapter VII. We begin by extending the last result to the class of all models if T is a countable ω-stable theory.

2.6 Theorem. *If T is a countable ω-stable theory then* $\mathbf{AT}$ *admits stationary* $\mathbf{AT}$*-strongly regular types.*

Proof. Let $M \subseteq N_1 \subseteq N$ with M properly contained in N. Choose a in $N - N_1$ to minimize the Morley rank, $R_M(t(a; M))$, and choose a formula $\phi(x; \overline{b}) \in t(a; M) = p$ with the same rank, say α, and degree as p. Clearly, the degree is one. Now if $a' \in N - M$ and $\models \phi(a'; \overline{b})$ then $R_M(t(a'; M)) \geq \alpha$ so $t(a'; \overline{b}) = p|\overline{b}$ and $\overline{a}' \downarrow_{\overline{b}} M$. Thus, $\phi(x, \overline{b})$ weakly isolates $t(a; M)$ in (N_1, N) and $(p, \phi(x, \overline{b}))$ is $\mathbf{AT}$-strongly regular as required.

The following exercise shows the necessity of assuming ω-stability for this theorem.

2.7 Exercise. Show that the theory of infinitely many independent unary predicates does not admit $\mathbf{AT}$-strongly regular types.

2.8 Exercise. Show, by forming a disjoint union of the last example with a strongly minimal set, that one can have (p, q) K-strongly regular and r a nonforking extension of q without (r, q) being K-strongly regular. (This differs from Theorem 1.8 as we require only $r \supseteq q$ rather than $r \supseteq p$.)

Two contrasting properties of Morley rank were used in the proof of Theorem 2.6. On the one hand, every type contains a formula with the same rank; on the other, a type has only finitely many distinct extensions of the same rank. Possessing both of these properties characterizes Morley rank. We can extend the scope of this result slightly by using different ranks to serve the two functions. However, we lose the ability to demand that p is stationary.

2.9 Exercise. Let T be a small superstable theory. Show T admits **AT**-strongly regular types. (Hint: First choose $a \in N - M$ to minimize the Shelah degree of $t(a; M) = p$. Then choose $\phi(x; \bar{b}) \in p$ with the same degree. Then choose a' satisfying $\phi(x; \bar{b})$ to minimize the Cantor-Bendixson rank in $S(\bar{b})$ of $t(a'; \bar{b})$. Show this is an appropriate choice.)

This exercise may seem illegitimate since our development of K-strongly regular types assumed that T admits K-prime models. However, the proof that weak isolation implies strong regularity in Theorem 1.7 does not require this hypothesis. The current proof of the converse does need the hypothesis.

The previous set of exercises shows that if T is a small superstable theory then T admits **AT**-strongly regular types. However, while one may find $p \in S(M)$ and ϕ over M with (p, ϕ) **AT**-strongly regular, there is no guarantee that there is a finite subset $A \subseteq M$ with $p|A$ stationary and $(p|A, \phi)$ **AT**-strongly regular. Indeed, Example VI.1.33 ii) shows this property may fail.

2.10 Exercise. Find an example to show the result of Exercise 2.9 can not be improved to require p to be stationary. (Hint: Consider the theory REF_{ω}.)

2.11 Exercise. Contrast the theories CEF_{ω} and REF_{ω} with respect to the existence of **AT**-strongly regular types.

The following characterization of K-minimal types requires the existence of regular types and so could not have been proved at the end of Section 1.

2.12 Theorem. *Let $M \in K$ and $p \in S(M)$. If K admits regular types then p is K-minimal if and only if for some K-strongly regular $q \in S(M)$, $p \underset{\sim}{\mapsto} q$.*

Proof. Suppose p is K-minimal. Since K admits regular types we can find a K-strongly regular q realized in $M[p]$. Thus, $p \mapsto q$. But since p is K-minimal, $q \mapsto p$. The converse follows quickly from Theorem 1.13.

We can combine this result with Theorem 1.14 to extend Corollary 1.15 and conclude that all K-minimal types yield isomorphic prime models.

2.13 Exercise. If $M \in K$, $p, q \in S(M)$ are K-minimal and $p \underset{\sim}{\mapsto} q$ then $M[p] \approx M[q]$.

The following construction provides another source of regular types as well as illustrating their connection with strongly minimal sets.

2.14 Definition. The type $p \in S(A)$ is *minimal* if for every formula $\phi(x)$, either $p \cup \phi$ or $p \cup \neg\phi$ has only finitely many solutions.

The following property of minimal types is straightforward.

2.15 Exercise. Every minimal type is regular.

This notion of minimality is quite distinct from K-minimality. Although by Theorem 2.12 and Exercise 2.15 every minimal type is K-minimal the converse is far from true.

2.16 Exercise. Find a regular type which is not minimal (say for **AT**).

2.17 Historical Notes. The existence arguments are derived from Theorem V.3.5 in [Shelah 1978], [Shelah 1982] and [Makkai 1984]. There is a considerably deeper treatment of minimal types in Section V.1 of [Shelah 1978]. They also play an important role in [Buechler 1984a].

3. Some Variants on Transitivity

At the outset of this chapter, we noted that the notion of regularity generalized two properties of strongly minimal sets. So far we have focused on the weak isolation aspect. Now, we consider the property, 'having a well defined dimension'. We showed in Section V.2 that the dimension of a set of indiscernibles is well defined if the set of indiscernibles has at least $\kappa(T)$ elements. To extend this result to sets with finite dimension depends, as in the theory of vector spaces, on a more detailed analysis of the independence notion. But in the theory of vector spaces, the extra care arises from the exchange principle which, in the form of the symmetry axiom, we always have. The extra difficulty arises here in guaranteeing that dependence is transitive. Accordingly we develop in this section, perhaps over pedantically, some variations on the notion of transitivity. Several of these will be seen in later sections to play an important role in our dimension theory. The notion of *transitivity of dependence* was formulated by Van der Waerden as follows.

If a depends on X and each $x \in X$ depends on Y then a depends on Y.

In Chapter II we extended the notion of dependence to deal directly with n-tuples (or, as we saw later, n-element sets). The proper generalization of this axiom is somewhat problematical. The most obvious generalization is

$$\text{If } (\overline{a} \not\downarrow B \cup C; A) \text{ and each } c \in C \text{ satisfies } c \not\downarrow_A B \text{ then } \overline{a} \not\downarrow_A B. \qquad (*)$$

Unfortunately, this version doesn't quite work. The basic problem is that if C is the union of the ranges of a collection of sequences which realize an n-type, we may not have enough information about individual members of C to carry through the proofs. This problem shows up in the proof of Theorem 4.1. The proper notion requires our conventions (see Convention X.1.29) about the treatment of n-tuples and the relation between $p^*(M)$ and $p(M)$.

3.1 Definition. Let p be a type over A. The relation of forking is *fully transitive* on $p(\mathcal{M})$ if for any sets B and C with $B \cup C \subseteq p(\mathcal{M})$ and any $\overline{a}$ realizing p, if $\overline{a} \not\downarrow_A (B \cup C)$ and each $\overline{c} \in C$ satisfies $\overline{c} \not\downarrow_A B$ then $\overline{a} \not\downarrow_A B$.

This would seem a perfectly natural extension of the first version of transitivity to n-tuples, but for one anamoly. Formally, the assertion $\overline{a} \not\downarrow_A (B \cup C)$ makes sense only when $B \cup C$ is thought of as a subset of $p^*(\mathcal{M})$ rather than $p(\mathcal{M})$. However, the symmetry axioms and Theorem II.10 guarantee the equivalence of $\overline{a} \downarrow_A p(M)$ with $\overline{a} \downarrow_A p^*M)$.

3.2 Exercise. Prove the notion of transitivity in Definition 3.1 implies that mentioned in the paragraph above the definition.

3.3 Exercise. Prove that if p is an n-type over A, $\overline{c} \in p^*(\mathcal{M})$, and $\lg(\overline{c}) < n$ then letting $q = t(\overline{c}; A)$, $q^*(\mathcal{M}) \subseteq p^*(\mathcal{M})$.

3.4 Exercise. Prove for p a type over A and $p \subseteq q \in S(A)$ that if forking is fully transitive on $p(\mathcal{M})$ then forking is fully transitive on $q(\mathcal{M})$.

Clearly, by the finite character of dependence we can reduce any question about transitivity to one about finite B and C. In the next lemma we observe that it is enough to check the case where C has one element to prove full transitivity.

3.5 Lemma. *Let $p \in S(A)$. Suppose for any $B \cup \overline{c} \subseteq p(\mathcal{M})$ and for any $\overline{a} \in p(\mathcal{M})$, if $\overline{a} \not\downarrow_A (B \cup \overline{c})$ and $\overline{c} \not\downarrow_A B$ then $\overline{a} \not\downarrow_A B$. Then forking is fully transitive on $p(\mathcal{M})$.*

Proof. Without loss of generality we assume $B \cup C$ is finite and induct on the cardinality of C. The hypothesis of the lemma is the case $|C| = 1$. Suppose $|C| = n + 1$. Let $\overline{d}$ denote the sequence $\langle \overline{c}_0, \ldots, \overline{c}_{n-1} \rangle$. We have $\overline{a} \not\downarrow_A B \cup \overline{d} \cup \overline{c}_n$ and $\overline{c}_n \not\downarrow_A B$. By monotonicity, $\overline{c}_n \not\downarrow_A B \cup \overline{d}$. Now $\overline{a} \not\downarrow_A (B \cup \overline{d})$ follows by using the hypothesis of the lemma with $B \cup \overline{d}$ playing the role of B. By induction, we have $\overline{a} \not\downarrow_A B$ as required.

There is another variant of transitivity which plays an important role.

3.6 Definition. Let $p \in S(A)$. The forking relation is *mildly transitive* on $p(\mathcal{M})$ if for each $\overline{a}$, $\overline{b}$, $\overline{c} \in p(\mathcal{M})$ whenever $\overline{a} \not\downarrow_A \overline{b}$ and $\overline{b} \not\downarrow_A \overline{c}$ then $\overline{a} \not\downarrow_A \overline{c}$.

This definition just asserts that for non-algebraic p forking is an equivalence relation on $p(\mathcal{M})$. Note that if $p(\mathcal{M})$ is mildly transitive then for any M we can assign an invariant to $p(M)$: the number of pairwise independent realizations of p in M. The exact properties of this invariant are unclear. It is unclear, for example, that if we defined a notion of dimension in terms of the maximal number of pairwise independent realizations of a type on which forking is mildly transitive whether the vitally important Theorem 4.4 (below) would still hold.

We have introduced these notions as variations of transitivity in this section to indicate one way in which they are related. Later we will revert to the more standard usage, p is regular if forking is fully transitive on $p(\mathcal{M})$; p has weight one if forking satisfies a condition slightly stronger than mild transitivity on $p(\mathcal{M})$. We will show that the term 'weight one', at least, is extremely descriptive. The next lemma formulates the minimal amount of transitivity necessary for a well defined dimension.

3.7 Definition. Let X be a set of n-tuples. The forking relation on X is *weakly transitive over A* if for any $\overline{a}, B, C$ contained in X the following condition holds. If B and C are each independent sets over A and if each $\bar{c} \in C$ satisfies $\bar{c} \not\downarrow_A B$ and $\bar{a} \not\downarrow_A (B \cup C)$ then $\bar{a} \not\downarrow_A B$.

If $A = \emptyset$ we just write 'weakly transitive'.

3.8 Exercise. Show that full transitivity implies weak transitivity implies mild transitivity.

We show now that weak transitivity implies that dimension is well defined. This proof is exactly as in linear algebra. We just note that the application of transitivity uses only weak transitivity. To ease readability we prove this theorem for X such that forking is weakly transitive over the empty set. The extension to an arbitrary A is routine.

3.9 Theorem. *Suppose the forking relation on X is weakly transitive over the empty set. For any Y and Z which are maximal independent subsets of X, $|Y| = |Z|$.*

Proof. By the finite character of forking the result is obvious unless one of Y and Z is finite. Suppose $|Z| < \omega$. We show by induction on $|A|$ for $A \subseteq Y$ that there exists a subset B of Z such that $|B| = |A|$ and, letting Z_0 denote $Z - B$, that $A \cup Z_0$ is a maximal independent set. This implies $|Y| \leq |Z|$. If $A = \emptyset$ there is nothing to prove. Suppose we have the result for $|A| \leq n$ and suppose $A = A' \cup \{a\}$ with $|A'| \leq n$. By induction, there exists B' with $|B'| = |A'|$ such that, letting Z_0' denote $Z - B'$, $A' \cup Z_0'$ is a maximal independent set. Thus $a \not\downarrow A' \cup Z_0'$. By the generalized symmetry principle (Corollary II.1.12), letting Z_b denote $Z_0' - \{b\}$, for any $b \in Z_0'$, $b \not\downarrow (A' \cup Z_b \cup \{a\})$. If $a \not\downarrow A' \cup Z_b$ then by weak transitivity, $b \not\downarrow A' \cup Z_b$ contradicting the induction hypothesis. (We don't need full transitivity since a and $A' \cup Z_b$ are separately independent.) Thus $A' \cup Z_b \cup \{a\}$ is independent as required.

The following example illustrates the difference between these variants on transitivity.

3.10 Example. Let T be the theory of the group $Z_4^{\aleph_0}$. Let r be the type over the empty set asserting x has order 4. Let M be a countable model of T and let N be $M \oplus Z_4 \oplus Z_4$. Consider the following three elements of N

	M	Z_4	Z_4
a	0	1	0
b	0	1	2
c	0	0	1

Now $c \not\downarrow_M a^\frown b$ and $b \not\downarrow_M a$ but $c \downarrow_M a$ so forking is not fully transitive nor even weakly transitive on $r(\mathcal{M})$. It is mildly transitive.

Our goal, of course, is to be able to attach a dimension to large subsets of models of T (large enough eventually so that they determine the model). The next lemma separates the part of our construction which is on the level of linear algebra from that which is model theory. We will apply this lemma in Section XIII.1.

3.11 Lemma. *Let X be a subset of $\mathcal{M}^n$. Suppose that there is an equivalence relation E on X such that*

i) *Forking over A is fully transitive on each equivalence class of E.*
ii) *If $I \subseteq X$ and $\bar{a} \not\downarrow_A I$ then $\bar{a} \not\downarrow_A I_{\bar{a}}$ where $I_{\bar{a}} = \{\bar{b} \subseteq I : \bar{b}E\bar{a}\}$.*

Then forking over A is weakly transitive on X.

Proof. Let $I, J \subseteq X$ be separately independent over A. Suppose $\bar{a} \not\downarrow_A (I \cup J)$ and for each $\bar{b} \subseteq I$, $\bar{b} \not\downarrow_A J$. By ii) $\bar{a} \not\downarrow_A (I_{\bar{a}} \cup J_{\bar{a}})$. Clearly, if $\bar{b} \in I_{\bar{a}}$ then $\bar{b}E\bar{a}$ implies $\bar{b} \not\downarrow_A J_{\bar{a}}$. Since forking is fully transitive on the equivalence class of $\bar{a}$, $\bar{a} \not\downarrow_A J_{\bar{a}}$ and the result is immediate by monotonicity.

With these notions of transitivity in hand we can draw some conclusions about the dimensions of types.

3.12 Definition. Let $p \in S(A)$ and $A \subseteq M$. Then $\dim(p, M)$ is the cardinality of any maximal independent set of realizations of p in M.

The following result is obvious from the considerations in Theorem 3.9.

3.13 Theorem. *If forking is weakly transitive on $p(\mathcal{M})$ then for any M which contains* $\operatorname{dom} p$, $\dim(p, M)$ *is well defined.*

3.14 Historical Notes. This section was motivated by Shelah's loose assertion in V.1 of [Shelah 1978] that on the realizations of regular types forking satisfies the axioms for vector spaces. The actual result proved is that the other axioms and weak transitivity (rather than full transitivity) are satisfied. The intent was that invariance of dimension holds. These distinctions were explored in [Baldwin 1984].

4. Strongly Regular Types and Compulsion

In this section we combine the two views of regularity discussed in Sections 1 and 3 to establish the role of orthogonality between K-strongly regular types in the ordering $\mapsto_K$. The first theorem establishes that we have been looking at the same notion from two viewpoints. Although the assumption that forking is fully transitive on $p(\mathcal{M})$ seems to be a purely local assumption (dealing only with realizations of p), it has consequences, e.g. those in Definition 1.1 of regularity, involving points which do not realize p.

4.1 Theorem. *Let $p \in S(A)$. The following are equivalent.*

i) *p is a regular type.*
ii) *Forking is fully transitive on $p(\mathcal{M})$.*

Proof. To show ii) implies i), it suffices, as in the proof of Theorem 1.7, to show that for any **S**-saturated M with $A \subseteq M$, if $\bar{a}$ realizes p then $\bar{a} \downarrow_A M$ or $t(\bar{a}; M) \perp p$. By Theorem VI.1.37, if $t(\bar{a}; M) \not\perp p$ there is a $\bar{b}$ realizing p such that $\bar{b} \downarrow_A M$ but $\bar{a} \not\downarrow_M \bar{b}$. By Axiom FI_3 and Proposition

X.1.37, $(\overline{a}\frown\overline{b}{\downarrow}M; A \cup p^*(M))$. Now if $\overline{a} \downarrow_A A \cup p^*(M)$ then $\overline{a} \downarrow_A M$ by transitivity of independence and we finish. But if $\overline{a} \not\downarrow_A A \cup p^*(M)$ note that since $(\overline{a}\frown\overline{b}{\downarrow}M; A \cup p^*(M))$ we have $(\overline{b}{\downarrow}M \cup \overline{a}; A \cup p^*(M) \cup \overline{a})$. Hence if $\overline{b} \downarrow_A A \cup p^*(M) \cup \overline{a}$, we conclude by transitivity of independence that $\overline{b} \downarrow_A M \cup \overline{a}$, contrary to the hypothesis that $\overline{a} \not\downarrow_M \overline{b}$. We are thus left with the case $\overline{b} \not\downarrow_A A \cup p^*(M) \cup \overline{a}$ and $\overline{a} \not\downarrow_A A \cup p^*(M)$ but then invoking the transitivity of dependence we have $\overline{b} \not\downarrow_A A \cup p^*(M)$. But $\overline{b} \downarrow_A M$ and this contradiction yields the first assertion.

To show i) implies ii) we show for any $B \cup \{\overline{a}, \overline{c}\} \subseteq p(M)$, if $\overline{c} \not\downarrow_A B$ and $\overline{a} \not\downarrow_A (B \cup \overline{c})$ then $\overline{a} \not\downarrow_A B$. To see this note that $\overline{c} \not\downarrow_A B$ implies by Definition 1.2 that $t(\overline{c}; B) \perp p$. Thus, if $\overline{a} \downarrow_A B$ (i.e. $t(\overline{a}; B) \parallel p$) we have $t(\overline{a}; B) \perp t(\overline{c}; B)$. In particular, $\overline{a} \downarrow_A (\overline{c} \cup B)$. We now deduce ii) from Lemma 3.5.

In the preceding proof we were careful to follow the notation of Chapter X which distinguishes between $p^*(M)$ and $p(M)$. Hereafter, we often take advantage of the remark after Definition and 3.1 and identify these two sets when speaking of independence. Note that this circumlocution is not formally correct if we speak of implication rather than independence. (Recall that if $a \downarrow_M b$ and $t(a; A)$ is stationary $t(a; A) \vdash t(a; A \cup b)$.)

The following version of regularity extends transitivity to domains which are not contained in $p(\mathcal{M})$.

4.2 Exercise. If $p \in S(A)$ is regular, $A \subseteq B$, $\overline{a} \cup X$ is a set of realizations of p such that $\overline{a} \not\downarrow_B X$ and for each $\overline{x} \in X$, $\overline{x}$ depends on B over A then $\overline{a} \not\downarrow_A B$. (Note this also holds if we replace $\overline{a} \not\downarrow_B X$ in the hypothesis by $\overline{a} \not\downarrow_A X \cup B$.)

We would like to have an analogue of Theorem 4.1 for strong regularity. Our success depends on the direction of the implication. Part i) of the following exercise verifies that one half of Theorem 4.1 (regularity implies transitivity) does go through for any acceptable K. This can be easily deduced from Theorem 1.12. In fact, this part of the proof of Theorem 4.1 applies to any class K such that if $p, q \in S(M)$ and $M \in K$ then $p \not\perp q$ implies $p \not\perp^a a$. The converse is harder; it is difficult to formulate a smooth characterization of strong regularity in terms of transitivity. The suggestion of Shelah embodied in part ii) of Exercise 4.3 seems to be the best.

4.3 Exercise. i) Show that if $p \in S(A)$ and (p, q) is K-strongly regular then forking is fully transitive on $p(\mathcal{M})$.

ii) Show (p, q) is K-strongly regular iff the dependence relation on $q^*(\mathcal{M})$ defined by '$\overline{a}$ depends on B over A if and only if $t(\overline{a}; A) \neq p$ or $(\overline{a} \not\downarrow B; A)$' is fully transitive.

Now we use both weak isolation and full transitivity to prove an important theorem about strongly regular types.

4.4 Theorem. *Let $M, N \in K$ and $p \in S(A)$ with $A \subseteq M \subseteq N$. Suppose that (p, p_0) is K-strongly regular. If C is a set of realizations of p_0 which* *is independent over M but $C \not\downarrow_A M$ then $C \not\downarrow_A p(M)$. Moreover,*

$$\dim(p,N) = \dim(p,M) + \dim(p^M,N).$$

Proof. (Fig. 3). If the first assertion fails, Proposition X.1.37 yields that $C \not\downarrow_A p_0(M)$ but $C \downarrow_A p(M)$. Without loss of generality, C is finite. Thus, for some finite $D \subseteq p_0(M) - p(M)$, $C \not\downarrow_{A\cup p(M)} D$. We show this is impossible by induction on the cardinality of the set of $\lg(p)$-tuples D. By the definition

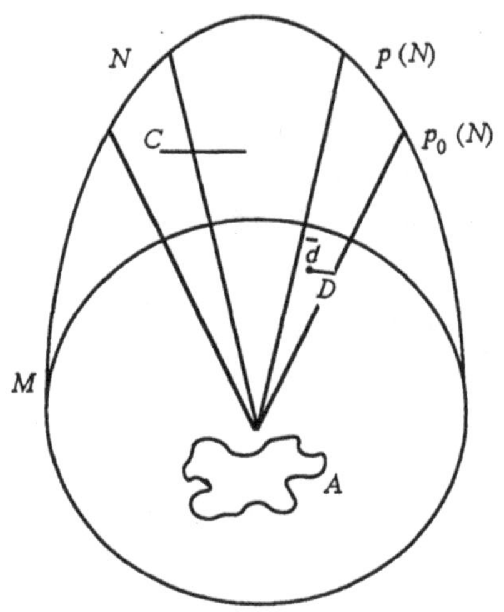

Fig. 3. Lemma XII.4.4.

of strong regularity, for each $\overline{d} \in D$, $t(\overline{d}; A \cup p(M)) \perp p^{A\cup p(M)}$. Invoking the triviality of orthogonality and the fact that C is independent over M, this establishes the result if $|D| = 1$. Suppose the conclusion holds when $|D| \leq k$ and consider D with $|D| = k+1$, say, $D = D' \cup \overline{d}$ and D' satisfies the induction hypothesis. Since $\overline{d}$ does not realize p, the definition of strong regularity implies that $\overline{d} \not\downarrow_A A \cup p(M) \cup D'$ and that $t(\overline{d}; A \cup p(M) \cup D') \perp p$. By the triviality of orthogonality we have $t(\overline{d}; A \cup p(M) \cup D') \perp t(C; A \cup p(M) \cup D')$ and we finish.

To prove the final statement, just extend a basis, X, for $p(M)$ to a basis, $X \cup Y$, for $p(N)$. We now show the new elements are a basis for $p^M(N)$. Since forking is fully transitive on $p(N)$, $Y \downarrow_A X$ implies $Y \downarrow_A p(M)$. By the first conclusion of the theorem, this implies $Y \downarrow_A M$ and we finish.

Theorem 1.13 showed that K-strongly regular types are K-minimal. Thus, two K-strongly regular types are either incomparable or equivalent modulo $\mapsto_K$. The next theorem and its corollaries provide a more syntactic description of this relation.

4.5 Theorem. *Let M be* **S***-saturated and suppose $p \in S(M)$ is a regular type. If $q \in S(M)$ and $q \not\perp p$ then $q \mapsto_S p$. Thus $q \rhd p$.*

Proof. Let $\bar{b}$ realize q and let N be **S**-prime over $M \cup \bar{b}$. Choose $A \subseteq M$ with $|A| < \lambda(I)$, such that p is strongly based on A and let $p_1 = p|A$. Since $p \not\perp q$, and over **S**-models $\perp^w$ implies $\perp$ (Theorem VI.1.40), there is $\bar{a}$ realizing p such that $\bar{a} \not\downarrow_M (M \cup \bar{b})$ and by monotonicity $\bar{a} \not\downarrow_M N$. By Axiom FI_3, $p|(A \cup p_1^*(N)) \vdash p$. So $(\bar{a} \not\downarrow A \cup p_1^*(N); A)$. Let I be a maximal independent subset of $p_1^*(N)$ such that $I \cap M$ is a maximal independent subset of $p_1^*(M)$. Since p (and therefore p_1) is regular, $\bar{a} \not\downarrow_A (A \cup I)$. If $I \cap (N - M) = \emptyset$ then $\bar{a} \not\downarrow_A M$, contrary to hypothesis. Thus there exists an $\bar{a}' \in I \cap (N - M)$. Now $\bar{a}' \downarrow_A (I \cap M)$ implies by the choice of $I \cap M$ and regularity again that $\bar{a}' \downarrow_A (A \cup p_1^*(M))$ and thus using FI_3 once more that $\bar{a}' \downarrow_A M$. Since p_1 is stationary, this yields $\bar{a}'$ realizes p as required.

Note that this argument gives a different proof (for the **S**-case) of Theorem 1.13: if $p \in S(M)$, M is **S**-saturated and p is regular then p is S-minimal.

The following exercise points out a slightly stronger version of Theorem 4.5 which we actually established in the preceding proof.

4.6 Exercise. Let M be **S**-saturated and suppose $p \in S(M)$ is a regular type, $q \in S(M)$, and $q \not\perp p$. If $\bar{a}$ realizes p, $\bar{b}$ realizes q, and $\bar{a} \not\downarrow_M \bar{b}$ then there is an $\bar{a}' \in M[\bar{b}] - M$ which realizes p and satisfies $\bar{a} \not\downarrow_M \bar{a}'$.

The next variant of Theorem 4.5 will be very useful in applications.

4.7 Exercise. Let $p \in S(A)$ and $q \in S(B)$. If $q \not\perp p$ and p is regular then $q \rhd^e p$.

Now comparing the models $M[p]$, $M[q_1]$, and $M[q_2]$ quickly yields

4.8 Corollary. i) *Suppose q_1 and q_2 are regular and $q_1 \not\perp q_2$. Then for any p, $p \perp q_1$ if and only if $p \perp q_2$.*

ii) *Nonorthogonality is the equivalence relation induced on the class of regular types by the relation $\mapsto\!\!\!\!\sim_S$.*

Proof. We prove only i) as ii) follows immediately from i). Without loss of generality we may assume p, q_1, and q_2 are in $S(M)$ for some **S**-saturated M. If $p \not\perp q_1$, then by Theorem 4.5, q_1 is realized in $M[p]$. Since $q_1 \not\perp q_2$, q_2 is realized in $M[q_1]$ which without loss of generality is a submodel of $M[p]$. But then by Exercise X.1.21 there are realizations $\bar{a}$, $\bar{b}$ of p and q_2 with $(\bar{a} \not\downarrow \bar{b}; M)$. Thus $p \not\perp q_2$.

4.9 Corollary. *If q is regular then for any p and r, $p \not\perp q$ and $r \not\perp q$ implies $p \not\perp r$.*

Proof. Suppose not. Without loss of generality, p, q and $r \in S(M)$ for some **S**-model M. Let $\bar{a}$, respectively $\bar{b}$, realize p and r. By Theorem 4.5, q is realized by some $\bar{c} \in M[\bar{a}] - M$ and some $\bar{c}' \in M[\bar{b}] - M$. Let α map $\bar{c}'$ to $\bar{c}$ fixing M. Then $\alpha(\bar{b})$ realizes r and $\bar{c} \in M[\bar{a}] \cap M[\alpha(\bar{b})]$. But, $p \perp r$ implies

$\alpha(\bar{b}) \downarrow_M \bar{a}$ which implies $M[\alpha(\bar{b})] \downarrow_M M[\bar{a}]$ which implies $M[\alpha(\bar{b})] \cap M[\bar{a}] = M$. This contradiction yields the result.

To extend these results to K-strongly regular types for arbitary acceptable classes K, we must extend the result that weak orthogonality implies orthogonality from types over **S**-models to types over K-saturated models. We do that here for stationary K-strongly regular types. This allows us to conclude the analog of Theorem 4.5 and Corollary 4.8.ii).

4.10 Theorem. *Suppose $M \in K$, $p \in S(M)$, and (p, p_0) is K-strongly regular. If $q \in S(M)$ is K-strongly regular, $q|\operatorname{dom}^*(q_0)$ is stationary, and $q \not\perp p$ then $q \mapsto_K p$ and $p \not\perp^w q$.*

Proof. Let $A = \operatorname{dom}^*(q_0)$. Since $p \not\perp q$, there is a $D \supseteq M$ and $\bar{a}$, $\bar{b}$ realizing nonforking extensions of p and q to $S(D)$ such that $(\bar{a} \not\downarrow \bar{b}; D)$. By Corollary X.1.22 there is a $\bar{b}' \in q_0(M[\bar{a}])$ such that $(\bar{b}' \not\downarrow \bar{a}; M)$. Let $r = t(\bar{b}'; M)$. Then r is a regular type, so either $r \perp q|A$ or r is a nonforking extension of the stationary type $q|A$. In the second case, $\bar{b}'$ realizes q and witnesses both $p \not\perp^w q$ and $p \leftrightarrows_K q$. But, $r \perp q|A$ is impossible. For, we have $r \not\perp p$ and $p \not\perp q|A$, since nonorthogonality is a parallelism invariant. Thus, by Corollary 4.8 i), $r \not\perp q|A$.

The contrapositive of this result yields: Suppose $p, q \in S(M)$ and p_0, q_0 are over A with (q, q_0) and (p, p_0) K-strongly regular and $q|A$ is stationary. If $p \perp^w q$ then $p \perp q$.

Note that we invoked Theorem 4.8 i) and thus, indirectly, Theorem 4.5 in the proof of Theorem 4.10. We are unable to give a uniform proof of this result for all acceptable classes K but must first prove the result for **S** and then transfer it to the other classes. This theme resounds through the rest of this book.

The proof of Theorem 4.10 did not require that K admit regular types. Thus, it applies for example to an arbitrary model and two stationary **AT**-strongly regular types in a superstable theory. Note, however, that in this situation the hypothesis that $q|A$ be stationary is not automatically satisfied. The extension of Theorem 4.10 to obtain the full analog of Theorem 4.5 depends on K admitting stationary regular types. With this hypothesis we will deduce the result for arbitrary types (with domain in K) in the next chapter and this will allow us to obtain the analog of Theorem 4.8 ii). Even without this hypothesis one can extend Theorem 4.8.ii) to any acceptable K.

4.11 Corollary. *Nonorthogonality is the equivalence relation induced on the class of K-strongly regular types by the relation $\leftrightarrows_K$.*

4.12 Exercise. Prove Corollary 4.11.

While our definition of a pair (p, ϕ) being **AT**-strongly regular allows p to fork over $\operatorname{dom}\phi$, the following exercise shows that has a powerful consequence. The exercise follows easily from Corollary 4.11 and the fact (established in Exercise 2.14 and Theorem 2.15 that in an ω-stable theory questions of orthogonality can be reduced to regular types.

4.13 Exercise (Buechler). Let T be ω-stable, $M \models T$, and $p \in S(M)$. Suppose that for some $\overline{c} \in M$ and some formula $\phi(x; \overline{c})$ the pair (p, ϕ) is **AT**-strongly regular. Show that if p forks over $\overline{c}$ then $p \dashv \overline{c}$.

4.14 Historical Notes. This entire chapter is an elaboration of Theorem V.1.19 of [Shelah 1978] which gives five equivalent forms of the definition of a regular type. We have separated these various notions to indicate the roles that the various formulations play. In addition, we have integrated the treatment of strongly regular types. The proof that weak orthogonality implies orthogonality for K-strongly regular types is derived from [Lascar 1982a]. We have not discussed the notion of a *semi-regular* type which is defined in Section V.4 of [Shelah 1978] and plays an important role in extending the study to strictly stable theories. Hrushovski [Hrushovski 1986] has extensively exploited this notion.

Chapter XIII

Decomposition Theorems and Weight

In this chapter we will show every type in a superstable theory can be decomposed in terms of a finite number of regular types. In fact, under a suitable operation we impose a structure on the stationary types which reflects the multiplicative structure of the natural numbers. The weight one types behave as primes in this representation. We first obtain a precise structure theory for finitely generated extensions of an **S**-model. We connect these structural results with the notion of weight. While we would like to develop such a decomposition theorem for models in an arbitrary acceptable class K, we can not do so uniformly. Rather, we first obtain the result for **S**-models in Section 1 then define weight in Section 2. We conclude Section 2 by invoking the notion of weight to prove Lachlan's theorem that a countable superstable theory has either 1 or infinitely many countable models. In Section 3 we show that in an ω-stable theory there are 'enough' **AT**-strongly regular types. With this tool we obtain an extension of the decomposition theorem to all models of an ω-stable theory in Section 4.

Except for a few results at the beginning of Section 2, we assume in this chapter that T is superstable.

1. The Decomposition Theorem For S-Models

In this section we restrict ourselves to the class of **S**-models. We show that each finitely generated **S**-model of a superstable theory has a well defined dimension. We will use this information to decompose all types in a superstable theory as a product of regular types. The results of this section provide one characterization of weight. In Section 4, we consider extending the results of this section to other classes K.

1.1 Definition. Let M be an **S**-model, $A \subseteq M$, and let $R(M, A)$ be the collection of points in M which realize stationary regular types over A. Then $\dim(R(M, A))$ is the cardinality of a maximal independent subset of $R(M, A)$.

Since we are dealing with **S** models in this section we will just refer to regular types (since any regular type over an **S** model is **S**-strongly regular). Note that we do not assert that the dimension defined here determines M. That will be true only in exceedingly special circumstances.

1.2 Exercise. Find **S**-models $M, M' \supseteq N$ which are not isomorphic although $\dim(R(M,N)) = \dim(R(M',N))$.

1.3 Lemma. *Forking is a weakly transitive relation on* $R(M,A)$. *In particular,* $\dim(R(M,A))$ *is well defined.*

Proof. By Theorem XII.3.13, we need only show the first statement. By Lemma XII.3.11, it suffices to define an equivalence relation on $R(M,A)$ such that i) for all $\overline{a}$, forking is fully transitive on $[\overline{a}]$, the equivalence class of $\overline{a}$ and ii) if $\overline{a} \not\downarrow_A I$ then $\overline{a} \not\downarrow_A I \cap [\overline{a}]$. We will show that nonorthogonality of stationary types is such an equivalence relation. To see i), suppose $\overline{c} \cup Z \cup \overline{b} \subseteq [\overline{a}]$, $(\overline{b} \not\downarrow \overline{c} \cup Z; A)$ and $(\overline{c} \not\downarrow Z; A)$. By Lemma XII.3.5 it suffices to show $(\overline{b} \not\downarrow Z; A)$. By regularity, we have $t(\overline{c}; A \cup Z) \perp t(\overline{c}; A)$. Since $\overline{b} \cup \overline{c} \subseteq [\overline{a}]$, $t(\overline{b}; A) \not\perp t(\overline{c}; A)$. Applying the contrapositive of the transitivity of nonorthogonality, we deduce $t(\overline{b}; A) \perp t(\overline{c}; A \cup Z)$. If $(\overline{b} \downarrow A \cup Z; A)$ then $t(\overline{b}; A \cup Z) \parallel t(\overline{b}; A)$. Thus, $t(\overline{b}; A \cup Z) \perp t(\overline{c}; A \cup Z)$. But this contradicts $(\overline{b} \not\downarrow \overline{c} \cup Z; A)$ and thus yields i).

Now suppose $\overline{a} \not\downarrow_A I$. We must show $\overline{a} \not\downarrow_A I_{\overline{a}}$ where $I_{\overline{a}} = I \cap [\overline{a}]$. Let $\hat{I}_{\overline{a}} = I - I_{\overline{a}}$. Assume for contradiction that $\overline{a} \downarrow_A I_{\overline{a}}$. Then certainly $\overline{a} \downarrow_A J_{\overline{a}}$ where $J_{\overline{a}}$ is a maximal independent subset of $I_{\overline{a}}$. Now, $t(\hat{I}_{\overline{a}}; A) \perp t(J_{\overline{a}}; A)$ and $t(\hat{I}_{\overline{a}}; A) \perp t(\overline{a}; A)$ so by the strong triviality of orthogonality (cf. Theorem VI.1.19) $\hat{I}_{\overline{a}} \downarrow_A I_{\overline{a}} \cup \overline{a}$. By monotonicity and transitivity of independence, we deduce $\overline{a} \downarrow_A I_{\overline{a}} \cup \hat{I}_{\overline{a}}$. From this contradiction we have condition ii) and the lemma.

The following result is crucial for our development. This initial version of the *three model theorem* is rather easy to prove for **S**-saturated models. We prove a difficult extension of it to arbitrary models of an ω-stable theory in Theorem 3.3. There are some further variants in the treatment of superstable theories which are not ω-stable [Shelah 198?].

1.4 Theorem. *Let* $N \subseteq M \subset M'$ *and suppose* M *and* N *are* **S**-*saturated. For every* $a \in M' - M$ *such that* $t(a; M)$ *is regular, either* $t(a; M) \dashv N$ *or there is some* $b \in M' - M$ *with* $t(b; M)$ *regular,* $t(b; N) \not\perp t(a; M)$, *and* $b \downarrow_N M$.

Proof. (Fig. 1). Suppose there is $a \in M' - M$ with $t(a; M) \not\dashv N$ and $p = t(a; M)$ regular. Then, by Proposition VI.2.5, for some $A \subset N$ with $|A| < \kappa(T)$, $p \not\dashv A$. Possibly enlarging A inside N, choose a finite set $B \subseteq M - N$ such that p is strongly based on $A \cup B$ and $B \downarrow_A N$. Now by the strong saturation of N choose $B' \subseteq N$, $stp(B'; A) = stp(B; A)$ and $B \downarrow_A B'$. Let p_B denote $p|B$ and $p_{B'}$ the image of p_B under an automorphism fixing A and mapping B to B'. Then $p_{B'}$ is regular and by Corollary VI.2.22 $p_B \not\perp p_{B'}$. Thus, by Theorem XII.4.5 the nonforking extension of $p_{B'}$ to M is realized in M' by some b which satisfies the theorem.

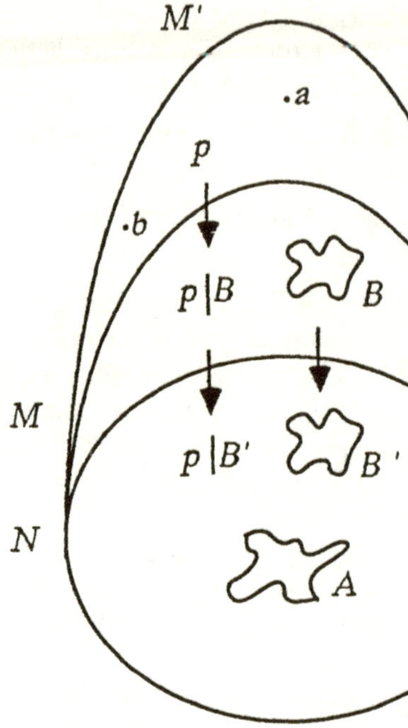

Fig. 1. The three model lemma (Strongly saturated version)

The last result would be more satisfying if it asserted either $t(M'; M) \dashv N$ or there is a $b \in M' - M$ with $b \downarrow_N M$ and $t(b; M)$ is regular. We obtain this stronger result in Lemma 4.3.

The following proof makes essential use of the fact that **S** is a powerful isolation relation. In Section 4 we will obtain a weaker version of this theorem for the class of **AT**-models.

1.5 Theorem (The Decomposition Theorem). *Let M be an* **S**-*saturated model of a superstable theory. For any finite sequence $\bar{c}$, if X is a basis for $R(M[\bar{c}], M)$ then $M[\bar{c}] \approx M[X]$.*

Proof. (Fig. 2). Let X be a basis for $R(M[\bar{c}], M)$. By Exercise X.1.21 $\bar{c}$ depends over M on each $x \in X$. So by Theorems VI.1.19 and II.2.16, X is finite. We show first that $X \rhd_M X^\frown\bar{c}$. Let $M_0 = M[X]$. We will construct a sequence of models $\langle M_i : i \leq \beta\rangle$ for some β and elements $\langle a_i : i < \beta\rangle$ such that $\bar{c} \in M_\beta$ and if $\bar{b} \in M_i$ for some $i < \beta$, then $X \rhd_M X^\frown\bar{b}$. Suppose we have constucted the sequence up to M_n and A_n. If $\bar{c} \notin M_n$, choose $a_{n+1} \in M_n[\bar{c}] - M_n$ to realize a regular type over M_n. Let $M_{n+1} = M_n[a_{n+1}]$. Then $t(\bar{c}; M_{n+1})$ forks over M_n for each n, so for some finite β, $\bar{c} \in M_\beta$. The maximality of X implies that there is no $b \in M - M_n$ such that $t(b; M_n)$ is regular and $b \downarrow_M M_n$. From Theorem 1.4, we deduce $t(a_{n+1}; M_n) \dashv M$. Thus for each n, $M_n \rhd_M M_{n+1}$. By the transitivity of dominance (Lemma

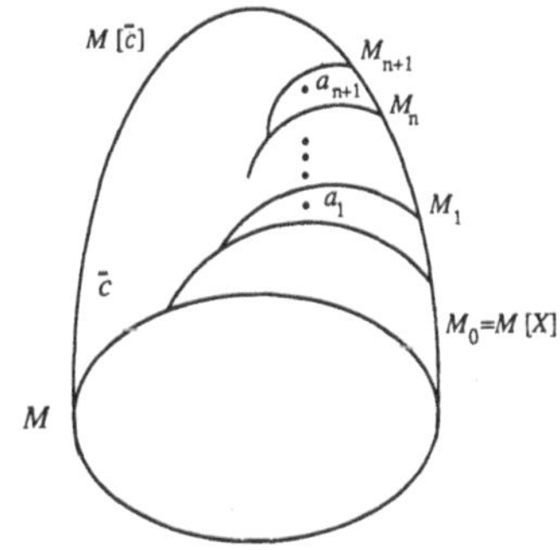

Fig. 2. The decomposition lemma

VI.3.10), $M_0 \rhd_M M_n$, for each n. But $X \rhd_M M_0$. So $X \rhd_M M_n$ for each n. In particular, $X \rhd_M X^\frown \bar{c}$. From Theorem X.2.3, we deduce there is a copy, M^0, of $M[X]$ which includes $\bar{c}$. From Theorem IX.4.1 and the fact (Theorem X.3.1) that all prime models are strictly prime, we deduce $M[X]$ is **S**-prime over $M \cup \bar{c}$ as required.

To see that the assumption that $\bar{c}$ is finite (or at least countable) is essential for Theorem 1.5, consider the theory of two refining equivalence relations. Let $M \models T$ and let C be an uncountable collection of points which are all in the same new E_1 class but are pairwise E_2 inequivalent. Now a subset of $M[C]$ which is maximal with respect to independence over M contains only a single element d and $M[d]$ contains only countably many E_2 classes. Thus $M[C]$ is not prime over a basis.

1.6 Exercise. Show that if X and Y are finite bases for $R(M, N)$ then $N[X] \approx N[Y] \approx M$.

1.7 Historical Notes. This section gives a more conceptual version of the argument in Theorem V.3.8 of [Shelah 1978].

2. Weight

We begin this section by defining weight and proving the rudimentary properties of weight which hold in any stable theory. Then we show that if T is superstable the weight of a type, p, can be thought of as the number of 'prime' factors of p under a suitable decomposition into regular types.

Intuitively, the weight of a type p is the maximal number of independent points that can be found such that a given realization of p depends upon each of them. A precise definition is somewhat more complicated. We have to take a supremum over various possibilites twice. In i) of the following definition we assign a 'preweight' to each type q by finding the maximal cardinality of an independent set C such that each element of C depends over $\operatorname{dom} q$ on some realization of q. In ii) we guarantee that the weight of a type p is a parallelism invariant by taking the supremum of the preweights of the nonforking extensions of p. Note that each q has a preweight since no C satisfying the conditions in ii) can have cardinality more than $\lambda \cdot \kappa(T)$ when q is a λ-type.

2.1 Definition. (Fig. 3).

i) Let $q \in S(B)$. The *pre-weight* of q (pwt(q)) is the supremum of the cardinalities of sets C such that C is an independent set of sequences over B and for some $\overline{a}$ realizing q, $\overline{a} \not\downarrow_B \overline{c}$ for each $\overline{c} \in C$.

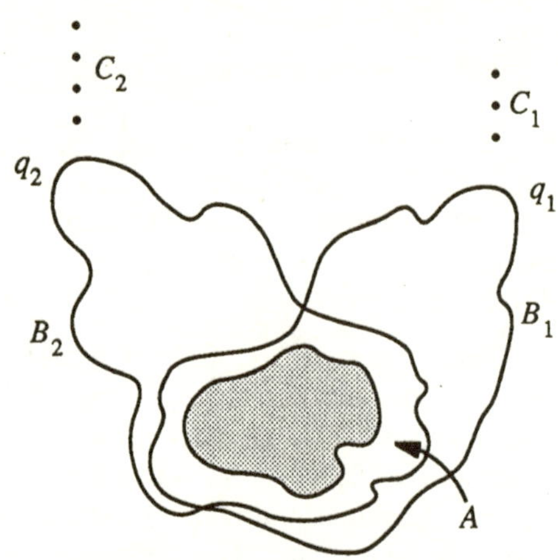

Fig. 3. $\mathrm{wt}(p) = \mathrm{pwt}(q_2) = 4$

ii) Let $p \in S(A)$. The *weight* of p (wt(p)) is the supremum over all $B \supseteq A$ and all nonforking extensions q of p to $S(B)$ of pwt(q).

To shorten notation we often write wt$(A; B)$ for wt$(t(A; B))$. There is no assumption in this definition that p is the type of a finite sequence. However, the proofs in this section are given for types of finite sequences. To extend to λ-types one must replace $\kappa(T)$ by $\kappa(T) + \lambda$.

It is clear from the definition that nonforking extensions can not increase weight. The following result then follows from Corollary II.2.11.

2.2 Exercise. Let p' be a nonforking extension of p; show $\mathrm{wt}(p) = \mathrm{wt}(p')$.

The following exercise shows that forking extensions can increase weight.

2.3 Exercise. Let T be the theory of two refining equivalence relations with infinite classes. Let $p \in S^2(\emptyset)$ be generated by $E_1(x, y) \wedge \neg E_2(x, y)$. Show that p has weight one and, in fact, is regular but that the extension p' of p generated by $E_1(x, a)$ has weight two.

2.4 Exercise. Show that if T is the theory of two crosscutting equivalence relations with all classes (and all intersections of classes from the two equivalence relations) infinite then the weight of the unique 1-type over the empty set is two.

2.5 Exercise. Let T be the theory of an infinite set, its three element subsets and a binary relation which holds between a three element set and each of its elements. Show the weight of a three element set is three.

2.6 Exercise. Show that it is necessary to pass through the notion of preweight by exhibiting a type with differing preweights depending on the choice of B. (Hint: Consider the theory of three crosscutting equivalence relations.)

2.7 Proposition. *Let $p \in S(A)$ and let M be a strongly $\kappa_r(T)$-saturated model containing A. There is a nonforking extension p' of p to $S(M)$ with* $\mathrm{wt}(p) = \mathrm{pwt}(p')$.

Proof. Consider any $B \supseteq A$, $\bar{b}$ realizing p, and $C = \langle \bar{c}_i : i < \alpha \rangle$ with $\alpha < \kappa(T)$ such that $\bar{b} \downarrow_A B$, for each i, $\bar{b} \not\downarrow_B \bar{c}_i$, and the $\bar{c}_i$ are independent over B. Using the local character of forking, there is a set $D \subseteq B$ with $|D| < \kappa_r(T)$ which satisfies the same conditions. Now choose $A_0 \subseteq A$ such that $\bar{b} \cup D \cup C \downarrow_{A_0} A$. By the strong $\kappa_r(T)$-saturation of M, we can choose $D' \subseteq M$, so that $t(D'; A_0 \cup \bar{b} \cup C) = t(D; A_0 \cup \bar{b} \cup C)$. Now choose $C' \cup \bar{b}'$ so that $t(C' \cup \bar{b}'; A_0 \cup D') = t(C \cup \bar{b}; A_0 \cup D')$, $\bar{b}'$ realizes $stp(\bar{b}; A_0)$, and $C' \cup \bar{b}' \downarrow_{A_0} M$. By Corollary II.2.10, we see that the required p' is $t(\bar{b}'; M)$.

Remember that if $\kappa(T)$ is regular and in particular if T is superstable then strong $\kappa_r(T)$-saturation is just $\mathbf{S}$-saturation.

2.8 Definition. Let $W(M, A)$ denote the set of $\overline{m} \in M$ such that $t(\overline{m}; A)$ has weight one.

Note that if $p \in S(A)$, the assertion that p has weight one implies that for any $\bar{a}$ realizing p and any sequences $\bar{b}$ and $\bar{c}$, if $(\bar{a} \not\downarrow \bar{b}; A)$ and $(\bar{a} \not\downarrow \bar{c}; A)$ then $(\bar{b} \not\downarrow \bar{c}; A)$. That is, we can pivot on $\bar{a}$. The weight one types play a very important role in the theory. They are a slight generalization of the regular types.

2.9 Theorem. *Supppose $A \subseteq M$ and $Z \subseteq W(M, A)$. Suppose $Y \subseteq M - A$ is independent over A and f mapping Y into Z satisfies $y \not\downarrow_A f(y)$. Then,*

i) *f is $1-1$.*
ii) *$f(Y)$ is independent over A.*

Proof. i) is obvious from the definition of a weight one type. For ii) we can assume by the finite character of forking that Y is finite. We show by induction on $|B|$ for $B \subseteq Y$, that $f(B) \cup (Y - B)$ is independent over A. When $B = Y$ we have the theorem. If $B = \emptyset$ there is nothing to show. Suppose $Y = Y_0 \cup B \cup y$ where we have by induction that $Y_0 \cup y \cup f(B)$ is independent. Then by hypothesis $y \not\downarrow_A f(y)$. So, if $f(y) \not\downarrow_A (Y_0 \cup f(B))$, since $f(y)$ has weight one, $y \not\downarrow_A (Y_0 \cup f(B))$ contrary to the induction hypothesis.

The argument for Theorem 2.9 can be rephrased as follows. Suppose p and q are complete weight one types over A which are weakly orthogonal. Let $R(a, b)$ hold if a realizes p, b realizes q and $a \not\downarrow_A b$. Now for every a

realizing p there is an a' realizing q with $R(a, a')$ and vice versa. Moreover, if $a_1, \ldots, a_n$ realize p, $a'_1, \ldots, a'_n$ realize q, and for each $i < n$, $R(a_i, a'_i)$ holds then exactly the same dependence relations hold among the $a_1, \ldots, a_n$ and the $a'_1, \ldots, a'_n$. This shows that weak non-orthogonality preserves any property of a type defined solely in terms of the properties of forking on realizations of the type. (Compare Chapter XVI.2 and [Baldwin 1984].)

For the next few results we assume T is superstable and tie together the definition of weight with the decomposition of **S**-saturated models in Section 1.

2.10 Theorem. *For any $A \subseteq M$, $R(M, A) \subseteq W(M, A)$. That is, every regular type has weight one.*

Proof. Suppose $t(\overline{m}; A)$ is regular, $\overline{b} \not\downarrow_A \overline{m}$ and $\overline{c} \not\downarrow_A \overline{m}$. By the definition of regularity, $t(\overline{m}; A \cup \overline{c}) \perp t(\overline{m}; A)$. From Exercise XII.4.7 we can conclude $t(\overline{b}; A) \rhd^e t(\overline{m}; A)$. Choose an **S**-model N with $\overline{m}^\frown\overline{b}^\frown\overline{c} \downarrow_A N$. Then $\overline{b} \rhd_N \overline{m}$. Now if $\overline{c} \downarrow_A \overline{b}$, we deduce in turn $\overline{c} \downarrow_N \overline{b}$, $\overline{c} \downarrow_N \overline{m}$, and, $\overline{c} \downarrow_A \overline{m}$ by the definition of domination and Corollary II.2.10. Thus, $\overline{c} \not\downarrow_A \overline{b}$ as required.

Suppose that $\overline{a}$ realizing $p \in S(M)$ and Y with $|Y| = \mu$ witness that $\mathrm{pwt}(p) = \mu$. Then, just rewording the definition, $t(Y; M) \rhd_M p$ and no subset of Y dominates p. With this observation we can link the weight of a type with the regular decomposition of a model prime over a realization of the type.

2.11 Theorem. *Let T be superstable.*

i) *Suppose M is **S**-saturated and $p \in S(M)$; then* $\mathrm{wt}(p) = \mathrm{pwt}(p, M) = \dim(R(M[p], M))$.
ii) *Thus, for any p,* $\mathrm{wt}(p) = n$ *if and only if there is an **S**-saturated M containing* $\mathrm{dom}\, p$ *and a nonforking extension p' of p to $S(M)$ such that* $\dim(R(M[p'], M)) = n$.
iii) $\dim(W(M, A)) = \dim(R(M, A))$. *Thus,* $\dim(W(M, A))$ *is well defined.*

Proof. i) The first equality holds by Propostion 2.7. For the second, let Y be an independent set which witnesses the preweight of p'. Since $t(Y; M) \rhd_M p$, applying Theorem X.2.5 p is realized by some $\overline{a} \in M[Y]$. But, $M[\overline{a}]$ must be $M[Y]$. Otherwise a proper subset of Y would dominate $\overline{a}$. Map Y into $R(M[p], M)$ choosing $f(y) \in M[y] - M$ to realize a regular type over M. By FI_1 for each y, $y \not\downarrow_M f(y)$. By Theorem 2.9, we deduce f is 1-1 and $f(Y)$ is independent. Thus $|Y| \leq \dim(R(M[p], M))$. But the other inequality is obvious by Theorem 2.10 so we have i).

Now ii) and iii) follow easily from Theorem 2.9 and i).

Now, we calculate the relation between the weight of a sequence and the weight of its components.

2.12 Theorem. i) *For any $\overline{a}$, $\overline{b}$, and A:*

$$\mathrm{wt}(\overline{a}^\frown\overline{b}; A) \leq \mathrm{wt}(\overline{a}; A) + \mathrm{wt}(\overline{b}; A \cup \overline{a}).$$

ii) *If* $\bar{a} \downarrow_A \bar{b}$,

$$\mathrm{wt}(\bar{a}^\frown\bar{b}; A) = \mathrm{wt}(\bar{a}; A) + \mathrm{wt}(\bar{b}; A).$$

Proof. Without loss of generality, we replace A by an **S**-saturated model, M. For i), we show:

$$\mathrm{wt}(\bar{a}^\frown\bar{b}; M) \leq \mathrm{wt}(\bar{a}; M) + \mathrm{wt}(\bar{b}; M \cup \bar{a}).$$

For this, suppose I is an independent set which witnesses the weight of $t(\bar{a}^\frown\bar{b}; M)$. Let I_0 be a maximal subset of I such that $I_0 \cup \bar{a}$ is independent. It is now straightforward to check a) for each $\bar{i} \in I_0$, $\bar{i} \not\downarrow_{M\cup\bar{a}} \bar{b}$ and so $|I_0| \leq \mathrm{wt}(\bar{b}; M \cup \bar{a})$ and b) if $\bar{i} \in I - I_0$ then $\bar{i} \not\downarrow_M \bar{a}$ so $|I - I_0| \leq \mathrm{wt}(\bar{a}; M)$. This establishes the first claim.

For ii), suppose C is a basis for $R(M[\bar{a}], M)$ and D is a basis for $R(M[\bar{b}], M)$. Now $C \downarrow_M D$ and for each $e \in C \cup D$, $\bar{a}^\frown\bar{b} \not\downarrow_M e$. Thus $\mathrm{wt}(\bar{a}^\frown\bar{b}; M) \geq \mathrm{wt}(\bar{a}; M) + \mathrm{wt}(\bar{b}; M)$. Since $\mathrm{wt}(\bar{b}; M \cup \bar{a}) = \mathrm{wt}(\bar{b}; M)$, we finish.

We can now read off a number of the characteristics of weight. First we make precise the decomposition of each type into regular types.

2.13 Definition. Let $\langle r_i : i < \alpha\rangle$ be a sequence of stationary types over a set A. Then the *product* of the r_i, denoted $\otimes r_i$, is the unique type of an independent sequence of tuples realizing the r_i. The r_i need not be distinct.

Note that if $\lg(r_i) = m_i$ then $\lg(\otimes r_i) = \sum_i m_i$.

2.14 Exercise. Show $\otimes r_i \perp \otimes r'_j$ if and only if $r_i \perp r'_j$ for each i and j.

2.15 Theorem. *Let T be superstable and q a stationary type. There exist regular types r_i, $i < \mathrm{wt}(q)$, such that $q \mathrel{\underset{\sim}{\mapsto}}_S \otimes r_i$. That is, $q \sqsupseteq^e \otimes r_i$.*

Proof. Choose an **S**-saturated M, $q' \in S(M)$ with $q' \parallel q$, and let the r_i be a list with appropriate repetitions of the types realized in a basis for $R(M[q'], M)$. We get the second statement because $\mathrel{\underset{\sim}{\mapsto}}_S$ is the same as $\sqsupseteq^e$.

This theorem shows that the quotient of the $\mapsto_S$ order on the class of stationary types in a countable superstable theory by the congruence relation $\mathrel{\underset{\sim}{\mapsto}}_S$ yields a structure similar to the natural numbers under the partial order of divisibility. The equivalence classes of the regular types correspond to the prime numbers.

The following exercises record some of the essential properties of weight. Exercise 2.18 is especially important. All of these exercises follow easily from the definitions.

2.16 Exercise. Show that if $A = \{\bar{a}_i : i < \mu\}$ then for any B,

$$\mathrm{wt}(A; B) \leq \sum_{i<\mu} \mathrm{wt}(\bar{a}_i; A_i \cup B).$$

2.17 Exercise. Show that $A \subseteq C$ implies $\mathrm{wt}(A; B) \leq \mathrm{wt}(C; B)$.

2.18 Exercise. Show that if $\bar{a} \rhd_M \bar{b}$ then $\mathrm{wt}(\bar{a}; M) \geq \mathrm{wt}(\bar{b}; M)$.

2.19 Exercise. Conclude from the previous exercise that if $\bar{a} \rhd_M \bar{b}$ and $\mathrm{wt}(\bar{a}; M) = 1$ then $\mathrm{wt}(\bar{b}; M) = 1$.

2.20 Exercise. Show that if p is K-strongly regular then $\mathrm{wt}(M[p]; M)$ is one.

2.21 Exercise. For any M and $\bar{a}$, $\mathrm{wt}(\bar{a}; M) = \mathrm{wt}(M[\bar{a}]; M)$.

The weight one types play a central role in the classification theory. Modulo $\sqsubseteq^e$ they are equivalent to regular types and thus share most of the important properties of regular types. The following results summarize some of their most important properties.

2.22 Theorem. *Suppose T is superstable. Let $p \in S(A)$. The following are equivalent.*

i) *p has weight 1.*
ii) *Suppose $\bar{a}$ realizes p, $\bar{a} \downarrow_A B$, $\bar{b} \not\downarrow_B \bar{a}$ and $\bar{c} \not\downarrow_B \bar{a}$. Then $\bar{b} \not\downarrow_B \bar{c}$.*
iii) *Let M be $\mathbf{S}$-saturated and $p' \in S(M)$ a nonforking extension of p. Then p' is $\mathbf{S}$-minimal.*
iv) *There is a regular type q with $p \sqsubseteq^e q$.*

Proof. ii) is just a restatement of the definition of weight one. If ii) holds and $\bar{a}$ realizes p' then for any $\bar{b}$ such that $\bar{a} \not\downarrow_M \bar{b}$, $\bar{b} \rhd \bar{a}$. Thus p is realized in $M[\bar{a}]$ and p is $\mathbf{S}$-minimal. (Indeed, we have shown that p' has weight one implies that if $p' \not\perp q$ then $q \mapsto_S p'$.)

Finally, suppose p' is $\mathbf{S}$-minimal. Let $\bar{a}$ realize p'. We want to show $\mathrm{wt}(p) = \mathrm{wt}(p') = 1$. If not, there exist $\mathrm{wt}(p') = n > 1$ independent elements in $M[\bar{a}] - M$. Choose any two, say $\bar{b}$ and $\bar{c}$. Since p' is $\mathbf{S}$-minimal there is an imbedding of $M[p']$ into $M[\bar{b}]$ and thus there are n independent elements contained in $M[p']$. But they are independent from $\bar{c}$ and all depend on $\bar{a}$, contradicting that $\mathrm{wt}(p') = n$. The equivalence of i) and iv) is immediate from Theorems 2.10 and 2.15.

2.23 Exercise. Show that if p has weight 1 then forking is mildly transitive on $p(\mathcal{M})$.

The $\mathbf{S}$ in part iii) of Theorem 2.10 cannot be replaced by an arbitrary acceptable K. Note that we have characterized $\mathrm{wt}(p) = 1$ in terms of dependence on two sets which need not realize p. Thus weight one is somewhat stronger than the assertion that forking is mildly transitive on $p(\mathcal{M})$ but, in view of Example XII.3.10, not as strong as asserting that forking is fully transitive on $p(\mathcal{M})$.

The following remarks are obvious from the assertion that each weight one type is $\sqsubseteq^e$ equivalent to a regular type.

2.24 Corollary. *Suppose T is superstable.*

i) *Nonorthogonality is transitive on weight one types.*
ii) *If q has weight one, $p \not\perp q$, and $r \not\perp q$ then $p \not\perp r$.*
iii) *If $p \not\perp A$, $q \not\perp p$ and p has weight one then $q \not\perp A$.*

2.25 Exercise. Show using Example XII.1.11 ii) that a type may have weight 1 without being **AT**-minimal.

2.26 Theorem. *If $A \subseteq M$ and $p \in S(A)$ has weight 1 then* $\dim(p, M)$ *is well defined.*

Proof. Suppose E and F are maximal independent subsets of $p(M)$ with $|E| = \mu < |F| = \mu'$. By Theorem 2.12, $\mathrm{wt}(E; A) = \mu$. But F is an independent set and each element of F depends on E. This contradicts the definition of weight one and yields the theorem.

We conclude this section by using weight to give a quick proof of an important result of Lachlan.

2.27 Theorem. *If T is a countable superstable theory then T has either* 1 *or infinitely many countable models.*

Proof. If T has fewer than $2^{\aleph_0}$ countable models then T is small and so admits prime models over finite sets. If T is not $\aleph_0$-categorical there is a finite set A with a nonprincipal type $p \in S(A)$. Let $E = \langle \overline{e}_i : i < \omega \rangle$ be an infinite set of indiscernibles based on p. Define models $\langle M_i : i < \omega \rangle$ so that M_i is prime over $A \cup E_i$. Suppose that k is the weight of p. By Theorem 2.12 and induction $\mathrm{wt}(E_i; A) = ik$. Note that for any $f \in p(M_i)$, $t(f; A \cup E_i)$ is isolated while $t(f; A) = p$ is not, so $f \not\downarrow_A E_i$. But then by the definition of weight, any independent (over A) set of realizations of p in M has cardinality at most ik. Thus, the models $\langle M_{ik} : i < \omega \rangle$ are pairwise nonisomorphic and we finish.

2.28 Exercise. Show that if any of $R_M(p), U(p), R_C(p)$ equals 1 then p is regular and thus has weight one.

The following exercises are taken from [Lascar 1984]. They follow easily from the U-rank inequalities discussed in Chapter VII.2.

2.29 Exercise. Suppose $p, q \in S(A)$, $U(p) \geq \omega^\alpha$ and $U(q) < \omega^\alpha$. Show $p \perp q$.

2.30 Exercise. If $U(p) = \omega^\alpha$ then p is regular.

From Theorem 2.22 iv) and Exercise 2.30 it is immediate that if $p \sqsubseteq q$ and $U(q) = \omega^\alpha$ then p has weight one. Lascar [Lascar 1984] establishes the converse, provided we work in T^{eq}. This proviso is essential; it is easy to find T and regular p with $U(p) = 2$ if we ignore T^{eq}.

2.31 Historical Notes. Shelah introduced the concept of weight in Section V.3 of [Shelah 1978]. We follow the somewhat simpler definition of [Makkai 1984]. The description of the relation of weight to regular types returns to Shelah's exposition. The product of types notation was introduced by Lascar in [Lascar 1976].

Theorem 2.27 has a long history. The first step in this direction is the proof by Baldwin and Lachlan [Baldwin & Lachlan 1971] that the conclusion holds for countable theories which are $\aleph_1$ categorical. This proof used many special properties of $\aleph_1$-categorical theories and yielded the additional infomation that all countable models of such a theory are homogeneous. Then Lachlan [Lachlan 1973] proved Theorem 2.27 by a complicated argument using rank. Lascar [Lascar 1976] simplified the proof by the use of U rank. The proof here is just a translation of his. Finally, Pillay [Pillay 1983] has given an even simpler proof and extended the result to what he calls the class of *normal theories.*

3. *Ubiquity of Regular Types*

This section is devoted to showing there are enough K-strongly regular types so that if a type $p \not\dashv M$ for $M \in K$ then $p \not\perp q$ for some K-strongly regular $q \in S(M)$. In the next section we will use this result to prove a decomposition theorem for finitely generated members of K. Throughout this section, we assume that T is superstable.

It is easy to deduce the following result from Theorems 2.15 and VI.1.19.

3.1 Proposition. *If $p \not\dashv M$ and M is* **S**-*saturated then for some regular $r \in S(M)$, $p \not\perp r$.*

Proof. For some $q \in S(M)$, $p \not\perp q$. Decompose q as $\otimes r_i$ with each r_i regular. We must have $p \not\perp r_i$ for some i.

It is much more difficult to show for an arbitrary acceptable class K, e.g **AT** in an ω-stable theory, that $p \not\dashv M$ implies $p \not\perp r$ for some K-strongly regular r. We require one easy lemma before the main assault.

3.2 Lemma. *If N is a model and M is* **S**-*saturated, $r \in S(M)$ is regular and $r \not\dashv N$ then for every $p \in S(M)$, if p is realized in $M[r]$ then $p \not\dashv N$.*

Proof. There is a $q \in S(N)$ with $r \not\perp q$. Since p is realized in $M[r]$, $p \not\perp r$. Thus, by Corollary XII.4.9 $p \not\perp q$, a fortiori, $p \not\dashv N$.

This next proof requires both the use of strong **I**-saturation as in Section 1 of this chapter and the consideration of indiscernibles to study orthogonality as in the last part of Section VI.1. The main thrust of this result is a *three model theorem.* Let $N \subseteq M \subseteq M_1$. Unless $t(M_1; M) \dashv N$, we want to find a $p \in S(M)$ which is realized in M_1, does not fork over N, and is K-strongly regular for an appropriate K. There are several variants on theorems with this general form. In the case at hand we make minimal assumptions on N and allow M to carry the burden. This is appropriate as the interesting application of this result is to arbitrary models N of an ω-stable theory.

3.3 Theorem (The three model theorem). *Let K admit regular types. Suppose $N \subseteq M \subseteq M_1$ and M is* **S**-*saturated. If for each $a \in M_1 - M$, $t(a; M) \not\dashv N$ then there is a $b \in M_1 - M$ such that $b \downarrow_N M$ and $t(b; M)$ is K-strongly regular.*

Proof. (Fig. 4). Since K admits regular types (Definition XII.2.1) we can choose $p \in S(M)$ and an **I**-formula p_0 over N such that p_0 weakly isolates $p|N$ in (M, M_1). By hypothesis, $p \not\nearrow N$. To complete the proof of the theorem, we establish the following claim.

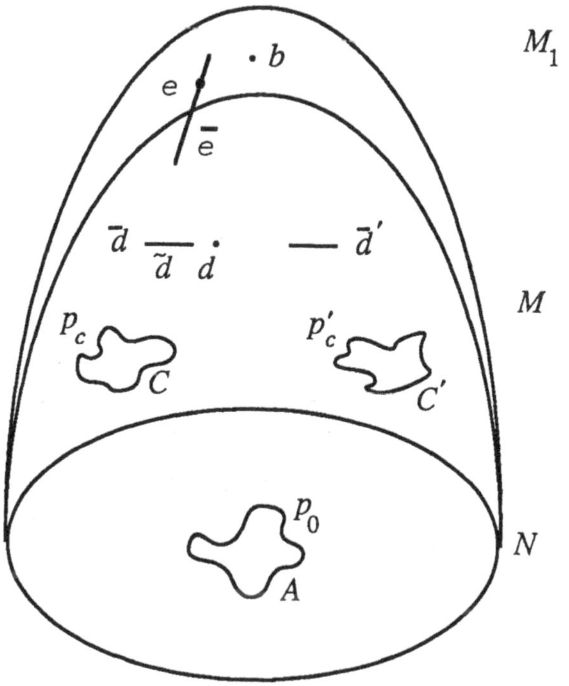

Fig. 4. The three model lemma (ω-stable version)

CLAIM. p does not fork over N.

Let $b \in M_1 - M$ realize p and choose $A \subseteq N$ and $C \subseteq M$ such that $|A| < \kappa(T)$, $|C| < \kappa(T)$, p_0 is over A and

$$b \underset{A\cup C}{\downarrow} M \text{ and } p_C = t(b; A \cup C) \text{ is stationary} \tag{1}$$

$$C \underset{A}{\downarrow} N \tag{2}$$

$$p_C \not\nearrow A\,. \tag{3}$$

We would be done if $b \downarrow_A A \cup C$, so we can assume for contradiction that $b \not\downarrow_A C \cup A$. We now invoke the test in Corollary VI.2.22 for $p_C \not\nearrow A$.

Choose $C' \subseteq M$ such that:

$$C' \underset{A}{\downarrow} C \tag{4}$$

$$stp(C; A) = stp(C'; A). \tag{5}$$

Let $p_{C'}$ denote the image of p_C under an automorphism which fixes A and maps C to C'. By Corollary VI.2.22, $p_C \not\perp p_{C'}$ so by Corollary VI.2.18 there is an n such that $p_C^n \not\perp^w p_{C'}^n$ and in particular $p_C^n \not\perp^a p_{C'}^n$. Let $\overline{d} = \tilde{d}^\frown d \in M^n$ with $\lg(\tilde{d}) = n-1$ realize p_C^n and let $\overline{d}' \in M$ realize $p_{C'}^n$. We can demand:

$$\overline{d} \downarrow_{A \cup C} A \cup C \cup C' \tag{6}$$

$$\overline{d}' \downarrow_{A \cup C} A \cup C \cup C' \tag{7}$$

$$\overline{d} \not\downarrow_{A \cup C \cup C'} \overline{d}' \tag{8}$$

and

$$C \cup \overline{d} \downarrow_N C'. \tag{9}$$

We will apply Axiom X.1.31 as in the proof of Theorem X.1.38. Since $(b \not\downarrow C \cup A; A)$ there is a formula $\psi(x,\overline{y}) \in F(A)$ and a sequence $\overline{c} \subset C$ such that:

$$\models \psi(b,\overline{c}) \text{ and if } \models \psi(b^*,\overline{c}^*) \text{ then } b^* \not\downarrow_A \overline{c}^* \text{ for any } \overline{c}^* \text{ and any } b^* \text{ realizing } t(b;A). \tag{$*$}$$

Without loss of generality we can take $\overline{c}$ as an enumeration of C. Similarly, (8) implies there is a formula $\theta(\overline{x},\overline{y};\overline{x}',\overline{y}') \in F(A)$ and $\overline{c}'$, (which without loss of generality we again take as an enumeration of C') such that:

$$\models \theta(\overline{d},\overline{c};\overline{d}',\overline{c}') \text{ and for any } \overline{e},\overline{f} \text{ if } \models \theta(\overline{d},\overline{c};\overline{e},\overline{f}) \text{ then } \overline{d} \not\downarrow_{A\cup\overline{c}} \overline{e}^\frown\overline{f}. \tag{$**$}$$

Let

$$r(\overline{d},\overline{c};\overline{x},\overline{c}') = \bigcup_{i<n} p_0(x_i) \cup \{\theta(\overline{d},\overline{c},\overline{x},\overline{c}')\} \cup \bigcup_{i<n} \{\psi(x_i;\overline{c}')\}.$$

Then $r(\overline{d},\overline{c};\overline{x},\overline{c}')$ is consistent since it is satisfied by $\overline{d}'$. Note that in the important **AT** case, r is a single formula. Applying (9) and the argument for Corollary X.1.13, we can choose a $\overline{c}'' \in N$ such that $r(\overline{d},\overline{c};\overline{x},\overline{c}'')$ is consistent. By (1) and the definition of p_C^n,

$$t(b;\tilde{d}^\frown\overline{c}^\frown\overline{c}' \cup A) = t(d;\tilde{d}^\frown\overline{c}^\frown\overline{c}' \cup A).$$

Thus, $r(\tilde{d},b,\overline{c};\overline{x},\overline{c}'')$ is consistent. Let $\overline{e} \in M_1$ realize this last type. In particular, $\models \theta(\tilde{d},b,\overline{c},\overline{e},\overline{c}'')$. We now show $\overline{e} \cap (M_1 - M) \neq \emptyset$. Suppose to the contrary, that $\overline{e} \subseteq M$. From $(**)$ we have $(\tilde{d}^\frown b \not\downarrow_{A\cup C} \overline{e}^\frown\overline{c}'')$. By monotonicity, this implies $b \not\downarrow_{A\cup C} \overline{e} \cup C'$ which contradicts (1). This yields $\overline{e} \cap (M_1 - M) \neq \emptyset$ as required.

Choose e from $\overline{e}$ with $e \in M_1 - M$. Then e realizes p_0 so $t(e;A) = t(b;A)$. But then since $\models \psi(e,\overline{c}'')$, $(*)$ yields $(e \not\downarrow_A \overline{c}''$, contrary to the weak isolation and we deduce the claim.

3.4 Corollary. *Let K admit regular types. If $N \in K$ and $q \not\perp N$ then there is a K-strongly regular type $r \in S(N)$ such that $q \not\perp r$.*

Proof. Without loss of generality, $q \in S(M)$ for a model $M \supseteq N$ which is **S**-saturated, **I**-saturated, and $|N|^+$-saturated, since we can take the nonforking extension of q to such a model. We can decompose q as a product of regular types r_i. Then one of the $r_i \not\perp N$ and so by Lemma 3.2 any type $p' \in S(M)$ which is realized in $M[r_i]$ is not orthogonal to N. By Theorem 3.3, there is a K-strongly regular $r \in S(N)$ whose nonforking extension to M is realized in $M[r_i]$. So $r \not\perp q$ and we finish.

3.5 Exercise (Saffe). Show that the assumption that N is a model is essential for Theorem 3.3 by considering the theory of two crosscutting equivalence relations.

Now we can obtain a factorization of any type into a product of K-strongly regular types whenever K admits regular types.

3.6 Theorem. *If T is superstable and K admits regular types then for any $N \in K$ and for any type $p \in S(N)$ there exists a family of K-strongly regular types $r_i \in S(N)$ for $i < n$ such that $p \sqsupseteq^e \otimes r_i$.*

Proof. By Theorem 2.15 $p \sqsupseteq^e \otimes r_i'$ for a sequence of r_i' each of which is regular. Now if $p \in S(N)$, each r_i' is not orthogonal to N. Thus, for each r_i' there is a strongly regular $r_i \in S(N)$ with $r_i' \not\perp r_i$. Then $\otimes r_i \sqsupseteq^e \otimes r_i'$ so $p \sqsupseteq^e \otimes r_i$.

Steinhorn has pointed out that Example XII.1.11 ii) shows that even for ω-stable theories we cannot strengthen Theorem 3.6 by replacing $\sqsupseteq^e$ by $\stackrel{\leftarrow}{\sim}_{\mathbf{AT}}$. We proved in Chapter VI that if M is **S**-saturated then for complete types over M weak orthogonality implies orthogonality. In Theorem XII.4.10 we removed the restriction on M but required both types to be **AT**-strongly regular and stationary. Now we remove the restriction that the types be strongly regular.

3.7 Corollary. *If T is superstable and K admits stationary regular types then for each $M \in K$, if $p, q \in S(M)$ and $p \perp^w q$ then $p \perp q$.*

Proof. By Theorem 3.6 we can let $q \sqsupseteq^e \otimes r_i$ where each r_i is K-strongly regular. It suffices to show $p \perp r_i$ for each i. But if $p \not\perp r_i$ then $p \not\perp^w r_i$ by Theorem XII.4.10.

The hypotheses of Corollary 3.7 express the properties required to prove the result. Nevertheless, they are unnatural. If K is **S** then the result is easy (Theorem VI.1.40). If K is **AT** then the only class of theories known to obey the hypotheses are the ω-stable theories. Thus, we have only a weak generalization of the following theorem of Lascar.

3.8 Corollary. *If T is ω-stable, $M \models T$, $p, q \in S(M)$ and $p \perp^w q$ then $p \perp q$.*

3.9 Exercise. Show the Example VI.1.33.i) is a small superstable theory so the assertion, 'For types over models, $\perp^w$ implies $\perp$' can not be extended from ω-stable to small superstable theories.

Two elements of a partial order are called *disjoint* if they have no common lower bound. We can now give a syntactic characterization of two types, p and q, being disjoint in the $\mapsto_K$-order. Namely, $p \perp q$. This observation follows quickly from the next theorem. It is unlikely to be an accident that Birkhoff's classic lattice theory text [Birkhoff 1949] uses $\perp$ to denote disjointness.

3.10 Theorem. *If the acceptable class K admits stationary regular types, $M \in K$, and $p \in S(M)$ is K-strongly regular then for any $q \in S(M)$ if $q \not\perp p$ then $q \mapsto_K p$.*

Proof. By Corollary 3.4 $q \sqsupseteq^e \otimes r_i$ where each $r_i \in S(M)$ is K-strongly regular. Thus $q \not\perp p$ implies $\otimes r_i \not\perp p$ implies for some i, $r_i \not\perp p$. By Theorem XII.4.10 $r_i \mapsto_K p$ and so $q \mapsto_K p$.

This theorem again improves Theorem XII.4.10 by removing the hypothesis that both of the types are K-strongly regular.

Now to see the equivalence of disjointness and orthogonality, let $M \in K$ and $p, q \in S(M)$. If p and q have a common lower bound, they have one which is regular. But then, by Corollary XII.4.9, $p \not\perp q$. Conversely, if $p \not\perp q$ then by Theorem 2.15 there exist regular factors r_1 of p and r_2 of q such that $p \not\perp r_1$, $r_1 \not\perp r_2$, and $q \not\perp r_2$. By Theorem 3.10, p and q are not disjoint. Of course, for $K = S$, we could prove this result already in Section XII.4.

There were actually three stages in the proof of Theorems 3.6, 3.10 and their assorted corollaries. We first have the result if M is an **S**-model and the types are arbitrary. Then we allow the model to be arbitrary but demand regularity of the types. Finally, we remove the regularity demand. This progression is obscured, if Theorem 3.10 itself is looked at, by the fact that the type p in Theorem 3.10 must be regular in all incarnations of that theorem.

Using Proposition 3.1 we can show that a weight one type which is not orthogonal to a set B shares many of the properties of a type actually over B. For example, we have the following generalization of Theorem VI.2.21.

3.11 Theorem. *Suppose* $\mathrm{wt}(p) = 1$, *$p \dashv A$, and $p \not\dashv B$. Then if $B \downarrow_A C$, $p \dashv C$.*

Proof. Without loss of generality, replace B by an **S**-model M. By Proposition 3.1, there is a regular $r \in S(M)$ with $p \not\perp r$. By Corollary 2.24, $p \dashv A$ implies $r \dashv A$. By Theorem VI.2.21, $r \dashv C$ whence by Corollary 2.24 again, now using that $\mathrm{wt}(p) = 1$, $p \dashv C$.

3.12 Exercise. Suppose T is stationary and K admits stationary regular types. Show that for any $M \in K$ each of the following properties implies its successor on the list: $p \mapsto_K q$, $p \rhd q$, $p \not\perp^a q$, $p \not\perp q$.

Consider the three implications in the preceding exercise. Corollary 3.7 shows the last can be reversed. We should not expect the second to be reversed for arbitrary p and q. For, we might factor have p as $r_1 \otimes r_2$ and q as $r_2 \otimes r_3$ where r_1, r_2, and r_3 are pairwise orthogonal regular types. The first implication is an equivalence for $K = \mathbf{S}$ (Corollary X.2.5); it cannot be reversed for $K = \mathbf{AT}$ ((Exercise X.2.6). Theorem 3.1 shows that all four conditions are equivalent if both p and q are K-strongly regular.

3.13 Historical Notes. The main result of this section, Theorem 3.3, first occurs in [Shelah 1982a]. Our proof is derived from comments of Buechler. Corollary 3.7 is due to Lascar.

4. *Linear Decomposition of Finitely Generated* **AT**-*Models*

In Section 1 we showed that in the category of **S**-models of a superstable theory any finitely generated **S**-model $N = M[\overline{c}]$ could be represented as $M[X]$ where X is a set of independent realizations of regular types over M. The cardinality of X is an invariant of the pair (N, M). In fact, it is the weight of $t(\overline{c}; M)$. We would like to provide a similar decomposition for finitely generated models (in the class **AT**) when T is ω-stable. The following examples show that an exact analog is impossible.

4.1 Example. i) Consider again Example XII.1.11. If we let M be a model of T and as before choose $a^\frown b \notin M$, not on the same Z-component but in the same equivalence class which does not intersect M, then it is impossible to choose an independent set X realizing **AT**-strongly regular types over M with $M[a,b] \approx M[X]$. Thus the best description of a 'basis' for $M[a,b]$ over M is to take the pair $\langle a, b\rangle$ as the basis and note that $t(b; M[a]) \dashv M$.

The next example shows the sensitivity of these notions to the proper choice of $M[a]$.

ii) Let T be the theory of a single unary function f satisfying axioms which assert $f^3(x) = f^2(x)$ but there are infinitely many elements with neither $f(x) = x$ nor $f^2(x) = f(x)$. Thus any countable model of T can be broken up into components where each component is a collection of sequences of integers of length at most two with $f(s) = s^-$ if $\lg(s) > 0$ and $f(s) = s$ if $\lg(s) = 0$. Now let M be a countable model of T and choose $a \in \mathcal{M}$ such that $f(a) \notin M$ but $f^2(a) \in M$. Choose $b \in \mathcal{M}$ such that $f(a) = f(b)$. Consider the model $M[a,b]$. Now $a \rhd_M M[a,b]$ so it is impossible to find a pair of elements in $M[a,b]$ which are independent over M. But there are copies of $M[a]$ with $b \notin M[a]$.

The first of these examples is fairly representative of the situation. The notion of a canonical resolution for N over M induces an invariant which characterizes the pair (M, N) up to isomorphism over M.

4.2 Definition. i) A *K-regular resolution* of $N = M[\overline{c}]$ is a sequence of models $M_0 = M, M_1, \ldots M_k$ such that each M_{i+1} is K-prime over a basis X_i for $R(N, M_i)$ and $M_k \approx M[\overline{c}]$.

ii) A *canonical K-regular resolution* of $M[\overline{c}]$ over M is a K-resolution of $M[\overline{c}]$ over M with minimal length.

We will apply the following lemma which is the promised strengthening of Lemma 1.4. Note that this result relies on the complicated Lemma 3.3.

4.3 Lemma. *Let $N \subseteq M \subset M'$ be models of the countable ω-stable theory T. Either $t(M'; M) \dashv N$ or for some $a \in M' - M$, $a \downarrow_N M$, and $t(a; N)$ is strongly regular.*

Proof. Suppose $t(M'; M) \not\dashv N$, by monotonicity for some $\overline{a} \in M' - M$, $p = t(\overline{a}; M) \not\dashv N$. By Theorem 3.4 there is a K-strongly regular $p' \in S(N)$ with $p \not\perp p'$. By Theorem XII.4.5, the nonforking extension of p' to $S(M)$ is realized in $M[\overline{a}] \subseteq M'$ which yields the result.

4.4 Lemma. *Let $N \subseteq M \subset M'$ be models of the countable ω-stable theory T. If N is finitely generated over M there is a finite canonical* **AT**-*regular resolution of N over M.*

Proof. We need only show there is a resolution as any shortest such will be canonical. Construct a sequence of submodels of $N = M[\overline{c}]$ according to the definition of a resolution. We can find such a sequence by Theorem 4.3. We need only show that the construction stops after finitely many steps. It suffices to show that for each i, $(\overline{c} \not\downarrow X_i; M_i)$, since such an infinite sequence would contradict the definition of $\kappa(T)$. So suppose for contradiction that $\overline{c} \downarrow_{M_i} X_i$. In particular, $\overline{c} \notin M_{i+1}$. Choose a strongly regular type q in $S(M_{i+1})$ realized by $b \in M_{i+1}[\overline{c}] - M_{i+1}$. By the maximality of X_i, we see $b \not\downarrow_M M_{i+1}$. But $\overline{c} \not\downarrow_{M_i} M_{i+1}$ so $\overline{c} \downarrow_{M_{i+1}} \overline{b}$ by the strong regularity of q. This contradicts the choice of $b \in M_{i+1}[\overline{c}]$.

Note that in forming a canonical resolution the choice of points or even of strongly regular types over M_i realized in X_i is not completely determined. But both the cardinality of X_i and the isomorphism type of M_{i+1} over M_i is fixed. Thus we justify the term 'canonical resolution'.

This uniqueness claim depends essentially upon the restriction in the definition of canonical resolution that k be minimal with $M_k \approx M[\overline{a}]$. For example, let T be the theory of an equivalence relation with infinitely many classes which are all infinite. If $M \models T$ and the equivalence class of a does not intersect M, then for any b which is equivalent to a, it is possible to choose $M[b] \subseteq M[a]$ which is isomorphic to but not equal to $M[a]$. In fact, we can choose arbitrarily long **AT**-resolutions of $M[a]$ over M.

4.5 Theorem. *Let $N \subseteq M \subset M'$ be models of a countable ω-stable theory T. If N is a finitely generated over M, there is finite sequence of models $M_0 = M, M_1, \ldots, M_k = N$ such that each M_i is determined up to isomorphism over M_{i-1}.*

Proof. Choose $\langle M_i : i \leq k \rangle$ inductively to satisfy the definition of an **AT**-resolution. Stop when some $M_i \approx N$. The uniqueness follows by induction from the following exercise.

4.6 Exercise. If X and Y are bases for $R(N, M)$ then $M[X] \approx M[Y]$. (Hint: Note that forking defines a $1-1$ correspondence between X and Y. Apply Corollary XII.1.15.)

4.7 Definition. Let $M_0 = M, M_1, \ldots, M_k = N$ be the canonical resolution of N over M. The *length* of this resolution is the sum of the dimensions of $R(M_{i+1}, M_i)$ for $i < k$.

Of course, both the integer k and the length of the resolution are invariants of the pair (M, N).

The following important reformulation of Lemma 4.3 provides a useful weakening of Theorem 1.5 for the class of **AT**-models of an ω-stable theory.

4.8 Exercise. If T is ω-stable, $M \subseteq N$, and $M, N \models T$ then $R(N, M) \triangleright_M N$.

4.9 Historical Notes. The results here refine the decompositions of models of ω-stable theories in [Lascar 1984] and [Makkai 1984].

Part D

The Number of Models

In the remainder of this book, we calculate the possible spectra for two classes. If T is a countable superstable theory we will give the possible functions $I(\aleph_\alpha, \mathbf{S})$. If T is a countable ω-stable theory we will give the possible functions $I(\aleph_\alpha, \mathbf{AT})$. A number of the theorems extend to more general classes of theories or to different classes of models. We have tried to build a framework which handles these more general cases. Thus, there are theorems and exercises referring to them. Some of the extensions we touch on are $I(T, \mathbf{AT}_\kappa)$ for any uncountable κ, uncountable T, and small countable superstable T.

The calculation proceeds by first classifying the theories and then computing the spectra in each class. We begin with the fact, proved in Section IX.6, that if T is not superstable then $I(\kappa, AT) = 2^{|\kappa|}$. (We gave the proof only for regular κ). Although we did not prove it here, it is shown in Chapter VII of [Shelah 1978], that $I(\kappa, \mathbf{S}) = 2^\kappa$. This justifies our assumption in the remainder of this book that T is superstable. Chapter XIV collects some of the main tools used in the computation. In Chapter XV we distinguish the bounded from the unbounded, or multidimensional, theories. We classify the spectra of bounded theories and compute a lower bound for the spectrum of an unbounded theory. Thereafter, we need only analyze unbounded theories. We also introduce in Chapter XV the notion of an eventually nonisolated type which is crucial for the study of countable models.

In Chapter XVI we introduce a major dividing line, the dimensional order property (DOP). We prove that if T has the dimensional order property then T has 2^κ $\mathbf{S}$-models in every power $\kappa \geq 2^{|T|}$. We also find a flaw in our classification of the classes of models to study. That is, we will prove that a theory T has the DOP for all of the classes K introduced earlier or for none of them. But there is another variant, the ENI-DOP which must be investigated to deal with countable models.

In Chapter XVII, we see that if T does not have the dimensional order property then every model can be decomposed into a tree of small models. If for each model the tree is well-founded, T is called 'shallow'; then we are able to assign invariants to the models in the manner suggested in Section

I.1. If not the theory is called 'deep' and again the theory has the maximal number of models.

Thus, we establish the 'main gap'. Either for all uncountable cardinals κ, $I(\kappa, K) = 2^\kappa$ or for all $\alpha > 0$, $I(\aleph_\alpha, K) \leq \beth_{\aleph_1}(\alpha)$. In Chapter XVIII we undertake a more detailed analysis of ω-stable theories. We establish both the Vaught and Morley conjectures for ω-stable T. That is, we show that if T is ω-stable then T has either countably many or $2^{\aleph_0}$ countable models and the spectrum function is increasing on uncountable cardinals. Shelah extends the second result to all theories in [Shelah 198?].

These computations of the spectrum functions have an interesting sidelight. It is by no means evident that the spectrum function of a theory T does not depend on the axioms for set theory. Indeed, for uncountable theories there are examples where it does so depend (cf. [Shelah 1978] Chapter IX). However, we will establish in this book equivalences (for countable theories) between various spectrum functions and certain 'syntactic' conditions which are clearly absolute.

We restrict our attention to countable first order theories. For the case of **S**-models this restriction is totally unnecessary. In order to extend the results to arbitrary models of an uncountable theory, one must generalize the notion of ω-stability. Shelah has done this (by generalizing Morley's definition via rank rather than the spectrum of stability) but we do not deal with this generalization here.

Throughout Part D, K denotes an acceptable class. Unless expressly asserted otherwise, we assume it admits stationary strongly regular types. This assumption provides the obvious obstruction to extending the results in Part D on countable ω-stable theories to countable superstable theories. Shelah surmounts this obstacle in [Shelah 198?].

Chapter XIV

The Construction of Many Nonisomorphic Models

We describe in this chapter the methods by which we later construct many nonisomorphic models of theories with certain specified properties. The basic technique is to construct for each graph G from a large family of graphs a model M_G in such a way that if M_G is isomorphic to M_H then G is isomorphic to H. The tool for this will be to encode the points of the graph and the edges between them by the sizes of certain indiscernible sets. In the first section of this chapter we review the construction of maximal sets of pairwise nonisomorphic graphs. In the second section we discuss how to identify the elements or, often, equivalence classes of elements that we want to use as the points of a graph. That is, we show how to construct models with the dimensions of types over certain base sets prescribed in advance. The methods of Section 2 are, in general, unable to prescribe dimensions below $\lambda(\mathbf{I})$. In Section 3, we make this limitation more precise and single out the exceptions to it.

1. Many Nonisomorphic Graphs

Many theorems showing that certain theories have many non-isomorphic models will be established by coding certain well known complicated classes, like the class of all graphs, into each theory satisfying a certain condition. In this section we justify this procedure by showing that the classes of graphs which we will interpret later do have many nonisomorphic models. We begin by establishing some nomenclature.

1.1 Definition. A *graph* is a set with a symmetric binary relation. A *directed graph* is a set with an asymmetric binary relation.

It is easy to construct 2^λ non-isomorphic directed graphs of power λ. For any $X \subseteq \lambda$, let G_X be the disjoint union of copies of the $\alpha \in X$ and use $<$ as the binary relation. It is somewhat more difficult to construct 2^λ

symmetric graphs of power λ, although the basic strategy is the same. To make it easier to identify the graph we impose one further condition.

1.2 Definition. A *triangle* in a graph G is a triple of points such that each pair is connected by an edge.

We will define another notion of triangle in Chapter XV. It will be clear from context which notion is meant.

1.3 Proposition. For every cardinal λ there are 2^λ non-isomorphic symmetric graphs which contain no triangles.

Proof. (Fig. 1). The key to the proof is to construct for each ordinal α a connected symmetric graph G_α which contains no triangles such that if $\alpha \neq \beta$ then $G_\alpha \not\approx G_\beta$. The universe of G_α will contain

$$\alpha \cup \{a_{\beta,\gamma} : \beta < \gamma < \alpha\} \cup \{b^0_{\beta,\gamma} : \beta < \gamma < \alpha\} \cup \{b^1_{\beta,\gamma} : \beta < \gamma < \alpha\}.$$

We will encode the usual ordering on α by means of the auxiliary elements.

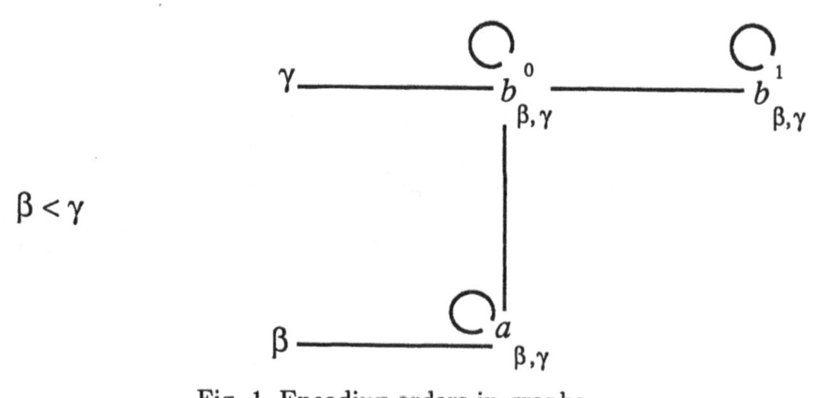

Fig. 1. Encoding orders in graphs

Let R be the symmetric closure of the following relation:

$$\begin{aligned}&\{\langle\beta, a_{\beta,\gamma}\rangle : \beta < \gamma < \alpha\} \cup \{\langle\gamma, b^0_{\beta,\gamma}\rangle : \beta < \gamma < \alpha\}\\ &\quad\cup \{\langle a_{\beta,\gamma}, b^0_{\beta,\gamma}\rangle : \beta < \gamma < \alpha\} \cup \{\langle a_{\beta,\gamma}, a_{\beta,\gamma}\rangle : \beta < \gamma < \alpha\}\\ &\quad\cup \{\langle b^1_{\beta,\gamma}, b^0_{\beta,\gamma}\rangle : \beta < \gamma < \alpha\} \cup \{\langle b^1_{\beta,\gamma}, b^1_{\beta,\gamma}\rangle : \beta < \gamma < \alpha\}.\end{aligned}$$

It isn't hard to verify that G_α is as required.

1.4 Exercise. Show that there are 2^λ nonisomorphic symmetric graphs of cardinality λ which contain no triangles but each point is connected to at least two others. (Hint: Put tails on the elements of the graph constructed for Proposition 1.3.)

1.5 Historical Notes. These results are all well-known. There is an elementary exposition in [Manaster 1972].

2. Models with Prescribed Dimensions

This section contains some of the most useful constructions in the entire book. We show how to construct models with specified dimensions for certain families of regular types. These constructions will be applied repeatedly to construct non-isomorphic models in the remainder of Part D.

The relation of orthogonality establishes a dichotomy on pairs of types. Roughly, if two types are not orthogonal, they have the same dimension; if they are orthogonal, their dimensions can vary arbitrarily. We devote this section to making this intuition precise.

We begin by showing that if the K-strongly regular type p is orthogonal to the K-strongly regular type q we can increase the dimension of p arbitrarily without increasing the dimension of q. Then we show that we can increase the dimension of a fixed q without increasing the dimension of any of a family of p's each orthogonal to q. In Theorem 2.4 we turn this argument on its head and show that we can fix the dimension of each type in a family X while increasing the dimensions of all types orthogonal to each $p \in X$. After some exercises illustrating these techniques, the remainder of the section considers sufficient conditions for the dimension of two types to be equal. Throughout this section we assume T is superstable.

2.1 Theorem. *Let T be superstable.*

i) *Suppose $p \perp q$ are K-strongly regular types which are each strongly based on subsets of $M \in K$. Then for any $\kappa \geq \max(|M|, \lambda_0(K))$ there is a model $N \in K$ with $\dim(p,N) = \dim(p,M)$ and $\dim(q,N) \geq \kappa$.*

ii) *Further, let X be a collection of K-strongly regular types each strongly based on a subset of M. Suppose $q \perp p$ for every $p \in X$. For each $\kappa \geq \max(|M|, \lambda_0(K))$ there is a model N with $|N| = \kappa$ such that $\dim(q,N) = \kappa$ and for each $p \in X$, $\dim(p,N) = \dim(p,M)$.*

Proof. (Fig. 2). i) Let E be an independent set of κ realizations of the nonforking extension of q to $S(M)$ and let N be K-prime over $M \cup E$. By XII.4.4, $\dim(p,N) = \dim(p,M) + \dim(p^M,N)$. But $\dim(p^M,N) = 0$ by Theorem X.4.6. Assertion ii) is easily shown by an increasing chain argument, repeatedly applying i).

2.2 Exercise. Prove Theorem 2.1 ii).

We now show how to combine these results to construct models with certain prescribed dimensions. The following definition codifies the situation of Theorem 2.1 ii).

2.3 Definition. i) If S is a family of stationary K-strongly regular types, the type q is *irrelevant* to S if $p \perp q$ for each $p \in S$.

Recall the following notation which is extremely useful to describe families of types.

ii) Suppose $p \in S(\overline{a} \cup A)$ and $t(\overline{b}; A) = t(\overline{a}; A)$. Then $p_{\overline{b}}$ denotes the image of p under an automorphism which maps $\overline{a}$ to $\overline{b}$ and fixes A.

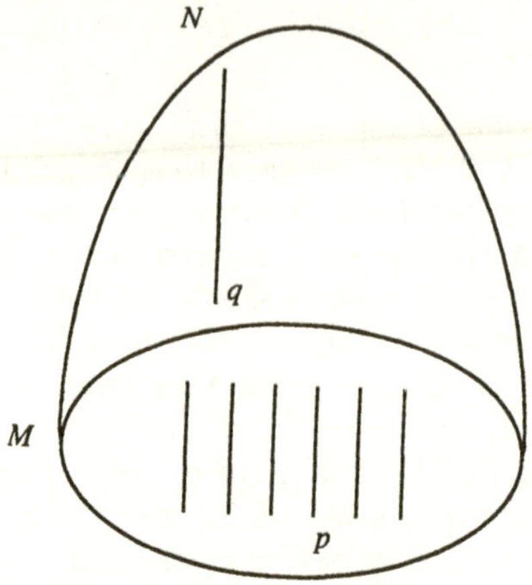

Fig. 2. Theorem XIV.2.1

It would be reasonable in i) to say q is orthogonal to S (and indeed Makkai in [Makkai 1984] does) but this seems to make the context bear too much burden as to which of three meanings of orthogonal is meant.

2.4 Theorem. *Let T be superstable. Let S be a family of pairwise orthogonal K-strongly regular stationary types over $A \subseteq M$. For any cardinal $\kappa \geq \max(\lambda_0(K), \kappa(T), |M|)$ there is a model $M_\kappa \supseteq M$ of T satisfying the following conditions.*

i) $|M_\kappa| = \kappa$.
ii) *If* $p \in S$, $\dim(p, M_\kappa) = \dim(p, M)$.
iii) *Suppose q is based on a subset of M_κ and q is irrelevant to S. Then* $\dim(q, M_\kappa)$ *is* κ.

Proof. Let $\langle q_i : i < \mu\rangle$ enumerate the K-strongly regular types which are strongly based on a subset of M and are irrelevant to S. Since T is superstable, $\mu \leq \kappa$. Define $\langle M_i : i < \mu\rangle$ by induction with $M_0 = M$. Take unions at limit ordinals and apply Theorem 2.1 ii) to choose M_{i+1} with $|M_{i+1}| = \kappa$, $\dim(q_i, M_{i+1}) = \kappa$, and for each $p \in S$, $\dim(p, M_{i+1}) = \dim(p, M_i)$. By induction, $\dim(p, M_i) = \dim(p, M)$ and by Theorem X.4.6 $\dim(p^{M_i}, M_{i+1}) = 0$. Let $N_0 = \bigcup_{i<\mu} M_i$. Construct N_i for $i < \omega$ by iterating this procedure. Then $M_\kappa = \bigcup_{i<\omega} N_i$ is the required model.

There are two uses of superstability in the preceding proof. The less important arises from the need to have $\mu \leq \kappa$. For this, one can assume that $\kappa \geq |M|^{<\kappa(T)}$ and $\kappa = \kappa^{<\kappa(T)}$. (The second of these requirements is needed for the iteration.) More essential for the application of the theorem is restriction to K-strongly regular types. The crucial use of regularity is to show (in the proof of Theorem 2.1) that

$$\dim(p, N) = \dim(p, M) + \dim(p^M, N).$$

2.5 Exercise. Let S be a family of stationary K-strongly regular types with each $p_{\overline{a}} \in S$ strongly based on $\overline{a} \in A$ and with $p_{\overline{a}} \perp t(A - \overline{a}; \overline{a})$.

Choose for each $p \in S$, a cardinal λ_p with $\lambda(\mathbf{I}) \leq \lambda_p \leq \kappa$. There is a model M_κ such that i) $\dim(p, M_\kappa) = \lambda_p$ if $p \in S$ and ii) $\dim(q, M_\kappa) = \kappa$ if q is irrelevant to S. (Hint: Let E_p be a set of λ_p independent realizations of p for each $p \in S$ and let $B = A \cup \bigcup_{p \in S} E_p$. Now if $p'_{\overline{a}}$ denotes the nonforking extension of $p_{\overline{a}}$ to $\overline{a} \cup E_{p_{\overline{a}}}$, $p'_{\overline{a}} \perp t(B - \overline{a}; \overline{a})$.)

The following exercise is easily proved by combining the techniques of Theorems 2.1 and 2.4.

2.6 Exercise. Let S be a family of stationary K-strongly regular types with each $p_{\overline{a}} \in S$ strongly based on $\overline{a} \in A$ and with $p_{\overline{a}} \perp t(A - \overline{a}; \overline{a})$. If λ_p, for each $p \in S$, denotes $\dim(p, M_p)$ where M_p is K-prime over the finite set on which p is strongly based and $\lambda_p \leq \kappa$, then there is a model M_κ such that i) $\dim(p, M_\kappa) = \lambda_p$ if $p \in S$ and ii) $\dim(q, M_\kappa) = \kappa$ if q is irrelevant to S.

Now we determine when two strongly regular types have the same dimension in a model N. There are several steps to this procedure. First we deal with an arbitrary pair of nonorthogonal K-strongly regular types over $M \in K$ with $M \subseteq N$. Then we restrict the types to be two copies of a type over an element $\overline{b}$. The relation between the dimensions of $p_{\overline{b}}$ and $p_{\overline{b}'}$ are seen to depend on properties of $\overline{b}$ and $\overline{b}'$. For this situation we deal first with the case that $\overline{b}$ and $\overline{b}'$ have the same strong type. Then we make the further assumption that the type of $\overline{b}$ is $\mathbf{I}$-isolated. With this assumption we no longer have to require the type to be stationary.

2.7 Theorem. *Suppose $p, q \in S(M)$, $M, N \in K$, $M \subseteq N$ and the pairs $(p, p_0), (q, q_0)$ are K-strongly regular. If $p \not\perp q$ then* $\dim(p, N) = \dim(q, N)$.

Proof. Let $E = \langle \overline{e}_\alpha : \alpha < \kappa \rangle$ be a basis for $p(N)$. Define by induction models N_α for $\alpha < \kappa$ with $N_0 = M$ and $N_{\alpha+1}$ a submodel of N which is K-prime over $N_\alpha \cup \overline{e}_\alpha$. By Exercise X.1.21, for each α, $N_\alpha \cap E = E_\alpha$. By Theorem XII.4.5 the nonforking extension of q to N_α is realized in $N_{\alpha+1}$. So $\dim(p, N) \leq \dim(q, N)$. Reversing the roles of p and q we finish.

This previous proof used in an essential way the hypothesis that p and q were types over K-models. If we relax this hypothesis we must weaken the conclusion somewhat. For cardinals κ μ, and λ, we write $\kappa = \mu \operatorname{mod}(\lambda)$ if $\kappa + \lambda = \mu + \lambda$.

2.8 Theorem. *If p, $q \in S(A)$ are K-strongly regular and $p \not\perp q$ then for any $N \in K$,* $\dim(p, N) = \dim(q, N) \ \operatorname{mod}(\lambda(\mathbf{I}))$.

Proof. Let $M \prec N$ be K-prime over A. By Theorem X.4.5 both $\dim(p, M)$ and $\dim(q, M)$ are less than or equal $\lambda(\mathbf{I})$. By Theorem XII.4.1 and Theorem XII.4.5 $\dim(p, N) = \dim(p, M) + \dim(p^M, N)$ and similarly for q. But $\dim(p^M, N) = \dim(q^M, N)$ by Theorem 2.7 so we finish.

We can remove the '$\operatorname{mod} \lambda(I)$' if we tighten our control on the types p and q.

2.9 Lemma. *Let $p \in S(A \cup \bar{b})$ be K-strongly regular and suppose $p \not\perp A$. Then for all $\bar{b}'$ realizing $stp(\bar{b}; A)$ and all $M \in K$ containing $A \cup \bar{b} \cup \bar{b}'$,* $\dim(p, M) = \dim(p_{\bar{b}'}, M)$.

Proof. (Fig. 3). Let p' denote $p_{\bar{b}'}$ and choose $N \prec M$ to be a K-prime model over $A \cup \bar{b} \cup \bar{b}'$. If $\bar{b} \downarrow_A \bar{b}'$ then $\{\bar{b}, \bar{b}'\}$ is a set of indiscernibles over A. Thus, N is K-prime over $A \cup \bar{b}' \cup \bar{b}$ (Note the change in order of $\bar{b}, \bar{b}'$.) and so $\dim(p, N) = \dim(p', N)$. By Theorem 2.7 and Theorem XII.4.4, we conclude $\dim(p, M) = \dim(p', M)$.

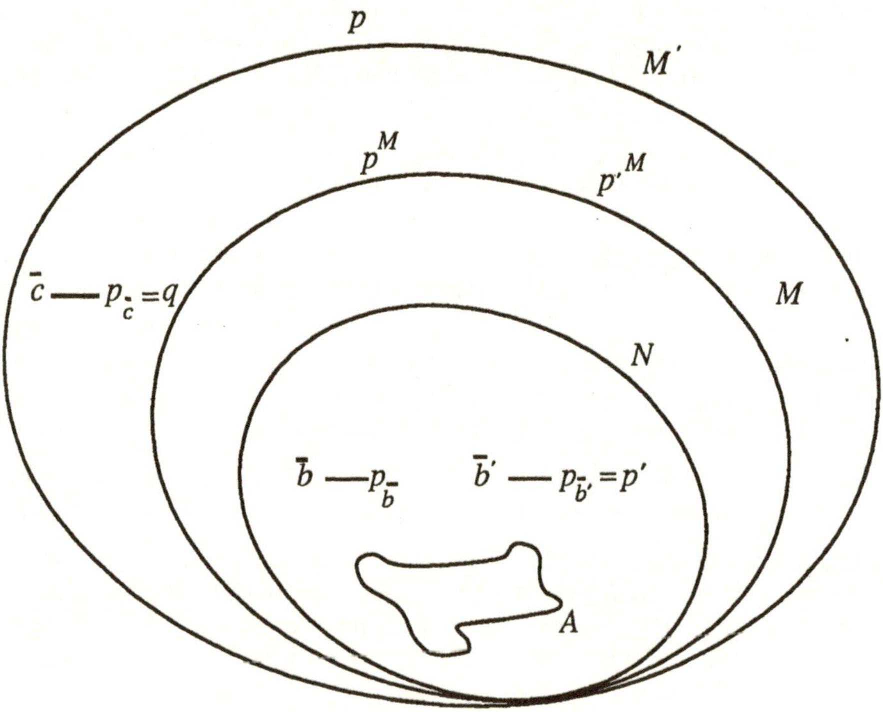

Fig. 3. Theorem XIV.2.9

Even if $(\bar{b} \not\downarrow \bar{b}'; A)$, we can reduce to the previous case as follows. Choose $\bar{c}$ realizing $stp(\bar{b}; A)$ with $\bar{c} \downarrow_A M$. Let $q = p_{\bar{c}}$ and choose M' which is K-prime over $M \cup \bar{c}$. Now let p^M and p'^M be the nonforking extensions to $S(M)$ of p and p' respectively. Since $p \not\perp A$, Theorem VI.2.22 implies $p \not\perp q$ and $p' \not\perp q$. By the transitivity of nonorthogonality on K-strongly regular types, $p^M \not\perp p'$. By Theorem 2.7 $\dim(p^M, M') = \dim(p'^M, M')$ and by the first case $\dim(p, M') = \dim(q, M') = \dim(p', M')$. By Theorem XII.4.4, we have

$$\dim(p, M') = \dim(p, M) + \dim(p^M, M')$$

and

$$\dim(p', M') = \dim(p', M) + \dim(p'^M, M').$$

By Exercise X.1.21, each realization, $\overline{a}$, of p^M in M' satisfies $\overline{a} \not\downarrow_M \overline{c}$. Thus, $\dim(p^M, M') < \kappa(T) = \omega$ so we can subtract to conclude $\dim(p, M) = \dim(p', M)$.

In this situation we can strengthen the criterion for nonorthogonality of Theorem VI.2.22.

2.10 Lemma. *Let $p \in S(A \cup \overline{b})$ be K-strongly regular and suppose $p \not\dashv A$. For any $\overline{b}'$ realizing $stp(\overline{b}; A)$, $p_{\overline{b}} \not\perp p_{\overline{b}'}$.*

Proof. By Lemma 2.9, for all $M \supseteq A$, $\dim(p_{\overline{b}}, M) = \dim(p_{\overline{b}'}, M)$. By the contrapositive of Theorem 2.1, we have the result.

If we add the requirement that $t(\overline{b}; A)$ is **I**-isolated then we can weaken the requirement in Lemma 2.10 that $\overline{b}'$ realize the same strong type as $\overline{b}$ to the requirement that it realize the same type. We obtain this generalization in Theorem 2.15. The following facts are needed in the argument.

2.11 Exercise. Show that if $t(\overline{a}; A)$ has finite multiplicity there is an E in $FE(A)$ such that $t(\overline{a}; A) \cup E(\overline{x}; \overline{a}) \vdash stp(\overline{a}; A)$.

2.12 Lemma. *Let T be stable and suppose $\overline{a}, \overline{b} \in M \models T$. Assume $t(\overline{a}\frown\overline{c}; \emptyset)$ has finite multiplicity and $t(\overline{c}; \overline{a})$ is **AT**-isolated. If $stp(\overline{a}; \emptyset) = stp(\overline{b}; \emptyset)$ then there exists $\overline{d} \in M$ with $stp(\overline{a}\frown\overline{c}; \emptyset) = stp(\overline{b}\frown\overline{d}; \emptyset)$.*

Proof. Choose $E \in FE(\emptyset)$ such that $t(\overline{a}\frown\overline{c}; \emptyset) \cup E(\overline{x}, \overline{a}\frown\overline{c}) \vdash stp(\overline{a}\frown\overline{c}; \emptyset)$. Let $\phi(\overline{y}; \overline{a}) \vdash t(\overline{c}; \overline{a})$. Since $stp(\overline{a}; \emptyset) = stp(\overline{b}; \emptyset)$, there is an $\overline{e} \in \mathcal{M}$ with $stp(\overline{a}\frown\overline{c}; \emptyset) = stp(\overline{b}\frown\overline{e}; \emptyset)$. Thus, $(\exists\overline{y})[\phi(\overline{y}; \overline{b}) \wedge E(\overline{a}\frown\overline{c}, \overline{b}\frown\overline{y})]$ is satisfiable. Choose $\overline{d}$ in M to witness this formula.

The following Lemma does not hold for an arbitrary acceptable class and I have not found an abstract formulation of a sufficient condition. Thus, I remark that it holds in the cases we are most interested in.

2.13 Lemma. *Let T be superstable and $K =$ **S** or T be ω-stable and $K =$ **AT**. Suppose p is **I**-isolated over A with $|A| < \lambda(\mathbf{I})$. For any $\overline{b}$ realizing p and any $M \in K$ with $A \subseteq M$, $stp(\overline{b}; A)$ is realized in M.*

Proof. This is obvious for the class of **S**-models. When T is a countable ω-stable theory we can use the fact that every type has finite multiplicity. For, if $p \in S(A)$ is an isolated type with multiplicity n the type of n points which realize p but realize different strong types over A is also isolated.

2.14 Exercise. Show that this result fails for an arbitrary acceptable class by considering the theory REF_ω and letting $K =$ **AT**.

2.15 Theorem. *Let T be superstable and $K =$ **S** or T be ω-stable and $K =$ **AT**. Let $A \subseteq M \in K$ and suppose $\overline{b}, \overline{b}'$ realize the **I**-isolated type, q, over A. Let $p \in S(A \cup \overline{b})$ be K-strongly regular, let p' denote $p_{\overline{b}'}$, and suppose $p \not\perp p'$. For any $M \supseteq A \cup \{\overline{b}, \overline{b}'\}$ with $M \in K$, $\dim(p, M) = \dim(p', M)$.*

Proof. Let $N \prec M$ be K-prime over $A \cup \overline{b}$ and choose by Lemma 2.13 $\overline{c} \in N$ realizing $stp(\overline{b}'; A)$. Let r denote $p_{\overline{c}}$. By Lemma 2.9 and Lemma 2.10, $\dim(p', M) = \dim(r, M)$ and $p' \not\perp r$. As N is also K-prime over $A \cup \overline{c}$, $\dim(p, N) = \dim(r, N)$. By the transitivity of nonorthogonality on regular types, $p \not\perp r$. Thus, $\dim(p^N, M) = \dim(r^N, M)$. By Theorem XII.4.4

$$\dim(p, M) = \dim(p, N) + \dim(p^N, M)$$

and

$$\dim(r, M) = \dim(r, N) + \dim(r^N, M).$$

So $\dim(p', M) = \dim(r, M) = \dim(p, M)$ as required.

2.16 Historical Notes. These kinds of construction originate in [Shelah 1978]. However, the emphasis on types rather than indiscernible sets is due to Lascar. Thus, the general outline of the first half of this section stems from Lascar via [Makkai 1984]. Theorem 2.7 through Theorem 2.13 comes fairly explicitly from [Bouscaren 1983] and [Bouscaren & Lascar 1983].

3. Tractable Types

We combine here the results of Section X.4 on the dimension of arbitrary indiscernible sets in K-prime models with the more refined results for strongly regular types described in Section 2 of this chapter. If a type is orthogonal to the empty set we are able to determine almost at will the dimension of various copies of the type. In this section we make precise 'almost at will'; in the next chapter we begin to apply these constructions. We are trying to develop a general framework to study both **S**-models and **AT**-models. In the second case, nonisolated types play a special role because they can have finite dimension.

3.1 Definition. We say the K-strongly regular stationary type p over A is (μ, K)-*tractable* if for any C with $t(C; A) \perp p$, if M is K-prime over $A \cup C$ then $\dim(p, M) < \mu + |A|$.

This definition is introduced primarily so that the following two situations can be treated uniformly in constructions. Recall from Definition VI.2.1 that a type p is unbounded if it orthogonal to the empty set.

3.2 Lemma. *Let $p \in S(B)$ be stationary. Suppose further that $B \subseteq A$, and $p \dashv \emptyset$. Regarded as a type over A,*

i) *p is $(\lambda(\mathbf{I})^+, K)$-tractable.*
ii) *If p is K-strongly regular and not $\mathbf{I}$-isolated then p is $(\lambda(\mathbf{I}), K)$-tractable.*

Proof. i) Let M be K-prime over $A \cup C$ and E a maximal independent set of realizations in M of p. By V.1.19 there is an $E_0 \subseteq E$ with $|E_0| < \kappa(T) + |A|$ such that, letting E' denote $E - E_0$, $E' \downarrow_B A$ and E' is a set of indiscernibles over A. Now if $\bar{e} \in E'$, $t(\bar{e}; A) \perp t(C; A)$ so, applying Theorem VI.1.19, $t(E'; A) \perp t(C; A)$. In particular, $E' \downarrow_A C$ and so E' is a set of indiscernibles over $A \cup C$. By Theorem X.4.5, $|E'| \leq \lambda(\mathbf{I})$. So $|E| \leq \lambda(\mathbf{I}) + \kappa(T) + |A|$. Since $\kappa(T) \leq \lambda(\mathbf{I})$, we have i).

ii) Note that $\bar{e} \in E'$ implies $\bar{e} \downarrow_B A \cup C$. Thus, by the open mapping theorem, $t(\bar{e}; A \cup C)$ is not **I**-isolated and so is not realized in M. That is, $E' = \emptyset$. Since $\lambda(\mathbf{I}) \geq \kappa(T)$, $|E| = |E_0| < \lambda(\mathbf{I})$ as required.

The following exercise is an immediate application of the lemma. Its solution is contained in the proof of the more complicated application in Theorem 3.5.

3.3 Exercise. Suppose $p_i \in S(A_i)$ is $\dashv \emptyset$ (and not **I**-isolated), $\{A_i : i < \kappa\}$ is an independent sequence of sets with each $|A_i| < \lambda(\mathbf{I})$, and M is K-prime over $A = \cup A_i$. Then for each i, $\dim(p_i, M) < \lambda(\mathbf{I})^+$ ($\dim(p_i, M) < \lambda(\mathbf{I})$.) (Hint: Apply Lemma 3.2 to each p_i taking A_i for both A and B and $\bigcup_{j \neq i} A_i$ for C).

3.4 Exercise. Show that if T is a countable ω-stable theory then $p \in S(A)$ is $(\aleph_0, AT)$-tractable if and only if p is nonprincipal.

We will repeatedly appeal to the construction described in the following theorem.

3.5 Theorem. *Suppose $\kappa > max(\lambda_0(\mathbf{I}), \lambda(\mathbf{I}))$. Fix a cardinal $\mu \geq \kappa(T)$. Let A_i for $i < \lambda$ be an independent sequence of sets with $|A_i| < \lambda(\mathbf{I})$. Let S_i be a set of (μ, K)-tractable types over A_i which are each orthogonal to $\emptyset$. Then there is a model N such that:*

i) $|N| = \kappa$.
ii) $\dim(p, N) < \mu$ *if* $p \in S = \bigcup_{i<\lambda} S_i$.
iii) $\dim(q, N) = \kappa$ *if q is over N and q is irrelevant to S.*

Proof. Let M be K-prime over $A = \bigcup_{i<\lambda} A_i$. If A^i denotes $A - A_i$, we have $A^i \downarrow_\emptyset A_i$. This implies by Theorem VI.2.21 that $p \perp t(A^i; A_i)$ for each i and each $p \in S_i$. Thus, by the definition of tractable, for each $p \in S$, $\dim(p, M) < \mu$. Now extend M to N to satisfy i) and iii) while preserving ii) by Theorem 2.4.

The following exercise indicates one way the sequence of independent sets A_i in the hypothesis of Theorem 3.5 can be found when applying that theorem.

3.6 Exercise. Let $p \in S(B)$ and $p \dashv \emptyset$. Suppose $t(C; B) \perp p$, $p \perp q$, $q \in S(C)$, and $q \dashv \emptyset$. Show that if E is an independent set of realizations of q then $p \perp t(E \cup C; B)$ and thus if M is K-prime over $B \cup C \cup E$, $\dim(p, M) \leq \lambda(\mathbf{I})$.

3.7 Historical Notes. Of course these notions originate with Shelah [Shelah 1978]. But this sort of construction is found more explicitly in [Makkai 1984] and [Bouscaren & Lascar 1983].

Chapter XV
The Width of a Theory

Our discussion of independence has focused on individual types. We have defined the dimension of a type p in a model M and if the type is regular or even has weight one then this dimension is always well defined. We have seen, however, that we may need to consider the dimension of more than one type in order to specify a model. In Section 1 of this chapter we define the width of a theory so that when this width is a cardinal number, it is the number of types whose individual dimensions must be specified to determine a model. Unfortunately, the situation is not always that simple.

There are three important cases. If all types have the same dimension, we call the theory unidimensional. Unidimensional theories are a natural generalization of $\aleph_1$-categorical theories. If there is a cardinal $\delta(T)$ such that we must specify the dimension of $\delta(T)$ types to determine each model, we call the theory bounded. If no such $\delta(T)$ exists, we call the theory unbounded. Section 2 begins the study of unbounded theories. We prove there lower bounds on the number of models for an unbounded theory. The bounded case is the comparatively straightforward generalization of vector space theory to allow several dimensions. In Section 3 of this chapter we derive a number of properties of the spectrum of a bounded theory.

The unbounded case is much more complicated. It is no longer possible to define a single unordered set as a basis for a model. Rather, there is a skeleton which is, roughly speaking, partially ordered by dependence. Chapter XVI is devoted to a detailed analysis of such skeletons.

We consider in Section 4 an extension of the notion of homogeneity to obtain some more detailed information about countable models of bounded theories. In particular, we solve Vaught's conjecture for ω-stable bounded theories.

1. Classifying Theories By Width

We begin this section by formally defining the width of theory. In some respects the arguments here just reformulate those in Chapter XIII.2.

1.1 Notation. Let $\mathcal{R}(A)$ denote the set of non-algebraic regular types in $S(A)$.

Note that in Definition XIII.1, $R(M, A)$ referred to a set of points; here $\mathcal{R}(A)$ is a set of types.

1.2 Definition. Let X be a set of regular types. For $X_0 \subseteq X$ and $p \in X$ we say p *depends* on X_0 if for some $q \in X_0$, $q \mapsto_K p$. $\delta(X)$ is the cardinality of a maximal independent set of types in X under this dependence relation. We will write $\delta(A)$ for $\delta(\mathcal{R}(A))$.

The treatment of this notion as a dependence relation is somewhat artificial. For most purposes, we could just describe the $\delta(X)$ defined here as the maximal number of pairwise orthogonal types in X.

We could just as easily define p to depend on X_0 if p is realized when all the members of X_0 are realized because the restriction of $\mapsto_S$ to K-strongly regular types is totally trivial. That is, whenever a pair of strongly regular types compel a third, one of them does. This observation makes the next theorem easy.

1.3 Theorem. *For any A, the cardinal $\delta(\mathcal{R}(A))$ is well defined.*

1.4 Exercise. For any set $M \in S$, $\delta(M)$ is the supremum of the cardinals κ such that there exist regular types $\{p_i \in S(M) : i < \kappa\}$ such that if $i \neq j$ then $p_i \perp p_j$.

1.5 Definition. The *width* of a superstable theory T, which we denote by $\delta(T)$, is $\sup\{\delta(M) : M \in \mathbf{S}\}$ if such a supremum exists; otherwise $\delta(T) = \infty$.

i) If $\delta(T) = 1$, T is called *unidimensional.*
ii) If $1 \leq \delta(T) < \infty$, T is called *bounded.*
iii) If $\delta(T) = \infty$, T is called *unbounded.*

We differ here from Shelah's terminology. He uses *multidimensional* to refer to what we call unbounded and *non-multidimensional* for our bounded.

Recall that throughout Part D we assume the class K admits stationary regular types. We defined the width of T in terms of the acceptable class S. Ostensibly, we could have defined a K-width of T $\delta_K(T)$ for each acceptable class K. In fact, for superstable T all such $\delta_K(T)$ are equal. To see this, choose $M \in S$. Extend M to an $M' \in K$ and a basis for $\mathcal{R}(M)$ to a set of pairwise orthogonal regular types in $S(M')$. By Theorem XIII.3.4 $\delta_K(M') \geq \delta_S(M') \geq \delta_S(M)$. Thus, $\delta_K(T) \geq \delta_S(T)$. But the converse is even easier noting that every K-strongly regular type is regular.

In the remainder of the book we concentrate on superstable theories. Our major aim is the calculation of the spectrum of each theory and we know T has 2^λ nonisomorphic models of power λ if T is not superstable so we need not investigate unstable and strictly stable theories. There are, however, important questions about the width of strictly stable theories ([Shelah 1978] V.4, [Buechler unk], [Hrushovski 1986]).

1.6 Exercise. Show that if every type has U-rank at most 1 then T is bounded.

The following theorem describes the fundamental properties of bounded theories.

1.7 Theorem. *The following are equivalent for a superstable theory T.*

i) $\delta(T) \leq 2^{|T|}$.
ii) *T is bounded.*
iii) *For every type p, $p \not\dashv \emptyset$.*
iv) *For every regular type p, $p \not\dashv \emptyset$.*

If T is ω-stable we can add a fifth equivalent.

v) *For every* **AT***-strongly regular type p, $p \not\dashv \emptyset$.*

Proof. Clearly i) implies ii). We show that ii) implies iii) and then that iii) implies i). Suppose iii) fails. Then, there is a type $p \dashv \emptyset$ based on a finite set $\overline{a}$ which is not (Exercise VI.2.4) in the algebraic closure of $\emptyset$. For any cardinal κ, let $\langle \overline{a}_i : i < \kappa \rangle$ be a sequence of independent elements realizing $stp(\overline{a}; \emptyset)$. Then if p_i is the image of p under an automorphism taking $\overline{a}$ to $\overline{a}_i$, by Theorem VI.2.22 the p_i are pairwise orthogonal and we contradict ii). Thus, ii) implies iii). To show iii) implies i), note that if there are κ pairwise orthogonal types with $\kappa > 2^{|T|}$, we can assume without loss that each is based on a finite set $\overline{a}_i$ and that all the $\overline{a}_i$ have the same length. Then by Theorem V.1.28, we can assume that the $\overline{a}_i$ are independent. Finally, since there are only $2^{|T|}$ strong types over a finite set, two of the types must be conjugate by a strong automorphism (Definition IV.3.9) mapping their base sets from one to the other. By Theorem VI.2.22 this contradicts iii) and yields the theorem. The equivalence of iii) and iv) is easy using Theorem XIII.2.15.

Now v) follows easily by decomposing an arbitrary type in an ω-stable theory as a product of **AT**-strongly regular types. (Apply Theorem XIII.3.6 (with $K =$ **AT**), Exercise XIII.2.14, and Corollary XIII.2.24.)

The following theorem shows that if T is unbounded this fact is witnessed by a particular type. The next definition picks out such a witness and provides a useful notation for discussing its conjugates.

1.8 Definition. If the regular type p is orthogonal to the empty set, we say p is an *unbounded type.* We say the unbounded type p is *based on the type* r if p is strongly based on the set A and $t(A; \emptyset) = r$.

Although for most of our applications T will be superstable and A a finite set this is not essential to the definition.

Examination of the argument for Theorem 1.7 shows that we have in fact proved.

1.9 Theorem. *If T is a bounded theory and $stp(\overline{a};\emptyset) = stp(\overline{b};\emptyset)$ then for each $p = p_{\overline{a}} \in S(\overline{a})$, $p_{\overline{a}} \not\perp p_{\overline{b}}$.*

The last theorem restates Theorem VI.2.22 in this important context.

1.10 Corollary. *If T is superstable and bounded then $\delta(T)$ is at most the number of strong types over $\emptyset$ times the number of types over a finite set. In particular, if T is ω-stable and bounded then $\delta(T) \leq \aleph_0$.*

It follows from [Cherlin, Harrington, & Lachlan 1985] that if T is $\aleph_0$-categorical, bounded, and ω-stable then $\delta(T)$ is finite.

If T is bounded the computation of $\delta(T)$ does not require the taking of suprema.

1.11 Theorem. *If T is bounded and admits K-strongly regular types then for all $M \in K$, $\delta(M) = \delta(T)$.*

Proof. If possible, choose $M \in K$ with $\delta(M) < \delta(T)$. Imbed M in N with $\delta(N) > \delta(M)$. Let X be a basis for $\mathcal{R}(M)$. Then X', the collection of nonforking extensions to N of the types in X, is an independent set of regular types over N but it is not maximal such. Let $q \in S(N)$ be regular and orthogonal to each member of X' and hence to each member of X. By Theorem 1.7 ii), $q \not\dashv \emptyset$ so $q \not\dashv M$. Therefore by Corollary XIII.3.4, $q \not\perp r$ for some regular $r \in S(M)$. Since X is a basis for $\mathcal{R}(M)$ this implies q is not orthogonal to some member of X. This contradiction yields the result.

1.12 Exercise. Let T be the theory of seven disjoint, infinite, unary predicates. Find a basis for the strongly regular types over the prime model of T.

1.13 Exercise. Let T be the theory of infinitely many disjoint, infinite, unary predicates. Find a basis for the strongly regular types over the prime model of T.

1.14 Historical Notes. The division of theories into multidimensional and non-multidimensional ones was made by Shelah in [Shelah 1978]. His definition is more complicated and applies to stable theories. The definition here is due to Lascar and agrees with Shelah's on superstable theories. The preparation of this chapter also depended heavily on the surveys by Pillay [Pillay 1983f] and Saffe [Saffe 1983].

2. Unbounded Theories

In this section we deal with unbounded or multidimensional superstable theories. At this time we will only provide lower bounds on the number of models in various classes of unbounded theories. In the next section we will determine the spectra of bounded theories. In later chapters we will return to a more detailed discussion of unbounded theories.

The simplest example of an unbounded theory is the theory of a single equivalence relation with infinitely many infinite classes. This theory is manifestly $\aleph_0$-categorical. There are $\aleph_0$ models of power $\aleph_1$. Each one is determined by the number of blocks of size $\aleph_1$ and the number of blocks of size $\aleph_0$. The first number can be any cardinal between 1 and $\aleph_1$; the second any cardinal between 0 and $\aleph_1$. By iterating this kind of argument one can show the following exercise.

2.1 Exercise. Let T be the theory of a single equivalence relation with infinitely many infinite classes. Show $I(\aleph_\alpha, \mathbf{AT}) = |\alpha + \omega|^{|\alpha+1|}$.

2.2 Exercise. Let T be the theory of an equivalence relation with infinitely many infinite classes each of which is a model of $\mathrm{Th}(Z, S)$. Show $I(\aleph_\alpha, \mathbf{AT}) = |\alpha + \omega|^{|\alpha+\omega|}$.

2.3 Exercise. Show that in these examples $I^*(\aleph_\alpha, \mathbf{AT}) = I(\aleph_\alpha, \mathbf{AT})$.

The calculation of the spectrum function given in this section specializes to a number of different cases. On first reading K should be taken to be $\mathbf{S}$ and we are just computing the number of strongly $\aleph_0$-saturated models of a countable superstable theory. To strengthen the conclusion take K to be your favorite acceptable class, e.g., the $\mathbf{SET}_{\aleph_3}$-saturated models of an uncountable theory. The argument also applies to give the number of models of an ω-stable theory but since we are proving lower bounds this reading provides no new information.

We strike here a theme of continuing importance. The number of models in a class K which have cardinality $\lambda = \aleph_\alpha$ is often computed not as function of λ but, since it depends on the number of cardinals less than λ, as a function of α. Recall that $\lambda_0(\mathbf{I})$ tells us the cardinality of the smallest $\mathbf{I}$-saturated model. Thus the calculation of the number of models in a class K depends also on β when $\lambda_0(\mathbf{I}) = \aleph_\beta$. When $K = \mathbf{AT}$, $\beta = 0$.

In the first major result of this section we show that the number of K-models with cardinality $\aleph_\alpha$ is at least as large as the number of functions from the set of cardinals between $\lambda_0(\mathbf{I})$ and $\aleph_\alpha$ to the set of cardinals less than $\aleph_\alpha$. Roughly speaking, for each dimension between $\aleph_\beta$ and $\aleph_\alpha$ we can fix the number of copies of a given type which have that dimension as any cardinal $< \aleph_\alpha$.

2.4 Theorem. *Suppose T is a countable superstable unbounded theory and $\lambda_0(\mathbf{I}) = \aleph_\beta$. Then $I(\aleph_\alpha, K) \geq |\alpha + \omega|^{|\alpha-\beta+1|}$.*

The proof will require several steps. We begin by fixing some notation. By Theorem 1.7, a theory is unbounded just if it contains a regular type which is orthogonal to $\emptyset$. Since T is superstable we can assume that this type is based on a finite set.

2.5 Notation. Let $p = p_{\overline{a}}$ be a K-strongly regular type which is strongly based on the finite sequence $\overline{a}$. Let q denote $stp(\overline{a};\emptyset)$. For any $M \in K$, let $D_{p,q}(M) = D(M) = \{p_{\overline{b}}: \models q(\overline{b}) \text{ and } \overline{b} \in M\}$

By Theorem 1.3, we know that $\delta(D(M))$, the maximal number of pairwise orthogonal copies of $p_{\overline{a}}$ based on M, is an invariant of M. That is, $\delta(D(M))$ does not depend on the choice of basis for $D(M)$. For the proof of Theorem 2.4 we refine this invariant somewhat.

2.6 Definition. For any $M \in K$ with $|M| = \aleph_\alpha$, any $\gamma < |\alpha - \beta + 1|$, and any basis J for $D(M)$, let $\delta^J_M(\gamma) = \delta(\gamma) = |\{p \in J : \dim(p, M) = \aleph_{\beta+\gamma}\}|$.

Ostensibly, the function δ depends on p, q, M, and J. The dependence on p and q is real. We simply keep these parameters constant throughout the discussion and suppress them for ease of reading. The next lemma shows δ does not really depend on J.

Note that δ maps the set of cardinals between $\aleph_\beta$ and $\aleph_\alpha$ into the set of cardinals less than $\aleph_\alpha$. Thus it can be viewed as a member of $|\alpha+\omega|^{|\alpha-\beta+1|}$. Recall that we do not distinguish in notation between the set of functions from one cardinal to another and the cardinality of that set of functions.

2.7 Lemma. *Let $|M| = \aleph_\alpha$ and suppose $M \in K$. For any I and J which are bases for $D(M)$, and for each $\gamma < |\alpha + \omega|$, $\delta^I_M(\gamma) = \delta^J_M(\gamma)$.*

Proof. (Fig. 1). There is a $1 - 1$ correspondence between I and J which assigns to each $p \in I$ the unique $q \in J$ such that $p \not\perp q$. By Theorem XIV.2.8, $\dim(p, M) = \dim(q, M) \bmod(\lambda(\mathbf{I}))$ and the result follows.

Thus we can simplify our notation by writing δ_M for δ^J_M. Now we return to the proof of Theorem 2.4.

Proof of Theorem 2.4. Let C be the set of cardinals less than or equal to $\aleph_\alpha$. Observe that $|C| = |\omega + \alpha|$. For each $f \in C^{|\alpha-\beta+1|}$ such that $f(\alpha - \beta) = \aleph_\alpha$ we will define a model M_f such that $\delta_{M_f} = f$. First, choose for each i with $\aleph_\beta \leq \aleph_i \leq \aleph_\alpha$ a set E_i of sequences realizing q such that $|E_i| = f(i)$ and $E = \bigcup_{i<\aleph_\alpha} E_i$ is an independent set. Then for each $\overline{e} \in E_i$ choose an independent set $A_{\overline{e}}$ of $\aleph_i$ realizations of $p_{\overline{e}}$. Let $A = \bigcup_{\overline{e}\in E} A_{\overline{e}}$. Let $p'_{\overline{e}}$ be the nonforking extension of $p_{\overline{e}}$ to $\overline{e} \cup A_{\overline{e}}$. Let $S = \{p'_{\overline{e}} : \overline{e} \in E\}$. Since each $p \in S$ is orthogonal to $\emptyset$, p is $(\lambda(\mathbf{I})^+, K)$-tractable by Lemma XIV.3.2. As $\{A_{\overline{e}} : \overline{e} \in E\}$ is an independent set, by Theorem XIV.3.5 there is a model N such that for each $\overline{e} \in E_i$, $\dim(p'_{\overline{e}}, N) \leq \lambda(\mathbf{I}) + \aleph_i$ and $\dim(q, N) = \aleph_\alpha$ if q is irrelevant to S. Since $\dim(p_{\overline{e}}, N) \geq \aleph_\beta \geq \lambda(\mathbf{I})$, $\dim(p_{\overline{e}}, N) = |A_{\overline{e}}| + \dim(p'_{\overline{e}}, N) = \aleph_i$. Extend $\{p_{\overline{e}} : \overline{e} \in E\}$ to a basis I for $D(N)$. Note that $\delta^I_N = f$ and the theorem follows.

The following exercises are just computations demonstrating the meaning of the theorem.

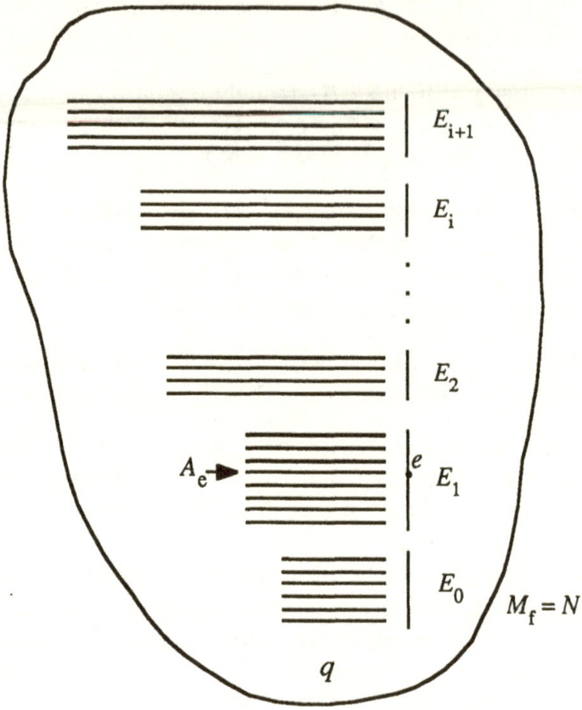

Fig. 1. Lemma XV.2.7: $|E_i| = f(i)$; $e \in E_i$ implies $A_e = \aleph_1$

2.8 Exercise. Show that if T is a countable ω-stable unbounded theory, then $I(\aleph_\alpha, \mathbf{AT}) \geq |\alpha + \omega|^{|\alpha+1|}$ if $\aleph_\alpha > \aleph_0$. In particular, T has infinitely many models in every uncountable power.

2.9 Exercise. Show that if T is a countable superstable unbounded theory, then $I(\aleph_\alpha, S) \geq |\alpha + \omega|^{|\alpha+1|}$ if $\aleph_\alpha \geq 2^{\aleph_0}$.

The next exercise outlines the means to extend the last sentence of Exercise 2.8 to the situation of Exercise 2.9.

2.10 Exercise (Saffe). Show that every countable superstable unbounded theory has infinitely many models in every uncountable power. Hint:

1) Note that T is small; otherwise it has $\geq 2^{\aleph_0}$ models in each power.
2) Thus T has prime models over countable sets of indiscernibles by Corollary V.1.26.
3) Let M be the prime model of T and choose $p \in S(M)$ and $\bar{e} \in M$ such that for some ϕ over $\bar{e}$, (p, ϕ) is $\mathbf{AT}$-strongly regular, $p \dashv \emptyset$, and p does not fork over $\bar{e}$. Let $q = t(\bar{e}; \emptyset)$. (Note that one can not guarantee that $p|\bar{e}$ is stationary.)
4) Let $E = \{\bar{e}_i : i < \omega\}$ be a strongly independent set of realization of q. Let N be prime over E and let p_i be a nonforking extension of the conjugate of p obtained by mapping $\bar{e}$ to $\bar{e}_i$. Thus the p_i are pairwise orthogonal.

5) Construct for each $n < \omega$ a model of T with cardinality $\aleph_\gamma$ such that exactly n nonorthogonality classes of the p_i are realized only countably many times. This construction is done by iterating the following observation. If N_1 is **L**-constructible over $N \cup \overline{a}$ where $\overline{a}$ realizes p_i then for $i \neq j$, $p_j|\overline{e}_j$ is omitted in N_1. (This is a little subtle since $p_j|\overline{e}_j$ is not stationary.)

It is not entirely clear how to extend this argument to get the full force of the Exercise 2.8. Saffe suggests that one can construct elementary submodels of the distinct **S**-models constructed in the last exercise so that the **S**-models are **S**-prime over the submodels. Then the uniqueness of prime models will guarantee the submodels are distinct.

It is tempting to think that an argument similar to that for Theorem 2.4 would show that if a countable ω-stable theory has a nonisolated type which is unbounded then the lower bound in Exercise 2.8 could be improved to $I(\aleph_\alpha, \mathbf{AT}) \geq |\alpha + \omega|^{|\alpha+\omega|}$. The difficulty is that if we assign to each of a family $\langle p_i : i < \omega \rangle$ of orthogonal copies of a nonisolated type a finite dimension (by saying $p_i \in S(\overline{a}_i)$ is realized by a set X_i of independent elements and taking the prime model M over all the $\overline{a}_i$ and X_i) then we know $\dim(p_i, M)$ is finite but we can not control its exact value. This problem is the fundamental difficulty of the Vaught conjecture (for ω-stable T).

The preceding argument shows in particular that an unbounded theory T has at least $2^{|\alpha|}$ models in power $\aleph_\alpha$. We will see later that this is a very weak lower bound. We show now that with one further requirement we can force T to have 2^λ models of power λ.

We now consider one consequence of assuming that forking is trivial on the realizations of a type p. We show here, by methods similar to the proof of the last theorem, that the existence of nontrivial types of a certain kind implies there are many models. In Section XVI.2 we will develop many useful properties of trivial types and connect this notion with the dimensional order property.

2.11 Definition. The stationary type $p \in S(A)$ is *trivial* if the dependence relation of forking is trivial on $p'(\mathcal{M})$ for any nonforking extension p' of p. That is, if $\overline{a}$, $\overline{b}$, $\overline{c}$ realize p' and are pairwise independent then they are independent. If $\overline{a}, \overline{b}, \overline{c}$ are pairwise independent but not independent we say they form a *triangle*.

The remainder of this section is devoted to the proof of the following theorem.

2.12 Theorem. *Let K be an acceptable class and p an unbounded (λ, K)-tractable type. Suppose p is based on $\overline{d}$ and $\overline{b} \rhd_\phi \overline{d}$ where $\overline{b}$ realizes a nontrivial, weight one type r. Then $I(\lambda, K) = 2^\lambda$ for every $\lambda \geq \lambda_0(\mathbf{I})$.*

The following special case includes the essential ideas shorn of the extra parameters.

Let T be ω-stable and suppose $p \in S(\overline{a})$ is an unbounded, stationary, nonprincipal type. If $t(\overline{a};\emptyset)$ has weight one and is nontrivial then T has $2^{\aleph_0}$ countable models.

We may assume in proving Theorem 2.12 that $\overline{d}$ realizes a nontrivial weight one type r. For, $\overline{b} \rhd_\emptyset \overline{d}$ implies $\mathrm{wt}(t(\overline{d};\emptyset)) = 1$. But if $t(\overline{b};\emptyset)$ and $t(\overline{d};\emptyset)$ both have weight one, domination and Theorem XIII.2.9 transfer nontriviality of $t(\overline{b};\emptyset)$ to nontriviality of $t(\overline{d};\emptyset)$.

To make the following argument easier to read, we use single letters such as a, b, c to refer to realizations of r and p rather than the vector notation $\overline{a}, \overline{b}, \overline{c}$ we have held to so far. We write $A^{(2)}$ for the set of two element subsets of A.

Proof of Theorem 2.12. (Fig. 2). Let r_1 be the type of a triangle of realizations of r. Let G be a graph of power λ which is a disjoint union of graphs $\langle G_i : i < \lambda\rangle$ such that for each i, $|G_i| < \lambda$, every element of G_i is connected to at two least others and G_i contains no triangles (i.e. no complete subgraph on three vertices). Let the vertices of G be a set $A = \bigcup_{i<\lambda} A_i$ of independent realizations of r. Apply Lemma II.2.26 to construct a set $C = \bigcup_{i<\lambda} C_i$ where $C_i = \{c_{\{a,b\}} : \langle a, b\rangle \in G_\mu\}$ such that, writing c_{ab} for $c_{\{a,b\}} = c_{\{b,a\}}$, each triple $\langle a, b, c_{ab}\rangle$ realizes r_1 and such that $D = A \cup C$ is independent relative to the partial ordering which has only the relations $a < c_{ab}$ and $b < c_{ab}$ for $a, b \in A$. Thus, $\langle a, b, c_{ab}\rangle$ is a triangle.

For each $d \in D$, let p_d denote the copy of p over d. To simplify notation, we write p_{ab} for $p_{c_{ab}}$. For each triple $\langle a, b, c_{ab}\rangle$ we call a and b *roots* of c_{ab}.

Now let $S = \{p_a : a \in A\} \cup \{p_{ab} : \langle a, b\rangle \in G\}$. Note that each p_{ab} is over some $A_i \cup C_i$ with $|A_i \cup C_i| < \lambda$. Take each of κ, λ, and μ as in Theorem XIV.3.5 and form a model M^G such that every type in S has dimension less than λ and all types irrelevant to S have dimension λ. We can apply Theorem XIV.3.5 since, if D^i denotes $\bigcup_{j<\lambda}(A_j \cup C_j) - (A_i \cup C_i)$ then by the independence relative to $<$ of D, $D^i \downarrow_\emptyset A_i \cup C_i$.

Since r has weight one, forking is an equivalence relation on $r(M^G)$. We will recover G from M^G by identifying the elements of G as equivalence classes mod $\not\downarrow$ of $r(M^G)$. For e realizing r, let $[e]$ denote the equivalence class of e under this relation. An equivalence class, $\mathcal{E}$, is *standard* if for some $e \in \mathcal{E}$, $\dim(p_e, M^G) < \lambda$. Note that $\mathcal{E}$ is standard just if for some $d \in D = A \cup C$, $[e] = [d]$. This follows because $p \dashv \emptyset$, so if $e \downarrow d$ for all $d \in D$, by Theorem VI.2.21 $p \dashv d$, and so p_e is irrelevant to S.

A *triangle*, $(\mathcal{E}_0, \mathcal{E}_1, \mathcal{E}_2)$, is a set of three equivalence classes such that some set of representatives (e_0, e_1, e_2) is pairwise independent but dependent. We showed in Theorem XIII.2.9 that a map $a \mapsto \hat{a}$ such that $a \not\downarrow_\emptyset \hat{a}$ preserves independence if its domain and range are contained in sets of realizations of weight one types. Thus, any set of representatives for a triangle of equivalence classes will be a triangle of elements. Note also that each equivalence class intersects D in at most one element. A *standard triangle* is one such that each $\mathcal{E}_i$ is standard.

The following proposition contains the main technical remarks necessary to finish the proof.

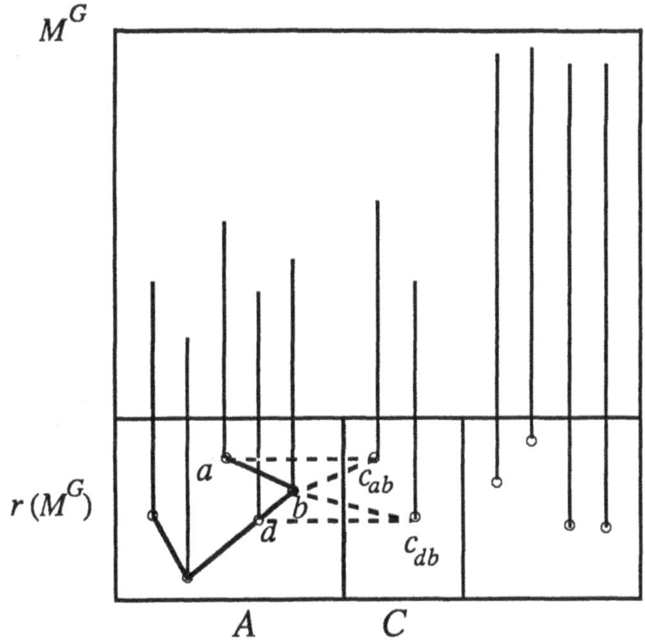

Fig. 2. Theorem XV.2.12.

2.13 Proposition. If $(\mathcal{E}_0, \mathcal{E}_1, \mathcal{E}_2)$ is a standard triangle then for some $a, b \in A$ and the associated $c_{ab} \in C$, $(\mathcal{E}_0, \mathcal{E}_1, \mathcal{E}_2) = ([a], [b], [c_{ab}])$.

The proof of the proposition relies on a number of observations about M^G which we record in the following exercises. The first depends only on the partial order defined on D.

2.14 Exercise. Show that if $\{a, b, d, e\} \subseteq A$ then i) if $\{a, b\} \cap \{d, e\} = \emptyset$ then $c_{ab} \downarrow_\emptyset c_{de}$ and ii) if $\{a, b\} \cap \{d, e\} = \{b\} = \{d\}$ then $c_{ab} \downarrow_b c_{de}$.

Now we can improve the second part of the last exercise using the fact that the triples $\langle a, b, c_{ab}\rangle$ were all required to be triangles.

2.15 Exercise. Show that if $\{a, b, d, e\} \subseteq A$ and $\{a, b\} \cap \{d, e\} = \{b\} = \{d\}$ then $c_{ab} \downarrow c_{de}$.

2.16 Exercise. Show that for each $c \in C$ there is a unique pair $\{a, b\}$ in $A^{(2)}$ such that $c \not\downarrow_\emptyset a^\frown b$.

Proof of 2.13. Since $(\mathcal{E}_0, \mathcal{E}_1, \mathcal{E}_2)$ is a standard triangle we know $(\mathcal{E}_0, \mathcal{E}_1, \mathcal{E}_2) = ([d_0], [d_1], [d_2])$ for some $d_0, d_1, d_2 \in D$. There are four cases.

Case 1: All the d_i are in A. Then the d_i do not form a triangle since A is independent.

Case 2: All the d_i are in C. Then the d_i do not form a triangle since by Exercises 2.14 and 2.15 C is an independent set.

Case 3: Two d_i, say d_0 and d_1, are in C, the third is in A. If d_2 is a common root of d_0 and d_1, we have (since C is independent relative to $<$) that $d_0 \downarrow_{d_2} d_1$ and since the elements of D are pairwise independent that $d_0 \downarrow_{\emptyset} d_1 {}^\frown d_2$. Thus the d_i do not form a triangle.

If d_2 is not a common root of d_0, d_1 then d_2 is not in the downward closure of $\{d_0, d_1\}$ under the order $<$ so $\{d_0, d_1, d_2\}$ is an independent set and again the d_i do not form a triangle.

Case 4: Thus we have two of the d_i, say d_0, d_1 in A and the third in C. By Exercise 2.16 $(d_0, d_1, d_2) = (a, b, c_{ab})$ for some $a, b \in A$.

This finishes the proof of the proposition.

Now we can finish the proof of the theorem. Note that by Exercise 2.16 and Proposition 2.13 no class $[c]$ with $c \in C$ can be in two standard triangles. Thus, we recover $\{[a] : a \in A\}$ as the standard classes which are contained in at least two standard triangles and we recover the relation G on A by $\langle [a], [b] \rangle$ is in G just if there is a triangle containing both $[a]$ and $[b]$.

2.17 Exercise. Let T be a superstable theory. Show that if $p \in S(\overline{a})$, $p \dashv \emptyset$ and $t(\overline{a}; \emptyset)$ is a nontrivial weight one type then for every $\lambda \geq 2^{|T|}$, T has 2^λ **S**-models of power λ.

2.18 Historical Notes. Theorem 2.4 is due to Shelah (Section V.4 of [Shelah 1978]), but the proof here has been influenced by some notes of Lascar, [Bouscaren & Lascar 1983], and [Makkai 1984]. Theorem 2.12 is part of the Harrington-Makkai proof [Harrington & Makkai 1985] of Shelah's theorem that a theory with the dimensional order property has 2^λ models of power λ. We place it in this section to emphasize that the proof does not involve the dimensional order property. However, we will show in Chapter XVI that the hypothesis of Theorem 2.12 does imply the dimensional order property.

3. *Bounded Theories*

In this section we solve the spectrum problem for theories that are bounded but not unidimensional. We also handle some cases of unidimensional theories which fit easily into the general framework. We want to simultaneously compute the spectrum function $I(\lambda, K)$ for a number of different classes K. The uniform proof requires more restrictive assumptions about the class K than we have previously made.

The following classes will fall under this rubric: The class of all models of a countable ω-stable theory with $\lambda \geq \omega$, the class of **S**-models of a superstable theory with $\lambda \geq 2^{|T|}$, and the class of μ-saturated models of a superstable theory for $\lambda \geq \sup(\mu, 2^{|T|})$.

The next result gives a normal form for any model of a bounded theory. To simplify the discussion we fix some notation at the beginning. We will continue to accumulate special notations throughout this section as we refine this normal form to make the calculation of the spectrum function more precise.

3.1 Notation. Let $\mathcal{P}$ denote the K-prime model of the theory T and let $\mathcal{B} = \langle r_i : i < \delta(T)\rangle$ be a basis for $\mathcal{R}(\mathcal{P})$, the K-strongly regular types over $\mathcal{P}$. Recall that for $p \in S(M)$, if p is strongly based on $\overline{a} \subseteq M$, we write $p_{\overline{a}}$ for the restriction $p|\overline{a}$ of p to $\overline{a}$. In this situation we will say $\overline{a}$ *sustains* $p_{\overline{a}}$.

3.2 Theorem. *Let $N \in K$. Then N is K-prime and in fact K-minimal over $\mathcal{P} \cup X$ where X is a maximal independent subset of N such that each $x \in X$ realizes some r_i.*

Proof. Let N_1 be K-prime over $M \cup X$. If $N_1 \neq N$ then since K admits regular types there is a $q \in S(N_1)$ which is K-strongly regular and realized in $N - N_1$. By the maximality of $\mathcal{B}$, $q \not\perp r$ for some $r \in \mathcal{B}$. But then the nonforking extension r' of r to $S(N_1)$ must be realized in $N_1[q] \subseteq N$. This contradicts the maximality of X.

3.3 Exercise. Show, by considering the theory of an equivalence relation with infinitely many infinite classes, that Theorem 3.2 may fail when T is unbounded.

Let $T' = \mathrm{Th}(\mathcal{M}, \mathcal{P})$ (i.e. name the elements of $\mathcal{P}$). Clearly the spectrum function for T' dominates that for T and we can now easily calculate an upper bound on the spectrum function for T'. Let C be the set of cardinals less than or equal $\aleph_\alpha$. If $\{r_i : i < \delta(T)\}$ enumerates a basis for $\mathcal{R}(\mathcal{P})$ there is a function from $C^{\delta(T)}$ onto $K_{\leq\alpha}$ given by $s \mapsto M$ if M is K-prime over $\mathcal{P} \cup X$ where X is an independent set containing $s(i)$ independent realizations of r_i, for $i < \delta(T)$ and $s \in C^{\delta(T)}$. Two refinements are necessary to convert this upper bound into an exact computation.

First, we replace $\mathcal{P}$ with a set B on which all the types in a basis for $\mathcal{R}(\mathcal{P})$ are based. The second refinement is motivated by the following simple example. Let T be the theory of an equivalence relation with two infinite classes. The theory T', which is obtained from T by naming the equivalence classes (that is, naming elements which sustain a basis for $\mathcal{R}(\mathcal{P})$) has four models of power at most $\aleph_1$, given by assigning the two equivalence classes the cardinalities $(\aleph_0, \aleph_0)$, $(\aleph_0, \aleph_1)$, $(\aleph_1, \aleph_0)$, or $(\aleph_1, \aleph_1)$. Of course, as models of T, the second and third choices are isomorphic. These two models are 'accidentally' isomorphic; there are two orthogonal types from the basis for $\mathcal{R}(\mathcal{P})$ which happen to be assigned the same dimension. To give a general description of this phenomenon we have to be more careful in our choice of a basis for the K-strongly regular types.

3.4 Notation. We impose some further constraints on $\mathcal{B} = \langle r_i : i < \delta(T)\rangle$, a basis for $\mathcal{R}(\mathcal{P})$. If $i \neq j$ then either r_i is a conjugate of r_j or r_i is orthogonal to every conjugate of r_j.

A basis satisfying this condition can easily be chosen by induction. Suppose $\langle r_i : i < k\rangle$ are chosen. If there is a type $r \in S(M)$ which is orthogonal to the r_i for $i < k$ and which is conjugate to one of them let $r_k = r$. If not, let

r_k be any regular type over $\mathcal{P}$ orthogonal to each of those already chosen. Note that in this case r_k is in fact orthogonal to all conjugates of the r_i for $i < k$. For, if $r_k \not\perp r_i'$ for some conjugate r_i' of r_i then by Lemmas XIV.2.13 and XIV.2.10 r_i' is nonorthogonal to some conjugate r_i'' of r_i which is based on $\mathcal{P}$ and orthogonal to the r_i for $i < k$.

With this choice, Theorem 1.9 implies that if r_i and r_j are conjugate the strong types of $\bar{b}_i$ and $\bar{b}_j$ are distinct. Thus, we have a set $B = \bigcup_{i<k} B_i$ where each B_i is a set of realizations of a type q_i over the empty set and each $p \in \mathcal{B}$ is based on some $\bar{b} \in B$.

Let G be the group of permutations $\hat{\alpha}$ of $\delta(T)$ such that for some α in $\mathrm{Aut}(\mathcal{M})$ there is a $B' = \alpha(B)$ such that for each $r_i \in \mathcal{B}$,

$$\alpha(r_i) \not\perp r_{\hat{\alpha}(i)}.$$

We have fixed a $1-1$ correspondence between $\mathcal{B}$ and $\delta(T)$ by our indexing so we can regard G as acting on either $\mathcal{B}$ or $\delta(T)$.

The following exercise justifies our replacing $\mathcal{P}$ by B.

3.5 Exercise. Modify the proof of Theorem 3.2 to show that if $M \in K$ and B and r_i are chosen according to 3.4 then M is a prime over B union a maximal independent set of realizations of the r_i.

With the discussion in Paragraph 3.4 in mind we make the following definition of a combinatorial structure to which we reduce the calculation of the spectrum problem.

3.6 Definition. Let μ and δ be cardinal numbers and let G be a group of permutations of δ. Then

$$(\mu^{\delta}/G)$$

denotes the result of partitioning the set of functions from δ into μ by the equivalence relation: $f \simeq g$ if there exists an $\alpha \in G$ with $g = f \circ \alpha$.

3.7 Exercise. Verify $\simeq$ is an equivalence relation.

Now we can calculate the spectra of $\mathbf{S}_\lambda$-saturated and $\mathbf{SET}_\lambda$-saturated models of a superstable theory. Calculation of the spectra of all models will require some further considerations. Note, however, that if T is countable and $\aleph_0$-categorical then every model of T is an $\mathbf{S}$-model and so this theorem applies.

3.8 Theorem. *Let T be a bounded superstable theory and suppose K is $\mathbf{SET}_\lambda$ or $\mathbf{S}_\lambda$ with $\lambda = \aleph_\beta \geq \lambda_0(\mathbf{I})$ (for the relevant $\mathbf{I}$). Then there is a group G of permutations of $\delta(T)$ such that for any α with $\aleph_\alpha \geq \lambda_0(\mathbf{I})$*

$$I^*(\aleph_\alpha, K) = |(|\alpha - \beta + 1|^{\delta(T)})/G|.$$

Proof. Fix B and G as in Notation 3.4. Let C be the set of cardinals κ with $\aleph_\beta \leq \kappa \leq \aleph_\alpha$. Clearly, $|C| = |\alpha - \beta + 1|$. For any $f \in C^{\delta(T)}$ define M_f to be K-prime over $B \cup X$ where $X = \bigcup_{i<\delta(T)} X_i$ and each X_i is an independent set of $f(i)$ realizations of r_i. By Exercise 3.5 each $M \in K$

must be isomorphic to some M_f. We must show $M_f \approx M_g$ if and only if $f \simeq g$. Suppose α is an isomorphism between M_f and M_g; write B' for $\alpha(B)$. Note that if $r_i \in \mathcal{B}$ is based on the empty set then α fixes r_i and $f(i) = g(i)$. Now consider an r_i based on some $\overline{a} \in B$. Then $\alpha(r_i)$ is based on $\overline{a}' = \alpha(\overline{a}) \in M_g$. Since $\mathcal{B}$ is a basis, $\alpha(r_i) \not\perp r_j$ for some $r_j \in \mathcal{B}$. For each i, let $\hat{\alpha}(i)$ be the j given by this procedure; then $\hat{\alpha} \in G$. We must show $f = g \circ \hat{\alpha}$. Since $\alpha(r_i) \not\perp r_j$, the careful selection of $\mathcal{B}$ guarantees that r_j is a conjugate of r_i and thus of $\alpha(r_i)$. Since α is an isomorphism and invoking Theorem XIV.2.15 we have

$$\dim(r_i, M_f) = \dim(\alpha(r_i), M_g) = \dim(r_j, M_g).$$

Thus, $f = g \circ \hat{\alpha}$ and $f \simeq g$. If $f \simeq g$ via a permutation γ then the automorphism of $\mathcal{M}$ which induces γ induces, via the uniqueness of K-prime models, an isomorphism of M_f and M_g.

We have reduced the calculation of the spectra of **S**-models to the calculation of $|\mu^\delta/G|$. Note that each equivalence class under $\simeq$ has at most $|G|$ elements. Thus for infinite μ if $\mu^\delta > |G|$ and especially if $\mu > |G|$, $|\mu^\delta/G| = |\mu|^\delta$. Thus G is relevant to the calculation of $I^*(\aleph_\beta, K)$ only for relatively small values of μ.

3.9 Exercise. Suppose that T is a countable superstable theory and that $\aleph_\alpha > 2^{\aleph_0} = \aleph_\beta$, then $I^*(\aleph_\alpha, S) = |\alpha - \beta + 1|^{\delta(T)}$.

The following exercise can be solved by analyzing the various possibilities for $\delta(T)$.

3.10 Exercise. Find all the possibilities for the function $I^*(\aleph_\alpha, S)$ when T is a countable superstable bounded theory.

One important fact simplifies the situation in Theorem 3.8. We have a lower bound of $\aleph_\beta$ on the dimension in a member of K of a strongly regular type over a finite set. When we deal with arbitrary models of a theory, some types may have only finite dimension in some models and we must distinguish this situation.

The following discussion is only relevant when T admits **AT**-prime models over finite sets and it will usually be applied when T is ω-stable. In fact, the analysis we now describe can be completely carried out only for a countable ω-stable theory. However, in view of the importance of these concepts for studying Vaught's conjecture for superstable theories we will make the definitions as general as possible and carry out as much of the analysis as we can for small superstable theories. The essential difficulty is that we can not guarantee that a small superstable theory admits *stationary* regular types.

We want to distinguish between those types which are '$\aleph_0$-categorical', that is, have infinite dimension in every model, and those to which we can assign finite dimensions. Roughly speaking, this is the difference between an isolated and a nonisolated type. However, noting that the only 1-type over the empty set in $\mathrm{Th}(Z, S)$ is isolated but every nonforking extension of it is not isolated, shows the situation is a little more complex. The following definition handles this difficulty.

3.11 Definition. Let $p_{\overline{a}} \in S(\overline{a})$ be **AT**-strongly regular.

i) $p_{\overline{a}}$ is *eventually nonisolated* if for some finite $\overline{b} \supset \overline{a}$, some nonforking extension $q_{\overline{b}}$ of $p_{\overline{a}}$ to $S(\overline{b})$ is not **AT**-isolated.
ii) $p_{\overline{a}}$ is *persistently isolated* if for every finite $\overline{b} \supset \overline{a}$, all nonforking extensions $q_{\overline{b}}$ of $p_{\overline{a}}$ to $S(\overline{b})$ are **AT**-isolated.

We extend this definition from types over finite sets to arbitrary types in the following way.

iii) We say an arbitrary type p is eventually nonisolated (persistently isolated) if it is a nonforking extension of an eventually nonisolated (persistently isolated) type.

These two concepts are the negations of each other and we have two separate names only to avoid dissonance. These notions are referred to in [Shelah, Harrington, & Makkai 1984] as eni and neni types. We may occasionally lapse into this usage. Contrary to [Shelah, Harrington, & Makkai 1984], we do not require that eventually nonisolated or persistently isolated types be stationary.

3.12 Exercise. If p is a nonisolated type over a finite set then every nonforking extension of p is also nonisolated.

3.13 Exercise. If $p \in S(\overline{a})$ is eventually nonisolated then every nonforking extension of p is eventually nonisolated.

3.14 Exercise. Determine the eventually nonisolated and the persistently isolated types in i) the theory of seven disjoint, infinite, unary predicates and ii) the theory of infinitely many disjoint unary predicates.

The next lemma shows the notion of eventual nonisolation is invariant under parallelism. The lemma after it shows that it is also invariant under nonorthogonality.

3.15 Lemma. *If $p \parallel q$ and p is eventually nonisolated then so is q.*

Proof. Suppose not and assume that p is eventually nonisolated but q is not. Without loss of generality we may assume that p is not isolated. Let $A = \operatorname{dom} p \cup \operatorname{dom} q$. Since p and q are parallel, there is a global type r which is a nonforking extension of both p and q. Since p is not isolated, the open mapping theorem guarantees that $r|A$ is not isolated. But since q is persistently isolated, $r|A$ is isolated. This contradiction yields the theorem.

In this case we have established that a concept defined for types which do not need to be stationary is invariant under parallelism.

3.16 Lemma. i) *If $p \in S(P)$ is persistently isolated and based on $\overline{a} \in P$ then for every $M \models T$, $\dim(p_{\overline{a}}, M) \geq \omega$.*

ii) *Suppose T is countable and ω-stable. If $p \in S(M)$ is persistently isolated and $q \not\perp p$ then q is persistently isolated.*

Proof. i) It suffices to show that $\dim(p_{\overline{a}}, \mathcal{P})$ is infinite. Clearly, $p|A$ is realized in $\mathcal{P}$ for any finite $A \subseteq \mathcal{P}$. But letting $A_0 = \overline{a}$ and choosing a_i to realize $p|A_i$ for $i < \omega$ we have the result.

ii) Suppose not. Without loss of generality we can replace p and q by parallel types over a single finite set A which are respectively isolated and not isolated. To simplify notation we call the resulting types p and q as well. Since T is ω-stable, we can assume p and q are stationary. Let M be prime over A. By i) $\dim(p, M) = \omega$ while by the omitting types theorem $\dim(q, M) = 0$. Let J be a proper subset of an infinite independent set I of realizations of p in M and choose N prime over $A \cup J$. By Corollary X.4.4, if $\overline{a} \in I - J$, $\overline{a}$ realizes $\mathrm{Av}(J, N)$, so $\overline{a} \downarrow_A N$. Thus, N is a proper elementary submodel of M. Since $p \not\perp q$, if q' is a nonforking extension of q to $S(N)$, $q' \not\perp t(\overline{a}; N)$. Thus q', and a fortiori q, is realized in $N[\overline{a}] \prec M$. This contradiction yields the theorem.

One of the major applications of these concepts will be to the proof of Vaught's conjecture for an ω-stable theory. It is natural to ask if we can extend that proof to superstable theories. For such an extension, we would like to prove the results here for small superstable theories. The assumption that T is ω-stable was used in two ways in the preceding proof: first, to guarantee that p and q were stationary and second, to find the model prime over $A \cup J$. It does not seem necessary to require p and q to be stationary. We could find a model prime over $A \cup J$ by invoking Corollary V.1.26 if T were a small superstable theory. In the next lemma we need again that a persistently isolated type is stationary. This implicit appeal to ω-stability seems harder to avoid.

The choice of $\overline{b}$ to witness that a type $p_{\overline{a}}$ is eventually nonisolated seems fairly arbitrary. In fact, it can be made in any model prime over $\overline{a}$.

3.17 Lemma. *Let T be a countable ω-stable theory and let $p_{\overline{a}}$ be a stationary eventually nonisolated type.*

i) *If $\overline{a} \subseteq M$, where M is prime over a finite set then* $\dim(p_{\overline{a}}, M) < \omega$.
ii) *There exists $\overline{b} \supseteq \overline{a}$ with $t(\overline{b}; \overline{a})$* **AT**-*isolated so that every nonforking extension of $p_{\overline{a}}$ to $S(\overline{b})$ is nonisolated.*

Proof. i) Suppose M is prime over the finite set $A' \supseteq \overline{a}$ and $\dim(p_{\overline{a}}, M) = \omega$. We will prove that for any finite $B \supseteq A'$ and any nonforking extension q of p to $S(B)$, q is isolated. Since any extension of $p_{\overline{a}}$ has a nonforking extension whose domain contains A', the open mapping theorem then yields that $p_{\overline{a}}$ is not eventually nonisolated. Assume for contradiction that q is nonisolated and let N be prime over B. Without loss of generality, $M \subseteq N$. Let I be an infinite independent set witnessing $\dim(p_{\overline{a}}, M) = \omega$. Since T is superstable there is a finite $I_0 \subset I$ such that $B \downarrow_{A \cup I_0} I - I_0$. Now, since q is stationary, any $c \in I - I_0$ realizes q. But if q is not isolated, $\dim(q, N) = 0$.

ii) Let M be **AT**-prime over $\overline{a}$. By i) $\dim(p, M) < \omega$. Let I_0 be a maximal independent subset of M realizing p. Then no nonforking extension of p to $S(\overline{a} \cup I_0)$ is realized in M. Thus, any such type is nonisolated and $I_0 \cup \overline{a}$ is the required $\overline{b}$.

3.18 Exercise. Show that Lemma 3.17 i) holds for a small superstable theory under the additional hypothesis that p has finite multiplicity.

3.19 Notation. Let T be a theory with dimension $\delta = \delta(T)$. Let δ_1 be the number of persistently isolated types in a basis $\mathcal{B}$ for $\mathcal{R}(\mathcal{P})$ and δ_2 the number of eventually nonisolated types. For simplicity, assume that $\mathcal{B} = \langle r_i : i < \delta \rangle$ is arranged with the persistently isolated types first. Let G_1 be the restriction of the group G of permutations of $\mathcal{B}$ defined in 3.4 to persistently isolated types and G_2 the restriction to eventually nonisolated types.

The next three theorems describe the spectra of a countable bounded ω-stable theory.

3.20 Theorem. *If T is a countable bounded ω-stable theory then for any uncountable $\aleph_\alpha$*

$$I^*(\aleph_\alpha, \mathbf{AT}) = ||\alpha + 1|^{\delta_1}/G_1| \times ||\alpha + \omega|^{\delta_2}/G_2|.$$

Proof. Let $\langle c_i : i < \alpha + \omega \rangle$ be an enumeration of the cardinals less than $\aleph_\alpha$. Map $|\alpha + 1|^{\delta_1} \times |\alpha + \omega|^{\delta_2}$ onto $I^*(\aleph_\alpha, \mathbf{AT})$ as follows. For each pair of functions $f \in (\alpha + 1)^{\delta_1}$ and $g \in (\alpha + \omega)^{\delta_2}$, choose $M_{f,g}$ such that $\dim(p_i, M_{f,g}) = c_{f(i)}$ for $i < \delta_1$ and $\dim(p_i, M_{f,g}) = c_{g(i)}$ for $\delta_1 \le i < \delta$. Just as in Theorem 3.8, we can see that $M_{f,g} \approx M_{f',g'}$ if and only if there exist $\alpha_1 \in G_1$ and $\alpha_2 \in G_2$ with $f' = f \circ \alpha_1$ and $g' = g \circ \alpha_2$.

The formula in Theorem 3.20 is misleading when $\delta(T)$ is infinite. In this case we can compute $I^*(\aleph_\alpha, K)$ without recourse to G. The computations in Theorems 3.20 and 3.21 are not contradictory. The group action is still present in the following situation; it just isn't reflected in the spectrum calculation.

3.21 Theorem. *Let T be a countable bounded ω-stable theory.*

i) *If $\delta_1(T) = \aleph_0$ then $I^*(\aleph_\alpha, \mathbf{AT}) \ge |\alpha + 1|^{\aleph_0}$.*
ii) *If $\delta_2(T) = \aleph_0$ then $I^*(\aleph_\alpha, \mathbf{AT}) \ge |\alpha + \omega|^{\aleph_0}$.*

Proof. The proof for the two parts is virtually identical so we do only i). Suppose first that there exists a type q and infinitely many r_i which witness that δ_1 is infinite and are all based on realizations $\overline{b}_i$ of q. Then, the r_i can be taken over a single realization $\overline{b}$ of q. For, since T is ω-stable, q has finite multiplicity so without loss of generality all the $\overline{b}_i$ realize the same strong type. But then for each i, j Theorem XIV.2.10 guarantees that $r_i \not\perp r_{ij}$ where r_{ij} denotes the conjugate of r_i over $\overline{b}_j$. So if $r_j \not\perp r_{ij}$ we obtain the contradiction $r_i \not\perp r_j$ by the transitivity of orthogonality on regular types. In this situation we obtain the required estimate by passing to the finite inessential extension of T obtained by naming $\bar{b}$. Thus, we can assume only finitely many of the r_i are based on realizations of any fixed type over $\emptyset$. Since each type has finite multiplicity, each set B_i defined in 3.4 is finite. Thus, there are infinitely many distinct B_i. For each $i < \omega$ choose an $r_i \in \mathcal{R}(\mathcal{P})$ which is strongly based on some $\bar{b} \in B_i$. Now, not only the r_i but any of their conjugates are pairwise orthogonal. Thus we can choose a model M fixing the dimensions of r_i in M independently and obtain the theorem.

The following exercise illustrates the loss of information inherent in describing structure results by spectrum functions.

3.22 Exercise. Show that the two spectrum functions in Theorem 3.21 have the same values.

3.23 Exercise. Show that by using the techniques of 2.6 and 2.7 we could obtain this result without recourse to the careful arrangement of $\mathcal{R}(\mathcal{P})$.

In Section 2 we showed that an unbounded ω-stable theory had infinitely many models in every uncountable power. The next exercise and following theorem give two successive weakenings of the hypothesis necessary to obtain this result.

3.24 Exercise. If T is a countable ω-stable theory and $\delta(T)$ is infinite then T has infinitely many models in every uncountable cardinal.

3.25 Theorem. *Let T be a countable ω-stable theory. If T is not unidimensional and $\delta_2(T) > 0$ then $I(\mu, \mathbf{AT}) \geq \omega$ for every uncountable μ.*

Proof. By Theorem 3.21 we can assume that $\delta(T)$ is finite. Let r_0 be an eventually nonisolated type and r_1 an orthogonal strongly regular type. Let M_n be prime over μ independent realizations of r_1 and n independent realizations of r_0. Now for each $m < \omega$ there are only finitely many models M_n such that $M_n \approx M_m$. For, such an M_n must have $\dim(r_i, M_n) = m$ for some i but there are only finitely many choices for i. Thus $I(\mu, \mathbf{AT})$ is infinite.

The following tables summarize some of the consequences for the spectrum problem of the work in this section. For $\mathbf{S}$, the spectrum depends on $\delta(T)$ and the group G. For $\mathbf{AT}$, we must also consider the role of δ_1 and δ_2.

The table compiles the results of Theorems 3.8, 3.20, and 3.21 with one minor adjustment. We have switched from describing the cumulative spectrum I^* to calculating the exactly the number of models in power $\aleph_\alpha$. This requires a modification of the definition of $(|\alpha + \omega|^\delta)/G$. Instead of considering all functions from δ into the set of cardinals less than or equal α, consider those functions which have $\aleph_\alpha$ in their range. Of course, (see the remark before Exercise 3.9) the group G is really interesting only when δ and α are both finite. In this case, it is smoother to use the original definition of (μ^δ/G) and calculate $I(\aleph_{m+1}, K)$ as $I^*(\aleph_{m+1}, K) - I^*(\aleph_m, K)$.

3.26 Theorem. *Let T be a countable bounded ω-stable theory. The following tables give the number of* **S***-models (models) of T with cardinality $\aleph_\alpha$ for $\alpha > 0$.*

Number of S-models

	$\delta(T) = 1$	$1 < \delta(T) < \omega$	$\delta(T) = \omega$
$I(\aleph_\alpha, \mathbf{S}_{\aleph_0})$	1	$\lvert\alpha + 1\rvert^{\delta}/G$	$\lvert\alpha + 1\rvert^{\omega}$

Number of models

$I(\aleph_\alpha, \mathbf{AT}_{\aleph_0})$	$\delta_1 = 0$	$\delta_1 = 1$	$1 < \delta_1 < \omega$	$\delta_1 = \omega$
$\delta_2 = 0$	impossible	1	$\lvert\alpha + 1\rvert^{\delta_1}/G$	$\lvert\alpha + 1\rvert^{\omega}$
$\delta_2 = 1$	1	$\lvert\alpha + \omega\rvert$	$\lvert\alpha + \omega\rvert$	$\lvert\alpha + 1\rvert^{\omega}$
$1 < \delta_2 < \omega$	$\lvert\alpha + \omega\rvert$	$\lvert\alpha + \omega\rvert$	$\lvert\alpha + \omega\rvert$	$\lvert\alpha + 1\rvert^{\omega}$
$\delta_2 = \omega$	$\lvert\alpha + \omega\rvert^{\omega}$	$\lvert\alpha + \omega\rvert^{\omega}$	$\lvert\alpha + \omega\rvert^{\omega}$	$\lvert\alpha + \omega\rvert^{\omega}$

In fact, [Cherlin, Harrington, & Lachlan 1985] shows that the case $\delta_2 = 0$ but δ_1 is infinite does not occur. We can read off from the second table and Lemma 3.28 (below) a number of the major early results on the spectrum problem.

3.27 Theorem. i) (Morley's Theorem) *If T is categorical in one uncountable power then T is categorical in every uncountable power.*

ii) (Baldwin-Lachlan Theorem) *If T is categorical in some uncountable power then T has either 1 or $\aleph_0$ countable models.*

iii) (Lachlan, Shelah) *If T is ω-stable and has finitely many models in some uncountable power then T is $\aleph_0$-categorical and there exists an integer m and a finite group G such that for $n < \omega$,*

$$I^*(\aleph_n, \mathbf{AT}) = |n^m/G|.$$

iv) (Shelah) *If T is a bounded ω-stable theory then T has 1, $\aleph_0$ or $2^{\aleph_0}$ countable models.*

Proof. For i), ii), and iii) we see by Theorem 2.4 that T must be bounded. In the first two cases, by Theorems 3.21 and 3.25, T must be unidimensional. In case iii), Theorem 3.21 implies T is finite dimensional. The remainder of the theorem can be read off from the chart and the following refinement of iv).

3.28 LEMMA. *The number of countable models of an ω-stable bounded theory is determined by the value of $\delta_2(T)$. More specifically,*

i) *If $\delta_2 = 0$ then T is $\aleph_0$-categorical.*
ii) *If $0 < \delta_2 < \omega$ then T has $\aleph_0$ countable models.*
iii) *If δ_2 is infinite then T has $2^{\aleph_0}$ countable models.*

Proof. For i) note that if there is a nonisolated type over some finite set, the preservation of eni by nonorthogonality guarantees that some member of the basis must be eventually nonisolated; that is $\delta_2 > 0$. Part ii) is immediate from Theorem 3.20 and part iii) from Theorem 3.21 ii).

We quote without proof Theorem IX.1.20 of [Shelah 1978].

3.29 Theorem. *Let T be superstable and λ the least cardinal in which T is stable. Then, for any μ with $\mu \geq \aleph_1 + |T|$, $I(T, \mathbf{AT}) \geq \min(2^\mu, 2^\lambda)$.*

Combining this result with Theorem 3.26 we have the following theorem of Lachlan.

3.30 Theorem. *Suppose T is a countable superstable theory which has finitely many models of power $\aleph_k$ for some $k < \omega$. Then, T is a bounded ω-stable theory. If T is not $\aleph_1$-categorical then T is $\aleph_0$-categorical and for $k < \omega$, $I^*(\aleph_k, \mathbf{AT}) = |(k+1)^m/G|$ for some m and G.*

Since we gave Theorem 3.29 without proof a crucial step in our argument for this theorem is missing: the move from T is superstable to T is ω-stable.

3.31 Historical Notes. Most of the results in this section are contained in the last theorem of the first edition of [Shelah 1978]. The discussion of eventually nonisolated and persistently isolated types is taken from [Shelah, Harrington, & Makkai 1984]. Our exposition depends very heavily on the treatments in [Bouscaren & Lascar 1983], [Saffe 1983], and especially [Pillay 1983f]. We thank David Kueker and the University of Maryland logic seminar for detecting some serious flaws in an earlier version of this section.

Morley proved not only Theorem 3.27 i) [Morley 1965], but also that an $\aleph_1$-categorical theory has at most $\aleph_0$ countable models which is subsumed in Theorem 3.27 ii). The other direction of Theorem 3.27 ii) is from [Baldwin & Lachlan 1971]. The results in Theorem 3.27 iii) appear in X.2.7 of [Shelah 1978] and, more identifiably, in [Lachlan 1975] and [Lachlan 1978].

There has been considerable further work on unidimensional theories. Kueker [Kueker 198?] conjectured that a countable theory whose every uncountable model is ω-saturated is either $\aleph_0$ or $\aleph_1$-categorical. Lachlan (unpublished) proved the conjecture for ω-stable theories. Buechler [Buechler 1984a] extended the result to superstable theories. Shelah [Shelah 1978] asked if every stable unidimensional theory is superstable. Hrushovski ([Hrushovski 1986], [Hrushovski 198?], [Hrushovski 198?a]) answered this question positively and also obtained Kueker's conjecture for stable theories. Shelah [Shelah 1986c] provided a rather straightforward argument that a countable unidimensional theory has either 1, 2^λ or $\min(2^\lambda, 2^{2^{\aleph_0}})$ models in power λ for each uncountable λ.

4. Almost Homogeneous Models

In this section we discuss work of Pillay, Bouscaren, and Lascar on a generalization of the notion of an ω-homogeneous model, an almost homogeneous model. As pointed out by Pillay, the most natural context for this concept is the class of those stable theories in which all types have finite multiplicity.

4.1 Definition. i) The theory T is *extra-stable* if T is stable and every type has finite multiplicity.

ii) The model M is *almost homogeneous* if for each pair of finite sequences $\overline{a}, \overline{b} \in M$ with $stp(\overline{a}; \emptyset) = stp(\overline{b}; \emptyset)$ and each $c \in M$ there exists a $d \in M$ with $stp(\overline{a}^\frown c; \emptyset) = stp(\overline{b}^\frown d; \emptyset)$.

4.2 Exercise. Show every ω-stable theory is extra-stable.

There is an inherently slippery feature of almost homogeneity. Namely, $stp(\overline{a}; \emptyset)$ is well defined only when we regard $\overline{a}$ as sitting in a model of T. It is this problem which makes condition i) of Theorem 4.10 necessary.

The first major result is that if T is extra-stable a countable model of T is determined by the types over the empty set which it realizes. The following more abstract situation allows us to combine two arguments from the original proof. We slightly generalize the notion of almost homogeneity. The purpose of this generalization is not so much to cover more cases as to emphasize the properties of almost homogeneity which dominate the following proofs.

4.3 Definition. Let $\mathcal{G}$ be a normal subgroup of $\text{Aut}(\mathcal{M})$ and for each submodel M of $\mathcal{M}$ denote by G_M the subgroup of $\text{Aut}(M)$ consisting of the restrictions of elements of $\mathcal{G}$ to automorphisms of M. We say $\overline{a}, \overline{b} \in M$ are *$\mathcal{G}$-conjugate* and write $\overline{a} \sim_{\mathcal{G}} \overline{b}$ if there is an $\alpha \in \mathcal{G}$ with $\alpha(\overline{a}) = \overline{b}$. We say $\overline{a}, \overline{b} \in M$ are *M-conjugate* and write $\overline{a} \sim_M \overline{b}$ if there is an $\alpha \in G_M$ with $\alpha(\overline{a}) = \overline{b}$. The structure M is *$\mathcal{G}$-homogeneous* if for each $\overline{a}, \overline{b} \in M$, $\overline{a} \sim_M \overline{b}$ if and only if $\overline{a} \sim_{\mathcal{G}} \overline{b}$.

There are two natural exemplars of this notion, ordinary ω-homogeneity where $\mathcal{G} = \text{Aut}(\mathcal{M})$ and almost homogeneity where $\mathcal{G} = \text{Saut}(\mathcal{M})$.

We say $\mathcal{G}$ is a *closed subgroup* if every model M and every automorphism α of M satisfy the following condition. If every restriction of α to a finite set can be extended to a member of $\mathcal{G}_M$ then $\alpha \in \mathcal{G}_M$. Both $\text{Aut}(\mathcal{M})$ and $\text{Saut}(\mathcal{M})$ are closed.

4.4 Exercise. Show that if M is countable and $\mathcal{G}$ is closed then M is $\mathcal{G}$-homogeneous if and only if for each pair of finite sequences $\overline{a}, \overline{b} \in M$ with $\overline{a} \sim_{\mathcal{G}} \overline{b}$ and each $c \in M$ there exists a $d \in M$ with $\overline{a}^\frown c \sim_{\mathcal{G}} \overline{b}^\frown d$.

The next lemma collects the conditions on $\text{Saut}(\mathcal{M})$ which are sufficient to prove most of the results presented here on almost homogeneous models.

4.5 Lemma. *Suppose $\mathcal{G}$ is* Saut($\mathcal{M}$) *and T is extra-stable.*

i) *For each $p \in S(\emptyset)$, the number of orbits of realizations of p under the action of G_M is finite.*
ii) *If $\{\overline{a}_0, \ldots, \overline{a}_n\}$ is a set of representatives for distinct orbits of $p(M)$ under $\mathcal{G}$ and $t(\overline{a}_0, \ldots, \overline{a}_n; \emptyset) = t(\overline{b}_0, \ldots, \overline{b}_n; \emptyset)$ then $\{\overline{b}_0, \ldots, \overline{b}_n\}$ is a set of representatives for distinct orbits of $p(M)$ under $\mathcal{G}$.*

Proof. Condition i) of Lemma 4.5 follows immediately from the definition of extra stable; condition ii) can be easily verified.

4.6 Exercise. Show that if T is extrastable and $\mathcal{G} = \mathrm{Saut}(\mathcal{M})$ then condition ii) holds.

4.7 Exercise. Show, in fact, that condition ii) depends only on the hypothesis that $\mathcal{G}$ is normal in $\mathrm{Aut}(\mathcal{M})$.

4.8 Exercise. Give an example of a type p and two models M and N of an extrastable theory such that the realizations of p split into a different number of strong types in M than in N.

4.9 Exercise. Show the phenomenon of the last exercise is impossible if p is an isolated type.

The following theorem yields that two countable almost homogeneous models which realize the same types are isomorphic.

4.10 Theorem. *Let T and $\mathcal{G}$ satisfy the conditions listed in Lemma 4.5.*

i) *If M and N realize the same types over the empty set, M is countable and N is $\mathcal{G}$-homogeneous then there is an elementary embedding of M into N.*
ii) *If M and N are countable, $\mathcal{G}$-homogeneous, and realize the same types over the empty set then they are isomorphic.*

Proof. i) Let $M = \{a_i : i < \omega\}$ and let $q_n = t(a_0, \ldots, a_{n-1}; \emptyset)$. We will construct by induction a set $S \subseteq \omega^{<\omega}$ and for each $s \in S$ an element $b_s \in N$. We write $\overline{b}_s$ for $\langle b_{s|1}, \ldots, b_s \rangle$ and let S_n denote the elements of S with length n. The construction will satisfy the following conditions.

a) S is closed under the formation of subsequences and each S_n is finite.
b) For each $s \in S_n$, $\overline{b}_s$ realizes q_n.
c) If $\overline{c} \in M$ realizes q_n, there is an $s \in S_n$ with $\overline{b}_s \sim_{\mathcal{G}} \overline{c}$.

Suppose for $m < n$ we have defined S_m and b_s for each $s \in S_m$; we must define S_n and for each $s \in S_n$, b_s. Let $C = \{\overline{c}_0, \ldots, \overline{c}_r\}$ be a complete list of representatives of the orbits of $q_n(N)$ under $\mathcal{G}$. Partition C by $\overline{c} \sim \overline{c}'$ if $\overline{c}|(n-1) \sim_{\mathcal{G}} \overline{c}'|(n-1)$. For each class D of this partition, choose by c) of the induction hypotheses an $s_D \in S_{n-1}$ such that for all $\overline{d} \in D$, $\overline{d}|(n-1) \sim_N \overline{b}_{s_D}$. By the $\mathcal{G}$-homogeneity of N, for each $\overline{d} \in D$ there exists a $b^{\overline{d}} \in N$ such that $\overline{d} \sim_N \overline{b}_{s_D} {}^\frown b^{\overline{d}}$. Fix an ordering of D and if $\overline{d}$ is the ith member of this ordering place $s_D {}^\frown i$ into S_n and set $b_{s_D {}^\frown i} = b^{\overline{d}}$.

Now S is an infinite finitely branching tree so by König's lemma it has an infinite branch, say η. Now, mapping a_i to $b_{\eta|(i+1)}$ yields the required imbedding of M into N.

ii) By i) we may assume $M \prec N$. We show that for $\overline{a} \in M$ and $\overline{b} \in N$, the relation $\overline{a} \simeq \overline{b}$ if and only if $\overline{a} \sim_{\mathcal{G}} \overline{b}$ is a 'back and forth' which yields the required isomorphism. The 'forth' is immediate so we prove only the 'back'. Suppose $\overline{a} \sim_{\mathcal{G}} \overline{b}$, $\lg(\overline{b}) = n$, and $d \in N$. We must find $c \in M$ with $\overline{a}^\frown c \sim_{\mathcal{G}} \overline{b}^\frown d$. Let $\overline{c} = \langle \overline{c}_0, \ldots, \overline{c}_r \rangle$ be a complete set of representatives for the orbits of $t(\overline{b}^\frown d; \emptyset)$ under $\mathcal{G}$ which intersect N and let $q = t(\overline{c}; \emptyset)$. Choose $\overline{c}' \in M$ to realize q. By condition 4.5 ii) $\overline{c}'$ is a set of representatives of distinct orbits in M of $t(\overline{b}^\frown d; \emptyset)$ under the action of $\mathcal{G}$. Since no more orbits can be represented in M than in N, for some $i < r$, $\overline{c}_i \sim_{\mathcal{G}} \overline{b}^\frown d$. In particular, $\overline{c}_i|(n-1) \sim_{\mathcal{G}} \overline{b}$. Since, by hypothesis, $\overline{a} \sim_{\mathcal{G}} \overline{b}$, we have $\overline{c}_i|(n-1) \sim_{\mathcal{G}} \overline{a}$. By the $\mathcal{G}$-homogeneity of M, there is a $c \in M$ with $\overline{a}^\frown c \sim_{\mathcal{G}} \overline{c}_i$. By transitivity $\overline{b}^\frown d \sim_{\mathcal{G}} \overline{a}^\frown c$ as required.

The remaining results in this section use the assumption that T is a countable ω-stable theory. This assumption is essential for the first result but it is open whether the hypothesis in the second case can be weakened to extra-stable.

4.11 Theorem. *Let T be countable, ω-stable, and bounded. Then every countable model of T is almost homogeneous.*

Proof. We first note that if T is ω-stable and bounded and $stp(\overline{a}; \emptyset) = stp(\overline{b}; \emptyset)$ then T satisfies the following condition.

$$\begin{array}{l}\text{For every } c \in M \text{ with } q = t(c; \overline{a}) \text{ strongly regular} \\ \text{there is a } d \in M \text{ with } stp(\overline{a}^\frown c; \emptyset) = stp(\overline{b}^\frown d; \emptyset).\end{array} \qquad (*)$$

For this, let $\alpha \in \operatorname{Saut}(\mathcal{M})$ map $\overline{a}$ to $\overline{b}$. Then since T is bounded, Lemma XIV.2.9 shows $\dim(q, M) = \dim(\alpha(q), M)$. In particular, $\alpha(q)$ is realized in M and we have the claim.

To see that the theorem follows from $(*)$, fix $\overline{a}, \overline{b}$, and $c \in M$. Since $\operatorname{Saut}(\mathcal{M})$ is closed, it suffices to find $d \in M$ with $stp(\overline{a}^\frown c; \emptyset) = stp(\overline{b}^\frown d; \emptyset)$. Choose $M_1 \subseteq M$ to be prime over $\overline{a}$ and choose $\overline{a}_1 \subseteq M_1$ with $\overline{a} \subseteq \overline{a}_1$ so that $t(c; M_1)$ is strongly based on $\overline{a}_1$. Moreover, let $t(c; M) \sqsupseteq^e \otimes r_i$ where each r_i is strongly regular and based on $\overline{a}_1$. By Theorem 3.2, we may suppose further that for some $\overline{e} \in M_1[c] \subseteq M$ realizing $\otimes r_i$, $t(c; M \cup \overline{e})$ is isolated over $\overline{a}_1 \cup \overline{e}$. This implies, in particular, that $t(c; \overline{a}_1 \cup \overline{e})$ is isolated by some $\phi(\overline{x}, \overline{a}_1, \overline{e})$. Since $\overline{a}_1 \subseteq M_1$, $t(\overline{a}_1; \overline{a})$ is principal so applying Lemma XIV.2.12, there is a $\overline{b}_1 \in M$ with $stp(\overline{a}^\frown \overline{a}_1; \emptyset) = stp(\overline{b}^\frown \overline{b}_1; \emptyset)$. Now if $\alpha \in \operatorname{Saut}(\mathcal{M})$ takes $\overline{a}_1$ to $\overline{b}_1$ there is, by $(*)$, for each e_i an $e_i' \in M$ realizing $\alpha(r_i)$. Since the e_i are independent, $\overline{e}'$ realizes $t(\alpha(\overline{e}); \overline{b})$. Now since $\phi(\overline{x}, \overline{b}_1, \overline{e}') \vdash stp(\alpha(c); \overline{b}_1)$ and $\overline{b}_1 \supseteq \overline{b}$, so M is almost homogeneous.

From Theorems 4.10 and 4.11 we have another proof of Theorem 3.27 iv).

4.12 Corollary. *Let T be countable ω-stable and bounded. Then T has $1, \aleph_0$ or $2^{\aleph_0}$ countable models.*

Proof. Each model of T is determined by the set of types over the empty set which are realized in it.

Finally, we show that for ω-stable theories, countable ω-homogeneous models are almost homogeneous. The following exercise shows where the difficulties of the theorem lie.

4.13 Exercise. Show that if M is a countable ω-saturated model of an extra-stable theory then M is almost homogeneous.

4.14 Theorem. *Suppose T is a countable ω-stable theory. Every countable ω-homogeneous model of T is almost homogeneous.*

Proof. Let $\overline{a}, \overline{b} \in M$ with $stp(\overline{a};\emptyset) = stp(\overline{b};\emptyset)$ and let $c \in M$. Choose $M_1 \subseteq M$ **AT**-prime over $\overline{a}$ and choose $\overline{a}_1 \in M_1$ with $t(c; M_1)$ strongly based on $\overline{a} \cup \overline{a}_1$. By Lemma XIV.2.12, there is a $\overline{b}_1 \in M$ so that $\overline{b}^\frown\overline{b}_1$ realizes $stp(\overline{a}^\frown\overline{a}_1;\emptyset)$. Since M is ω-homogeneous, there is a $d \in M$ with $t(\overline{a}^\frown\overline{a}_1{}^\frown c;\emptyset) = t(\overline{b}^\frown\overline{b}_1{}^\frown d;\emptyset)$. Since $stp(\overline{a}^\frown\overline{a}_1;\emptyset) = stp(\overline{b}^\frown\overline{b}_1;\emptyset)$ there is an $\alpha \in \text{Saut}(\mathcal{M})$ which maps $\overline{a}^\frown\overline{a}_1$ to $\overline{b}^\frown\overline{b}_1$. If $q = stp(c;\overline{a}^\frown\overline{a}_1)$, d realizes $\alpha(q)$. Thus, by Lemma IV.3.20, $stp(\overline{a}^\frown\overline{a}_1{}^\frown c;\emptyset) = stp(\overline{b}^\frown\overline{b}_1{}^\frown d;\emptyset)$ and we finish.

There are a number of interesting questions raised by this section. We adopted an abstract exposition of Theorem 4.10 to emphasize that it is an assertion about certain families of group representations with little model theoretic content. In contrast, Theorem 4.11 makes heavy use of the stability machinery. This use seems intrinsic to the proof but we have no example of any model which is homogeneous without being almost homogeneous. There are several further speculations and insightful examples in [Pillay 1982a].

4.15 Historical Notes. This section is a mild rewrite of [Pillay 1982a] and one theorem from [Bouscaren & Lascar 1983]. Bouscaren and Lascar proved a more general version of Theorem 4.11 by induction on U-rank. Their argument applies when α_T is finite without assuming that T is bounded.

Chapter XVI

The Dimensional Order Property

We discuss in this chapter the main dividing line between superstable theories with and without a good structure theory; this dividing line is called, 'the dimensional order property' or just DOP. If a theory has the dimensional order property then it is possible to interpret an arbitrary binary relation into T by considering the dimensions of sets. This 'nonstructure' result leads to the conclusion that T has 2^{λ} models in every cardinality λ greater than $2^{|T|}$. The great significance of this concept stems from the perhaps even more remarkable consequences of its negation. Essentially, the negation of DOP (NDOP) is the assertion that the relation $p \dashv M$ is a trivial dependence relation. This hypothesis allows one to decompose every model as a tree of small models. There are a number of equivalent formulations of DOP which are useful in various contexts. The name is suggested by the following variant. If T has DOP then there exists a two parameter family $\{p_{\overline{a}\overline{b}}\}$ of copies of a type p such that for any choice of infinite cardinals $\lambda_{\overline{a},\overline{b}}$, there is a model M with $dim(p_{\overline{a}\overline{b}}, M) = \lambda_{\overline{a},\overline{b}}$. This leads to the construction of 2^{κ} models with cardinality $\kappa \geq 2^{|T|}$ by constructing $\overline{a}_i$, $\overline{b}_i$ for $i < \kappa$ and using $dim(p_{\overline{a}_i\overline{b}_j})$ to encode an arbitrary binary relation on κ.

In the first section of this chapter we discuss the notion of free amalgamation of models in a class K. This leads to the formalization of DOP as an assertion about the triviality of $\dashv$. We develop in Section 2 some technical properties of trivial types which are extremely useful in Section 3 and Chapter XVII. In Section 3, we show that the DOP implies T has many models. In the following chapter we will continue the discussion of theories without the DOP.

1. *Avatars of the Dimensional Order Property*

We first formally define K-NDOP in a way that will be useful in Chapter XVII and beyond. Then we show that the K in K-NDOP was superfluous; the dimensional order property does not, in fact, depend on the class K. We show that NDOP can be described in terms of finitely generated models and we develop the form of the DOP which makes the name appear most natural.

Throughout this chapter we assume that T is superstable. This is not really necessary and Shelah develops much of the machinery for arbitrary stable theories in [Shelah 1982]. We sacrifice the greater generality for the convenience of a ready supply of regular types. The most immediate advantage of Shelah's added generality is the ability to prove such theorems as, 'A stable but not superstable theory with the dimensional order property has 2^λ models which are λ-saturated and have power λ'. Despite this restriction we often write $\kappa(T)$ when by this convention we mean ω.

Most of the work in Chapter XIV and before was one-dimensional in the sense that we could order our constructions linearly. We now want to consider model constructions where the relations between the building blocks are inherently of a higher dimension. In this book we extend primarily from 1-dimensional diagrams to 2-dimensional diagrams. The full development of classification theory, particuarly the study of infinitary languages [Shelah 1983a], requires the study of n-dimensional diagrams for all finite n.

Much of the work in the remainder of this book is concerned with partially ordered families of sets or, more often, of models. We introduced in Chapter II the notion of an independent system. We denoted the universe of such a system by an upper case Roman letter and the elements with lower case Roman letters, even though it was understood that the elements of the independent system might be subsets of $\mathcal{M}$. In some of the ensuing definitions (specifically a normal family in Section XVII.2) we have to deal with elements of the sets in an independent family. In these more complicated situations we have adopted the convention of referring to the systems as $\bar{A}$, $\bar{B}$, etc., the elements of the system as A, B, etc. and the elements of the elements of the system as a, b, etc. However, we retain the practice of denoting the system by upper case Roman letters and the members (subsets of models) by lower case Roman letters when their elements are not required.

When discussing independent families of models we introduce the following systematic ambiguity.

1.1 Notation. Let $(A, <)$ be a partial order. Let $I \subseteq A$ be an ideal (i.e. downward closed, cf. Definition II.2.24). Then we denote $\bigcup I \subseteq \mathcal{M}$ by I.

It should be clear from context when I denotes a set of subsets of $\mathcal{M}$ and when it denotes the union of that set. Similarly, we use A ambiguously to denote $\cup A$. For example, we will frequently speak of a model M which is prime over (the system) A. This really means M is prime over $\cup A$.

This notation seems to be less cumbersome than that of [Shelah 1982] and [Harrington & Makkai 1985]. They let $\langle \mathcal{I}, < \rangle$ be a partially ordered set and let $\langle A_i : i \in \mathcal{I} \rangle$ be a collection of subsets (of the monster model) indexed by $\mathcal{I}$. For $I \subseteq \mathcal{I}$, A_I denotes $\bigcup_{j \in I} A_j$. Letting subsets of $\mathcal{M}$ act as their own indices allows us to reduce the number of levels of notation by one.

We begin by considering the simplest kind of nonlinear diagram, an independent triangle.

1.2 Definition. i) A triple $\mathcal{N} = \langle N_0, N_1, N_2 \rangle$ with $N_1 \downarrow_{N_0} N_2$, N_0 contained in $N_1 \cap N_2$, and each $N_i \in K$ is called a *free K-amalgam.*

ii) A quadruple $\mathcal{N} = \langle N_0, N_1, N_2, N_3 \rangle$ such that $\langle N_0, N_1, N_2 \rangle$ is a free amalgam, and N_3 is K-prime over $N_1 \cup N_2$ is called a *full (free) K-amalgam.*

Since we deal only with free amalgams we will often omit the adjective free. We say that N_3 *completes* the amalgam $\mathcal{N}$. If the only restriction on the N_i is that they be subsets of the monster model we refer to $\mathcal{N}$ as a set-amalgam.

1.3 Definition. The theory T *does not have the K-dimensional order property,* written T satisfies K-NDOP, if for any full amalgam $\langle N_0, N_1, N_2, N_3 \rangle$ of K-models and type p, if $p \dashv N_1$ and $p \dashv N_2$ then $p \dashv N_3$.

If this condition fails we say T has the K-DOP. We show in Theorem 1.7 that we can drop the K and speak just of the *DOP*.

The simplest example of a theory with the dimensional order property is the theory of two crosscutting equivalence relations. (Fig. 1). A model, M, of this theory is easily visualized as a checkerboard. Suppose M is extended to M_1 and M_2 by adding to M a new row for M_1 and a new column for M_2, thus producing a checkerboard with one corner missing. Then the type

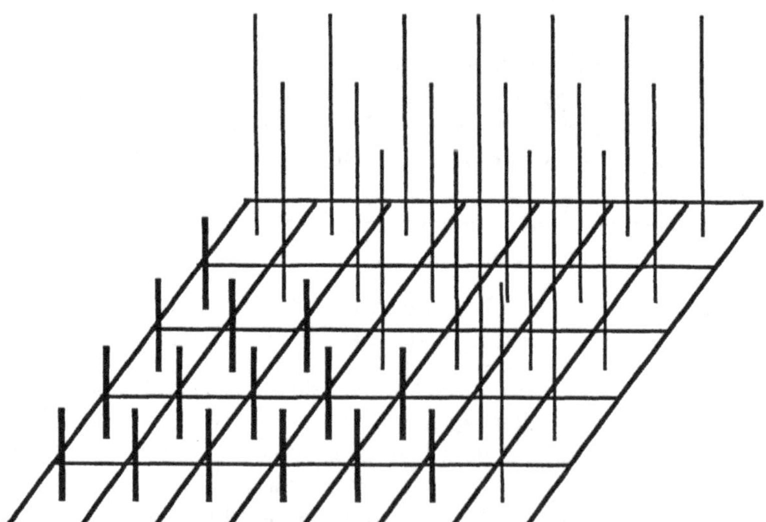

Fig. 1. The dimensional order property (DOP)

of a point in the missing corner square is orthogonal to each of M_1 and M_2 but is realized in every model containing both of them.

The next result sharpens the conditions on DOP and provides a version which is more useful for proving nonstructure results. The three parts of the theorem successively strengthen the result.

1.4 Theorem. *If the full free* **S**-*amalgam* $\mathcal{N}$ *and the type* p *witness the* **S**-*DOP then we may assume*

i) $p \in S(N_3)$ *and* p *is regular.*
ii) N_1 *and* N_2 *are finitely generated over* N_0.
iii) *For* $i = 1, 2$, $N_i = N_0[a_i]$ *where* a_i *realizes a regular type over* N_0.

Proof. i) Since $p \not\perp N_3$, there is a regular $q \in S(N_3)$ with $q \not\perp p$. Write p as $\otimes r_j$ where each r_j is regular. Then, for some j, $r_j \not\perp q$. Now, by the transitivity of nonorthogonality on regular types, if $q \not\perp N_i$ then $r_j \not\perp N_i$ and, a fortiori, $p \not\perp N_i$. Thus we can take q as the desired p.

ii) We assume p satisfies condition i). There is a finite $B \subseteq N_3$ with $|B| < \kappa(T)$ such that p is strongly based on B. Then $t(B; N_1 \cup N_2)$ is **S**-isolated by a finite $A \subseteq N_1 \cup N_2$. Let $\overline{a}_i = A \cap (N_i - N_0)$ for $i = 1, 2$. Now, $\langle N_0, N_0[\overline{a}_1], N_0[\overline{a}_2], N_0[\overline{a}_1, \overline{a}_2]\rangle$ and $p|N_0[\overline{a}_1, \overline{a}_2]$ meet the requirements.

iii) We show that if **S**-NDOP holds for models generated by one element realizing a regular type then it holds for all finitely generated models. Let $\mathcal{N}$ be a free **S**-amalgam with N_1 and N_2 finitely generated over N_0. Let X_1 and X_2 be bases for $R(N_i; N_0)$. The proof is by induction on the sum of the cardinalities of the X_i. Choose $x_i \in X_i$ for $i = 1, 2$ and let $X_i' = X_i - \{x_i\}$. Since $X_1 \cup X_2$ is independent over N_0,

$$\langle N_0[X_1' \cup X_2'], N_0[X_1' \cup X_2], N_0[X_1 \cup X_2'], N_0[X_1 \cup X_2]\rangle$$

is a full **S**-amalgam. Thus, NDOP implies either $p \not\perp N_0[X_1 \cup X_2']$, or $p \not\perp N_0[X_1' \cup X_2]$. Without loss of generality, we may assume the first case occurs. But now, note that $\langle N_0, N_0[X_1], N_0[X_2'], N_0[X_1 \cup X_2']\rangle$ is a full **S**-amalgam so by induction $p \not\perp N_0[X_1]$ or $p \not\perp N_0[X_2']$. In either case we finish, either directly or by applying upward monotonicity for $\not\perp$.

The next lemma helps to explain the name dimensional order property. It shows that in the presence of DOP, if N_3 is **S**-prime over $N_1 \cup N_2$ then N_3 contains an infinite set of indiscernibles over $N_1 \cup N_2$. The dimension of this set of indiscernibles can be fixed arbitrarily to code binary relations into models of T. In particular, consider again the theory of two crosscutting equivalence relations; we are free to place any infinite number of points on each square of the checkerboard.

1.5 Lemma. *Suppose the full amalgam* $\mathcal{N}$ *witnesses that* T *possesses the* **S**-*DOP. Then* N_3 *contains an infinite set of indiscernibles over* $N_1 \cup N_2$.

Proof. (Fig. 2). Let $p \in S(N_3)$ witness the **S**-DOP. Choose $B \subseteq N_3$ with p strongly based on B and $|B| < \kappa(T)$. We will show the following

1.6 Claim. *For any* C *containing* B *with* $|C| < \kappa(T)$ *there is a* $D \supseteq C$ *with* $|D| < \kappa(T)$ *such that* $p|D \vdash p|(N_1 \cup N_2 \cup D)$.

With this fact in mind it is easy to construct the set of indiscernibles. Namely, choose e_i for $i < \omega$ so that e_i realizes $p|(E_i \cup N_1 \cup N_2)$. We can choose the $e_i \in N_3$ since the claim entails that p_i is implied by a type over a set with cardinality $< \kappa(T)$.

We now prove the claim. Using the definition of $\kappa(T)$ and monotonicity, we can extend C to D with $|D| < \kappa(T)$ so that, letting $D_i = D \cap N_i$ for $i < 3$,

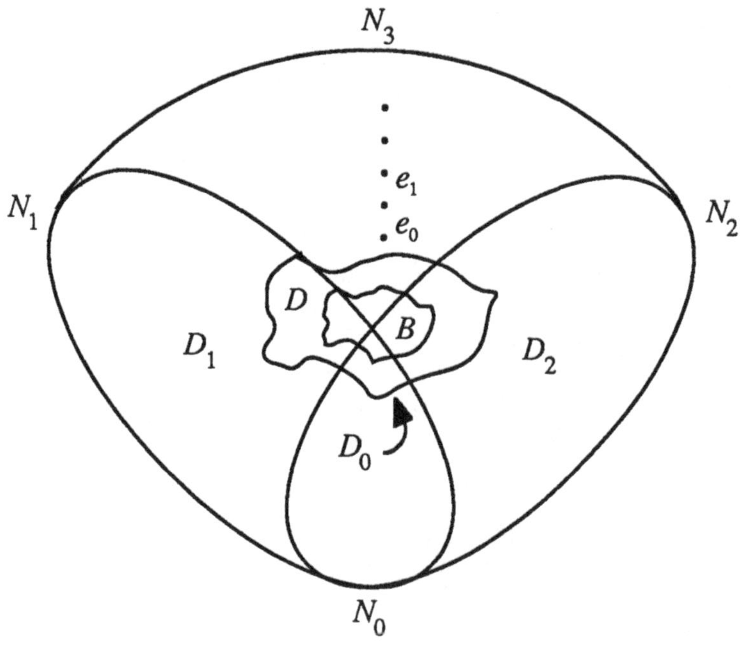

Fig. 2. Lemma XVI.1.5: Simplifying DOP

$$D \underset{D_1 \cup D_2}{\downarrow} (N_1 \cup N_2) \tag{1}$$

and

$$D \underset{D_0}{\downarrow} N_0, D \underset{D_1}{\downarrow} N_1, \text{ and } D \underset{D_2}{\downarrow} N_2. \tag{2}$$

By the definition of a free amalgam we have

$$N_1 \underset{N_0}{\downarrow} N_2. \tag{3}$$

Now, from (1) and the monotonicity properties of forking we can deduce $N_2 \downarrow_{N_1 \cup D_2} N_1 \cup D$. Similarly, (3) implies $N_2 \downarrow_{N_0 \cup D_2} N_1 \cup D_2$. Now, transitivity of independence yields $N_2 \downarrow_{N_0 \cup D_2} N_1 \cup D$. By monotonicity again, we conclude $N_1 \downarrow_{N_0 \cup D} N_2$. The downward preservation of $\dashv$ in the second coordinate implies $p \dashv N_0 \cup D_1$ and $p \dashv N_0 \cup D_2$. Now, using (2) we have $t(N_1; N_0 \cup D) \perp p|(N_0 \cup D)$ and $t(N_2; N_0 \cup D) \perp p|(N_0 \cup D)$. From the strong triviality of orthogonality (Theorem VI.1.19) we conclude $p|(N_0 \cup D) \perp t(N_1 \cup N_2; N_0 \cup D)$. Applying a similar argument to N_0 and D_0 we have $p|D \perp t(N_0; D)$. Since $p|D$ and $p|(N_0 \cup D)$ are stationary $\perp$ implies weak orthogonality (cf. Exercise VI.1.23) so we have $p|D \vdash p|(N_0 \cup D) \vdash p|(N_1 \cup N_2 \cup D)$ as required.

In the light of Theorem X.4.9, Lemma 1.5 asserts that in the presence of DOP prime models over free amalgams are not minimal. The next lemma provides a converse to this result by showing that NDOP implies **S**-prime models over free amalgams are **AT** minimal.

1.7 Lemma. *Let T be a countable ω-stable theory. If T satisfies* **S***-NDOP and $\mathcal{N}$ is a full* **S***-amalgam then N_3 is* **AT***-minimal over $N_1 \cup N_2$. In particular, N_3 is* **AT***-prime over $N_1 \cup N_2$.*

Proof. (Fig. 3). Suppose for contradiction that there is a proper submodel N_3' with $N_1 \cup N_2 \subseteq N_3' \subseteq N_3$. Then some strongly regular $p \in S(N_3')$ is realized in $N_3 - N_3'$. By NDOP, p is not orthogonal to one of N_1, N_2, say $p \not\dashv N_1$. Applying a consequence of the Three Model Theorem, Theorem XIII.4.3, to N_3, N_3', and N_1, there is a strongly regular type $q \in S(N_3')$ with $q \not\perp p$ such that q does not fork over N_1. Since $N_3 = N_1[N_2]$, for any

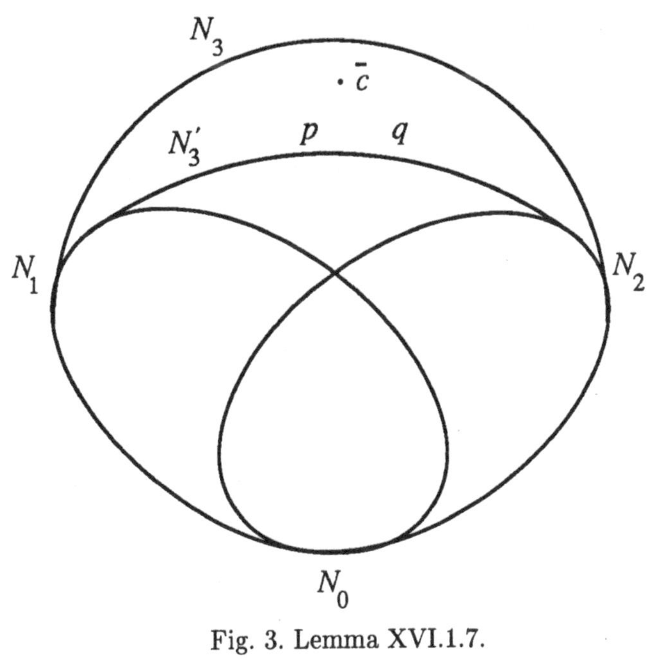

Fig. 3. Lemma XVI.1.7.

$a \in N_3$ $a \not\downarrow_{N_1} N_2$ and a fortiori $a \not\downarrow_{N_0} N_3'$. By Lemma XIII.4.3, $t(N_3; N_3')$ and therefore q is orthogonal to N_0. Since **AT**-strongly regular types are **AT**-minimal some $\bar{c} \in N_3$ realizes q. The choice of q yields $\bar{c} \downarrow_{N_0} N_3'$. Thus, by the definition of $\dashv$, $N_1 \downarrow_{N_0} N_2$ implies $\bar{c} \downarrow_{N_1} N_2$. But as $N_2 \rhd_{N_1} N_3$, this is impossible and we conclude the theorem.

Corollary XVII.2.4 and the exercise following it give a more satisfying proof of the last result.

The next theorem provides the most intuitive description of the negation of the dimensional order property, the assertion that $\dashv$ is a trivial dependence relation. The proof of the equivalence of this characterization with the original definition is from [Lascar 1985]. One consequence of this result is that we can refer to just the DOP rather than the K-DOP. We discuss another approach to this result in Theorem XVII.1.21.

1.8 Definition. We say $\dashv$ is *trivial* if for any three sets A, B, and C with $B \downarrow_A C$ and any type p, if $p \dashv B$ and $p \dashv C$ then $p \dashv B \cup C$.

The following exercises make the equivalence between NDOP and triviality of $\dashv$ more plausible.

1.9 Exercise. i) Show that $\dashv$ is trivial if and only if for every regular p, if $p \dashv B$ and $p \dashv C$ then $p \dashv B \cup C$.

ii) Suppose $p \dashv M \cup A$ implies $p \dashv M[A]$. Show that for any K, the K-NDOP is equivalent to the assertion that $\dashv$ is trivial.

The hypothesis of Exercise 1.9 ii) does not hold in general. The following more complicated argument of Lascar shows that the conclusion does. We employ the following notation in the proof.

1.10 Notation. Let $\mathcal{N} = \langle N_0, N_1, N_2 \rangle$ be a free K-amalgam. We call the K'-amalgam $\mathcal{N}' = \langle N'_0, N'_1, N'_2 \rangle$ a *parallel amalgam* to $\mathcal{N}$ if the following diagram (Fig. 4) is independent with respect to the partial order indicated by the arrows.

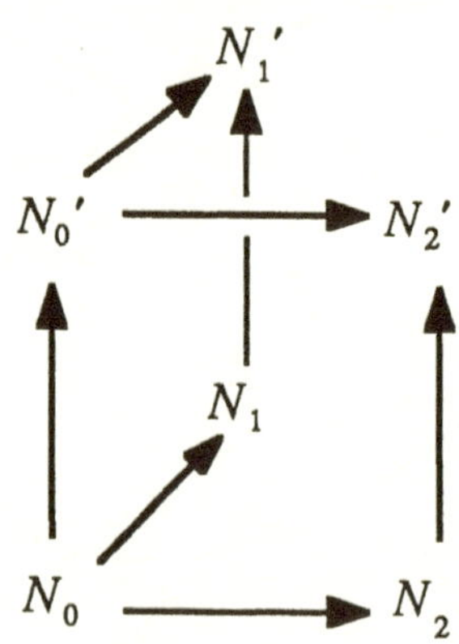

Fig. 4. A parallel amalgam

1.11 Theorem. *The following are equivalent.*

i) *T has the* **S**-*NDOP.*

ii) *For some acceptable class K, T has the K-NDOP.*

iii) *$\dashv$ is trivial.*

Proof. That i) implies ii) is obvious. To see that ii) implies iii), fix a set-amalgam $\mathcal{N}=\langle N_0, N_1, N_2\rangle$ and assume $p \not\perp N_1 \cup N_2$. Without loss of generality, we may assume p is regular. Extend $\mathcal{N}$ to a parallel amalgam $\langle M_0, M_1, M_2\rangle$ of K-models and let M_3 be K-prime over $M_1 \cup M_2$. Since $p \not\perp N_1 \cup N_2$, $p \not\perp M_3$. By Theorem XIII.3.3, there is a K-strongly regular $q \in S(M_3)$ with $p \not\perp q$. By the K-NDOP, q is not orthogonal to one of M_1 or M_2, say $q \not\perp M_1$. But $M_1 \downarrow_{N_1} N_1 \cup N_2$ implies by Lemma XIII.3.11 that $p \not\perp N_1$ and we finish.

We now show that iii) implies i) (Fig. 5). Let $\mathcal{N} = \langle N_0, N_1, N_2, N_3\rangle$ be a full $\mathbf{S}$-amalgam and suppose $p \in S(N_3)$. We must show $p \not\perp N_1$ or $p \not\perp N_2$. This would be immediate from the definition of triviality but we don't know $p \not\perp N_1 \cup N_2$.

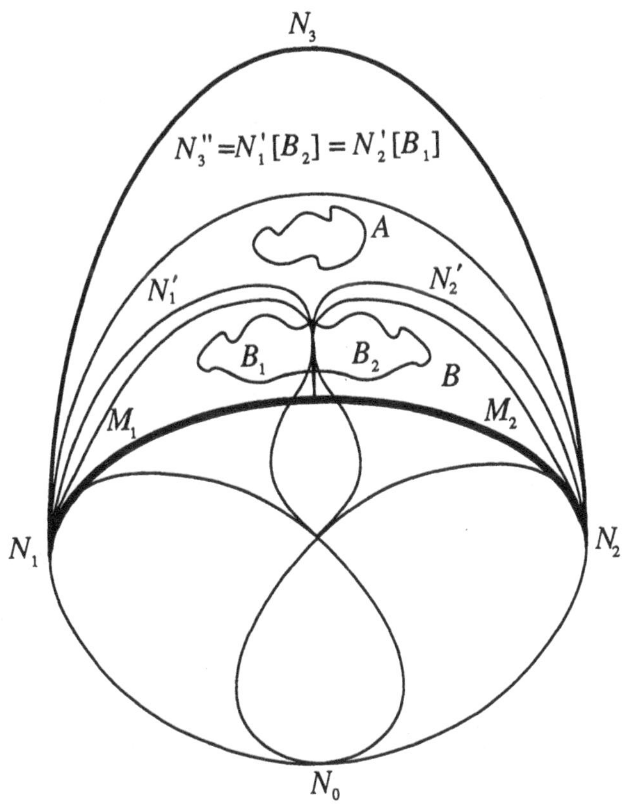

Fig. 5. Theorem XVI.1.11: iii) → i)

Let p be strongly based on the finite subset A of N_3. The proof will be by induction on the U-rank of $t(A; N_1 \cup N_2)$. Without loss of generality, p is regular.

If $U(A;N_1 \cup N_2) = 0$ then p does not fork over $N_1 \cup N_2$ so $p \not\perp N_1 \cup N_2$ and we finish by the triviality of $\dashv$.

Suppose $U(A;N_1 \cup N_2) = \alpha$ and if $\langle M_0, M_1, M_2, M_3\rangle$ is a full amalgam and $p \in S(M_3)$ with p strongly based on the finite subset B of M_3 with $U(B;M_1 \cup M_2) < \alpha$, then $p \not\perp M_1$ or $p \not\perp M_2$. Choose N_3' to be **S**-prime over $N_1 \cup N_2$ with $A \downarrow_{N_1 \cup N_2} N_3'$. Let N_3'' be $N_3'[A]$. Let B be a basis for the realizations of regular types over N_3' in N_3''. Since $A \downarrow_{N_1 \cup N_2} N_3'$, for each $b \in B$, $t(b;N_3') \not\perp t(A;N_3')$ which implies $t(b;N_3') \not\perp N_1 \cup N_2$. Since $\dashv$ is trivial this implies each $t(b;N_3')$ is nonorthogonal to one of N_1 or N_2. By Theorem XIII.3.3 we may without loss of generality replace each $b \in B$ such that $t(b;N_3') \not\perp N_1$ ($\not\perp N_2$) by a b' with $b' \downarrow_{N_1} N_3'$ ($b' \downarrow_{N_2} N_3'$). Thus, we can assume B is the disjoint union of B_1 and B_2 where the elements of B_1 (B_2) are independent from N_3' over N_1 (N_2).

By Theorem X.1.28 we can choose copies N_1' and N_2' of $N_3'[B_1]$ and $N_3'[B_2]$ such that $N_3'' = N_1'[B_2] = N_2'[B_1]$. Moreover, we can choose copies M_1 and M_2 of $N_1[B_1]$ and $N_2[B_2]$ respectively so that $N_3'' = N_1'[M_2] = N_2'[M_1]$.

Now let p' be a nonforking extension to $S(N_3'')$ of $p|(A \cup N_1 \cup N_2)$. Note i) $U(A;M_1 \cup N_2') < U(A;N_1 \cup N_2)$ and ii) $M_1 \downarrow_{N_1} N_2'$. The first of these assertions is obvious; the second is a routine application of the forking calculus. Thus we may apply the induction hypothesis to conclude: $p' \not\perp M_1$ or $p' \not\perp N_2'$.

In the first case, recall that $N_3 = N_1[N_2]$ and $A \subset N_3$ so $N_2 \downarrow_{N_1} B_1$ implies successively that $N_2 \downarrow_{N_1} M_1$, $N_3 \downarrow_{N_1} M_1$, and $A \downarrow_{N_1} M_1$. But $p \not\perp A$ and $p \not\perp M_1$ implies by Theorem XIII.3.11 that $p \not\perp N_1$ as required.

In the second case we consider the representation of N_3'' as $N_2'[M_1]$ and deduce first that $p \not\perp M_2$ or $p \not\perp N_1'$. From the first of these alternatives we deduce as in the previous paragraph that $p \not\perp N_2$ and finish. If not, we have $p \not\perp N_1'$ and $p \not\perp N_2'$. But $N_1' \downarrow_{N_3'} N_2'$ implies, again by Theorem XIII.3.11, that $p \not\perp N_3'$. But $A \downarrow_{N_1 \cup N_2} N_3'$ yields $p \not\perp (N_1 \cup N_2)$ and thus by triviality of orthogonality $p \not\perp N_1$ or $p \not\perp N_2$ and we finish.

1.12 Exercise. Deduce from the previous theorem that if T is a countable ω-stable theory then T has **S**-NDOP if and only if T has the **AT**-NDOP.

We have shown that the dimensional order property depends only on T and not on a class of models K. If we vary the definition by restricting the type p which can witness the nontriviality of $\dashv$, we do obtain a different notion. The remainder of this section is a first step towards proving Vaught's conjecture for ω-stable theories. To properly count countable models, just as in Chapter XIV, we must distinguish eventually nonisolated types.

1.13 Definition. The theory T *does not have the ENI-dimensional order property,* if for any free amalgam $\langle N_0, N_1, N_2, N_3\rangle$ of **S**-models and ENI type p, if $p \dashv N_1$ and $p \dashv N_2$ then $p \dashv N_3$. We write T satisfies ENI-NDOP.

Naturally, the negation of this property is referred to as the ENI-DOP. Notice that the theory CER_2 of two cross-cutting equivalence relations is an $\aleph_0$-categorical, ω-stable theory with DOP but which does not have ENI-DOP.

1.14 Exercise. Give an example of an ω-stable but not $\aleph_0$-categorical theory with DOP but without ENI-DOP.

1.15 Exercise. Show that if T has the ENI-DOP then T has the DOP.

Since we showed in Section XV.2 that nonorthogonality preserved ENI types, the following result can be deduced immediately from Theorem 1.4.

1.16 Theorem. *If the full* **S**-*amalgam* $\mathcal{N}$ *and the ENI-type* p *witness the ENI-DOP then we may assume*

i) $p \in S(N_3)$.
ii) N_1 *and* N_2 *are finitely generated over* N_0.
iii) *For* $i = 1, 2$, $N_i = N_0[a_i]$ *where* $t(a_i; N_0)$ *is strongly regular.*

1.17 Historical Notes. The dimensional order property was first defined and applied in [Shelah 1982]. Our treatment depends greatly on later expositions by [Harrington & Makkai 1985] and [Lascar 1985]. In addition to proving that NDOP is equivalent to the triviality of $\dashv$, Lascar introduces in that paper another interesting notion. He says a type p is bounded by a set A if when writing p as $\otimes r_i$, where each r_i is regular, all of the r_i are not orthogonal to A. Lascar calls a theory without the dimensional order property presentable. This terminology is justified by the decomposition theorem in Chapter XVII.

2. *Triviality of Forking*

In this section we develop some technical consequences of the assumption that forking is trivial on the realizations of a type p. In Section XV.3 we showed that if there exists a type $p \in S(\overline{a})$, p is orthogonal to the empty set and $t(\overline{a}; \emptyset)$ is a nontrivial weight one type then there exist many non-isomorphic models. In this section we expound the properties of trivial types. We will rely on the results here in the proof that DOP implies a theory has the maximal number of models and in the proof of the Vaught and Morley conjectures for ω-stable countable theories.

Recall from Section XV.2 the definition of a trivial type.

2.1 Definition. The stationary type $p \in S(A)$ is *trivial* if for any nonforking extension p' of p, any three pairwise independent realizations of p' are in fact independent.

We call a triple of points which form a counterexample to triviality a *triangle*. Clearly, if p is trivial so is any nonforking extension of p. Given a triple $\langle \overline{a}, \overline{b}, \overline{c} \rangle$ of realizations of $p \in S(A)$ and $B \supseteq A$, we can choose $\langle \overline{a}', \overline{b}', \overline{c}' \rangle$ realizing $t(\overline{a}^\frown \overline{b}^\frown \overline{c}; A)$ with $\overline{a}'^\frown \overline{b}'^\frown \overline{c}' \downarrow_A B$. Since the first triple is a triangle over A if and only if the second is a triangle over B we deduce the invariance of triviality under nonforking extensions.

2.2 Proposition. *If p' is a nonforking extension of the stationary type p, then p is trivial if and only if p' is trivial.*

Using this fact, it is easy to prove the following proposition.

2.3 Proposition. *If p is a trivial type and I is a set of pairwise independent realizations of p then I is independent.*

When p is regular we can extend the triviality property to sets which do not realize p.

2.4 Lemma. *Let $p \in S(A)$ be trivial and regular. If $\overline{a}$ realizes p and for some B_1, B_2 with $B_1 \downarrow_A B_2$, $\overline{a} \not\downarrow_A B_1 \cup B_2$ then $\overline{a} \not\downarrow_A B_1$ or $\overline{a} \not\downarrow_A B_2$.*

Proof. Without loss of generality we can replace A by a strongly $\kappa(T)$-saturated model M. Thus, p denotes $t(\overline{a}; M)$. For $i = 1, 2$, let $D_i = p(M[B_i])$. Then, $p \perp t(M[B_i]; M[D_i])$. For, if not, p would be realized in $M[B_i] - M[D_i]$ contradicting the choice of the D_i. Now since $M[B_1] \downarrow_M M[B_2]$, $M[B_1] \downarrow_{M[D_1] \cup M[D_2]} M[B_2]$, and for each i $t(M[B_i]; M[D_1] \cup M[D_2]) \parallel t(M[B_i]; M[D_i])$. Since orthogonality is preserved by parallelism and is trivial we have $t(M[B_1] \cup M[B_2]; M[D_1] \cup M[D_2]) \perp p$. Now, if $\overline{a} \downarrow_M (D_1 \cup D_2)$, then $\overline{a} \downarrow_M M[D_1] \cup M[D_2]$. By the definition of orthogonality we conclude

$$\overline{a} \underset{M[D_1] \cup M[D_2]}{\downarrow} M[B_1] \cup M[B_2].$$

Thus, $\overline{a} \not\downarrow_M B_1 \cup B_2$ implies $\overline{a} \not\downarrow_M D_1 \cup D_2$. So by triviality of p, $\overline{a} \not\downarrow_M D_1$ or $\overline{a} \not\downarrow_M D_2$. But the first implies $\overline{a} \not\downarrow_M B_1$ and the second implies $\overline{a} \not\downarrow_M B_2$ so we finish.

The following corollary is almost immediate.

2.5 Corollary. *Suppose that I is a set of realizations of the trivial regular type $p \in S(A)$ and that I is independent over A. If each $a \in I$ satisfies $a \downarrow_A B$ then I is an independent set over B.*

Proof. Since p is trivial, it suffices to show that I is pairwise independent over B. Thus, it is more than enough to show that if $a, b \in I$ then $a \downarrow_A B \cup b$. But by Lemma 2.4, the last assertion follows from $a \downarrow_A B$, $b \downarrow_A B$, and $a \downarrow_A b$, which all hold.

We would like to extend Lemma 2.4 from trivial regular types to trivial weight one types. In fact, we will have to strengthen the hypothesis on $B \cup C$. Before proving the extension we must show that triviality is a property of the nonorthogonality class of a weight one type.

2.6 Theorem. *If p and q are nonorthogonal weight one stationary types and q is trivial then p is trivial.*

Proof. Without loss of generality, assume that p and q are in $S(M)$ for some strongly $\kappa(T)$-saturated model M. Since p and q are nonorthogonal, they are not almost orthogonal. Thus we can define a map f from $p(\mathcal{M})$ into $q(\mathcal{M})$ such that $a \not\downarrow_M f(a)$. Since q has weight one, if a_1 and a_2 are independent so are $f(a_1)$ and a_2. Now we can see that f preserves pairwise independence. For if a_1 and a_2 were independent but $f(a_1)$ and $f(a_2)$ were not, the assumption that $t(f(a_2); M)$ has weight one would be contradicted. Now, suppose that $A = \{a_1, a_2, a_3\}$ are pairwise independent realizations of p. Note that f is 1-1 on A. Then f^{-1} satisfies the conditions of Theorem XIII.2.9 so A is independent.

Another easy consequence of Lemma 2.4 strengthens the triviality of forking to include all realizations of trivial regular types.

2.7 Corollary. *If p and q are trivial regular types then forking is trivial on $p(\mathcal{M}) \cup q(\mathcal{M})$.*

Now we can obtain the promised extension of Lemma 2.4 to weight one types. The following proof relies on superstability by assuming the existence of regular types. I don't know whether this assumption can be avoided.

2.8 Theorem. *If $a \not\downarrow_M c_1{}^\frown c_2$ with $c_1 \downarrow_M c_2$, $t(a; M)$ is trivial, and the type of each of a, c_1, c_2 over M has weight one then $a \not\downarrow_M c_1$ or $a \not\downarrow_M c_2$.*

Proof. Without loss of generality, we assume that M is strongly $\kappa(T)$-saturated. Since $p \not\perp t(c_1{}^\frown c_2; M)$, p is realized by some $a' \in M[c_1, c_2]$. Form $M[a'] \prec M[c_1, c_2]$ and choose $d \in M[a']$ to realize a regular type q over M. Then $p \not\perp q$ so by Lemma 2.6, q is trivial. Since $d \in M[c_1, c_2]$, $d \not\downarrow_M c_1{}^\frown c_2$. Applying Lemma 2.4, we can assume without loss of generality that d depends on c_1 over M. Since q has weight one, d and c_2 are independent. If d depends on a over M, transitivity of forking on weight one types yields the result. But if a and d are independent, applying Lemma 2.4 again, we have $d \downarrow_M a{}^\frown c_2$. Thus, $d \downarrow_{M \cup c_2} a$. But $a \not\downarrow_M c_1{}^\frown c_2$ and $a \downarrow_M c_2$ implies $a \not\downarrow_{M \cup c_2} c_1$. Similarly, $d \not\downarrow_M c_1{}^\frown c_2$ and $d \downarrow_M c_2$ implies $d \not\downarrow_{M \cup c_2} c_1$. The last three assertions contradict the hypothesis that $t(c_1; M \cup c_2)$ has weight one so we finish.

The following two exercises show the true strength of Theorem 2.8. The first one shows that if an element a depends on an independent set I of realizations of a type then over an appropriate base it depends on a subset I_0 with $|I_0| \leq 2$. Combining this observation with Theorem 2.8 yields Exercise 2.10.

2.9 Exercise. Suppose p and q are nonalgebraic stationary types over A. Show that if a realizes p and I is a minimal set of realizations of q such that $a \not\downarrow_A I$ and $|I| \geq 2$ then there is an S-saturated model M and realizations c_1, c_2 of q such that $a \not\downarrow_M c_1{}^\frown c_2$ but $a \downarrow_M c_1$, $a \downarrow_M c_2$.

2.10 Exercise. Suppose I and J are independent sequences realizing trivial weight one types $p, q \in S(A)$ such that each $\overline{a} \in I$ satisfies $\overline{a} \not\downarrow_A J$ and each $\bar{b} \in J$ satisfies $\bar{b} \not\downarrow_A I$. Show $\bar{a} \not\downarrow_A \bar{b}$ establishes a 1-1 correspondence between I and J.

Another version of this result continues the program of finding sufficient conditions to make orthogonality and almost orthogonality equivalent. We began by proving the equivalence if the two types are over a strongly saturated model (Theorem VI.1.40); later we extended the result to arbitrary models of a countable ω-stable theory (Corollary XIII.3.8.) Here we move in another direction by imposing no condition on the set but requiring that the types be trivial. [Buechler 1986], extending [Cherlin, Harrington, & Lachlan 1985], shows that trivial can be replaced by modular. We will rely heavily on the next two results in Chapter XVIII.

2.11 Corollary. *If $p, q \in S(A)$ are stationary trivial weight one types then $p \not\perp q$ implies $p \not\perp^a q$.*

Proof. By Corollary VI.2.18, $p \not\perp q$ implies there are finite independent sequences E, F of realizations of p, q, respectively, so that $t(E; A) \not\perp^a t(F; A)$. But since p and q are trivial, Theorem 2.9 yields that for some $e \in E, f \in F$, $e \not\downarrow_A f$ and we finish.

2.12 Lemma. *Let $p, q \in S(A)$ be stationary trivial types with $p \not\perp q$. Suppose p is regular and for some ϕ, (q, ϕ) is strongly regular. Then for every b realizing p and every $M \supseteq A \cup b$, there is a $c \in M$ which realizes q and $b \not\downarrow_A c$.*

Proof. By Theorem 2.11, $p \not\perp^a q$. Let b realize p and choose d realizing q with $b \not\downarrow_A d$. Then there is a formula $\psi(x, y)$ such that for any e, $\models \psi(b, e)$ implies $b \not\downarrow_A e$. Fix $M \supseteq A \cup b$ and choose $c \in M$ so that $\models \phi(c) \wedge \psi(b; c)$. Since (q, ϕ) is strongly regular either c realizes q and we finish or $t(c; A) \perp q$. But $t(c; A) \not\perp p$ and $p \not\perp q$ so, since p is regular, $t(c; A) \not\perp q$.

The important Theorems 2.9 and 2.11 require only that p and q have weight one. The next attribute holds for regular trivial types but not for general weight one trivial types.

2.13 Definition. The type $p \in S(A)$ is *totally trivial* if for any $\langle \bar{a}, \bar{b}, \bar{c} \rangle$ realizing p, if $\bar{a} \not\downarrow_A \bar{b}\frown\bar{c}$ then $\bar{a} \not\downarrow_A \bar{c}$ or $\bar{a} \not\downarrow_A \bar{b}$.

Using the transitivity characterization of regularity, it is easy to see

2.14 Proposition. *Every regular trivial type is totally trivial.*

The next example shows the necessity of regularity for the last proposition.

2.15 Example. (Fig. 6). Let the language of T contain unary predicates R (for regular) and W (for weight one), a binary relation E, and a ternary relation S. For simplicity, in the following description we identify each relation symbol of L with its interpretation in a model of T. Roughly, a model of T consists of two sets R and W. There is an equivalence relation with infinitely many infinite classes on W. Each class is named by an element of R. There is a graph defined on the elements of W. Each component of the graph lies in a single equivalence class. The graph is symmetric and contains no cycles. Finally, each edge is labeled by an element of R. For each element a of W and each element b of R, exactly one other element of W is connected to a by an edge labeled by b.

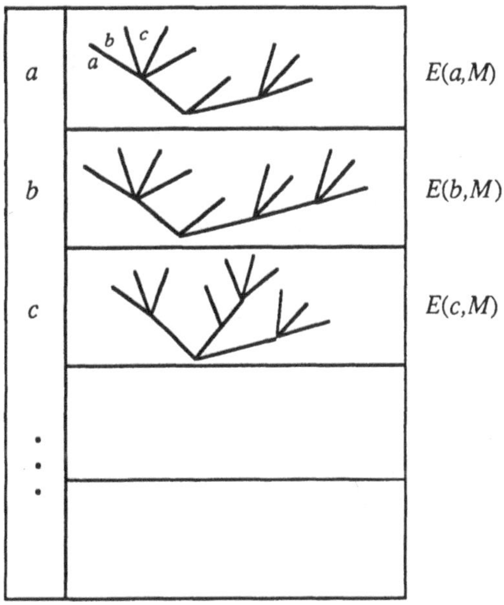

Fig. 6. Example XVI.2.15.

More formally, $E \subseteq R \times W$; each element of R is connected via E with infinitely many elements of W but each element of W is connected to exactly one element of R. Moreover, $S \subseteq R \times W \times W$. If $\langle r, w_1, w_2 \rangle \in S$ then $\langle r, w_2, w_1 \rangle \in S$ and for some r' both $\langle r', w_1 \rangle$ and $\langle r', w_2 \rangle$ are in E. Thus, the projection of S on $W \times W$ defines a symmetric graph on W which has no cycles and connects only points in the same equivalence class defined by R and E. The first coordinate of S determines a labeling of the edges of this graph by elements of R. Namely, each point will be related to infinitely many other points by the relation $(\exists x)R(x) \wedge S(x, y, z)$. There is a $1-1$ correspondence between the points in R and the points related to an element a. That is, for each $a \in W$, f_a is a $1-1$ map from R to the neighbors of a in the graph given by $f_a(x)$ is the unique z such that $S(x, a, z)$.

Now we claim, first, T is ω-stable. To verify this one must first check that (after adding predicates which indicate the distance between pairs of points in the graph and the possible equalities among the labels of the edges of a path connecting the pair) T admits elimination of quantifiers and then count types. Second, each 1-type containing $R(x)$ is regular and trivial. Here are two key observations to support this claim. Any permutation of R which preserves the number of components in the graph associated with a point of R extends to an automorphism of the model. Any permutation of an equivalence class which is a homomorphism for the expanded language extends to an automorphism of the model.

But, some 1-types which contain $W(x)$ have weight one and are trivial but are not totally trivial. If $W(x) \in q \in S(M)$ then the Morley rank (or U-rank) of q is determined by the 'distance' of a realization of q from M. It is easy to check that the rank of q is at most $\omega + 1$ and this rank is attained by the type $\hat{q}$ which asserts that x is not related by E to any element of M. Now, if $a, b \in W$ realize $\hat{q}$, they are independent over M if and only if they are in distinct classes of the partition. Thus, forking is trivial on $\hat{q}$. But, suppose a and b are in the same class of the partition and there exists a $c \in R$ with $S(c, a, b)$ and $\neg E(c, a)$. Then for any $d \in W$ with $E(c, d)$, d is independent from each of a and b but depends on $a ^\frown b$.

The following lemma connects the dimensional order property with the existence of non-trivial types. For convenience, we say a type r has depth at least one if there is a nonforking extension p of r with the following properties. For some $\mathbf{S}$-model M $p \in S(M)$, and there is a realization $\overline{c}$ of p, and a regular type $q \in S(M[\overline{c}])$ with $q \dashv M$. This is a special case of the definition of the depth of a type in Chapter XVII.

2.16 Lemma. *If T does not have the dimensional order property then T has no nontrivial regular type p with depth one or more.*

Proof. Without loss of generality assume $p \in S(M)$ and M is S-saturated. Suppose $\langle \overline{a}, \overline{b}, \overline{c} \rangle$ form a triangle which witnesses the nontriviality of p and that p has depth at least one. Choose $A \subseteq M$ with $|A| < \kappa(T)$ such that $t(\overline{a} ^\frown \overline{b} ^\frown \overline{c}; M)$ is strongly based on A. Form $M[\overline{a}, \overline{b}]$. Now $t(\overline{c}; A \cup \overline{a} ^\frown \overline{b})$ is realized in $M[\overline{a}, \overline{b}]$ by some $\overline{c}'$. If $\overline{c}' \not\downarrow_A M$ then $t(\overline{c}'; M) \perp p$. But since $\overline{a}$ and $\overline{b}$ are independent realizations of p, the only regular types realized in $M[\overline{a}, \overline{b}]$ are those which are not orthogonal to p. Thus, $\overline{c}' \downarrow_A M$. Suppose for contradiction that $\overline{c}' \not\downarrow_M \overline{a}$. Then $t(\overline{c}'; M \cup \overline{a}) \perp p$. Hence $\overline{c}' \downarrow_{M \cup \overline{a}} \overline{b}$. But, since $\overline{b} \downarrow_A M \cup \overline{a}$, transitivity of independence yields $\overline{c}' \downarrow_A \overline{b}$. This contradicts the choice of $\overline{c}'$. Now, since $\overline{a} \downarrow_M \overline{c}'$ and $\overline{b} \downarrow_M \overline{c}'$, $p_{\overline{c}'}$ is orthogonal to both $M[\overline{a}]$ and $M[\overline{b}]$ which gives an example of the DOP.

2.17 Historical Notes. This section appeared in [Baldwin & Harrington]. Lemma 2.16 is from [Saffe 1983]. Many of the other results were implicit in the main gap papers [Shelah 1982]. We mentioned after Lemma V.2.6 and in Section X.4 the similarity between an infinite set of indiscernibles and a model. The same phenomenon occurs in [Shelah 1986c] where Shelah proves a version of Lemma 2.12 replacing M by the algebraic closure of an infinite set of indiscernibles.

3. DOP Implies Many Nonisomorphic Models

In this section we prove any superstable theory with DOP has 2^λ models in every $\lambda \geq 2^{|T|}$ and an ω-stable countable theory with ENI-DOP has $2^{\aleph_0}$ countable models. To obtain a precise common statement of these two results, recall from Section XIV.2 that a type p over a set A is (μ, K)-tractable if there are fewer than μ realizations of p in a K-prime model over A.

3.1 Theorem. *Let T be a superstable theory with the DOP. Suppose the (λ, K)-tractable type p witnesses the DOP and $\lambda \geq \lambda_0(\mathbf{I})$. If $\mu \geq \lambda_0(\mathbf{I})$ then $I(\mu, K) = 2^\mu$.*

Remember that for $K = \mathbf{S}$, $\lambda_0(\mathbf{S}) = 2^{|T|}$ while for countable ω-stable T, $\lambda_0(\mathbf{AT}) = \aleph_0$. Our argument follows that of [Harrington & Makkai 1985]. Shelah's original proof [Shelah 1982] was more along the line discussed in Section I.5.

We begin by translating the DOP into a collection of data about a finite set of finite sequences. We will use these finite sequences to code a family of graphs into models of T with power λ and thus deduce the theorem.

The following reduction depends on several ideas. The first is that because of superstability, we can reduce the relations between the models which witness DOP according to Definition 1.3 to relations between finite sequences. The second is that because N_1 and N_2 were K-minimal (Theorem 1.4), we can demand that these finite sequences have weight one. Finally, since N_1 and N_2 were prime over K-strongly regular types we can demand either that $(N_1 \perp N_2; N_0)$ or, invoking Corollary XII.1.15, that N_1 is isomorphic to N_2 over N_0.

To apply the following lemma to a countable ω-stable theory recall that for such a theory a nonprincipal type is $(\aleph_0, \mathbf{AT})$-tractable.

3.2 Reduction. Suppose T is superstable and a (λ, K)-tractable type witnesses that T has DOP. Then, there exist finite sequences $\overline{n}_0$, $\overline{n}_1$, $\overline{n}_2$, and $\overline{n}_3$ such that $\overline{n}_0 \subseteq \overline{n}_1 \subseteq \overline{n}_3$, $\overline{n}_0 \subseteq \overline{n}_2 \subseteq \overline{n}_3$,

$$\overline{n}_1 \underset{\overline{n}_0}{\downarrow} \overline{n}_2,$$

$$\overline{n}_1 \rhd_{\overline{n}_2} \overline{n}_3,$$

$$\overline{n}_2 \rhd_{\overline{n}_1} \overline{n}_3,$$

and

$$\overline{n}_1 \cup \overline{n}_2 \rhd_{\overline{n}_0} \overline{n}_3.$$

Furthermore, there is a (λ, K)-tractable type $p \in S(\overline{n}_3)$ with $p \dashv \overline{n}_1$ and $p \dashv \overline{n}_2$. Finally, if $q_1 = t(\overline{n}_1; \overline{n}_0)$ and $q_2 = t(\overline{n}_2; \overline{n}_0)$ we can guarantee that q_1 and q_2 are stationary weight one types and either $q_1 = q_2$ or $q_1 \perp q_2$.

Proof. (Fig. 7). By Theorem 1.4 (or Theorem 1.16 for the ω-stable case) we can find a full amalgam of K-models, $\langle N_0, N_1, N_2, N_3 \rangle$ and a (λ, K)-

tractable type $r \in S(N_3)$ such that $r \dashv N_1$, $r \dashv N_2$. Moreover, we can insist that for $i = 1, 2$, $N_i = N_0[\overline{a}_i]$ where $t(\overline{a}_i; N_0)$ is K-strongly regular. If $t(\overline{a}_1; N_0)$ and $t(\overline{a}_2; N_0)$ are not orthogonal, Corollary XII.1.15 implies that we can find an isomorphism which fixes N_0, takes $\overline{a}_1$ to $\overline{a}_2$ and N_1 to N_2.

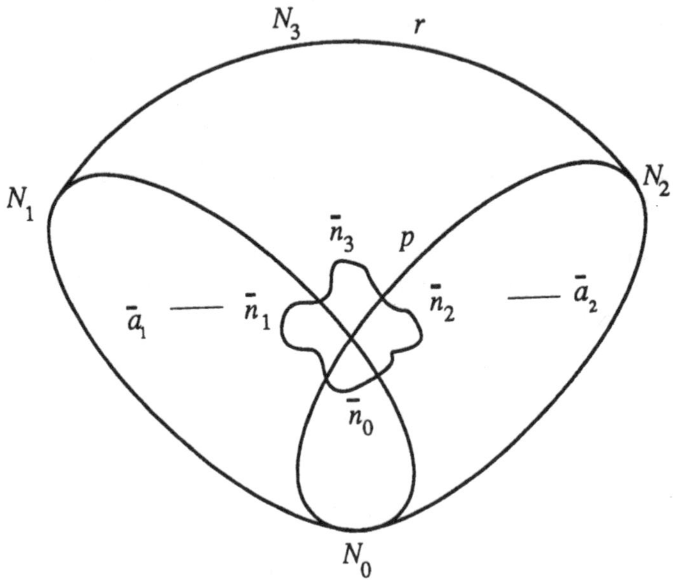

Fig. 7. Reducing to finite sets

Choose $\overline{n}_3 \subseteq N_3$ so that r is strongly based on $\overline{n}_3$. Let $p = r|\overline{n}_3$. Then choose $\overline{n}_i$, for $i = 1, 2$, so that $t(\overline{n}_3; N_i)$ is strongly based on $\overline{n}_i$. If $t(\overline{a}_1; N_0) \perp t(\overline{a}_2; N_0)$, clearly $t(\overline{n}_1; N_0) \perp t(\overline{n}_2; N_0)$. If not, replace $\overline{n}_2$ by $\overline{n}_2$ union the image of $\overline{n}_1$ under the isomorphism mapping N_1 to N_2 and similarly for $\overline{n}_1$. This replacement guarantees that $q_1 = q_2$. Now choose $\overline{n}_0 \subseteq N_0$ so that $t(\overline{n}_1 {}^\frown \overline{n}_2; N_0)$ is strongly based on $\overline{n}_0$. Since for $i = 1, 2$, $N_i = N_0[\overline{a}_i]$ and $t(\overline{a}_i; N_0)$ is K-strongly regular, each $t(\overline{n}_i; N_0)$ has weight one. Since N_3 is K-prime over $N_1 \cup N_2$, $\overline{n}_1 \rhd_{\overline{n}_2} \overline{n}_3$ and $\overline{n}_2 \rhd_{\overline{n}_1} \overline{n}_3$. Similarly, $\overline{n}_1 \cup \overline{n}_2 \rhd_{\overline{n}_0} \overline{n}_3$.

We can make the following simplifications in notation.

3.3 Notation. We will replace the vector notation by using individual letters, a, b, c to refer to finite tuples. Specifically we denote $\overline{n}_i$ as n_i for i=0,1,2,3. We will assume that the sequence $\overline{n}_0$ is named in the language. In particular this means that q_1, q_2 are stationary types over the empty set. For any model M we denote by $Q(M)$ the set of all tuples which realize either q_1 or q_2.

We are going to build many models by coding graphs on realizations of q_1 and q_2. The next result will simplify picking out the graph in the model. It generalizes Corollary 2.7 (since the q_i are only weight one rather than regular). We are able to prove the stronger result because of the additional information that the q_i are either equal or orthogonal.

3.4 Proposition. i) *q_1 and q_2 are trivial types.*
ii) *Forking is trivial on $Q(\mathcal{M})$.*

Proof. i) We have that $n_1 \rhd_{n_2} n_3$. Since $p \dashv n_1$ and $wt(t(n_1; n_2)) = 1$, by Theorem XV.2.12, either we are finished or $t(n_1; n_2)$ is trivial. Since $n_1 \downarrow_{n_0} n_2$, this implies q_1 is trivial. The argument for q_2 is entirely symmetric.

ii) If $q_1 = q_2$ then ii) is just a restatement of i). So suppose $q_1 \perp q_2$ and $I \subseteq Q(\mathcal{M})$ is pairwise independent but not independent and $|I|$ is minimal among all such sets. Then we can write I as $C \cup D \cup \{b\}$ where $C \subseteq q_1(\mathcal{M})$ and $D \cup \{b\} \subseteq q_2(\mathcal{M})$. By the triviality of q_2, $b \downarrow D$ and by orthogonality of q_1 and q_2, $b \downarrow C$. To get $b \downarrow C \cup D$, it suffices by Lemma II.2.5 to show $b \downarrow_D C$. This holds because $t(b; \emptyset) \perp t(C; \emptyset)$, $C \downarrow D$, and by induction $b \downarrow D$.

To reduce eyestrain at the possible cost of straining the reader's short term memory, we introduce a number of notions which are used only in this proof. Here is the first set.

3.5 Notation. A *germane pair* $\langle a, b \rangle$ is a realization of $t(n_1, n_2; \emptyset)$; a *germane triple*, $\langle a, b, e_{ab} \rangle$ is a realization of $t(n_1, n_2, n_3; \emptyset)$. If $\langle a, b, e_{ab} \rangle$ is a germane triple, we denote by p_{ab} a copy of $p \in S(e_{ab})$ which is orthogonal to both a and b.

Note that if $\langle a, b \rangle$ is a germane pair then $a \downarrow b$. Although we write e_{ab} and p_{ab}, there are many choices for this set and type for any given a, b.

3.6 Exercise. If $\langle a, b \rangle$ is a germane pair and $q_1 = q_2$ then $t(a^\frown b; \emptyset) = t(b^\frown a; \emptyset)$.

The principal technical difficulty of this proof centers around the following problem. Given p_{ab} and p_{cd} with $p_{ab} \not\perp p_{cd}$; show $a \not\downarrow c$ and $b \not\downarrow d$. This does not hold in general; the next few lemmas lay out special conditions where it does hold. We will then arrange the construction so we can obtain those conditions.

First suppose that $b = d$ or $a = c$.

3.7 Proposition. i) *If $\langle a, b \rangle$ and $\langle a', b \rangle$ are germane pairs and $p_{ab} \not\perp p_{a'b}$ then $a \not\downarrow a'$.*
ii) *If $\langle a, b \rangle$ and $\langle a, b' \rangle$ are germane pairs and $p_{ab} \not\perp p_{ab'}$ then $b \not\downarrow b'$.*

Proof.

i) If $a \downarrow a'$, since forking is trivial on $Q(\mathcal{M})$, $\{a, a', b\}$ is an independent set. As $a \rhd_b e_{ab}$ and $a' \rhd_b e_{a'b}$, we can conclude $e_{ab} \downarrow_b e_{a'b}$. Since $p_{ab} \dashv b$, this implies $p_{ab} \perp p_{a'b}$.
ii) If $q_1 = q_2$, ii) is a restatement of i). If not, it still follows by a symmetric proof.

To consider the case when a, b, c, d are distinct we introduce some further notation.

3.8 Notation. A 4-tuple $\langle a, b, c, d \rangle$ is in *normal position* if $\langle a, b \rangle$ and $\langle c, d \rangle$ are germane pairs and $a \downarrow d$ and $b \downarrow c$.

Note that $q_1 \perp q_2$ implies that any two germane pairs are in normal position. The notion of normal position can be visualized by regarding a, b, c, d as the vertices of a square. We will connect two vertices of the square to indicate the points are independent. The germane pairs guarantee the top and bottom are independent (draw the top and bottom). (Fig. 8). The additional requirement of normal position is that the diagonals are independent (draw the diagonals). (Fig. 9). Thus, if $\langle a, b, c, d \rangle$ is in normal

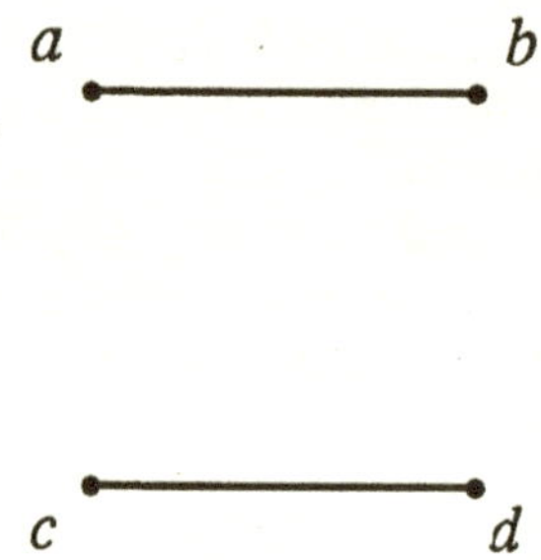

Fig. 8. Germane pairs

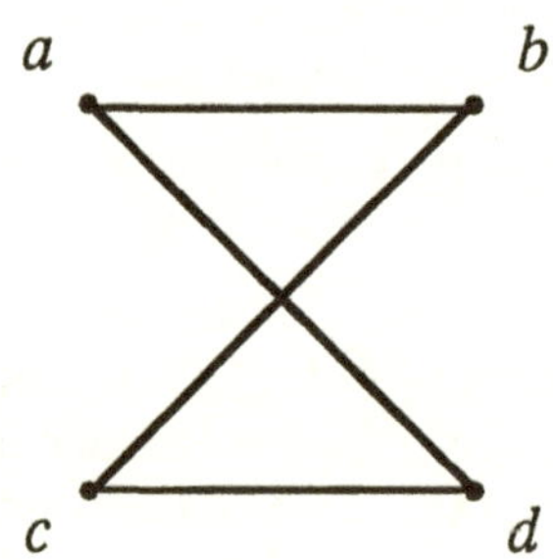

Fig. 9. Normal position

position, a and c realize q_1 while b and d realize q_2.

3.9 Exercise. If $\langle a, b, c, d \rangle$ is in normal position and $q_1 = q_2$ then $d \downarrow a^\frown b$.

The next proposition concerns whether we can draw the sides.

3.10 Proposition. *If $\langle a, b, c, d \rangle$ is in normal position with $a \downarrow c$ and $b \downarrow d$ then $p_{ab} \perp p_{cd}$.*

Proof. (Fig. 10). The hypotheses and the triviality of forking on $Q(\mathcal{M})$ yield that $\{a, b, c, d\}$ is an independent set and, in particular, $ab \downarrow cd$. Since $ab \rhd e_{ab}$ and $cd \rhd e_{cd}$, we deduce using Theorem VI.2.21 from $p_{ab} \dashv \emptyset$ that $p_{ab} \dashv e_{cd}$ and thus $p_{ab} \perp p_{cd}$.

Note that the argument for Proposition 3.10 depends only on the properties of the set $\{a, b\}$, not the ordered pair $\langle a, b \rangle$. Thus, we have

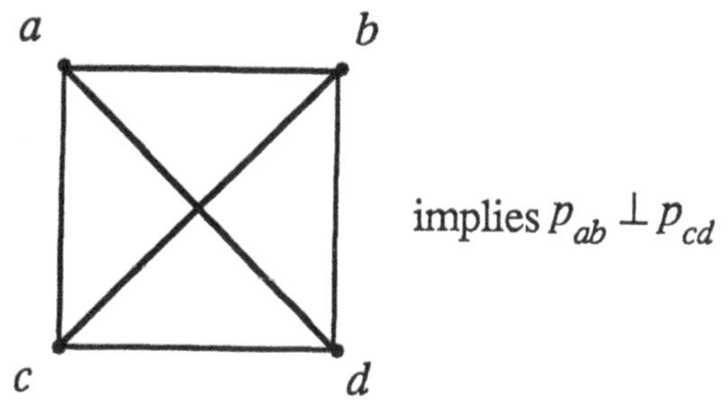

Fig. 10. Proposition XVI.3.10.

3.11 Exercise. Assume $\langle a, b, c, d\rangle$ is in normal position. Then $p_{ba} \perp p_{cd}$, $p_{ba} \perp p_{dc}$, $p_{ab} \perp p_{cd}$, and $p_{ab} \perp p_{dc}$.

This shows if $\langle a, b, c, d\rangle$ is in normal position and $p_{ab} \not\perp p_{cd}$ then at least one of $a \not\downarrow c$ and $b \not\downarrow d$ holds. We want them both to hold; this requires some further hypotheses.

3.12 Proposition. (Fig. 11). *Suppose $\langle a, b, c, d\rangle$ is in normal position and $p_{ab} \not\perp p_{cd}$.*

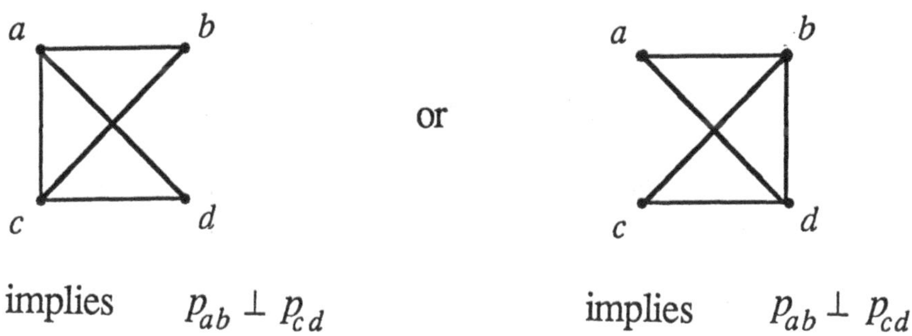

Fig. 11. Proposition XVI.3.12.

i) *If $q_1 = q_2$ then $a \not\downarrow c$ and $b \not\downarrow d$.*
ii) *If $q_1 \perp q_2$ and $ac \downarrow bd$ then $a \not\downarrow c$ and $b \not\downarrow d$.*

Proof. i) By Proposition 3.10, we know $a \not\downarrow c$ or $b \not\downarrow d$. Suppose $a \not\downarrow c$ and $b \downarrow d$. Since $q_1 = q_2$, $t(a^\frown b; \emptyset) = t(c^\frown d; \emptyset)$. Let α be an automorphism which interchanges a and b. Choose $\langle \hat{c}, \hat{d}\rangle$ to realize $t(\alpha(c), \alpha(d); a^\frown b)$ and with $\hat{c}\hat{d} \downarrow_{ab} cd$. We have $p_{cd} \not\perp p_{ab}$ and $p_{ab} \not\perp p_{\hat{d}\hat{c}}$ so $p_{cd} \not\perp p_{\hat{d}\hat{c}}$. As germane pairs, we have $c \downarrow d$ and $\hat{c} \downarrow \hat{d}$. We chose $\hat{d}\hat{c} \downarrow_{ab} cd$ and were given $d \downarrow ab$ so $d \downarrow \hat{d}$ and $d \downarrow \hat{c}$. Now $c \downarrow b$, $b = \alpha(a)$, and $\hat{c}$ realizes $t(\alpha(c); b)$ implies $\hat{c} \downarrow a$. Similarly,

$\hat{d} \downarrow a$. But $c \not\downarrow a$ so since q_1 has weight one $c \downarrow \hat{c}$ and $c \downarrow \hat{d}$. By Exercise 3.11 we conclude $p_{cd} \perp p_{\hat{d}\hat{c}}$ and this contradiction yields part i).

ii) Suppose for contradiction that $b \downarrow d$. By monotonicity from $ac \downarrow bd$ we conclude

$$c \underset{a}{\downarrow} b \tag{1}$$

and

$$a \underset{c}{\downarrow} d. \tag{2}$$

From $ac \downarrow bd$ and $b \downarrow d$, Corollary II.2.10 yields

$$b \underset{ac}{\downarrow} d. \tag{3}$$

Since $b \rhd_a e_{ab}$, (1) and Lemma VI.3.12 imply:

$$b \rhd_{ac} e_{ab}. \tag{4}$$

Similarly $d \rhd_c e_{cd}$, (2), and Lemma VI.3.12 imply:

$$d \rhd_{ac} e_{cd}. \tag{5}$$

Now (3), (4), and (5) imply $e_{cd} \downarrow_{ac} e_{ab}$. Using $b \rhd_a e_{ab}$ again, we conclude from (1) that $ac \downarrow_a e_{ab}$. From transitivity of independence, we deduce $e_{cd} \downarrow_a e_{ab}$. Now $p_{cd} \dashv a$ implies $p_{cd} \perp p_{ab}$. This contradiction implies $b \not\downarrow d$.

Note that the roles of bd and ac are completely symmetric in this argument. Thus repeating the argument with reversed roles yields $a \not\downarrow c$.

Beyond the additional hypothesis there is a certain asymmetry between the arguments for i) and ii). In i) we use the fact that one side of the square is dependent in proving that the other is. In ii), with no hypothesis about either side, we prove by contradiction that a given side is dependent.

If we were able to conclude that an arbitrary 4-tuple $\langle a, b, c, d\rangle$ in normal position, with $p_{ab} \not\perp p_{cd}$ satisfied $a \not\downarrow c$ and $b \not\downarrow d$ then we could code a graph G inside a model M by requiring $dim(p_{ab}, M)$ to be small iff $\langle a, b\rangle \in G$. The necessity of the extra hypothesis in Proposition 3.12 ii) forces us to be more subtle and replace a single link by an infinite set of links. Moreover, in the case when $q_1 \neq q_2$ we must make a further reduction which follows from Proposition 3.12 i) when $q_1 = q_2$. From Proposition 3.10 we know that if four points are in normal position and the associated types are nonorthogonal then either the left side or the right side must be dependent. We show now that, in fact, this result holds uniformly.

3.13 Proposition. *Either*

i) *for every* $\langle a, b, c, d\rangle$ *in normal position with* $p_{ab} \not\perp p_{cd}$, $a \not\downarrow c$ *or*
ii) *for every* $\langle a, b, c, d\rangle$ *in normal position with* $p_{ab} \not\perp p_{cd}$, $b \not\downarrow d$.

Proof. If $q_1 = q_2$ this follows from Proposition 3.12 i). If $q_1 \perp q_2$ let $\langle a, b, c, d\rangle$ be in normal position with $p_{ab} \not\perp p_{cd}$ and suppose $a \not\downarrow c$. If i) fails we can, by applying an automorphism, assume it fails for $\langle a, b, c', d'\rangle$ for some choice of c', d'. That is, $p_{ab} \not\perp p_{c'd'}$ but $a \downarrow c'$ and $b \not\downarrow d'$. Choose $\langle \hat{c}, \hat{d}\rangle$ to realize $t(c', d'; ab)$ but with $\hat{c}\hat{d} \downarrow_{ab} cd$. Since $q_1 \perp q_2$ $\langle c, d, \hat{c}, \hat{d}\rangle$ is in normal position. Triviality of forking yields $c' \downarrow ab$ and so $\hat{c} \downarrow ab$. By transitivity of independence $\hat{c} \downarrow c$. Moreover $d \downarrow \hat{d}$. For, if not, we have $\hat{d} \not\downarrow d$ and (by choice of d' and $\hat{d}$) $\hat{d} \not\downarrow b$. But then $d \not\downarrow b$. Since $p_{ab} \not\perp p_{cd}$ and $p_{ab} \not\perp p_{\hat{c}\hat{d}}$, transitivity of nonorthogonality for regular types yields the required contradiction.

With the proposition in mind we can assume the following

3.14 Convention. Condition 3.13 ii) holds.

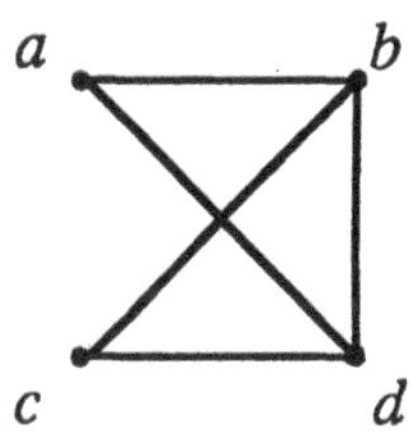

implies $p_{ab} \perp p_{cd}$.

Fig. 12. Convention XVI.3.14.

We will not appeal to this convention until the proof of the last proposition in this section.

We next describe the family of graphs which we will be able to recover, then the properties of the model erected on each graph, and finally the recovery process.

3.15 Construction (The graph H). Let G be a connected symmetric graph on a set B. Add to B a single element $\hat{b}$ and let A be an infinite set with a distinguished element $\hat{a}$. Define a graph H on $A \cup B \cup \{\hat{b}\}$ as follows. Each element of B is connected to $\hat{a}$. Each element of A is connected to $\hat{b}$. If $b, b' \in B$ and $\langle b, b'\rangle \in G$ then each of b, b' is connected in H to every member of a countable subset $A_{b,b'}$ of A. The subsets of A associated with distinct pairs from B are disjoint. Rename $B \cup \{\hat{b}\}$ as B.

Thus the main feature relating G and H is that two elements, b_1, b_2, of B are connected in G if and only if there is an infinite subset of A such that each element of that subset is connected in H to both b_1 and b_2.

Take the set $A \subseteq q_1(\mathcal{M})$ and the set $B \subseteq q_2(\mathcal{M})$ such that $A \cup B$ is an independent set. Now, by Lemma II.2.26 extend $A \cup B$ to $D = A \cup B \cup E$ by choosing for each germane pair $\langle a, b\rangle$ from $A \times B$ an element e_{ab} such that $\langle a, b, e_{ab}\rangle$ is a germane triple and so that D is independent with respect to the partial order whose only relations are $a < e_{ab}$ and $b < e_{ab}$. If $q_1 = q_2$, choose $e_{ab} = e_{ba}$. Fix for $\langle a, b\rangle \in A \times B$, a copy p_{ab} of p in $S(e_{ab})$. If $q_1 = q_2$, let $p_{ab} = p_{ba}$.

3.16 Exercise. Show $\langle \hat{a}\hat{b}\rangle$ is a germane pair.

We will use the following abbreviation only in the next proposition. For $\langle a, b\rangle \in A \times B$, let $D_{ab} = \{a, b, e_{ab}\}$.

3.17 Proposition. *For* $\langle a, b\rangle \in A \times B$, $p_{ab} \perp t(D - D_{ab}; D_{ab})$.

Proof. By the choice of $<$, $(D - D_{ab}) \downarrow_{ab} D_{ab}$. But, $(D - D_{ab}) \downarrow ab$ so $(D - D_{ab}) \downarrow (D_{ab})$. Since $p_{ab} \dashv \emptyset$ we finish.

With this in hand we can define from H (and thus from G) a model M^G such that $M^G \approx M^{G'}$ implies $G \approx G'$.

3.18 Construction (The model M^G). Let $S = \{p_{ab} : \langle a, b\rangle \in H\}$. Form by Theorem XIV.3.5 a model M^G with $|M^G| = \lambda$ and containing D so that $dim(p, M^G) < \lambda$ if $p \in S$ and $dim(q, M^G) = \lambda$ if q is irrelevant to S.

We will recover H and thus G from M^G by recovering the equivalence classes under forking of the elements of $A \cup B$ and then seeing which of them are related by H.

3.19 Notation. We will refer to a pair $\langle a, b\rangle \in A \times B$ as a *standard pair*. For any $d \in Q(M^G)$ we denote the equivalence class of d (for the equivalence relation of forking restricted to $Q(M^G)$) by $[d]$. Since M^G is fixed, we write Q for $Q(M^G)$.

Note that if $c \in Q$, $[c] \cap (A \cup B)$ has at most one element. The remainder of the proof depends on the crucial observation that for any germane pair $\langle a, b\rangle$, if $dim(p_{ab}, M^G) < \lambda$ then for some standard pair $\langle a', b'\rangle$, $p_{ab} \not\perp p_{a'b'}$.

The first step is to recover the domain of the graph H from M^G.

3.20 Proposition (Recover domain).

i) *If* $c \in Q$ *then* $[c] = [a]$ *for some* $a \in A$ *iff* $dim(p_{c\hat{b}}, M^G) < \lambda$.
ii) *If* $c \in Q$ *then* $[c] = [b]$ *for some* $b \in B$ *iff* $dim(p_{c\hat{a}}, M^G) < \lambda$.

Proof. i) The construction guarantees that the condition is fulfilled whenever $[c] = [a]$ for some $a \in A$. If $dim(p_{c\hat{b}}, M^G) < \lambda$ then for some standard pair $\langle a, b\rangle$, $p_{c\hat{b}} \not\perp p_{ab}$. If $b \not\downarrow \hat{b}$ then $b = \hat{b}$ so by Proposition 3.7 i) we finish. Thus, $b \downarrow \hat{b}$. If $q_1 \neq q_2$, $q_1 \perp q_2$ and $\langle c, \hat{b}, a, b\rangle$ are in normal position so by Proposition 3.10, $a \not\downarrow c$. If $q_1 = q_2$ then $\langle b, a\rangle$ is a germane pair. Suppose for contradiction that $a \downarrow c$. Then $\langle c, \hat{b}, b, a\rangle$ is in normal position. By Proposition 3.12 i) $c \not\downarrow b$ and $\hat{b} \not\downarrow a$. But $a \in A$ and $\hat{b} \in B$ so this last condition is impossible and we finish.

ii) The proof of ii) is entirely analogous.

It remains to recover the graph H. We must introduce one last bit of notation.

3.21 Notation. For $b \in B$ and an infinite $X \subseteq A$, we write $X \mathcal{F} b$ if for all but finitely many elements x of X, $\langle x, b\rangle \in H$.

We now show that $\mathcal{F}$ can be recovered from M^G. It is clear that we can recover G from this $\mathcal{F}$.

3.22 Proposition. *For any $X \subseteq A$ and any $b \in B$, $X\mathcal{F}b$ if and only if*

$$\begin{array}{l}\textit{for some } \tilde{b} \in [b] \textit{ and all but finitely many } x \in X, \\ \textit{there is an } x' \in [x] \textit{ with } dim(p_{x'\tilde{b}}, M^G) < \lambda.\end{array} \qquad (*)$$

Note that we are able to choose a fixed $\tilde{b} \in [b]$. This strengthens the natural version which would make $\tilde{b}$ depend on x.

Proof. Clearly $X\mathcal{F}b$ implies $(*)$. Suppose $(*)$ holds for X and b but $X\mathcal{F}b$ fails. Then there is an infinite subset $X_0 \subseteq X$ such that for each $x \in X_0$ there is an $a_x \in [x]$ with $dim(p_{a_x\tilde{b}}, M^G) < \lambda$ but $dim(p_{xb}, M^G) = \lambda$. The first condition implies that for each $x \in X_0$ there is an $\hat{x} \in A$ and a $b_x \in B$ such that $p_{a_x\tilde{b}} \not\perp p_{\hat{x}b_x}$. Since $A \cup B$ is independent, $A \cap B = \emptyset$, and forking is transitive on $Q(M^G)$, the tuple $\langle a_x, \tilde{b}, \hat{x}, b_x\rangle$ is in normal position for each $x \in X_0$. By Convention 3.14 for each x, $\tilde{b} \not\downarrow b_x$. Since each $b_x \in B$, this implies all the b_x are equal. Call the common value $\bar{b}$.

Thus we have a single pair $\langle \tilde{b}, \bar{b}\rangle$ and an infinite set of pairs $\langle a_x, \hat{x}\rangle$ such that for each x, $p_{a_x\tilde{b}} \not\perp p_{\hat{x}\bar{b}}$.

Let $X_0' = \{a_x : x \in X_0\}$ and $\hat{X}_0 = \{\hat{x} : x \in X_0\}$. Clearly X_0' is infinite. To see that $\hat{X}_0$ is infinite we show the map $x \mapsto \hat{x}$ is $1-1$. If not, for some $x \neq y \in X_0$ we have $p_{a_x\tilde{b}} \not\perp p_{\hat{x}\bar{b}}$ and $p_{a_y\tilde{b}} \not\perp p_{\hat{x}\bar{b}}$ so $p_{a_x\tilde{b}} \not\perp p_{a_y\tilde{b}}$ which implies $a_x \not\downarrow a_y$, contrary to the hypothesis that $x \downarrow y$. Now choose $X_1 \subseteq X_0' \cup \hat{X}_0$ such that X_1 is finite and $\tilde{b}\bar{b} \downarrow_{X_1} X_0' \cup \hat{X}_0$. Then choose $w \in X_0$ such that neither a_w nor $\hat{w}$ is in X_1. Since A is independent, this implies $a_w\hat{w} \downarrow \tilde{b}\bar{b}$. Since $\langle a_w, \tilde{b}, \hat{w}, \bar{b}\rangle$ is in normal position and $p_{a_w\tilde{b}} \not\perp p_{\hat{w}\bar{b}}$, this contradicts Proposition 3.12 ii) and completes the proof of the proposition.

Since there are 2^λ nonisomorphic graphs of the requisite sort and we have proved that $M_G \approx M_{G'}$ implies $G \approx G'$, we conclude Theorem 3.1.

3.23 Historical Notes. The proof presented in this section is based on that in [Harrington & Makkai 1985]. This proof differs significantly from Shelah's [Shelah 1982]. Here we use the forking technology to construct an interpretation of the theory of graphs into models of T. Shelah makes another application of the general method discussed in Section I.5. That is, he constructs generalized Ehrenfeucht-Mostowski models over the graphs and argues on general grounds that they are not isomorphic. This general argument requires a detailed combinatorial analysis in Chapter VIII of [Shelah 1978]. We avoid that analysis. On the other hand, Shelah applies his method again to show that theories with the 'omitting types order property' [Shelah 198?] have the maximal number of models in each uncountable power. There is no known extension of the method here to that case. Shelah's method will be expounded in [Hodges 198?]. Saffe [Saffe 1982] also proved this result by a method similar to that used here. This result leaves one lacuna in the general program. Let T be a countable superstable theory which is not ω-stable and suppose the continuum hypothesis fails.

Then for an uncountable cardinal, κ, below the continuum we have not proved that the dimensional order property implies that T has 2^κ models of power κ. This problem is resolved by Theorem IX.1.20 of [Shelah 1978].

Chapter XVII

NDOP: Theories Without the Dimensional Order Property

In this chapter we investigate theories without the dimensional order property. As in the previous chapter, our main concerns will be with **S**-models of superstable theories and arbitrary models of ω-stable theories. We will show that for any acceptable class K, if T does not have the DOP, each K-model of T can be decomposed as a K-prime model over a skeleton which is an independent tree of 'small' K-models. If the model has cardinality λ, this tree will be isomorphic to a subset of $\lambda^{<\omega}$.

The following simple example is a good prototype for the kind of counting done in this chapter. The theory in the example is *deep* in the sense made precise in Section 2 and does not have the dimensional order property. Thus by Theorem 4.8, it has 2^λ models of power λ for each uncountable λ. We sketch here how this result can be seen directly.

Let T be the theory with a single function symbol f such that there is a unique point which is mapped to itself by f and all points have infinitely many preimages. T admits elimination of quantifiers. The models of T are best regarded as unions of components where two elements a and b are in the same component if for some m and n, $f^m(a) = f^n(b)$. There are, in fact, 2^λ possibilities for the unique component which has a base point so the others can be ignored in calculating the number of models of power λ. The component with a base point can be thought of as a subset of the tree $\lambda^{<\omega}$ which is closed under predecessor. Thus our task reduces to showing there are 2^λ such trees. To see this, we will show that each ordinal $\alpha < \lambda$ can be coded by a tree, $T_\alpha \subseteq \lambda^{<\omega}$. The T_α are constructed similarly to the coding of an arbitary countable ordinal by a subtree of $\omega^{<\omega}$. The difference is that rather than placing a single copy of T_β, for each $\beta < \alpha$, on the first level to code α, we place λ copies of each such T_β. This allows each point to have infinitely many preimages without disturbing the effect of the coding. There is a series of exercises at the end of Section 3 which explores the meaning of the notions defined in this chapter for this example.

Section 1 contains some preliminary results on the type of tree we use for a skeleton. In Section 2 we discuss representations of models and prove the decomposition theorem. In Section 3 we define the depth of a theory and use the decompostion theory to produce an upper bound on $I(\aleph_\beta, K)$ in terms of β and the depth of T. This requires that we calculate the number of (labeled) trees of each depth contained in $\lambda^{<\omega}$. In Section 4, we begin the proofs of the lower bounds on the spectrum function. We are able to complete the computation for theories with high depth and when $\aleph_\beta \geq \aleph_\omega$. The remaining cases are considered in the next chapter.

Since we only defined the dimensional order property for superstable theories, there is a de facto assumption in this section that the theory is superstable. In fact, this hypothesis is irrelevant to the discussion in Section 1 of normal trees. It becomes important when we assume the existence of **S**-prime models and in Section 2 where we show the existence of a decomposition. Shelah [Shelah 1982] discusses the dimensional order property for arbitrary stable theories.

1. *Normal Trees*

This section contains technical results on a generalization of an independent tree, namely a normal tree. We also introduce the notion of a stable system and prove some preliminary results about such systems. They play a small role in this book and a much larger role in Shelah's study of superstable but not ω-stable theories. We assume familiarity with the properties of independent systems which were discussed at the end of Section II.2.

We begin by establishing some notation for the trees that will form the skeletons of the decomposition. Informally, a tree is a partially ordered set which is isomorphic to a subset of $\langle \lambda^{<\omega}, \subseteq \rangle$ for some infinite λ. We give a more formal definition below, but these are the structures to remember. In the following notations for the 'cone above an element' or 'set of predecessors of an element' note that we form a union in our definition. This simplifies the notation when these concepts are applied.

1.1 Notation. i) Recall that in a discrete partial order, we denote by a^- the predecessor of a. If $b^- = a$ we say 'a precedes b' or 'b succeeds a'.

ii) We now define by induction the *cone above* a. Let $a^1_> = \{b : b^- = a\}$ and for each n, $a^{n+1}_> = \bigcup\{b^n_> : b^- = a\}$. The cone (strictly) above a is $a_> = \bigcup_{n<\omega} a^n_>$. Finally $a_\geq = a_> \cup \{a\}$. Similarly we can define $a_< = \bigcup\{b : b < a\}$. Recall that $a_\#$ denotes the set of elements incomparable to a.

iii) A partial order $(A, <)$ is a *tree* if no pair of incomparable elements has a common upper bound. Thus in a tree the set of predecessors of each element is linearly ordered by the induced ordering.

iv) A *well-founded* tree is a tree $(A, <)$ such that for each $a \in A$, $a_<$ is wellordered. The ordinal of this wellordering is called the *height* of a, and denoted $\mathrm{ht}(a)$. If each element has at most λ successors we call λ the *width* of the tree. Note that the cardinality of the tree is $\lambda^{<\omega} = \lambda$ as λ is always infinite. A node with no successor is called a *leaf.*

1.2 Exercise. Show $a_> = \bigcup\{b : b > a\}$

The next definition specifies the form we want the skeleton of a model to have.

1.3 Definition. (Fig. 1). $(A, <, A_{\langle\rangle})$ is a *normal tree* if

i) The partial order $(A, <)$ is isomorphic to a downward closed subset of the tree $(\lambda^{<\omega}, \subseteq)$ for some cardinal λ.
ii) $a < b$ implies $a \subseteq b$.
iii) $A_{\langle\rangle}$ denotes the element mapped to the empty node.
iv) For every $a \in A$, $a^1_>$ is an independent set over a.
v) For every $a \in A$ with $a^- \neq A_{\langle\rangle}$, $t(a; a^-) \dashv a^{--}$.

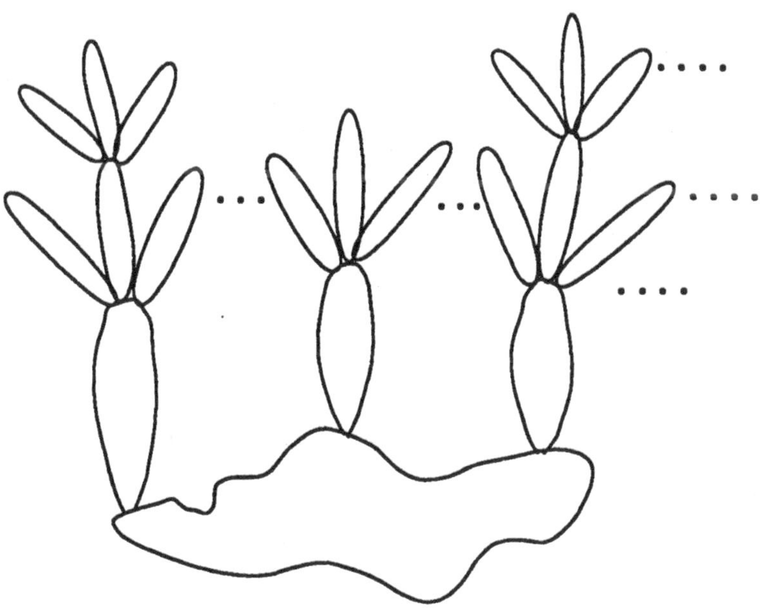

Fig. 1. A normal tree

Note that ii) implies that for each $a \in A$, $a^- = a_<$. Condition v) implies that the partial order $<$ on A is the same as the partial order $<'$ defined by $a <' b$ if a dominates b over $A_{\langle\rangle}$.

Our first task is to verify that any normal tree is independent with respect to $<$. We could, of course, have added that property to the definition but the version used reflects more clearly the way in which normal trees are constructed. The proof proceeds via two claims.

1.4 Claim 1. *For every n, $t(a^n_>;a) \dashv a^-$. Hence, $t(a_>;a) \dashv a^-$.*

Proof. If $n = 1$, the result follows from the triviality of orthogonality. Specifically, this is Exercise VI.2.5. Suppose we have this result for m and $n = m+1$. Then, by induction, for each $b \in a^1_>$, we have $t(b^m_>;b) \dashv a$. We also have $t(b;a) \dashv a^-$. Thus, by the transitivity of $\dashv$, (compare Lemma VI.3.10 and Exercise VI.3.11), $t(b^m_>;a) \dashv a^-$. Since $a^1_>$ is an independent set, by Theorem VI.2.21 $\{b^m_\geq : b \in a^1_>\}$ is an independent set. Now as in the case $n = 1$, $t(a_>;a) \dashv a^-$.

The following remark is actually contained in the proof of Claim 1.

1.5 Exercise. Show that for any a in a normal tree, $\{b_\geq : b^- = a\}$ is an independent set.

1.6 Claim 2. *For every a in a normal tree, $a \downarrow_{a_<} a_\#$.*

Proof. We induct on the height of a. If $\mathrm{ht}(a) = 0$, there is nothing to prove. Suppose $a^- = b$ and we have shown by induction that $b \downarrow_{b_<} b_\#$. To show $a \downarrow_{a_<} a_\#$, note that $a_\# = b_\# \cup X$ where $X = \{c : c \geq b \wedge a \not\leq c\}$. Since A is normal, $b = a_<$ and $b^- = b_<$. By Claim 1, $t(b_>;b) \dashv b^-$; we conclude $t(b_>;b) \perp t(b_\#;b^-)$. Since $b_> = a_\geq \cup X$, we have, in particular, that $b_\# \downarrow_b (a_\geq \cup X)$. By monotonicity, this yields $(b_\# \cup X) \downarrow_{X \cup b} a_\geq$. But, the last exercise shows $(X \cup b) \downarrow_b a_\geq$ so by transitivity $(b_\# \cup X) \downarrow_b a_\geq$. That is, $a_\# \downarrow_{a_<} a_\geq$ which is more than is required.

1.7 Claim 3. *If $(A, <, A_{\langle\rangle})$ is a normal tree then for any $a, b \in A$ with neither equal to $A_{\langle\rangle}$ and $a^- \neq b^-$, $t(a;a^-) \perp t(b;b^-)$.*

Proof. As $a^- \cap b^-$ is a proper subset of $a \cap b$, $t(a;a^-) \dashv (a^- \cap b^-)$ and, since $a^- \downarrow_{a^- \cap b^-} b^-$ by Theorem VI.2.21, the result follows easily.

One of the major tasks in the remainder of this text is the reduction of problems about the class of all models to the consideration of problems about **S**-saturated models. Relative **I**-saturation (Section X.1) is an important tool for this purpose. To apply it we need Theorem 1.8.

In some of the ensuing definitions we have to deal with elements of the sets in an independent family. Recall that in Chapter XVI we adopted the convention of referring in such situations to the systems as $\bar{A}$, $\bar{B}$, etc.,the elements of the system as A, B etc. and the elements of the elements of the system as a, b, etc. We often identify a system $\bar{A}$ or an ideal J with its union and depend on the context to indicate which is meant. Sometimes for emphasis we insert the union sign.

The proof of the following theorem contains an important idea masquerading as a technical device. Let $\bar{A}$ be a partially ordered set of sets and $\bar{a} \in \cup\bar{A}$. Then $\bar{a}$ can be partitioned as $\langle \bar{a}_0, \ldots, \bar{a}_k \rangle$ for some k so that there is a set $\langle A_0, \ldots, A_k \rangle$ of elements of $\bar{A}$ such that the A_i are incomparable with respect to the partial order and $\bar{a} \cap A_i = \bar{a}_i$. The crucial point is that when $(\bar{A}, <)$ is an independent partial order, this partition breaks $\overline{a}$ down into smaller independent pieces. This reduction permits one to carry out an induction on the length of $\overline{a}$. This idea also shows up in, e.g. [Shelah 1975a] and [Baldwin & Shelah 1985].

We prove the following result in the context of an admissible class K. Thus the sequence $\overline{a}$ may be infinite (e.g. $K = \mathbf{S}_{\aleph_1}$.) However, for the major applications in this book $\lg(\overline{a})$ and so γ are finite.

1.8 Theorem. *Let $(\bar{A}, <)$ be an independent family of models from a class K and J an ideal in $\bar{A}$ such that*

for every $A \in \bar{A}$ there is a maximal element B of J which is less than A. (∗)

Then $J \leq_I \bar{A}$.

Proof. Suppose $q(\overline{x}; \overline{b})$ is an I-formula with parameters $\overline{b} \in J$. Suppose for $\overline{a} \in \bar{A}$, $\models q(\overline{a}; \overline{b})$. Partition $\overline{a}$ as $\langle \overline{a}_i : i < \gamma \rangle$ where each $\overline{a}_i = A_i \cap \overline{a}$ for some $A_i \in \bar{A}$ and the A_i for $i < \gamma$ are incomparable with respect to the partial ordering. For each $i < \mu$, choose using $(*)$ B_i to be a maximal member of J which is contained in A_i.

Now we show by induction on $i < \gamma$ that there exists a function f from $\langle \overline{a}_i : i < \gamma \rangle$ into J such that for each i

$$\models q(f(\overline{a}_0), f(\overline{a}_1), \ldots, \overline{a}_i, \overline{a}_{i+1}, \ldots; \overline{b}).$$

Suppose we have defined $f(\overline{a}_l)$ for $l < i$. Let $J_1 = \{A \in \bar{A} : A \leq A_i\}$ and let J_2 be the ideal generated by J and the A_j for $j \neq i$. Since $(\bar{A}, <)$ is an independent partial order, $J_1 \downarrow_{J_1 \cap J_2} J_2$. By monotonicity $\overline{a}_i \downarrow_{B_i} \overline{a}' \frown \overline{b}$ where $\overline{a}'$ is the sequence containing $f(\overline{a}_l)$ for $l < i$, and the $\overline{a}_l$ for $l > i$. Since B_i is strongly I-saturated there is an $\overline{a}'_i = f(\overline{a}_i) \in B_i$ such that

$$\models q(f(\overline{a}_0), f(\overline{a}_1), \ldots, \overline{a}'_i, \overline{a}_{i+1}, \ldots; \overline{b}).$$

The condition $(*)$ imposed on I in Theorem 1.8 seems to be the abstract content (at least in this context) imposed on a partial order by Shelah's concrete representation of a *stable system* as a family of sets indexed by finite subsets of some other set. For more information see [Shelah 198?].

1.9 Exercise. Show that if $(\bar{A}, <)$ is a normal tree and J is an ideal in $\bar{A}$, then J satisfies $(*)$.

We deduce the following immediately from Corollary X.1.18.

1.10 Theorem. *Suppose $\lambda(I)$ is regular, J is an ideal in a normal tree and M_J is I-prime over J. Then $M_J \downarrow_J \bar{A}$.*

The following result which is applied in Section XVII.4 also follows easily from Corollary X.1.18. If J is an ideal in a normal tree $\bar{A}$ of I-models M_J denotes the I-prime model over J.

1.11 Corollary. *If $J_0 \subseteq J_1$ are a pair of ideals in a normal tree $\bar{A}$ then $M_{J_0} \downarrow_{J_0} J_1$ so M_{J_1} is constructible over M_{J_0}. If $\langle J_n : n < \omega \rangle$ is an increasing* *sequence of ideals and M_n is I-prime over J_n then $\bigcup M_n$ is I-prime over $\bigcup J_n$.*

Recall that the notion of parallel free amalgams was introduced in Notation XVI.1.10. Since two parallel amalgams are one of the simplest examples of an independent partial order we can immediately conclude

1.12 Corollary. *Let the* **AT**-*amalgam $\mathcal{N} = \langle N_0, N_1, N_2\rangle$ be parallel to the set-amalgam $\mathcal{A} = \langle A_0, A_1, A_2\rangle$. Then $N_1 \cup N_2 \leq_{\mathbf{AT}} A_1 \cup A_2$.*

1.13 Exercise. Show that if $\mathcal{A}$ is a set-amalgam, there is a parallel amalgam of **S**-models.

Using Theorem 1.8 we can construct a full amalgam of **S**-models, $\mathcal{N}'$, parallel to a given full amalgam of models, $\mathcal{N}$. We will use this result to give another proof that the dimensional order property depends on T rather than on a particular acceptable class K.

1.14 Theorem. *If $\mathcal{N} = \langle N_0, N_1, N_2, N_3\rangle$ is a full* **AT**-*amalgam then there is a full* **S**-*amalgam parallel to it.*

Proof. Choose N_0', an **S**-model containing N_0 and independent from $N_1 \cup N_2$ over N_0. Extend $N_1 \cup N_0'$ to an **S**-model N_1' with $N_1' \downarrow_{N_0' \cup N_1} N_2$ and hence by transitivity with $N_1' \downarrow_{N_0'} N_2$. Then extend $N_2 \cup N_0'$ to an **S**-model N_2' with $N_2' \downarrow_{N_0' \cup N_2} N_1'$ and hence by transitivity with $N_2' \downarrow_{N_0'} N_1'$. By Theorem 1.8, $N_1 \cup N_2 <_{\mathbf{AT}} N_1' \cup N_2'$. Let $\langle a_i : i < \alpha\rangle$ be a construction of N_3 over $N_1 \cup N_2$. By Lemma X.1.18 ii), for each i,

$$N_1 \cup N_2 \cup A_i \leq_{\mathbf{AT}} N_1' \cup N_2' \cup A_i.$$

Invoking Theorem X.1.17 this implies $t(a_i; N_1' \cup N_2' \cup A_i)$ is isolated. Thus, N_3 is constructible over $N_1' \cup N_2'$. So N_3 can be embedded in any **S**-prime model over $N_1' \cup N_2'$ which completes the theorem.

Now we give a second argument that the notion of DOP does not really depend upon a class of models but is a property of a theory. For simplicity, we show only the equivalence for the two most important cases.

1.15 Theorem. *Let T be a countable ω-stable theory. Then T has the* **S**-*NDOP if and only if T has the* **AT**-*NDOP.*

Proof. It is immediate that the **AT**-NDOP implies the **S**-NDOP. Suppose T has the **S**-NDOP. If $\mathcal{N}$ is a full **AT**-amalgam we can by Theorem 1.14 extend $\mathcal{N}$ to a parallel full **S**-amalgam. Now let p be an arbitrary type in $S(N_3)$ and p' a nonforking extension to $S(N_3')$. By the **S**-NDOP, p' and hence p is not orthogonal to, say N_1'. By the choice of a parallel amalgam, $N_1' \downarrow_{N_1} N_2$. As N_3 is **AT**-prime over $N_1 \cup N_2$, we deduce from Corollary X.1.20 that $N_1' \downarrow_{N_1} N_3$. Since $p \not\perp N_1'$, Corollary VI.2.22 yields $p \not\perp N_1$. Thus T satisfies **AT**-NDOP.

1.16 Historical Notes. Shelah introduced the notion of a normal tree in [Shelah 1982]. The treatment here relies both on the account in [Makkai

1984] and [Harrington & Makkai 1985] and on [Shelah 198?]. The proof of Theorem 1.8 given here is based on that in [Shelah 198?] which is considerably simpler than that in [Shelah 1982] which was reproduced in [Makkai 1984].

2. Decompositions of Models

In this section we show that if T is a countable superstable theory without the dimensional order property then every **S**-model of T admits a decomposition into models of power the continuum. Moreover, if T is ω-stable we can require the constituent models to be countable.

The difficult extension of this result to get countable models in the decomposition of a model of an arbitrary superstable theory has been obtained by both Hart and Saffe [Hart 1986].

2.1 Definition. By a *K-representation* of a model M we mean a normal tree $(\bar{A}, <, A_{\langle\,\rangle})$ of subsets of M such that:

i) For each $A \in \bar{A}$, $A \in K$.
ii) For every $A \in \bar{A}$ except $A_{\langle\,\rangle}$, there is an $a_A \in A$ such that $t(a_A; A^-)$ is K-strongly regular and $A = A^-[a_A]$.
iii) M is K-prime over $\bigcup \bar{A}$.

We write p_A for $t(A; A^-)$ and abbreviate $(\bar{A}, <, A_{\langle\,\rangle})$ by $\bar{A}$. Note that for each A, p_A is a stationary type.

Note that each p_A is a weight one type so $\not\perp$ is an equivalence relation on the set of p_A for $A \in \bar{A}$. Moreover, by 1.7 (Claim 3), each equivalence class consists entirely of successors of a single node. There may be several equivalence classes above any particular node. We often refer to a K-representation of M as K-decomposition of M. The difference is approximately that between an internal and external direct sum.

Before showing that under favorable conditions each K-model has a K-decomposition we deduce a few properties which are implied by the existence of such a decomposition. The first extends the triviality of $\dashv$ in a theory without DOP by replacing the pair of independent sets by an independent tree.

2.2 Lemma. *Let T be a superstable theory which satisfies NDOP. If $\bar{A}$ is a K-decomposition of the model M and $p \not\dashv M$ then for some $A \in \bar{A}$, $p \not\dashv A$.*

Proof. First, we show that we can assume $\bar{A}$ is finite. Let $q \in S(M)$ with $p \not\perp q$. Since $M \in K$, there is a finite subset $B \subseteq M$ with q strongly based on B. Thus, $p \not\perp q|B$ and $p \not\dashv B$. For some finite $\bar{A}_0 \subseteq \bar{A}$, B is **I**-atomic over $A_0 = \cup \bar{A}_0$ and so we can choose M_0 which contains B, is contained in M and is K-prime over A_0. Renaming M_0 as M and $\bar{A}_0$ as $\bar{A}$, we have the required reduction.

Now, we work by induction on the finite cardinality of $\bar{A}$. If $\bar{A}$ is linearly ordered the result is evident. If not, $\bar{A} = \bar{A}_1 \cup \bar{A}_2$ where the $\bar{A}_i$ are ideals and $\bar{A}_1 \cap \bar{A}_2$ is linearly ordered and has a maximal element A. Since $\bar{A}$ is an independent tree, $\bar{A}_1 \downarrow_A \bar{A}_2$. Moreover if, for $i = 1, 2$, we choose N_i to be K-prime over $\bigcup \bar{A}_i$, $N_1 \downarrow_A N_2$. If M_3 is K-prime over $N_1 \cup N_2$, there is, as M is prime over $\bigcup \bar{A}$, an embedding of M into M_3. Thus $p \not\perp M_3$. But then by NDOP, $p \not\perp$ one of N_1, N_2, say N_1. By induction, $p \not\perp A'$ for some $A' \in \bar{A}_1$ and we finish.

From this result we can deduce that M is tied even more tightly to an $\bar{A}$ which represents it.

2.3 Definition. The K-model M is *K-minimal* over $\bar{A}$ if $\bigcup \bar{A} \subseteq M$ and there does not exist a K-model N with $\bigcup \bar{A} \subseteq N \subseteq M$ and $N \neq M$.

2.4 Corollary. *Let T be a superstable theory which satisfies NDOP. If $\bar{A}$ is a K-representation of M then M is K-minimal over $\bar{A}$.*

Proof. Let $N \prec M$ be K-prime over $\bigcup \bar{A}$. If N is a proper subset of M, there is a $p \in S(N)$ which is realized in $M - N$ and is K-strongly regular. By Lemma 2.2, $p \not\perp A$ for some $A \in \bar{A}$. Thus, there is a K-strongly regular $q \in S(A)$ such that the nonforking extension q^N of q to $S(N)$ satisfies $q^N \not\perp p$. So q^N is realized in $M - N$ by some $\bar{b}$. But $\bar{A} \rhd_A M$ implies $\bar{b} \downarrow_A M$ and this contradiction yields the theorem.

2.5 Exercise. Show that if T is ω-stable and $\bar{A}$ is an **S**-decomposition of the **S**-model M then, in fact, M is **AT**-minimal over $\bar{A}$.

The next two results are further technical properties of a decomposition which we will need in the next section. To simplify the statements of these results we fix some more notation. Recall that $A^+ = \{B : B^- = A\}$. In a K-representation we define A_+ as $\{a_B : B^- = A\}$. Now we describe A_+.

2.6 Corollary. *If $\bar{A}$ is a K-representation of M then for each $A \in \bar{A}$, A_+ is a maximal independent set of realizations in M of strongly regular types over A.*

Proof. Note that $A_+ \rhd_A \bar{A}$. If $t(\bar{c}; A)$ is K-strongly regular and $\bar{c} \downarrow_A A^+$ then $\bar{c} \downarrow_A \bar{A}$. But we can choose $\bar{c} \in M = A[\bar{A}]$ so this is impossible.

We want to extend the descriptions of the successors of a point in a normal tree to a description of the successors of an ideal. Thus, we write I^+ for $\{A : A^- \in I\}$ and I_+ for $\{a_A : A^- \in I\}$. As usual we systematically confuse I^+ with $\bigcup I^+$. In order to study I_+ it is important that we find a common domain for the types in I_+. The natural candidate is M_I.

2.7 Lemma. *If I is an ideal in the K representation, $\bar{A}$, of M then I_+ is a maximal independent set of realizations in M of K-strongly regular types over M_I.*

Proof. Let $\bar{a} \in M$ realize p, a K-strongly regular type over M_I and suppose $\bar{a} \downarrow_{M_I} I_+$. Since T satisfies NDOP, $p \not\perp A$ for some $A \in I$. Thus, for some regular $q \in S(A)$, $q \not\perp p$. But then, $p \not\perp q^{M_I}$ so q^{M_I} is realized by some $\bar{c} \in M_I[\bar{a}]$. Since $\bar{a} \triangleright_{M_I} M_I[\bar{a}]$, $\bar{a} \downarrow_{M_I} I_+$ implies $\bar{c} \downarrow_{M_I} I_+$. By transitivity of independence, $\bar{c} \downarrow_A I_+$. But, $\bar{A} \downarrow_I M_I$ implies $A_+ \subseteq I_+$ so this contradicts the previous corollary.

This result implies that I^+ is a set of models each of which is K-prime over one of a maximal independent set of realizations in M of K-strongly regular types over M_I. Note that A^+ (I^+) and A_+ (I_+) dominate each other over A (M_I). Now we obtain the most important consequence of the NDOP.

2.8 Theorem. *If T is superstable and does not have the dimensional order property, then every K-model of T has a K-representation.*

Proof. (Fig. 2). Let $M \in K$ and suppose $|M| = \lambda$. We define a partial isomorphism, $f : \eta \mapsto M_\eta$ from $\lambda^{<\omega}$ into the set of K-submodels of M such that the image of f is the required K-representation. Let $M_{\langle\,\rangle}$ be any

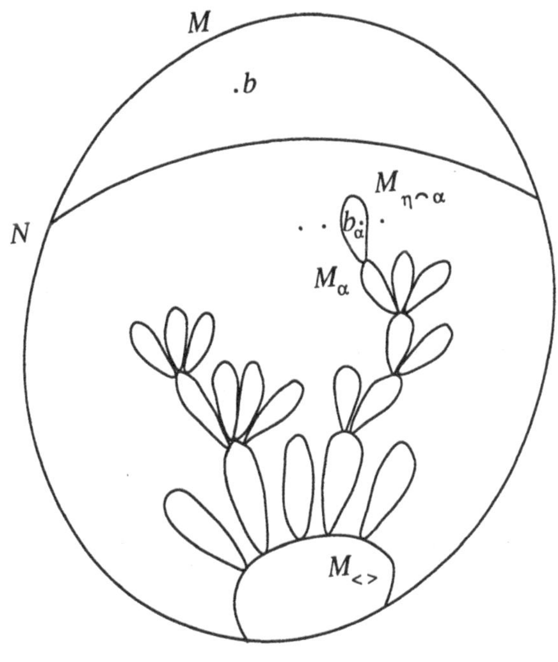

Fig. 2. Theorem XVII.2.8: The Decomposition Lemma

copy of the K-prime model of T in M. If $\eta \in \lambda^{<\omega}$ and $f(\eta)$ is defined, let $\{b_\alpha : \alpha < \mu \leq \lambda\}$ be a maximal subset of M such that:

i) $\{b_\alpha : \alpha < \mu\}$ is independent over M_η.
ii) For each α, $t(b_\alpha; M_\eta)$ is K-strongly regular.
iii) If $\eta \neq \langle\,\rangle$, for each α, $t(b_\alpha; M_\eta) \dashv M_{\eta^-}$.

Now, for $\alpha < \mu$, define $f(\eta^\frown\alpha) = M_\eta[b_\alpha]$. We must show that if $\bar{A}$ is the range of f and $<$ is $\subseteq$ then $(\bar{A}, <, M_{\langle\,\rangle})$ is a K-representation of M. Clearly conditions i)-ii) of the definition are satisfied. We must show M is K-prime over $\bar{A}$. Suppose $N \prec M$ is K-prime over $\bar{A}$. If $N \neq M$ then, as in the proof of Lemma 2.2, there is an $M_\eta \in \bar{A}$ and a $q \in S(M_\eta)$ such that some $b \in M - N$ realizes q^N. Moreover, we can choose η minimal so that $q \dashv M_\nu$ for any $\nu \subsetneq \eta$ or $\eta = \langle\,\rangle$. In particular, $b \downarrow_{M_\eta} N$. Thus if $\{b_\alpha : \alpha < \mu\} \cup \{b\}$ is the set of independent elements associated with M_η in the construction, $\{b_\alpha : \alpha < \mu\} \cup \{b\}$ is independent. If $\eta = \langle\,\rangle$, we clearly contradict the maximality of the b_α. If $\eta \neq \langle\,\rangle$, we must also check that $t(b; M_\eta) \dashv M_{\eta^-}$. But this follows by the minimal choice of η.

2.9 Historical Notes. Most of this material is from [Shelah 1982] and [Harrington & Makkai 1985]. The proof of Theorem 2.8 is from [Harrington & Makkai 1985] and greatly simplifies Shelah's construction (which applies in more general situations). Shelah's argument for the existence of a representation requires an approximation from both above and below. The simplest form of this argument which has no need for prime models appears in [Baldwin & Shelah 1985]. Construction of $A^-[a_A]$ in the general setting but without appealing to prime models is explained in [Hart 1986] and [Shelah 1982], [Shelah 198?]. Lemmas 2.6 and 2.7 which simplify the arguments in Section XVII.4 are taken from [Baldwin & Harrington].

3. Trees, Labeled Trees and Upper Bounds

The major goal of this section is to compute upper bounds on the number of models of various theories which do not have the K-dimensional order property. In the last section we saw that each K-model of such a theory could be represented by a tree. Thus estimating the number of such trees will give this upper bound. In fact, we will be able to refine our estimates by assigning another invariant to the theory, its depth. We begin by discussing the notion of depth and counting the number of trees with a given depth in a purely combinatorial context. Then we will show how to extend the definitions and apply the results in a more general model theoretic context.

3.1 Definition. Let $(A, <, A_{\langle\,\rangle})$ be a tree.

i) We define by induction the *κ-depth* of $a \in A$, denoted $\mathrm{dp}_\kappa(a)$.
For every $a \in A$, $\mathrm{dp}_\kappa(a) \geq 0$.
For any β, $\mathrm{dp}_\kappa(a) \geq \beta$ if for every $\alpha < \beta$ there exists at least κ successors, c, of b with $\mathrm{dp}_\kappa(c) \geq \alpha$.

ii) The κ-depth of $(A, <, A_{\langle\,\rangle}) = \mathrm{dp}_\kappa(A_{\langle\,\rangle})$.

We may omit the subscript κ if the value is clear from context. Note that when $\kappa = 1$ we are dealing with the usual foundation rank of the tree.

Our next step is to calculate the number of trees of width λ and depth α. Of course, the calculation of dp_κ actually also depends on κ. We deal here with the case $\kappa = 1$; other κ are treated in Lemma 4.11. The function which enumerates such trees has a curious discontinuity at depth ω.

3.2 Definition. i) Let $t(\gamma,\alpha)$ denote the following function.

- If $1 \leq \alpha < \omega$, $t(\gamma,\alpha) = \beth_{\alpha-1}(|\gamma|)$.
- If $\alpha \geq \omega$, $t(\gamma,\alpha) = \beth_{\alpha+1}(|\gamma|)$.

ii) Let $T(\gamma,\alpha) = \min(t(\gamma,\alpha), 2^{\aleph_\gamma})$.

The proof of the following lemma is a straightforward induction based on the following observation. For each α, the number of trees of depth α and width $\aleph_\beta$ is the same as the number of functions from the set of trees with depth less than α into the number of cardinals less than $\aleph_\beta$. The most delicate point is to notice that this observation forces the jump at $\alpha = \omega$. We include the proof because several variants of it are discussed later.

3.3 Lemma. *For any α and β, the number of trees $(A, <, A_{\langle\rangle}) \subseteq \aleph_\beta^{<\omega}$ with depth at most α is $T(\beta + \omega, \alpha)$.*

Proof. Let $g(\beta,\alpha)$ denote the number of trees $(A, <, \langle\rangle) \subseteq \aleph_\beta^{<\omega}$ with depth at most α. As an aid to showing $T(\beta+\omega,\alpha) = g(\beta,\alpha)$, let $T^*(\beta,\alpha)$ denote $\sum_{\gamma<\alpha} T(\beta,\gamma)$ and $g^*(\beta,\alpha)$ denote $\sum_{\gamma<\alpha} g(\beta,\gamma)$. Now for fixed β, we prove the result by induction on α.

If $\alpha = 1$, the number of trees of power at most $\aleph_\beta$ is the number of cardinals less than $\aleph_\beta$, that is, $\beta + \omega = T(\beta + \omega, 1)$. For $\alpha > 1$, each tree, A, of power at most $\aleph_\beta$ with depth at most α is determined by a function, f, from the set of trees of depth $< \alpha$ into the set of cardinals $\leq \aleph_\beta$. Namely, if B is a tree with depth $< \alpha$, $f(B) = \kappa$ if and only if there are κ elements $a \in A$ with height 1 such that $(a)_\geq \approx B$. There are $|\beta + \omega|^{g^*(\beta,\alpha)}$ such functions.

Now if $\alpha < \omega$, $g^*(\beta,\alpha) = g(\beta, \alpha - 1)$ so $g(\beta,\alpha) = |\beta + \omega|^{g(\beta,\alpha-1)} = |\beta+\omega|^{T(\beta+\omega,\alpha-1)} = T(\beta+\omega,\alpha)$. While if $\alpha \geq \omega$, $g^*(\beta,\alpha) = T^*(\beta,\alpha)$ so $g(\beta,\alpha) = |\beta+\omega|^{T^*(\beta,\alpha)} = T(\beta+\omega,\alpha+1)$.

In order to generalize this remark to calculate an upper bound on the number of K-models of a theory without the DOP, we apply the decomposition theorem to extend the notion of depth from trees to models of such theories. We must study a slightly more complicated combinatorial object, a labeled tree.

If we decompose a model of a theory without the dimensional order property by the procedure in Theorem 2.8 we can assign a depth to each element of the model according to its position in the representing tree. In order to make the discussion uniform over the various models of T we define the depth of a type by induction. There are several minor variations of the definition of depth in the literature. The definition here agrees with that in [Shelah 1982] and [Saffe 1983] but disagrees with [Harrington & Makkai 1985] and [Lascar 1985].

3.4 Definition. Suppose T is a theory without the dimensional order property. Let M be an **S**-model and $p = t(\overline{a}; M)$ be regular and let N be **S**-prime over $M \cup \overline{a}$.

i) We define the *depth* of p by induction as follows:

- $\mathrm{dp}(p) \geq 0$ for all such p.
- If α is zero or a successor ordinal, $\mathrm{dp}(p) \geq \alpha + 1$ if there is some $q \in S(N)$, $q \dashv M$ and $\mathrm{dp}(q) \geq \alpha$.
- If β is a limit ordinal, $\mathrm{dp}(p) \geq \beta + 1$ if for every $\alpha < \beta$, $\mathrm{dp}(p) > \alpha$.

Finally, $\mathrm{dp}(p)$ is the least β such that $\mathrm{dp}(p) \not\geq \beta + 1$ or ∞ if there is no such β.

ii) If $p \in S(A)$ is stationary and regular, let M be **S**-prime over A and set $\mathrm{dp}(p) = \mathrm{dp}(p^M)$.

iii) $\mathrm{dp}(T)$ is one more than the supremum of the $\mathrm{dp}(p)$ for all regular p if this supremum exists. If the supremum exists we say T is *shallow*. If not, T is *deep*.

If q has weight one, let $\mathrm{dp}(q) = \mathrm{dp}(r)$ where r is a strongly K-regular type which is not orthogonal to q. We deduce from Theorem XIII.2.22 and Theorem 3.14 (below) that this assignment does not depend on the choice of r.

We define the *eni-depth* of a type or a theory by a similar induction but with the added condition that the type p has eni-depth ≥ 1 if and only if there is an eni type $q \in S(M[a])$ with $q \dashv M$.

Note that a theory has depth 1 just if it is bounded. Observe that neither a type nor a theory can have limit depth. When computing the depth of a theory remember the type p in clause i) of the definition of depth cannot be algebraic.

3.5 Exercise. Show that the theory of an infinite set has depth 1.

3.6 Exercise. Let $p \in S(M)$ and suppose that a realization $\overline{a}$ of p occurs in the **S**-decomposition $(\bar{A}, <)$ of a model N. Show that $\mathrm{dp}(\overline{a})$ in $\bar{A}$ is at most $\mathrm{dp}(p)$. Show there is a model N and a **S**-decomposition $(\bar{A}, <)$ where equality holds.

The last exercise showed the relation between the depth of a type and the depth of a tree which decomposes a model. Ostensibly, there should be a different notion of depth for each acceptable class K. However, using Lemma XV.1.7 it is possible to prove (cf. [Harrington & Makkai 1985]) that for a countable ω-stable theory computing depth with respect to **S**-decompositions or with respect to **AT**-decompositions gives the same result. Thus, for our purposes the only distinguished case is the eni-depth.

3.7 Exercise. Show that for any shallow theory, T, $\text{eni-dp}(T) \leq \mathrm{dp}(T)$ and the inequality may be strict.

3.8 Examples. Recall from Section XV.3 that there were essentially four types of nonconstant spectra for bounded ω-stable countable theories. The four spectrum functions assigned to $I^*(\aleph_\alpha, AT)$ one of $\alpha+1, \alpha+\omega, (\alpha+1)^\omega$, and $(\alpha+\omega)^\omega$. By choosing $\alpha+1$ to represent the spectra in the finite dimensional case, we are exploiting the group action discussed in Section

XV.3 to simplify the function. We now give four examples of depth two theories with simple spectrum functions. They arise by placing a model of one of the depth one theories in each equivalence class of an equivalence relation with infinitely many classes.

i) Let T_0 be the theory of an equivalence relation with infinitely many infinite classes.
ii) Let T_1 be the theory of an equivalence relation with infinitely many infinite classes, each of which is a model of $\mathrm{Th}(Z, S)$.
iii) Let T_2 be the theory of an equivalence relation with infinitely many infinite classes, each of which contains a model of the theory of infinitely many disjoint unary predicates.
iv) Let T_3 be the theory of an equivalence relation with infinitely many infinite classes, each of which contains a model of the theory of infinitely many disjoint unary predicates where each of the unary predicates contains a model of $\mathrm{Th}(Z, S)$.

These theories are the simplest examples of the following four kinds of theory. T_0 has a finite number of independent isolated depth 0 1-types. T_1 has a finite number of independent depth 0 1-types and at least one is not isolated. T_2 has infinitely many independent depth 0 1-types which are all isolated. T_3 has infinitely many independent nonisolated depth 0 1-types.

The following table indicates the spectrum functions of these four theories. The second column shows the function directly as it is computed from α. The third and fourth columns show that under particular assumptions on α some of the functions coincide.

Theory	$I(\aleph_\alpha, AT)$	*finite* α	*infinite* α
T_0	$\lvert\alpha+\omega\rvert^{\lvert\alpha+1\rvert}$	$\lvert\omega\rvert$	$2^{\lvert\alpha\rvert}$
T_1	$\lvert\alpha+\omega\rvert^{\lvert\alpha+\omega\rvert}$	$2^{\lvert\omega\rvert}$	$2^{\lvert\alpha\rvert}$
T_2	$\lvert\alpha+\omega\rvert^{\lvert\alpha+1\rvert^{\lvert\omega\rvert}}$	$2^{2^{\lvert\omega\rvert}}$	$2^{\lvert\alpha\rvert^{\lvert\omega\rvert}}$
T_3	$\lvert\alpha+\omega\rvert^{\lvert\alpha+\omega\rvert^{\lvert\omega\rvert}}$	$2^{2^{\lvert\omega\rvert}}$	$2^{\lvert\alpha\rvert^{\lvert\omega\rvert}}$

There is one other fundamental example. It is described in detail in Section XVIII.4. The following example shows there are theories without NDOP with each depth up to $|T|^+$.

3.9 Example. Let EER_β denote the theory of β *expanding equivalence relations*. That is, the language contains binary relations E_γ for $\gamma < \beta$ and EER_β asserts that each E_γ is an equivalence relation with infinitely many infinite classes and each $E_{\gamma+1}$ class contains infinitely many E_γ classes. Then EER_β is an ω-stable theory without the dimensional order property. Moreover, $\mathrm{dp}(\mathrm{EER}_\beta) = \beta + 1$ if $\beta < \omega$ and $\mathrm{dp}(\mathrm{EER}_\beta) = \beta + 2$ if $\beta \geq \omega$.

3.10 Exercise. Let T be EER_β and suppose $\beta < \omega$. Show that

$$I^*(\aleph_\alpha, \mathbf{AT}) = \beth_{\beta-1}((\alpha+\omega)^{|\alpha+1|}).$$

3.11 Exercise. Let T be EER_β and suppose $\beta \geq \omega$. Show that

$$I^*(\aleph_\alpha, \mathbf{AT}) = \beth_{\beta+1}(|\alpha + \omega|^{|\alpha+1|}).$$

The divergence of results between these two exercises shows the necessity for the peculiar cases in Definition 3.2.

3.12 Exercise. Consider the effect of using the other three cases of Example 3.8 as the 'ground' step of a theory of depth β.

3.13 Exercise. Find an example of a theory with depth $\omega + 1$. (Hint: Consider the disjoint union of theories with finite depth.)

We now show the notion of depth is truly well defined by showing the depth of a strongly regular type p depends only on the parallelism class, indeed, only on the nonorthogonality class of p.

3.14 Theorem. *Let $p \in S(M)$ and $p' \in S(N)$ be regular types where M and N are* **S***-models. If i) p' is a nonforking extension of p or ii) $p' \not\perp p$ then* $\text{dp}(p) = \text{dp}(p')$.

Proof. We prove simultaneously by induction on α that supposing either i) or ii), $\text{dp}(p) \geq \alpha$ if and only if $\text{dp}(p') \geq \alpha$. Suppose we have proved the result for all β less than α. For i), we first show $\text{dp}(p) \geq \alpha$ implies $\text{dp}(p') \geq \alpha$. Choose a realization $\bar{c}$ of p'. Then $M[\bar{c}] \downarrow_M N$. Let $q \in S(M[\bar{c}])$ be orthogonal to M. Then by Theorem VI.2.21, $q \dashv N$. If $\text{dp}(q) < \alpha$, then by induction $\text{dp}(q) = \text{dp}(q^{N[\bar{c}]})$ and $q^{N[\bar{c}]} \dashv N$ so $\text{dp}(p') \geq \text{dp}(q)$. Since we can pick q with each depth $< \alpha$ ($< \beta$ if $\alpha = \beta + 1$ and β is a limit ordinal), $\text{dp}(p') \geq \alpha$. To see $\text{dp}(p') \geq \alpha$ implies $\text{dp}(p) \geq \alpha$, fix $q \in S(N[\bar{c}])$ with $\text{dp}(q) < \alpha$ and $q \dashv N$. Now $N[\bar{c}]$ can be viewed as $N[M[\bar{c}]]$ and $M[\bar{c}] \downarrow_M N$ so by NDOP, $q \not\dashv M[\bar{c}]$ and thus $q \not\perp q_0$ for some regular $q_0 \in S(M[\bar{c}])$. By induction, $\text{dp}(q) = \text{dp}(q_0)$ so $\text{dp}(p) \geq \alpha$.

For part ii), we can invoke i) to assume that $M \prec N$. Then if $\bar{c}$ realizes p', we can find by XII.1.15, using the nonorthogonality and regularity of p and p', a $\bar{d}$ realizing p^N with $N[\bar{c}] \approx N[\bar{d}]$. But then $\text{dp}(p') = \text{dp}(p^N)$ whence by part i) $\text{dp}(p) \geq \alpha$ if and only if $\text{dp}(p^N) \geq \alpha$ if and only if $\text{dp}(p') \geq \alpha$ and we finish.

The following theorem is an immediate corollary of Definition 3.4 and Theorem 2.8.

3.15 Theorem. *If T has the NDOP then every model of T has a normal K-decomposition $(A, <, A_{\langle\,\rangle})$ with* $\text{dp}(A) \leq \text{dp}(T)$.

We would like to obtain an upper bound on the number of K-models of T by saying that Theorem 3.15 defined a function from the set of trees with width at most λ and depth $\text{dp}(T)$ onto the K-models of T with cardinality at most λ. Unfortunately, this isn't quite true. While every K-model of T is K-prime over a normal tree, the model is determined not just by the shape of the tree but by the types that are realized. Thus we have a relation on the class of ordered pairs of trees and models which is not a function. By adding additional structure on the trees we can make this relation into a function.

By the uniqueness of K-prime models, we see that two normal trees $(\bar{A}, <, \bar{A}_{\langle\,\rangle})$ and $(\bar{B}, <', \bar{B}_{\langle\,\rangle})$ represent isomorphic K-models M and N if i) the partial orderings $(\bar{A}, <, \bar{A}_{\langle\,\rangle})$ and $(\bar{B}, <', \bar{B}_{\langle\,\rangle})$ are isomorphic by some isomorphism α and ii) for each $A \in \bar{A}$ and each strongly regular type $p \in S(A)$ such that a successor of A has the form $A[a]$ for some a realizing p, the number of successors of A that are prime over a realization of p is the same as the number of successors of $\alpha(A)$ that are prime over a realization of $\alpha(p)$. We reflect this property in the tree structure by labeling each node $A[a]$ by a unary predicate representing $t(a; A)$.

3.16 Definition. i) A *κ-labeled* tree is a tree $(A, <, A_{\langle\,\rangle})$ such that for each $a \in A$, there is a family $\{U_{a,i} : i < \kappa\}$ of unary predicates and each successor of a satisfies one of these predicates.

ii) A *κ-partially labeled* tree is a tree $(A, <, A_{\langle\,\rangle})$ such that for each $a \in A$ with $\mathrm{dp}_\rho(a) = 1$ there is a family $\{U_{a,i} : i < \kappa\}$ of unary predicates and each successor of a satisfies one of these predicates.

Thus, in a partially labeled tree we label only the 'leaves' or top nodes. Usually the trees can be pruned without loss of generality so that the ρ in this definition can be taken to be 1. Thus, we did not clutter the notation by recording the dependence on ρ.

We will show that it suffices to study partially labeled trees. The following notations make the statement of the result simpler.

iii) Let $L_{\beta,\alpha,\kappa}$ denote the number of κ-labeled trees of depth at most α and power at most $\aleph_\beta$.

iv) Let $P_{\beta,\alpha,\kappa}$ denote the number of κ-partially labeled trees of depth at most α and power at most $\aleph_\beta$.

Note that since all models in, for example, an **S**-representation of a model of a countable theory have cardinality at most $2^{\aleph_0}$, we will always be able to find a κ (depending on K) such that every K-representation corresponds to a κ-labeled tree.

Now an induction like that to prove Lemma 3.3 shows that whether we label all the nodes or just those on the top has no effect on the number of trees. That is,

3.17 Lemma. *For any κ, $L_{\beta,\alpha,\kappa} = P_{\beta,\alpha,\kappa} = T(|\beta+\omega|^\kappa, \alpha)$.*

We tie up the results of this section with the following theorem.

3.18 Theorem. *Let T be a superstable theory without the DOP and with $\mathrm{dp}(T) = \alpha$, then for any β with $\aleph_\beta \geq \lambda_0(\mathbf{I})$*

$$I^*(\aleph_\beta, K) \leq T(|\beta+\omega|^{\lambda_0(I)}, \alpha).$$

Proof. We decompose each model by a depth α tree of models, each with cardinality at most $\lambda_0(I)$. Thus we have a function from the collection of labeled trees with depth α and width at most $\aleph_\beta$ onto the models with cardinality at most $\aleph_\beta$.

3.19 Corollary. *If T is a countable ω-stable theory with infinite depth α*

$$I^*(\aleph_\beta, \mathbf{AT}) \leq T(|\beta+\omega|, \alpha).$$

Proof. Since $\lambda_0(I) = \aleph_0$, the theorem implies: $I^*(\aleph_\beta, \mathbf{AT}) \leq T(|\beta+\omega|^\omega, \alpha)$. But if $\alpha \geq \omega$, $T(|\beta+\omega|^\omega, \alpha) = T(|\beta+\omega|, \alpha)$.

The following exercises refer to the example discussed in the introduction to Chapter XVII. The theory T has a single unary function symbol f; there is a unique point which is mapped to itself by f; all others have infinitely many preimages.

3.20 Exercise. Show that T is ω-stable. (Hint: Show that for any $A \subseteq \mathcal{M}$ and any $m \in \mathcal{M}$ the orbit of m under $\mathrm{Aut}_A(\mathcal{M})$ is determined by a pair $\langle n, a\rangle$ where $\langle n, a\rangle \in \omega \cup \{\infty\} \times A \cup \{*\}$ and $\langle n, a\rangle$ ($\langle \infty, *\rangle$) is assigned to m if $f^n(m) = a$ (there is no such n and a).)

3.21 Exercise. Let $A \subseteq B$ and $a \in M \models T$. If $\mathrm{cl}(a) \cap (\mathrm{cl}(B) - \mathrm{cl}(A)) = \emptyset$ show $a \downarrow_A B$.

3.22 Exercise. Suppose $f(a_0) = f(a_1)$ and $f(a_1) = a_2$. Show $t(a_0; a_1) \dashv a_2$.

3.23 Exercise. Let a be the unique element in $\mathcal{M}$ with $f(a) = a$. Let q_0 denote $t(a; \emptyset)$. For each n find types $p_1 \ldots p_n$ which witness that $\mathrm{dp}(q_0) \geq n$.

3.24 Exercise. Repeat Exercise 3.23 for $q_1 = t(b; a)$ where $f(b) = a$ and $f(a) = a$. Conclude that T is deep.

3.25 Exercise. Show the U-rank of the type q_0 from Exercise 3.23 is ω. Prove no type has higher U-rank.

3.26 Exercise. Show the Morley rank of $x = x$ is ω.

3.27 Exercise. What is the effect on the last four exercises of adding to T the axiom $f^7(x) = f^6(x)$?

3.28 Historical Notes. These results were first proved in [Shelah 1982]. Our presentation was greatly influenced by [Harrington & Makkai 1985] and [Saffe 1981]. The function we have labeled T plays a similar major role in [Baldwin & Shelah 1985]. The proof of Theorem 3.14 by a double induction is due to Saffe [Saffe 1981].

4. *Quasi-Isomorphisms and Lower Bounds*

In this section we want to compute lower bounds on $I^*(\aleph_\beta, K)$ for a theory T which does not have the DOP and has depth δ. In fact, we will show this function is eventually bounded below by the same function $T(\beta, \alpha)$ which we saw was an upper bound in Section 3. The basic strategy behind such a lower bound argument is simple. We would like to show that if both $(\bar{A}, <, \bar{A}_{\langle\,\rangle})$ and $(\bar{B}, <', \bar{B}_{\langle\,\rangle})$ are K-representations of M then the two trees are isomorphic. We do not need to worry about labeling the trees as in Section 2 if we can achieve the desired lower bound with unlabeled trees. It turns out that for small depths and cardinals less than $\aleph_\omega$ we must take note of the labeling.

Unfortunately, we are unable to guarantee that the map from models to skeletons is actually a function. There may be non-isomorphic normal representations $\bar{A}$ and $\bar{B}$ of a model M. We show, however, that two trees which represent the same model are quasi-isomorphic and that there are sufficiently many non-quasi-isomorphic trees to calculate the spectrum. We can not even attain this result for all models. But we will obtain it for a large enough subset to calculate the number of models when $\aleph_\beta \geq \aleph_\omega$ or the depth of T is infinite. After some preliminary analysis, we defer to the next chapter the exact computation of the lower part of the spectrum for theories with finite depth.

A quasi-isomorphism is simply a correspondence between two trees which is $1-1$ and preserves order but is defined almost everywhere, rather than everywhere. More formally, we say

4.1 Definition. Two trees $A = (A, <, A_{\langle\,\rangle})$ and $B = (B, <, B_{\langle\,\rangle})$ are *μ-quasi-isomorphic*, denoted $A \approx_{q,\mu} B$, if there is a 1-1 relation $h \subseteq A \times B$ such that:

i) If $\langle a, b\rangle$ and $\langle a', b'\rangle$ are in h then $a < a'$ if and only if $b < b'$.
ii) For all $a \in A$, the number of successors of a which are not in $\operatorname{dom} h$ is less than μ.
iii) For all $b \in B$, the number of successors of b which are not in $\operatorname{rng} h$ is less than μ.
iv) If $\operatorname{dp}(a) = 1$ and $\langle a, b\rangle \in h$ then for each successor a' (b') of a (b) there is a successor b' (a') of b (a) with $\langle a', b'\rangle \in h$.

We write $A \approx_q B$ if the μ-ample (see below) trees A and B are μ quasi-isomorphic. In the relevant cases (μ-ample) we will be able to deduce from condition i) that the b referred to in condition iv) also has depth 1.

First, we construct from two K-representations of the same model a pair of $\lambda_0(I)^+$-quasi-isomorphic trees. Then we turn to the question of finding enough trees which are not quasi-isomorphic. We require several lemmas for the first task.

For any tree $\bar{A}$, we denote by $\bar{A}^1$ the subtree consisting of those nodes with nonzero depth. Given representations $\bar{A}$ and $\bar{B}$ of a properly constructed model M, we will first define a bijection between between $\bar{A}^1$ and $\bar{B}^1$ by sending A to $\hat{A}$ if the type of some successor of A over A is nonorthogonal to the type of some successor of $\hat{A}$ over $\hat{A}$ We will show that for an appropriate choice of a model $M_{A\hat{A}}$ containing A and $\hat{A}$, forking determines the same bijection (between 'most' of the successors of A and 'most' of the successors of $\hat{A}$.) To see that this bijection preserves order, we first note that Theorem VI.2.21 shows that order is preserved on the two levels immediately above A and $\hat{A}$. Then we extend this result to the cones above A and $\hat{A}$ by induction on height above A or $\hat{A}$ and the properties of normal representations established in Section 2.

The following notation allows us to define a tree 'lying above' an ideal I. Note that the new tree $\bar{A}_I$ is not a subtree of $\bar{A}$ but is obtained by shrinking an intial segment of $\bar{A}$ to a point. Lemma 4.3 relies heavily on the fact obtained in Theorem 1.10: $M_I \downarrow_I A$.

4.2 Notation. Let I be an ideal in the normal tree $\bar{A}$. We will define a new (normal) tree $\bar{A}_I$. The base node in the new tree is M_I. The other nodes will be B' for $B \in \bar{A} - I$ where B' is defined by the following induction. If $B \in I^+$, $B' = M_I[B]$. If $B \in I^{n+1}$ (The $n + 1$st predecessor of B is in I) then $B' = (B^-)'[B]$. $\bar{A}_I = \{B' : B \in \bar{A} - I\}$.

4.3 Lemma. *$\bar{A}_I$ is a normal tree.*

Proof. The first three conditions of Definition 1.3 are clear. It is easy to conclude iv) and v) by induction. We give only the first stage of the induction since it contains all the ideas. For iv), note that if $A \in I^+$, then $J = \{B : B^- = A^- = C\}$ is independent over C and independent from I over C. Thus, $\{M_I[B] : B \in J\}$ is independent over M_I.

For v), let $A^{--} \in I$. Then $A \cup A^- \downarrow_{A^{--}} I$. Thus $A \cup A^- \downarrow_{A^{--}} M_I$ and so $A \downarrow_{A^-} M_I[A^-]$. Now $t(A; A^-) \dashv A^{--}$ implies $t(A; A^-) \dashv M_I$. Thus, since orthogonality is preserved by nonforking extensions, $t(A; M_I[A^-]) \dashv M_I$. Since $A \rhd_{M_I[A^-]} M_I[A^-][A]$, we conclude $t(M_I[A^-][A]; M_I[A^-]) \dashv M_I$.

In the following theorem we construct a quasi-isomorphism between the nodes of positive depth in two K-repesentations of a K-model M. [Harrington & Makkai 1985] constructs a quasi-isomorphism between the entire representing tree. However, the argument here is much simpler and suffices for much of the spectrum computation. For the remainder we must label the leaves on the trees.

A quasi-isomorphism can fail to be an isomorphism in two ways. First, it is not defined everywhere. Second, it does not preserve height. The first of these problems can be circumvented by dealing with sufficiently bushy trees. The second requires several special tricks. We begin with the first problem. Recall from Definition 3.1 the concept of κ-depth ($\mathrm{dp}_\kappa(a)$) of a node in a tree.

4.4 Definition. A tree $(A, <)$ is called *μ-ample* if for every $a \in A$, $\mathrm{dp}_1(a) = \mathrm{dp}_\mu(a)$.

The following easy result shows that there are the maximal number of non-quasi-isomorphic μ-ample trees. To prove it, we can, for example, specify the set of depths of elements of height 1 in each tree.

4.5 Lemma. *For every $\lambda \geq \mu$, there are 2^λ non-isomorphic, μ-ample trees of width λ.*

4.6 Exercise. Show that if h is a μ-quasi-isomorphism between the μ-ample trees $\bar{A}$ and $\bar{B}$ with $h(A) = B$, the μ depth of A in $\bar{A}$ is the same as the μ depth of B in $\bar{B}$.

We show now that by specifying some easily fulfilled conditions on the representing trees we can guarantee that if two trees represent the same model then the trees obtained by peeling off the leaves are quasi-isomorphic. The key to applying this result is that if $\bar{A}$ has infinite depth, the derived tree has the same depth. Another important idea in this proof is the fact that for each $A \in I^+$, A and a_A are bidominant over M_I. Thus, we can use the triviality of $t(a_A; M_I)$ to deduce properties of A.

4.7 Theorem. *Suppose $\mu \geq \lambda_0(\mathbf{I})^+$ and that $\lambda(\mathbf{I})$ is regular. Let the μ-ample trees $\bar{A}$ and $\bar{B}$ be K-representations of the K-model M. Suppose further that only one nonorthogonality class of K-strongly regular types over A is represented in $\bar{A}$, for each $A \in \bar{A}$ (and similarly for $\bar{B}$). Let $\bar{A}^1$ ($\bar{B}^1$) denote the nodes in $\bar{A}$ ($\bar{B}$) which have depth at least one. Then $\bar{A}^1 \approx_q \bar{B}^1$.*

Proof. (Fig. 3). We first define a map $A \mapsto \hat{A}$ which is a bijection between $\bar{A}^1$ and $\bar{B}^1$ but may not preserve order. Then we show that by throwing away from the domain (range) of the map less than μ successors of any node, we can restrict our map to one which preserves order. Thus, the restricted map is a quasi-isomorphism on the nodes of positive depth. Define $h: A \mapsto \hat{A}$ if for some successor A_1 of $A \in \bar{A}$ and some $B_1 \in \bar{B}$, with $B_1^- = \hat{A}$, $t(A_1; A) \not\perp t(B_1; \hat{A})$. By Lemma 2.8, the transitivity of nonorthogonality on regular types, and the fact that only one nonorthogonality class over A ($\hat{A}$) is realized in M, this map is a bijection.

Fix $A, \hat{A}$. Choose a model $M_{A\hat{A}}$ with $|M_{A\hat{A}}| \leq \mu$ such that $A, \hat{A} \in M_{A\hat{A}}$ and for some ideal $I \subseteq \bar{A}$ and some ideal $J \subseteq \bar{B}$, $M_{A\hat{A}}$ is prime over each of I and J. This model can be chosen by a back and forth. First choose a model, M_0, prime over an ideal of $\bar{A}$ which contains $\hat{A}$ and then an M_1 prime over an ideal of $\bar{B}$ which contains $A \cup M_0$ and so on. Since the property of being an elementary submodel has finite character, this process terminates in ω steps. By Corollary 1.11 the union of the resulting chain satisfies the conditions.

Less than μ successors of either A or $\hat{A}$ are contained in $M_{A\hat{A}} = M_I$. Prune $\bar{A}^1$ and $\bar{B}^1$ by discarding for each A ($\hat{A}$) with depth at least three in $\bar{A}$ ($\bar{B}$) its successors in $M_{A\hat{A}}$. To preserve symmetry discard $h(A)$ if A is discarded and vice versa. In order to show that h preserves order on the pruned trees it suffices to show the following condition. If A_1 was not pruned where $A_1^- = A$ and $A_2 > A_1$ then $h(A_2) > h(A_1)$.

By Theorem 1.10, $\bar{A} \downarrow_I M_I$. Since $A_1 \in \bar{A}$, $A_1 \downarrow_I M_I$. As A_1 does not depend on I over A, we have $A_1 \downarrow_A M_I$. The analogous result holds for the successors of $\hat{A}$. Now by Lemma 2.7, J^+ is a maximal independent subset of realizations of K-strongly regular types over M_J. Since $M_J = M_{A\hat{A}} = M_I$ and J^+ dominates M over M_J, each $A_1 \in I^+$ depends on J^+ over $M_{A\hat{A}}$. By triviality and weight one, there is a unique $\tilde{A}_1 \in J^+$ such that $A_1 \not\downarrow_{M_{A\hat{A}}} \tilde{A}_1$. Now for any other $B \in J^+$ we have $B \downarrow_{B^-} M_{A\hat{A}}$. By Lemma 4.3 we have normal trees $\bar{A}_I$ and $\bar{B}_J$ each with initial node $M_{A\hat{A}}$. Let $(A_1)_{\geq}$ denote

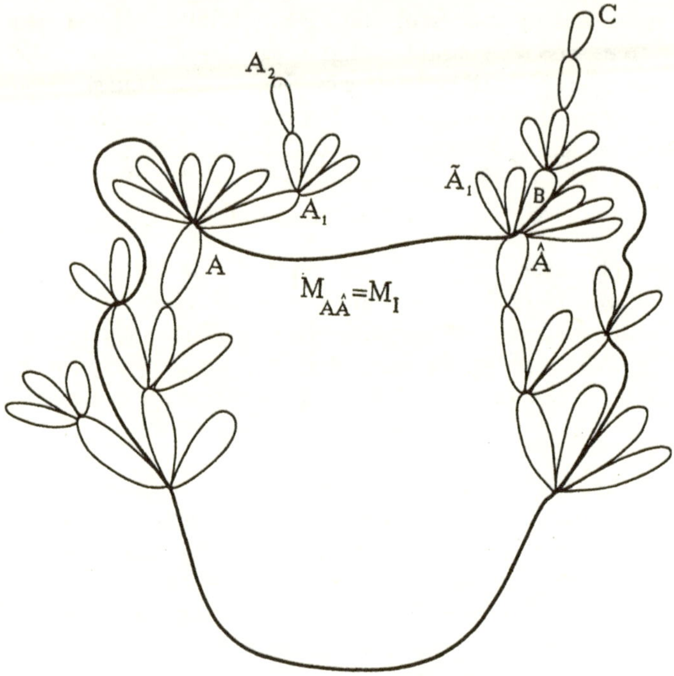

Fig. 3. Theorem XVII.4.7: The construction of quasi-isomorphisms

the cone above A_1 in $\bar{A}_I$ and $(B)_{\geq}$ denote the cone above B in $\bar{A}_J$. Since $A_1 \downarrow_{M_{A\hat{A}}} B$, we have a normal tree $M_{A\hat{A}} \cup (A_1)_{\geq} \cup (B)_{\geq}$. By Lemma 1.7 we see that for any $A_2 > A_1$, and any $C > B$, $t(A_2; A_2^-) \perp t(C; C^-)$. Thus, $h(A_1) = \tilde{A}_1$ and $h(A_2) \geq \tilde{A}_1$. This shows that h preserves order almost everywhere.

We must still verify condition iv) of Definition 4.1 by showing that if A_2 has depth one in $(\bar{A})^1$ and A_2 is in the restricted domain of h, then h is a bijection between the successors of A_2 and the successors of $h(A_2)$. If A_2 is in the restricted domain of h, neither A_2 nor $h(A_2)$ was pruned in the construction. Then, for any successor A_1 of A_2, the argument in the last paragraph shows $h(A_1) > h(A_2)$. Note that for any $A \in \bar{A}$ there is a chain of length n above $A \in \bar{A}$ if and only if there is chain of length n above $h(A)$ in $\bar{B}$. Thus, $h(A_2)$ has depth two in B. Since $h(A_1)$ has depth at least one in B, $h(A_1)$ is an immediate successor of $h(A_2)$.

The fact that a quasi-isomorphism preserves depth is implicit in the argument for condition iv). In general, a quasi-isomorphism need not preserve height.

Now we can obtain the first precise result on the spectrum problem of NDOP theories.

4.8 Theorem. *If T is superstable and has the NDOP but is deep then for every uncountable λ, $I^*(\lambda, K) = 2^\lambda$.*

Proof. Since T has infinite depth, for any $\lambda_0(I)$-ample tree, $\bar{A}$, of depth less than λ^+, we can construct a model M with cardinality λ which is K-prime over a tree isomorphic to $\bar{A}$. If two such trees represent the same model they are quasi-isomorphic by a quasi-isomorphism constructed as in Theorem 4.7. Without loss of generality we can assume that all height 1 types in $\bar{A}$ and $\bar{B}$ are based on a single finite set D. Naming D does not affect the number of models of T in an uncountable cardinal. But since the quasi-isomorphism is given by nonorthogonality and all types in $\bar{A}$ ($\bar{B}$) except those of elements with height 1 are orthogonal to D, we see h must map elements of height 1 to elements of height 1. But since it is easy to construct trees $\bar{A}_X$ for $X \subseteq \lambda$ so that for $A \in \bar{A}_X$ with height 1, $\mathrm{dp}(A)$ in $\bar{A}_X$ is η if and only if $\eta \in X$, we conclude there are 2^λ nonisomorphic models.

The next exercise generalizes one of the major ideas from the previous theorem, the importance of a quasi-isomorphism preserving height. Although this result is crucial for the rest of the argument, its straightforward but tedious proof is omitted.

4.9 Exercise. Show that if there is μ-quasi-isomorphism between two μ-ample trees which preserve height then they are isomorphic.

Thus, we need to find some trick to make quasi-isomorphisms preserve height. We will use two different devices. In, e.g. the ω-stable case, the first works when computing the number of models of power κ for $\kappa \geq \aleph_\omega$. If this trick is not sufficient we will use the second trick and some further analysis.

For the first trick, we restrict the tree $\bar{A}$ by choosing for each $n < \omega$ a set of cardinals, X_n, and requiring that the number of successors of a node of height n and depth 1 be a cardinal in X_n. The next definition describes the kind of trees we will map onto and the function which counts them. We introduce the new parameter γ here to handle the superstable and ω-stable cases simultaneously. We will need to choose γ so that $\aleph_\gamma = \lambda_0(I)$. Thus, for a countable superstable theory, we need $\aleph_\gamma = 2^{\aleph_0}$ and for a countable ω-stable theory, $\gamma = 0$.

4.10 Definition. Let $\bar{X} = \langle X_n : n < \omega \rangle$ be a family of sets of cardinal numbers which lie between $\aleph_\gamma$ and $\aleph_\beta$ such that the X_n are pairwise disjoint $\{\aleph_{\gamma+1}, \aleph_\beta\} \subseteq X_1$, and for each n $|X_n| \geq 2$. The tree $(A, <)$ is *constrained* by $\bar{X}$ if for each $a \in A$ with $\mathrm{ht}(a) = n$ and $\mathrm{dp}(a) = 1$, $|\{b : b^- = a\}| \in X_n$.

The usual kind of induction allows us to count the number of constrained trees of power $\aleph_\beta$. In the ensuing lemma we fix $\aleph_\gamma$ as $\lambda_0(\mathbf{I})$. The constraint on the ordinal β is satisfied by all infinite β when $\lambda_0(\mathbf{I}) = \aleph_0$.

4.11 Lemma. *If $\beta \geq \gamma + \omega$ and α is infinite there are $T(\beta, \alpha)$ trees of power $\aleph_\beta$ which are $\bar{X}$-constrained and $\aleph_{\gamma+1}$-ample but have depth at most α.*

Proof. Although the statement of the theorem is for infinite α, the key to the proof is the analysis for finite α. Since $\beta \geq \gamma + \omega$, we can find at least two cardinals between $\aleph_\gamma$ and $\aleph_\beta$ to put in each X_n. Note, then, that there are at least two of the desired trees with depth one, $|\beta + \omega|$ of depth two, and $T(|\beta + \omega|, n-1)$ of depth n for each finite $n \geq 2$. Thus, there are $T(\beta, \omega)$ of depth ω. The rest of the induction is routine.

The proof of Lemma 4.11 illustrates an important point in the actual calculation of the number of models. While one may not be able to guarantee that every cardinal less than $\aleph_\beta$ is possible as the number of successors of an arbitrary node (either because of ampleness or some defect in coding), this problem can be surmounted if the tree has infinite depth. The following two exercises emphasize this point.

4.12 Exercise. Complete the proof of Lemma 4.11. (Hint: Consider the depth two case. Suppose $\{\kappa_0, \kappa_1\} \subseteq X_1$. Each function f from 2 to the set of $\kappa < \aleph_\beta$ with $\aleph_\beta \in \operatorname{rng} f$ determines an appropriate tree by assigning for i equal 0 or 1 κ_i successors to $f(i)$ nodes of height one.)

4.13 Exercise. Show that for the case of an ω-stable theory, where $\gamma = 0$, the definition of constrained can be altered to require each $|X_n| = |\beta|$. With this alteration Lemma 4.11 holds for finite α but the change does not suffice to prove Theorem 4.14 for finite α.

Now we can compute the spectrum when both β and α are sufficiently large. Note that if T is a countable ω-stable theory, γ in the following theorem is 0 and so we are just requiring β and α to be infinite. The arithmetic of the following argument depends on two observations. If $\alpha \geq \lambda_0(\mathbf{I})$ then $T(|\beta + \omega|^{\lambda_0(\mathbf{I})}, \alpha) = T(\beta, \alpha)$. If a tree $\bar{A}$ has infinite depth then $\bar{A}^1$ has the same depth.

4.14 Theorem. *Let $\aleph_\gamma = \lambda_0(\mathbf{I})$. Suppose $\beta \geq \gamma + \omega$. If T is a superstable theory, the NDOP holds and* $\mathrm{dp}(T) = \alpha$ *is infinite then $I^*(\aleph_\beta, K) = T(\beta, \alpha)$.*

Proof. From Lemma XVII.3.17, we have the upper bound $T(|\beta + \omega|^{\aleph_\gamma}, \alpha)$. If two $\aleph_{\gamma+1}$-ample trees $\bar{A}$ and $\bar{B}$ represent the same model and we denote by $\bar{A}^1$, $\bar{B}^1$ the nodes with depth at least one then by Theorem 4.7 $\bar{A}^1$ and $\bar{B}^1$ are quasi-isomorphic. If we insist the trees are constrained the quasi-isomorphism preserves height. Applying the second observation before the theorem we see that a lower bound is the number of $\bar{X}$-constrained, $\aleph_{\gamma+1}$-ample trees of power $\aleph_\beta$ and depth α. By Lemma 4.11, this number is $T(\beta, \alpha)$. Since $\alpha \geq \aleph_\gamma$, this equals $T(|\beta + \omega|^{\aleph_\gamma}, \alpha)$ and we finish.

It remains to compute the spectrum for powers $\aleph_\beta$ when $\beta < \gamma + \omega$. For concreteness we restrict the theorems to counting all models of a countable ω-stable theory. We will not obtain the full solution until Section XVIII.5. However, we can solve the problem for theories of infinite depth with the methods we have now. We need the following sharper estimate on the number of ample trees.

4.15 Lemma. *For any $1 \leq m < \omega$ there are $2^{\aleph_m}$ partially 2-labeled trees of power $\aleph_m$ which are $\aleph_1$-ample and have depth less than $m + 2$.*

Proof. For any $k > 0$, there are $\aleph_0$ partially 2-labeled trees of cardinality $\aleph_k$ and depth one. Namely, let the nth tree have n nodes of height one labeled by U_0 and $\aleph_k$ labeled by U_1. Since any uncountable tree of depth one is $\aleph_1$-ample, each T_n is. By forming for any subset $X \subseteq \omega$ a tree M_X such that if $a \in M_X$ has height one then $(a_{\geq}) \approx T_n$ for some $n \in X$ and such that each such subtree occurs $\aleph_1$ times, we create $\beth_1(\aleph_0)$ partially 2-labeled trees which are $\aleph_1$-ample and have depth two. Continuing inductively, there are $\beth_m(\aleph_0)$ partially 2-labeled trees which are $\aleph_1$-ample, have cardinality $\aleph_k$ and depth at most $m+1$. Since $\beth_{m+2}(\aleph_0) \geq 2^{\aleph_m}$ this yields the lemma.

4.16 Theorem. *If T is an ω-stable countable theory without the dimensional order property and* $\mathrm{dp}(T) \geq \omega$ *then for $\beta < \omega$, $I^*(\aleph_\beta, \mathbf{AT}) = 2^{\aleph_\beta}$.*

Proof. Fix $\beta = m$ with $1 \leq m < \omega$. Choose $n = m + 2$. For each partially 2-labeled $\aleph_1$-ample tree S of height n we build a normal tree $\bar{A}_S$ of height $n+1$ with $(\bar{A}_S)^1$ isomorphic to S. Each A in $\bar{A}_S$ will have only a single type based on it (to guarantee the uniformity condition in the hypothesis of Theorem 4.7). If the leaf $A \in S$ is labeled by U_0, A will have $\aleph_0$ successors in $\bar{A}_S$; if the label is U_1 the node will have $\aleph_k$ successors in $\bar{A}_S$. We require that if $\mathrm{ht}(A) = k$, then $\mathrm{dp}(t(A; A^-)) = n - k$. Then the quasi-isomorphism between two representations of the same model will preserve height. By Theorem 4.7 if $\bar{A}_{S_0}$ and $\bar{A}_{S_1}$ both represent the same model, $(\bar{A}_{S_0})^1 \approx_q (\bar{A}_{S_1})^1$. Thus, they are isomorphic as trees. To see that the isomorphism preserves labels we need only note that if $h(A_0) = B_0$ then there exist A_0' and B_0' with $t(A_0'; A_0) \not\perp t(B_0'; B_0)$. Since nonorthogonal types in an ω-stable theory have the same dimension modulo $\aleph_0$, A_0 and B_0 have the same label. By Lemma 4.15 we obtain a lower bound of $2^{\aleph_m}$ on the number of models of power $\aleph_m$. Since this is the crudest upper bound, we complete the proof of the theorem.

4.17 Exercise. Let T be a superstable countable theory without the dimensional order property and $\mathrm{dp}(T) \geq \omega$. Show that if $\aleph_\gamma = \lambda_0(\mathbf{S})$ then for $\gamma < \beta < \gamma + \omega$, $I^*(\aleph_\beta, \mathbf{S}) = 2^{\aleph_\beta}$.

There are two difficulties with the method of proof of the last theorem. First, the requirement of considering trees of depth $m + 2$ makes it inherently unsuitable if we are trying to find precise estimates for the number of models of a theory with depth m. Secondly, the methods used fail if we try to label finite dimensions. Both of these problems are resolved in the complete solution of the spectrum problem for finite depth theories in Section XVIII.5. The key to the solution is to continue to build $\aleph_1$-ample trees but to preserve labels for finite dimensions. In addition to the methods described here, that proof relies on Bouscaren's analysis in [Bouscaren 1984].

4.18 Historical Notes. These results were originally proved by Shelah in [Shelah 1982]. Our account owes a great deal to the expositions of [Harrington & Makkai 1985] and [Saffe 1983]. The present proof of Theorem 4.7 on the existence of a quasi-isomorphisms is from [Baldwin & Harrington]. The trees discussed in this section play an important role in [Baldwin & Shelah 1985] and in that paper there are other applications of the counting functions discussed here. The case treated in Theorem 4.16 was not discussed explicitly in either [Harrington & Makkai 1985] or [Shelah 1982], but was in [Saffe 1981].

Chapter XVIII

Vaught and Morley Conjectures for ω-Stable Countable Theories

In this chapter we complete the proof of two of the most important results about the spectrum problem. We prove Vaught's conjecture for countable ω-stable T: A countable ω-stable theory has either countably many or $2^{\aleph_0}$ countable models. Furthermore, we prove Morley's conjecture for countable ω-stable T: If $\aleph_1 \leq \kappa < \lambda$, $I(\kappa, AT) \leq I(\lambda, AT)$.

For Vaught's conjecture the assumption that T is ω-stable is made because it allows us to prove the theorem. The conjecture has not been resolved for any other class in the stability hierarchy although there are partial results by [Lascar 1981] and [Shelah 1978a]. In contrast, for Morley's conjecture the assumption that T is ω-stable is part of a systematic program. If T is not superstable then T has 2^λ models in power λ for all uncountable λ. Although we did not prove this result for singular λ in Section IX.6, we did prove the function was increasing for stable but not superstable T. Thus, the only omissions in our treatment of Morley's conjecture for countable theories are the unstable case and the superstable but not ω-stable case. The first of these cases is handled in the first edition of [Shelah 1978] while both are solved in the second edition.

For each problem we know by Section XVI.3 that T does not have the appropiate version of the dimensional order property. That is, if T has the DOP then T has 2^λ models of power λ for $\lambda > \aleph_0$. In addition, if the ω-stable theory T has the ENI-DOP then T has $2^{\aleph_0}$ countable models. In Chapter XVII, we completed the solution of the spectrum problem in uncountable cardinalities for ω-stable T except for shallow theories with finite depth. For theories with finite depth we did not resolve the difficulties which arise when there are ENI-types on the leaves of the representing tree of a model. The resolution of this difficulty is closely connected to the analysis of depth two types which is necessary to prove Vaught's conjecture for countable ω-stable T.

In Section 1 of this chapter we begin the analysis of types of low depth by considering what it means for one type to support another and some transfer properties of this notion. We justify this study of types with low depth in Section 2 by showing that if there is an ω-stable counterexample to

Vaught's conjecture then it has depth two. We also compute lower bounds on the number of models in uncountable powers under certain conditions on the theory T. In Section 3 we give a detailed analysis of types of depth two which meet the conditions we discussed in Section 2 and prove Vaught's conjecture for ω-stable T. Section 4 contains an example of Shelah which shows the necessity of some of the more complicated steps in the argument. Finally, in Section 5 we complete the proof of Morley's conjecture for ω-stable T.

There are many partial results on Vaught's conjecture. The first significant general result was Morley's proof that a sentence in $L_{\omega_1,\omega}$ has either $2^{\aleph_0}$ or at most $\aleph_1$ countable models. That proof is essentially a result in descriptive set theory. Most arguments in that field would extend to pseudoelementary classes in $L_{\omega_1,\omega}$. But Morley's result does not. Most of the other work not reported here concentrates on particular theories, usually with an ordering of some kind. Thus, Rubin [Rubin 1974] showed that any complete first order theory of a linear order with monadic predicates had either finitely many or $2^{\aleph_0}$ countable models. Further work along this line in [Miller 1981] and [Marcus 1980] culminated in the proof by Steel [Steel 1978] that Vaught's conjecture holds for any $L_{\omega_1,\omega}$ sentence all of whose models are trees. Shelah [Shelah 1978a] shows that if the theory T has Skolem functions and admits a linear order of an infinite set then T has $2^{\aleph_0}$ countable models. Lascar [Lascar 1981] proved Vaught's conjecture for stable theories with Skolem functions. Lascar and Boucaren [Bouscaren & Lascar 1983], [Bouscaren 1983] made substantial progress on the ω-stable case before it was completed by Shelah [Shelah, Harrington, & Makkai 1984].

For ease of reading we adopt the convention of writing single letters to represent finite sequences rather than littering the text with bars.

In this chapter T is a *countable ω-stable theory with NDOP. Moreover, T is shallow and when dealing with Vaught's conjecture T has ENI-NDOP.* Thus, throughout the chapter we consider only the acceptable class **AT** and all notions (e.g. prime, isolation) are with respect to that class.

1. Supportive Types

In this section we work with a type $p \in S(A)$ and a realization a of p. The major task is to analyze the types which 'depend' on $A \cup a$ either directly by being over $A \cup a$ or in the slightly less direct fashion discussed in this section. The next definition describes in more detail the situation that we must analyze. We then prove a number of technical lemmas which allow us to pursue this analysis.

1.1 Definition. The type p *needs* the finite sequence a over the set A, written p needs a/A, if

i) $p \dashv A$ and $p \not\dashv A \cup a$.
ii) p is stationary and strongly regular.
iii) $t(a; A)$ is stationary and has weight one.

Of course, condition i) is the active part of the definition. We differ from [Harrington & Makkai 1985] in not requiring that p be ENI. They make this requirement because for the Vaught conjecture only the ENI case is required. For the Morley conjecture we must discuss persistently isolated types which need specific a. The type $q \in S(A)$ is *supportive* if for some a realizing q and some type p, p needs a/A. In this situation we sometimes say q *supports* p. We say p needs $q \in S(A)$ if for some realization b of q, p needs b/A. The type $q \in S(A)$ is *ENI-supportive* if for some a realizing q and some ENI type p, p needs a/A. We extend these notions to global types by saying a type $p \in S(\mathcal{M})$ is ENI-supportive if for some A, p does not fork over A and $p|A$ is ENI-supportive. Note that if T is an ENI-depth two theory, every ENI-supportive type has depth one and the ENI-types which need supportive types have depth zero. This notion of needing is somewhat more restrictive than what the word need might bring to mind. In particular, needing is not transitive. That is, if $q \in S(c)$, b realizes $p \in S(a)$, q needs b/a, and p needs $a/\emptyset$ it does not follow that q needs $a/\emptyset$.

1.2 Exercise. Show that if p needs a/A then for some finite $A_0 \subseteq A$, p needs a/A_0. (Remember that throughout this chapter we assume that T is ω-stable.)

1.3 Exercise. Show that when T is a depth two theory, every supportive type is not orthogonal to the empty set.

1.4 Exercise. Find the supportive types in the theory of an equivalence relation with infinitely many infinite classes. Do the same for ENI-supportive types if each class contains a model of $\mathrm{Th}(Z, S)$. Show the ENI-depth of the first example is one (the minimum for any theory) while the second has ENI-depth two.

The next two exercises explore the preservation of needing to supersets of A which are independent from a.

1.5 Exercise. Show that if q needs c/A and $c \downarrow_A B$ then q needs c/B.

If a theory does not have the ENI-DOP use the triviality of $\dashv$ to show

1.6 Exercise. If T has ENI-NDOP, q is ENI, q needs c/B and $c \downarrow_A B$ then q needs c/A. Contrast this exercise with Lemma XIII.3.11.

We prove the following results simultaneously for various properties which may or may not be modified by ENI, e.g. NDOP and ENI-NDOP. As usual, this dual purpose is indicated by parentheses around 'ENI'. For Vaught's conjecture we are interested in the ENI-NDOP case while for Morley's conjecture we need both versions. Our principal goal in the following few paragraphs is the analysis of supportive types of depth one. However, the following notation slightly generalizes that situation to the study of the relation between pairs of types whose depth differs by one. This added generality is needed for the study of the Morley conjecture in Section 5.

1.7 Notation. A type s is ENI-constrained by q if eni-dp$(s) \leq$ eni-dp$(q) + 1$.

If the definition is made for depth rather than eni-depth one just says q constrains s. Note that r needs p does not imply that r constrains p but there will be some s which constrains p and needs p. If the (ENI)-depth of s is less than or equal one, s is (ENI)-constrained by any q which needs it.

1.8 Theorem. *Suppose T satisfies the (ENI)-NDOP, q is a stationary (ENI)-type, and q (ENI)-constrains s. For any c realizing s if $q \dashv A \cup c$ and $c \rhd_A d$ then $q \dashv A \cup d$.*

Proof. Choose a strongly $|A| + \kappa(T)$-saturated model M with $M \downarrow_A c^\frown d$. Then $d \downarrow_{A \cup c} M \cup c$ so we can choose a copy $M[c]$ of the **S**-prime model over $M \cup c$ with $d \downarrow_{A \cup c} M[c]$. Assume for contradiction that $q \not\dashv A \cup d$. Now, $q \not\dashv A \cup c \cup d$ implies by Theorem XIII.3.11 that $q \dashv M[c]$. Then, clearly, $q \not\dashv M[c][d]$. The models M, $M[c]$, $M[c,d]$ give us the situation of a three model theorem. Since $q \dashv M[c]$, $M[c]$ is a proper subset of $M[c,d]$. If $t(M[c,d]; M[c]) \dashv M$, the depth of $t(c; M)$ is at least two more than the depth of q contrary to hypothesis. Thus, by Theorem XIII.3.4 there is a $b \in M[c,d] - M[c]$ which is independent from c over M. But since $c \rhd_M d$ this implies $b \downarrow_M c^\frown d$. But this is absurd since $b \in M[c,d]$.

We draw a useful conclusion from this result.

1.9 Corollary. *Suppose T is an (ENI)-depth 2 theory which satisfies the (ENI)-NDOP and q is a stationary, (ENI)-type. If $q \dashv M \cup c$ then $q \dashv M[c]$.*

The next exercise provides a simpler proof of a weak version of Theorem XVI.1.11.

1.10 Exercise. Suppose T is an (ENI)-depth two theory satisfying the (ENI)-NDOP and q is a stationary (ENI)-type. Show that if $C = A \cap B$, $A \downarrow_C B$, $q \dashv A$, and $q \dashv B$ then $q \dashv A \cup B$.

1.11 Exercise. Show that requiring eni-dp$(s) \leq$ eni-dp$(q) + 1$ is needed to prove Theorem 1.8. (Hint: Consider the theory of two refining equivalence relations with an additive structure on the classes of the coarser equivalence relation.)

We now analyze in more detail the situation when a type q needs b/B and $t(b; B)$ has (ENI)-depth one. We can 'slightly' enlarge both b and B to find a base for a strongly regular type which is not orthogonal to s. The next lemma makes 'slightly' precise and gives a normal form for depth one types.

1.12 Lemma. *Suppose $s \in S(B)$ is (ENI)-constrained by q and q needs s over B. There is a finite set D which is atomic over B and a pair of types $\hat{s}$ and s_1 in $S(D)$ which satisfy the following conditions. If b realizes s there is a d realizing s_1 with $b \subseteq d$ and a $c \subseteq d$ which realizes $\hat{s}$ such that*

i) $t(c; D)$ *is strongly regular.*
ii) $t(d; D \cup c)$ *is isolated.*
iii) *For some strongly regular* $\hat{q} \in S(D \cup d)$, $q \not\perp \hat{q}$ *and* $\hat{q}$ *needs* d/D.
iv) $c \triangleright_D d$.

Proof. Suppose q needs b/B. Choose M prime over B with $M \downarrow_B b$. Since q has weight one Lemma XIII.3.11 implies $q \dashv M$. Applying monotonicity, $q \not\dashv B \cup b$ implies $q \not\dashv M[b]$. Choose $c \in M[b]$ with $t(c; M)$ strongly regular. By Exercise XIII.4.8 and since b has weight one, $c \triangleright_M M[b]$. By Lemma 1.8 $q \not\dashv M \cup c$. Thus, for some strongly regular $q' \in S(M[c])$, $q \not\perp q'$. Choose $D \subseteq M$ and $d \subseteq M[c] - M$ with $b \cup c \subseteq d$ so that q' is strongly based on $D \cup d$. Let $\hat{q} = q' | D \cup d$. Setting $\hat{s}$ as $t(c; D)$ and s_1 as $t(d; D)$, conditions ii) and iii) are satisfied. We can invoke Lemma VI.3.12 and possibily increase the size of D to guarantee condition iv). This completes the proof.

This result refines Corollary XIII.3.4. There, we showed that if $p \not\dashv M$ and $M \models T$ then $p \not\perp q$ for some strongly regular $q \in S(M)$. While we can not relax the requirement that M be a model to show this for an arbitrary set B we find the strongly regular type by only slightly extending B. Buechler [Buechler 1986] shows that if T is superstable and '1-based' then we can find the regular type in $S(B)$. In the light of Lemma 1.12 we can establish a normal form for each depth one type. Actually, this normal form will depend on q. In this section we exploit the normal form separately for each choice of q. Later we will reduce to the case that s supports only finitely many types. Then we can choose a single s_1 to work for all of them.

1.13 Notation. For any depth one type s and any q which needs s we fix a pair of types $\hat{s}$ and s_1 satisfying Lemma 1.12. Without loss of generality we may assume each of $s, \hat{s}, s_1$ have domain D. Now, $\hat{s}$ is strongly regular and if d realizes s_1, there is a $c \subseteq d$ realizing $\hat{s}$ such that $t(d; D \cup c)$ is atomic.

We want to show that if q is nonorthogonal to one realization of s it is nonorthogonal to all realizations which depend on the first. This kind of transfer requires some relation between the depth of s and that of q; for example, q constrains s. An easy counter example to the conclusion of the following lemma but with $\text{dp}(t(c; A)) = 2$ and $\text{dp}(q) = 0$ can be found in the theory REF_2.

1.14 Lemma. *Suppose* T *satisfies the (ENI)-NDOP and* q *is a stationary (ENI)-type. Suppose both* $t(c; A)$ *and* $t(c'; A)$ *are stationary with weight one and that* q *(ENI)-constrains* $t(c'; A)$. *If* $q \dashv A$ *and* $c \not\downarrow_A c'$ *then* $q \not\dashv A \cup c$ *implies* $q \not\dashv A \cup c'$.

Proof. Let $B = \text{dom}\, q$ and choose an **S**-model M with $B \cup c \cup c' \downarrow_A M$. Then $q \dashv M$, $q \not\dashv M \cup c$, and $c \not\downarrow_M c'$. Choose an independent set X of realizations of strongly regular types over M so that both $X \cup c$ and $X \cup c'$ are maximal independent sets of realizations in $M[c, c']$ of weight one types. The existence of X is guaranteed by $c \not\downarrow_M c'$. Now, $q \not\dashv M \cup c$ implies two facts. First, $q \not\dashv M[c, c']$. Second, since $c \downarrow_M x$ for each $x \in X$, we have $c \downarrow_M M[X]$. Applying Theorem XIII.3.11, we conclude $q \dashv M[X]$. By the (ENI)NDOP, $q \dashv M[X, c']$ if and only if $q \dashv M[c']$. Now suppose $q \dashv M[X, c']$. Consider the three models $M[X] \subseteq M[X, c'] \subseteq M[c, c']$. If $t(M[c, c']; M[X, c']) \dashv M[X]$ the (ENI)-depth of $t(c'; M[X])$ which equals the (ENI)-depth of $t(c'; M)$ is greater than the depth of q plus two. Thus, Theorem XIII.4.3 yields a $b \in M[c, c'] - M[c', X]$ which is independent from $M[c', X]$. Because this contradicts the choice of X, $q \not\dashv M[X, c']$; whence, $q \not\dashv M[c']$. By Theorem 1.8, this implies $q \not\dashv M \cup c'$ and thus, since $B \downarrow_{A \cup c'} M[c']$, $q \not\dashv A \cup c'$.

The next lemma is a further transfer property for needing.

1.15 Lemma. *Let T be an ω-stable theory with the (ENI)-NDOP. Suppose $t(b; \emptyset)$ and $t(b'; \emptyset)$ have (ENI)-depth ≤ 1. If p needs c/b, p' needs c'/b', and $t(c; b) \perp t(c'; b')$ then $p \perp p'$.*

Proof. Suppose first that $c \downarrow_b b'$ and $c' \downarrow_{b'} b$. Then, $p \dashv b \cup b'$ and since $t(c; b) \perp t(c'; b')$, $c \downarrow_{b \cup b'} c'$. Thus, by Theorem XIII.3.11, $p \dashv b \cup b' \cup c'$. But, if $p \not\perp p'$ then $p' \not\dashv b' \cup c'$ implies by Theorem XIII.2.24 iii) that $p \not\dashv b' \cup c'$. So $p \not\dashv b \cup b' \cup c'$ and we have the result. Thus, the major task of the proof is to replace b, c, b', c' by $\hat{b}, \hat{c}, \hat{b}', \hat{c}'$ which satisfy the following conditions.

i) $stp(c \frown b; \emptyset) = stp(\hat{c} \frown \hat{b}; \emptyset)$ and $stp(\hat{c}' \frown \hat{b}'; \emptyset) = stp(c' \frown b'; \emptyset)$.
ii) $t(\hat{c}; \hat{b}) \perp t(\hat{c}'; \hat{b}')$.
iii) p needs $\hat{c}/\hat{b}$ and p' needs $\hat{c}'/\hat{b}'$.
iv) $\hat{c} \downarrow_{\hat{b}} \hat{b}'$ and $\hat{c}' \downarrow_{\hat{b}'} \hat{b}$.

For this, choose first $\hat{b}, \hat{b}'$ so that $\{b \frown c \frown b', \hat{b}, \hat{b}'\}$ is independent over the empty set. Let $q = t(c; b)$ and $q' = t(c'; b')$. Since $t(b; \emptyset)$ has (ENI)-depth at most one, $q \not\dashv \emptyset$. Writing $\hat{q}$ for $q_{\hat{b}}$, we have by Corollary VI.2.22 that $q \not\perp \hat{q}$. Let M be an **S**-model containing $\{b, \hat{b}', \hat{b}\}$ with $c \downarrow_{b\hat{b}'\hat{b}} M$; witness the nonorthogonality of q and $\hat{q}$ by choosing $\hat{c}$ realizing $\hat{q}$ with $\hat{c} \downarrow_{\hat{b}} M$ and $\hat{c} \not\downarrow_M c$. By transitivity of independence, $c \downarrow_b M$. Thus, p needs c/b implies p needs c/M. By Lemma 1.14 since $\hat{c} \not\downarrow_M c$, p needs $\hat{c}/M$ and thus p needs $\hat{c}/\hat{b}$. Since $\hat{b}' \in M$, $\hat{c} \downarrow_{\hat{b}} \hat{b}'$. Similarly, we can choose $\hat{c}'$ realizing the nonforking extension to $S(M)$ of $q'_{\hat{b}'}$ and verify that p' needs $\hat{c}'/\hat{b}'$ and $\hat{c}' \downarrow_{\hat{b}'} \hat{b}$. Then we complete the proof as outlined.

The following theorem plays a key role in the proof of Vaught's conjecture (cf. Lemma 3.3).

1.16 Theorem. *Let $A \subseteq M$ and suppose $s \in S(A)$ is a trivial (ENI)-depth one type. If $q \in S(M)$ needs s then there is a $b \in M$ such that q needs b/A.*

Proof. By definition of needing s, q needs d/A for some realization d of s. Suppose for contradiction that $d \downarrow_A M$. By Theorem XIII.3.11, this implies $q \dashv M \cup d$ contrary to the choice of d. Thus, $d \not\downarrow_A M$. Applying Theorem 1.12, choose $c' \subseteq d$ which realizes the strongly regular type $\hat{s} \in S(A)$ and so that d is atomic over $A \cup c$. Since nonorthogonality preserves triviality, $\hat{s}$ is also trivial. By Theorem XVI.2.12 and the triviality of $\hat{s}$, $\hat{s}$ is realized in M by some c with $c \not\downarrow_A c'$. Now, choose $b \subseteq M$ so that $t(b; A \cup c) = t(d; A \cup c')$. Since $b \not\downarrow_A d$, Lemma 1.14 implies q needs b/A.

1.17 Historical Notes. Most of this material is from the first two sections of [Shelah, Harrington, & Makkai 1984]. We have benefited greatly from discussions about this material with Buechler, Harrington, and Saffe.

2. Toward the Vaught and Morley Conjectures

In this section we obtain sufficient conditions on the type structure of a theory to compute certain lower bounds on the number of models of T. We prove first that if the countable ω-stable theory T has fewer than $2^{\aleph_0}$ countable models then it has ENI-depth at most two. We refine this analysis to show that T has only finitely many nonorthogonality classes of types in each of the following categories.

i) R^e: ENI-types which are orthogonal to the empty set.
ii) S^e: Types which support ENI-types.
iii) Q^e_b: ENI-types which need a fixed realization b of a supportive type.

As a step toward the Morley conjecture, we then make a similar analysis of the types which depend on an (ENI)-depth two type p over a set X. We introduce the notion of a frame $\langle R_p, S_p, Q_{p,s}\rangle$ for p. A frame is a triple of sets of types which form a basis for those types which need a specific realization of p over X or need some type which needs that realization over X. We prove a lower bound for the number of models of T in an uncountable power if for some p and s, R_p, S_p, or $Q_{p,s}$ is infinite.

We begin now to restrict the possibilities for a countable ω-stable theory which has fewer than $2^{\aleph_0}$ countable models. At first glance the following theorem seems to be too weak. It is easy to see that what appears to be a typical example of an ENI-type which is orthogonal to the empty set (an equivalence relation with infinitely many classes, each a model of (Z, S)) (cf. XVII.3.8) leads immediately to $2^{\aleph_0}$ models. Thus, one might hope to prove that any countable ω-stable theory with fewer than $2^{\aleph_0}$ models has ENI-depth one. However, the example in Section 4 shows this intuition is incorrect.

2.1 Theorem. *Let T be a countable ω-stable theory without ENI-DOP. If T has ENI-depth greater than 2 then T has $2^{\aleph_0}$ countable models.*

Proof. If T has ENI-depth greater than 2 then there is a type q with ENI-depth 2. Without loss of generality, we may assume q is over the empty set and choose a to realize q. Then there exist finite sequences $a \subsetneq b \subsetneq c$ and a nonisolated $p \in S(c)$ such that $p \dashv b$ and $t(c; b) \dashv a$; moreover p can be chosen stationary and strongly regular while $r = t(b; a)$, and $t(c; b)$ are stationary weight one types. Thus, both p and r have weight one. Without loss of generality we can absorb a into the language and work over the empty set.

For each $X \subseteq \omega$ we define a model M_X as follows. Let $B = \{b^n : n \in X\}$ be an independent sequence of realizations of $t(b; \emptyset)$. For each $n \in X$, choose a sequence $C^n = \{c_i^n : i < n\}$ of sequences which are independent over b^n such that $c_i^n \frown b^n$ realizes $t(c \frown b; \emptyset)$. Since $t(c; b) \dashv \emptyset$, $\{C^n : n \in X\}$ is an independent sequence over $\emptyset$. Let p_i^n denote $p_{c_i^n}$ and $\mathcal{P}$ denote the collection of all the p_i^n.

Use the choice of the C^n, the fact that $t(c; b)$ is stationary and Corollary VI.2.22 to show

2.2 Exercise. For $n \neq m$ or $i \neq j$, $p_i^n \perp p_j^m$.

Returning to the proof of Theorem 2.1, since the c_i^n for $i < n$ are independent over b^n, if $\tilde{p}_i^n$ denotes the nonforking extension of p_i^n to $S(C^n)$ then for any $M \supset C^n$, $\dim(\tilde{p}_i^n, M) = \dim(p_i^n, M)$. By Theorem XIV.2.4, there is a model M_X such that for each $n \in X$ and each $i < n$, $\dim(\tilde{p}_i^n, M_X) < \omega$ and if q is ω-irrelevant to $\mathcal{P}$ then $\dim(q, M_X)$ is infinite.

Since $r = t(b; \emptyset)$ has weight one, forking is mildly transitive on realizations of this type. Thus forking is an equivalence relation on $r(\mathcal{M})$. By a class $\mathcal{B}$ or $[b]$ we mean an equivalence class under this relation. To show that $M_X \approx M_Y$ implies $X = Y$ we will assign an integer $m([b])$ to each class $[b]$ and prove that $\{m([b]) : b \in r(M_X)\} = X$. Namely, $m(\mathcal{B})$ is the supremum over all $b' \in \mathcal{B}$ of the $m \leq \omega$ such that there exists an independent sequence c'_i for $i < m$ over b' with $t(c'_i \frown b'; \emptyset) = t(c \frown b; \emptyset)$ and with $\dim(p_{c'_i}, M_X) < \omega$.

The following extension of Exercise 2.2 is the key to rest of the argument. It is an immediate consequence of Lemma 1.15

2.3 Exercise. Consider any b_0, b_1, e_0, e_1 such that b_0, b_1 realize $t(b; \emptyset)$ and $t(e_i \frown b_i; \emptyset) = t(c \frown b; \emptyset)$ for $i = 0, 1$, and $b_0 \downarrow_\emptyset b_1$. Then $p_{e_0} \perp p_{e_1}$.

2.4 Exercise. Prove the preceding exercise directly rather than deriving it from Lemma 1.15.

We use this fact to compute $m(\mathcal{B})$ in several cases. If $\hat{c} \frown \hat{b}$ realizes $t(c \frown b; \emptyset)$ and $\dim(p_{\hat{c}}; M_X) < \omega$ then for some p_i^n, $p_{\hat{c}} \not\perp p_i^n$. Applying the contrapositive of Exercise 2.3, $(\hat{b} \not\downarrow b^n; \emptyset)$ for some n. Thus, if $\mathcal{B} \neq [b^n]$ for some $n \in X$, we have $m(\mathcal{B}) = 0$.

Now we show $m([b^n]) = n$. By the construction, $m([b^n]) \geq n$. Suppose $\hat{b} \in [b^n]$ and for some k that $\langle \hat{c}_i : i < k \rangle$ is an independent sequence over $\hat{b}$ with $t(\hat{c}_i \frown \hat{b}; \emptyset) = t(c \frown b; \emptyset)$. For every $m \neq n$, $\hat{b} \downarrow_\emptyset b^m$ so by Exercise 2.3, letting $\hat{p}_i$ denote $p_{\hat{c}_i}$, $\hat{p}_i \perp p_i^m$ if $m \neq n$. But $\dim(\hat{p}_i, M_X) < \omega$ implies there is some $p \in \mathcal{P}$ with $\hat{p}_i \not\perp p$. So for each $i < k$ there is an $\hat{i} < n$ with $\hat{p}_i \not\perp p_{\hat{i}}^n$. Since the p_i^n are pairwise orthogonal and nonorthogonality is transitive on strongly regular types, the map $i \mapsto \hat{i}$ must be $1-1$ and we conclude $m([b^n]) = n$.

Thus, the isomorphism type of M_X determines X and we finish the proof of Theorem 2.1.

We have shown any ω-stable theory with fewer than $2^{\aleph_0}$ countable models has ENI-depth at most two. We need a slightly stronger observation than the assertion that R^e is finite. The following lemma is an immediate consequence of Theorem XV.3.21 ii).

2.5 Lemma. *If T has less than $2^{\aleph_0}$ countable models then for any finite set A, there are only finitely many nonorthogonality classes of ENI-strongly regular types which are not orthogonal to A.*

This shows that R^e is finite. Moreover, for each $s \in S^e$ and each realization b of s, Q^e_b is finite. The following theorem shows we may assume S^e is finite.

2.6 Theorem. *Let T have the ENI-NDOP. If there are infinitely many pairwise orthogonal ENI-supportive types then T has $2^{\aleph_0}$ countable models.*

Proof. If there are infinitely many ENI-supportive types $\langle s_n : n < \omega\rangle$ which are pairwise orthogonal then we will normalize them as in the proof of Theorem XV.3.25. Let s_n be strongly based on b_n, let c_n realize s_n and suppose $q_n \in S(d_n)$ needs c_n/b_n. Without loss of generality, we can assume that all conditions of Lemma 1.12 are satisfied. In particular, $b_n \subsetneq c_n \subsetneq d_n$ and q_n is not isolated. Construct an independent sequence $\langle b_n{}^\frown c_n{}^\frown d_n : n < \omega\rangle$ realizing these types. As in Paragraph XV.3.4 we can require that for $i \neq j$ either b_i and b_j have the same type over the empty set or every conjugate of s_i is orthogonal to s_j. (If some conjugate $\alpha(s_i)$ is not orthogonal to s_j, then α^{-1} of s_j could be chosen as s_j.) Since none of the s_i are orthogonal to the empty set and $s_i \perp s_j$, the first alternative implies b_i and b_j realize distinct strong types over the empty set (Corollary VI.2.23). But, since T is ω-stable $t(b_i;\emptyset)$ has finite multiplicity. Thus, we can find a subsequence $S = \langle s_i : i < \omega\rangle$ such that if $i \neq j$ then every conjugate of s_i is orthogonal to s_j and $t(b_i;\emptyset) \neq t(b_j;\emptyset)$.

By Theorem XIV.2.4, for any $X \subseteq \omega$ there is a model M_X such that if $n \in X$, $\dim(q_n, M_X) < \omega$ and if $n \notin X$, $\dim(q_n, M_X) = \omega$. We claim $n \in X$ if and only if there is a triple $\langle b, c, d\rangle$ realizing $t(b_n{}^\frown c_n{}^\frown d_n;\emptyset)$ and with $\dim(q_d, M_X) < \omega$, where q_d is a copy of q_n over d. Clearly, if $n \in X$, we have the claim by construction. Fix n such that $\langle b, c, d\rangle$ realizes $t(b_n{}^\frown c_n{}^\frown d_n;\emptyset)$. For any $m \neq n$, the choice of S guarantees $t(c;b) \perp t(c_m;b_m)$. Since q_d needs c/b and q_m needs c_m/b_m, Lemma 1.15 implies $q_d \perp q_m$. Thus, if $n \notin X$, $q_d \perp q_m$ for every $m \in X$. By Theorem XIV.2.4, $\dim(q_d, M_X) = \omega$ as required. Since we can recover X from the isomorphism type of M_X, if $X \neq X'$ then $M_X \not\approx M_{X'}$. We have proved the theorem.

This completes the contribution of this section to Vaught's conjecture. We turn to the Morley conjecture. We now want to provide lower bounds for the number of models in power κ of an ω-stable theory which has finite depth. We describe these bounds in terms of the height of a type.

2.7 Definition. Let $p \in S(B)$ be strongly regular. If $p \not\perp \emptyset$ then $\mathrm{ht}(p) = 0$. If for some a and A, p needs a/A with $p' = t(a; A)$ and $\mathrm{ht}(p') = n$ then $\mathrm{ht}(p) \geq n + 1$. Finally, $\mathrm{ht}(p)$ is the least n such that $\mathrm{ht}(p) \not\geq n + 1$. We say p' witnesses that the height of p is at least $n + 1$.

Note that without loss of generality one may assume that $A \cup a$ is contained in $\mathrm{dom}\, p$. Thinking of p as occuring in a representation of a model M, the height of p is the height of a realization of p, not the height of the domain of p.

A major technical difficulty in establishing lower bounds arises from the problem of ensuring that nonorthogonal types in different representations of the same model have the same dimension. We solved this problem in Chapter XVII by using $\aleph_1$-ample trees to guarantee that, at least often enough, the types we were comparing could be based on the same set. The trouble is that at the top levels we may need to preserve finite dimensions but in a model prime over an $\aleph_1$-ample tree of height m there may be no types with finite dimension. The solution is to construct trees which are $\aleph_1$-ample up to height $m - 2$ or $m - 1$ and to label the top nodes of these trees with the dimensions that we want to preserve at the top level. Using the fact that the shorter trees are $\aleph_1$-ample, we obtain an isomorphism between the shorter trees. By applying the results of Chapter XIV and Section 1 of this chapter, we are able to preserve the labels on the top nodes.

Thus, the combinatorial principal underlying the counting below is the following lemma, which is proved by induction as similar results were earlier. In order to treat the problem uniformly in $\aleph_\beta$, we must allow the number of labels to depend on the subscript β.

2.8 Lemma. *Let $\lambda = \lambda(\beta)$. If $\beta \geq 1$, there are at least $\beth_m(\lambda)$ partially λ-labeled trees of power $\aleph_\beta$ which have height m and are $\aleph_1$-ample.*

Although Lemma 2.8 is proved by a routine induction, there is one important feature. In the first step, we do not try to calculate how many times the various labels are realized. Rather, we observe that there are 2^λ subsets of λ and that any set X of labels can have the property that each member of X is realized at least once and therefore at least $\aleph_1$ times while labels which are not in X are not realized. This is essentially the same idea as the proof of Lemma XVII.4.11

The following definition defines the context for the detailed analysis of depth two types. We formalize a situation analogous to the discussion of R^e, S^e, and Q^e_b earlier in this section. For the study of Morley's conjecture we must consider types which are not ENI; thus, the superscript e's disappear. The word set in this definition is not used in our technical sense (smaller than $|\mathcal{M}|$). In fact, if $\tilde{S}_a$, for example, is infinite then it will be a class in the technical sense since it contains all depth one types based in $\mathcal{M}$ which need a/X.

2.9 Definition. Let $p \in S(X)$ have depth two and let a realize p. A *frame* at a for p consists of three sets of types: $\tilde{R}_a$ $\tilde{S}_a$ and $\tilde{Q}_a$:

i) $\tilde{R}_a$ is the set of depth zero types which need a/X.
ii) $\tilde{S}_a$ is the set of depth one types which need a/X.
iii) For each $s \in \tilde{S}_a$, and any realization b of s, $\tilde{Q}_{a,b}$ is the set of types that need $b/\operatorname{dom} s$. $\tilde{Q}_{a,s}$ denotes an arbitrary conjugate over $X \cup a$ of $\tilde{Q}_{a,b}$. We write $\tilde{Q}_a$ for $\bigcup\{\tilde{Q}_{a,s} : s \in \tilde{S}_a\}$
iv) Let R_a, S_a, and for each $s \in S_a$ and each b realizing s, $Q_{a,b}$ be maximal pairwise orthogonal subsets of $\tilde{R}_a$, $\tilde{S}_a$, and $\tilde{Q}_{a,b}$ respectively.

Although this definition refers to a specific realization a of p, there is clearly an automorphism fixing X and taking the frame at a to the frame at b for any other realization b of p. To emphasize the independence of these notions from the specific realization of p, we write $\langle \tilde{R}_p, \tilde{S}_p, \tilde{Q}_p \rangle$ to denote an arbitrary conjugate over X of $\langle \tilde{R}_a, \tilde{S}_a, \tilde{Q}_a \rangle$ and $Q_{p,s}$ to denote an arbitrary conjugate over X of $Q_{a,s}$.

We can simplify the study of the frame for p by replacing $\tilde{R}_p$, $\tilde{S}_p$, and for each $s \in \tilde{S}_p$, $\tilde{Q}_{p,s}$ by maximal pairwise orthogonal subsets R_p, S_p, and for each $s \in S_p$, $Q_{p,s}$. As in the original case we write Q_p for the union of the $Q_{p,s}$ over $s \in S_p$. We may sometimes fix a realization a of p and speak, for example, of S_a. In this section, since we are looking for lower bounds we can work directly with $\langle R_p, S_p, Q_p \rangle$. In Section 5, to compute upper bounds, we must rely on the results of Section 1 to show that controlling $\langle R_p, S_p, Q_p \rangle$ allows one to control $\langle \tilde{R}_p, \tilde{S}_p, \tilde{Q}_p \rangle$.

In the next definition we distinguish three kinds of depth two type. In the following theorem we give, for each kind, lower bounds on the number of models of T if T has a depth two type with height r of that kind.

2.10 Definition. Let $p \in S(A)$ have depth two.

The type p has *kind I* if there is an $s \in S_p$ such that $Q_{p,s}$ is infinite.
The type p has *kind II* if S_p is infinite.
The type p has *kind III* if R_p is infinite.

Recall that in the construction of models over trees for a countable ω-stable theory we are always considering types over countable sets. Thus, the A in the following theorem can be taken to be countable.

2.11 Theorem. *Let $p \in S(A)$ have height r and depth two.*

i) *If p has kind I then $I(\aleph_\beta, AT) \geq \beth_{r+1}(|\beta + 1|^{|\omega|})$.*
ii) *If p has kind II then $I(\aleph_\beta, AT) \geq \beth_{r+1}(|\beta + \omega|)$.*
iii) *If p has kind III then $I(\aleph_\beta, AT) \geq \beth_r(|\beta + 1|^{|\omega|})$.*

Proof. For each of the three cases we will describe a $1-1$ correspondence between $\aleph_1$-ample labeled trees of height r or $r+1$ (depending on the case) and models of T. The exact labeling depends upon the kind of p, but the construction of a core tree is the same in each case.

For any $\aleph_1$-ample tree of height r, X, form a tree of models, $\bar{A}_X$, as follows. Choose p_i for $i < r$ by induction so that $p = p_0$ and p_{i+1} witnesses that the height of p_i is $r - i$. Note that these types are pairwise orthogonal.

Now let the elements of $\bar{A}_X$ with height i be prime models over realizations of conjugates of p_{r-i}. Depending upon the kind of p, we describe below the method for completing the tree $\bar{A}_X$ to a tree $\hat{A}_X$ by adding elements of height $r+1$ and $r+2$. In discussing kind I, we will label the nodes of height $r+1$. For kinds II and III we will label the top nodes (i.e. nodes of height r) of X.

Let M_X be prime over $\hat{A}_X$. Now, suppose $M_X \approx M_Y$. Then, we know by Theorem XVII.4.7 that $X \approx_q Y$ and since X and Y are $\aleph_1$-ample that $X \approx Y$, as trees. It remains to show that this isomorphism preserves the labels. However, this will be clear when we explain the assignment of labels in each case. Then in each case we complete the proof of the theorem by counting the labels and thus the labeled trees.

i) When p has kind I, we make the trees of height $r+1$ be $\aleph_1$-ample as well. Thus, if $M_X \approx M_Y$ there is an isomorphism α between the subtrees of elements with height at most $r+1$ of $\hat{A}_X$ and $\hat{A}_Y$. Since p has kind I, there is an element B of $\hat{A}_X$ which has height $r+1$ and has infinitely many pairwise orthogonal strongly regular types in $S(B)$. That is, B is $X[a][b]$ where b is a realization of the $s \in S_p$ which supports infinitely many types. In fact, without loss of generality, we can require all models of height $r+1$ to have this form. But now, for each $q_i \in S(B)$, $\dim(q_i, M_X) = \dim(\alpha(q_i), M_Y)$. Thus, we can label a node B of height $r+1$ by $f \in |\beta+1|^{|\omega|}$ if for each $i < \omega$, there are i mutually orthogonal types in $S(B)$ which are orthogonal to B^- and satisfy: $\dim(q_i, M_X) = f(i)$. We force this last condition to hold by putting $f(i)$ independent realizations of q_i into $\hat{A}_X$. By Lemma 2.8, we have $\beth_{r+1}(|\beta+1|^{|\omega|})$ distinct models of power $\aleph_\beta$ as required.

To see the labels are preserved, suppose $\hat{A}_X$ and $\hat{A}_Y$ represent M. Then there is an isomorphism between $(\hat{A}_X)^1$ and $(\hat{A}_Y)^1$ induced by nonorthogonality as in Theorem XVII.4.7. Moreover, continuing the notation of that theorem, if $q \in S(B)$ and $\mathrm{ht}(B) = r+1$ then $\alpha(q) \not\perp q$ and both are strongly based on $M_{B,\alpha(B)}$. But then $\dim(q, M) = \dim(\alpha(q)M)$.

ii) If p has kind II, S_p is infinite. We will label the nodes of height r in $\bar{A}_X$. Each such node corresponds to a realization a of p. We label the node by telling for each $s \in S_a$, the number of realizations of s which support a type of dimension $\aleph_\gamma$ for $\omega \leq \gamma \leq \beta$. There are then

$$(|\beta+\omega|^{|\beta+1|})^{\omega} = |\beta+\omega|^{|\beta+\omega|}$$

labels. If α is an isomorphism between the elements of height at most r in $\hat{A}_X$ and $\hat{A}_Y$, then a and $\alpha(a)$ have the same label. Applying Lemma 2.8, there are at least $\beth_r(|\beta+\omega|^{|\beta+\omega|}) = \beth_{r+1}(|\beta+\omega|)$ models of power $\aleph_\beta$ as required.

iii) If p has kind III, for each a realizing p and for each $r \in R_a$, there are at least $|\beta+1|$ possible dimensions for r. Since R_a is infinite, there are $|\beta+1|^{\omega}$ possible labels for a. Thus, by Lemma 2.8 again, there are at least $\beth_r(|\beta+1|^{|\omega|})$ models of T as required.

Apply Theorem XVII.3.19 to show the following exercise.

2.12 Exercise. If T has depth $r+2$ and some p of height r has kind I then $I(\aleph_\beta, AT) = \beth_{r+1}(|\beta+1|^{|\omega|})$.

In this section we have computed lower bounds on $I^*(\aleph_\beta, \mathbf{AT})$ if one of R_p, S_p, or $Q_{p,s}$ is infinite. When studying Vaught's conjecture, β is zero. By Theorem 2.1, we may assume that T has ENI-depth two. In that case we found that if one of R^e, S^e or Q^e_s is infinite the lower bound is 2^ω. Thus, for further analysis of Vaught's conjecture in Section 3 we may assume each of R^e, S^e, and Q^e_s is finite. In Section 5 we improve the lower bounds in uncountable cardinalities for this case and consider the upper bounds for all cases.

2.13 Historical Notes. The discussion of Vaught's conjecture relies on [Shelah, Harrington, & Makkai 1984] and [Bouscaren 1984]. The material on Morley's conjecture appeared in [Baldwin & Harrington]. I was greatly aided by discussions with Buechler and with Saffe, who gave the first proof of the finite depth case of the Morley conjecture [Saffe 1983].

3. The Vaught Conjecture for ω-Stable Countable Theories

In this section we complete the proof of Vaught's conjecture for an ω-stable countable theory. With this in hand we finish in Section 5 the finite depth case of the proof, outlined in Chapter XVII, that the spectrum function of a countable ω-stable theory is increasing on uncountable cardinals. Throughout this section we assume T is a countable ω-stable theory with fewer than $2^{\aleph_0}$ countable models. The following notion of a frame for T formalizes the ideas introduced at the beginning of Section 2 but since T has ENI-depth two it is less complex than the analogous notion of a frame for p introduced in Definition 2.9.

3.1 Definition. A *frame* for T is a triple $\langle R^e, S^e, Q^e\rangle$ satisfying the following conditions. R^e is a maximal set of pairwise orthogonal, ENI-depth zero, ENI-types which are each nonorthogonal to the empty set. S^e is a nonorthogonality basis for the ENI-supportive types. For each $s \in S^e$ and each b realizing s, Q^e_b is a nonorthogonality basis for the ENI-types which need b over dom s. We write Q^e_s for an arbitrary representative of the conjugacy class of the $Q^e_{s,b}$ and Q^e for the union of the Q^e_s over $s \in S$.

Let M_0 be the prime model of T. Then, by Theorem XIII.3.4 each $p \in S^e \cup R^e$ is nonorthogonal to some strongly regular type which is strongly based on M_0. By Lemma 2.5 and Theorem 2.6, R^e and S^e are finite. Thus, by choosing the right members of each nonorthogonality class we can assume all members of $R^e \cup S^e$ are based on the same finite subset A of M_0. Moreover, we can assume the members of R^e are nonisolated and strongly regular.

By Lemma 2.5 again there are only finitely many mutually orthogonal ENI-types which are not orthogonal to A. Thus, we can choose a finite set D with $A \subseteq D \subseteq M$ so that if $r \not\perp A$, there is an $\tilde{r} \in S(D)$ which is strongly regular and not orthogonal to r. Perhaps enlarging our original choice, we let R^e denote a nonorthogonality basis for the set of all ENI-depth zero types which are not orthogonal to D. For each realization b of $s \in S^e$, we denote by Q_b a nonorthogonality basis for the types which need b/A. As in the previous two cases, we can assume that all members of Q_b are in $S(A \cup b')$ where b' is a finite sequence which is atomic over $A \cup b$. There is one slippery point in this specialization. There may be realizations of some $s \in S^e$ which depend on D over A. Since there are only finitely many such points we will see, in the proof Theorem 3.13, that they are harmless.

We now fix a particular model M and describe a 'basis' for M with respect to a specific frame $\langle R^e, S^e, Q^e\rangle$ for T. Recall that each $s \in S^e$ is associated with a strongly regular $\hat{s}$ given in Notation 1.13.

3.2 Notation. (Fig. 1). An ENI-*extended basis* or *set of reference points* for M relative to $\langle A, D, R^e, S^e, Q^e\rangle$ consists of

i) A basis J_r of $r(M)$ for each $r \in R^e$.
ii) A basis I_s of $s(M)$ for each $s \in S^e$.
iii) For each $b \in I_s$ and each $q \in Q^e_b$, a basis $J_{b,q}$ for $q(M)$.

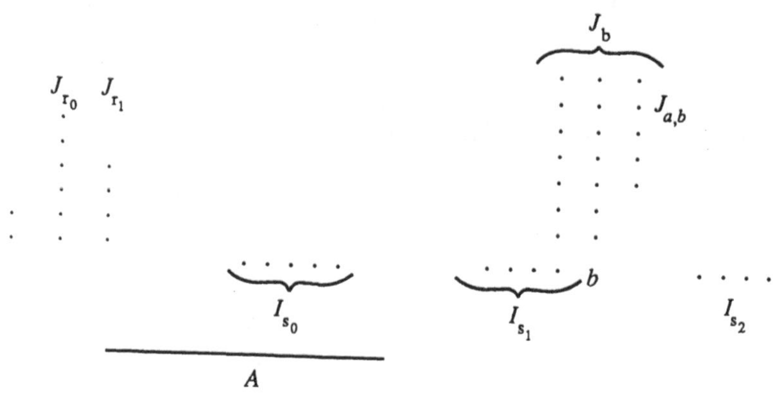

Fig. 1. An extended basis

Since Q^e_s is finite we are able to identify the s_1 from Notation 1.13 with s. Associated with I_s is a set $I_{\hat{s}}$, a basis for $\hat{s}(M)$, such that for each $b \in I_s$ there is an $a \subseteq b$ with $a \in I_{\hat{s}}$ and $t(b; A \cup a)$ is isolated.

We write J_b for $\bigcup_{q \in Q^e_b} J_{q,b}$ and J_s for $\bigcup_{b \in I_s} J_b$; we call J_b a basis for Q^e_b in M. We write I^e for $\bigcup_{r \in R} J_r \cup \bigcup_{s \in S^e} I_s \cup \bigcup_{s \in S^e} J_s$; we call I^e an extended basis for M.

Of course, some J_r and I_s may be empty. If I_s is empty, then no Q_b^e for b realizing s is even defined. Note that I is independent over A relative to the partial ordering which places c above b if $c \in J_b$ and otherwise the elements of I^e are incomparable.

The next lemma justifies the terminology 'extended basis'.

3.3 Lemma. *If I^e is an ENI-extended basis for M then M is atomic over $D \cup I^e$. Thus, if M is countable, M is prime over $D \cup I^e$.*

Proof. Let N be a maximal atomic model with $I^e \cup D \subseteq N \subseteq M$. The existence of N is guaranteed by Zorn's lemma since the union of atomic models is atomic and if $E \subseteq N$ is atomic over $D \cup I^e$ then there is a model N' which is atomic over $D \cup I^e$ with $I^e \cup D \cup E \subseteq N' \subseteq M$.

We claim $N = M$. In the ensuing two paragraphs we eliminate each possibility for a strongly regular type $q \in S(N)$ which might be realized in $M - N$. The definition of a frame provides a classification of the possible q.

First, we show that no type $q \in S(N)$ which satisfies either of the following two conditions can be realized in $M - N$. i) q is ENI. ii) q is nonorthogonal to a type which has a realization $b \in M$ such that b supports an ENI-type. For, if so, either $q \not\perp r$ for some $r \in R^e$ or $q \not\perp s$ for some $s \in S^e$ or q needs b/A for some $b \in I_s$. By Theorem 1.16 if q needs b/A, $q \not\perp q_0$ for some $q_0 \in Q_b^e$. Now in any case, since the appropropriate r, s, or q_0 is strongly regular, and not orthogonal to q, one of r^N, s^N, or ${q_0}^N$ is realized in $M - N$ contradicting the maximality of I^e.

Now suppose some persistently isolated $q \in S(N)$ is realized by some $c \in M - N$. We will show $N \cup c$ is atomic over $I^e \cup D$ and thus contradict the maximality of N. Let E be an arbitrary finite subset of N and without loss of generality assume q is strongly based on E and that $D \subseteq E$. (That is, if necessary, expand E a little.) Then $q|E$ is isolated. Since $c \downarrow_E N$ and $t(c; E)$ is stationary, $t(c; E) \vdash t(c; N)$. Applying transitivity of atomicity (Lemma IX.1.11) twice, $N \cup c$ is atomic over $E \cup I^e$ and then over $D \cup I^e$.

The next step in the proof is a solution of the difficulty we encountered in Chapter XVII. The decomposition tree describes J_b for only one member b from each of the dependency classes under forking of the realizations of the supportive types. But J_b and $J_{b'}$ may vary drastically even when b and b' are dependent. Rather than choosing a representative from each such class (by choosing a basis), it is necessary to describe J_b for each member of the class. Since in this case we are dealing with depth one types, this is not too difficult.

For the present purposes of Vaught's conjecture the β in the following definition and lemmas can be set at zero. However, it doesn't clutter the definition or excessively complicate the proofs. So since we apply the more general notion in Section 5 we obtain its properties at this time.

3.4 Definition. Let $s \in S(A)$ have depth one and let $A \subseteq M$ with $M \models T$ and $|M| = \aleph_\beta$.

i) For $c \in s(M)$ the *local β-configuration* of s at c is $\overline{v}_{c,M}$, the vector of dimensions in M of the types $q \in Q^e_c$.

ii) For $b \in s(M)$, the *β-configuration* of s at b is $\overline{v}_{[b],M} = \{\overline{v}_{c,M} : c \not\perp_A b\}$. If β is fixed we speak of an s-configuration.

The following definitions will clarify the effect of a particular type $s \in S^e$.

iii) The model M is *s-simple* if M is prime over A, a basis for s in M and for each b in this basis, a basis for Q^e_b in M.

Here is an even simpler model.

iv) The model M is *s-trite* if M is prime over $A \cup b \cup J_b$ where b realizes s and J_b is a basis for Q^e_b.

A β-configuration of s at b in M is said to be trite or simple if M is trite or simple.

Each β-configuration is thus a set (not a sequence) of countable (finite in the Vaught conjecture case) sequences of cardinals less than or equal $\aleph_\beta$. We write configuration for 0-configuration. Vectors denoted by $\overline{s}$, $\overline{t}$ range over local configurations; vectors denoted by $\bar{S}$, $\bar{T}$ range over configurations.

In Section 5 we will rely on an obvious variant on the notion of an s-configuration to include types $s \in S_a - S^e_a$. This variant is obtained by allowing the types q in the definition of a local β-configuration to range over Q_c rather than just Q^e_c (i.e. we include isolated types.)

Given the dimension of each type in R^e and for each $s \in S^e$ and for each vector $\overline{t}$, the number of members, b, of a basis for s for which $\overline{v}_{b,M}$ should equal $\overline{t}$, we may determine a model M of T. However, there may be more than one such sequence which yields the same model. For, if $\overline{v}_{b,M} = \overline{t}$ and $\overline{v}_{c,M} = \overline{s}$ and $b \not\perp c$, we *may* obtain the same model by choosing a basis element with associated dimensions $\overline{s}$ as by choosing one with associated dimensions $\overline{t}$. This situation is illustrated by the example in Section XVIII.4.

We now show that the β-configurations which occur in trite models are either equal or disjoint and then that any configuration is a union of trite β-configurations. This analysis is based on [Bouscaren 1984].

3.5 Proposition. Let b, c realize s and suppose $b \not\perp_A c$. For any $q \in Q_c$, there is a $\tilde{q} \in Q_b$ with $q \not\perp \tilde{q}$.

Proof. Since q needs c/A, Lemma 1.14 implies q needs b/A. But, Q_b is a nonorthogonality basis for the types which need b/A so we finish.

The following lemma employs the notation of 3.2. It is here that we make crucial use of the choice of the strongly regular type $\hat{s}$.

3.6 Lemma. *Let M be s-simple. Suppose $b \in s(M)$ and J_b is a basis for Q^e_b in M. Let $M_1 \subseteq M$ be prime over $A \cup b \cup J_b$. For any $c \in s(M)$ with $c \not\perp_A b$, there is a $c' \in M_1$ with $\overline{v}_{c,M} = \overline{v}_{c',M_1}$.*

Proof. We first show that for each $c \in s(M_1)$ which depends on b over A, $\overline{v}_{c,M} = \overline{v}_{c,M_1}$. Clearly, each component of $\overline{v}_{c,M_1}$ is less than the corresponding component of $\overline{v}_{c,M}$. For the converse, choose any $q \in Q^e_c$. By Proposition 3.5 there is a $q' \in Q^e_b$ with $q \not\perp q'$. Now if q is realized in $M - M_1$, so is $(q')^{M_1}$. But this contradicts the assumption that J_b is a basis for Q^e_b in M.

Note that $\overline{v}_{[b],M_1}$ is the same for every M_1 prime over $A \cup b \cup J_b$ (by the uniqueness of prime models). Thus, it suffices to show that for every $c \in s(M)$ with $c \not\downarrow_A b$, there is a copy of M_1 which contains a c' with $\overline{v}_{c',M_1} = \overline{v}_{c,M}$. By the first paragraph of this proof, we need only find an M_1 and a $c' \in M_1$ so that $\overline{v}_{c',M} = \overline{v}_{c,M}$. Recall that by Lemma 1.12 there is a strongly regular type $\hat{s}$ such that there is an $a \subseteq b$ which realizes $\hat{s}$ and such that i) $a \rhd_A b$ and ii) $t(b; a \cup A)$ is isolated. Let $d \subseteq c$ realize $\hat{s}$ and satisfy that i) $d \rhd_A c$ and ii) $t(c; d \cup A)$ is isolated.

We now use the s-simplicity of M to find a copy of M_1 which contains an appropriate c'. Let I'_s denote $I_s - \{a\}$, $I'_{\hat{s}}$ denote $I_{\hat{s}} - \{b\}$, and J'_s denote $J_s - \{J_b\}$. Making essential use of the strong regularity of $\hat{s}$ we show

$$d \underset{A \cup b \cup J_b}{\downarrow} I'_{\hat{s}}.$$

For, $c \not\downarrow_A b$ implies $c \not\downarrow_A b \cup J_b$. Since $d \rhd_A c$, $d \not\downarrow_A b \cup J_b$. Now, by regularity, $t(d; A \cup b \cup J_b) \perp \hat{s}^{A \cup b \cup J_b}$. Thus, $d \downarrow_{A \cup b \cup J_b} I'_{\hat{s}}$.

Recall that for each $f \in I_s$, there is an $g \in I_{\hat{s}}$ such that $g \rhd_A f$ and that $f \rhd_A J_f$. This implies that $I'_{\hat{s}} \rhd_A (I'_s \cup J'_s)$. The displayed formula and the independence of $I'_{\hat{s}}$ from $b \cup J_b$ imply $d \cup b \cup J_b \downarrow_A I'_{\hat{s}}$. From domination we get $d \cup b \cup J_b \downarrow_A I'_s \cup J'_s$. Now monotonicity yields $d \downarrow_{A \cup b \cup J_b} I_s \cup J_s$. Since M is atomic over $A \cup I_s \cup J_s$, by the open mapping theorem $t(d; A \cup b \cup J_b)$ is isolated. Choose a copy of M_1 prime over $A \cup b \cup J_b$ and containing d and choose $c' \in M_1$ to realize $stp(c; A \cup d)$.

Now, both Q^e_c and $Q^e_{c'}$ form nonorthogonality bases for the ENI types which need d/A. Thus, $\not\perp$ defines a $1-1$ correspondence between Q^e_c and $Q^e_{c'}$ sending, say, q to $\tilde{q}$. Since any member of Q^e_c or $Q^e_{c'}$ is not orthogonal to $A \cup d$ while c and c' realize the same isolated type over $A \cup d$, Theorem XIV.2.9 yields $\dim(q, M) = \dim(\tilde{q}, M)$. Thus $\overline{v}_{c',M} = \overline{v}_{c,M}$.

3.7 Corollary. i) *Every simple s-configuration is a trite s-configuration.*
ii) *If $\bar{S}$ and $\bar{T}$ are trite s-configurations and $\bar{S} \cap \bar{T} \neq \emptyset$ then $\bar{S} = \bar{T}$.*

Proof. In Lemma 3.6 we showed that any configuration realized in the s-simple model M is also realized in the s-trite model M_1. Thus we have i). For ii), suppose two trite configurations, $\bar{S}$ and $\bar{T}$, intersect in a tuple $\bar{t}$. There are trite models M and N and an element b with $\overline{v}_{b,M} = \overline{v}_{b,N} = \bar{t}$, but $\overline{v}_{[b],M} = \bar{S}$ and $\overline{v}_{[b],N} = \bar{T}$. Let the model M_1 be prime over $A \cup b \cup J_b$. By Lemma 3.6, $\bar{S} = \overline{v}_{[b],M} = \overline{v}_{[b],M_1} = \overline{v}_{[b],N} = \bar{T}$.

Two easy lemmas complete our analysis of s-configurations. It remains open whether, assuming for instance that T is ω-stable with fewer than $2^{\aleph_0}$ countable models, one can prove that every s-configuration is trite.

3.8 Lemma. *If $\bar{S}$ is a β-configuration for s then $\bar{S}$ is a union of trite β-configurations for s.*

Proof. Let $M \supset A$ and suppose $\overline{v}_{[b],M} = \bar{S}$. For any $c \subset s(M)$ which depends on b, let M_c be prime over $c \cup J_c$ where J_c is a basis for Q^e_c in M. Then, since s has weight one, if $d \in s(M_c)$, $d \not\downarrow_A b$, so $\overline{v}_{[c],M_c} \subseteq \bar{S}$ as required.

3.9 Lemma. *For every depth one type $s \in S$ with $|Q^e_s| < \omega$, there are either $|\beta + \omega|$ or for some natural number k, $|\beta + k|$ distinct trite β-configurations of s.*

Proof. Let C be the set of cardinals less than $\aleph_\beta$ and suppose $|Q^e_s| = n$. Then each trite configuration $\bar{W}$ is a subset of C^n. For each $i < n$, there is at most one infinite cardinal λ such that for some $v \in \bar{W}$, $v_i = \lambda$. For, $v_i = \dim(q, M)$ for some $q \in Q_b$. But, if $c \not\downarrow_A b$ and $\tilde{q} \in Q_c$ with $q \not\perp \tilde{q}$, $\dim(q, M) = \dim(\tilde{q}, M) \bmod \aleph_0$. Thus, if $w = \overline{v}_{c,M}$, $w_i = \lambda$ as well. Thus, there are at least $|\beta + 1|$ distinct trite β-configurations. There are at most $|\beta + \omega|$ since this number is attained when all finite dimensions can be discriminated.

Since each β-configuration is a union of trite configurations the number of configurations will be $2^{|\beta+k|}$ or $2^{|\beta+\omega|}$. When β is finite the first case of Lemma 3.9 gives $2^{|\omega|}$ configurations; the second finitely many. The following definition isolates the difficult case. In the next section we show that such pathology actually exists.

3.10 Definition. For any B and any $s \in S(B)$ with depth one, let Q^e_s be the collection of depth zero types which are ENI and need some realization of s. Suppose Q^e_s is non-empty. If there are $|\beta + \omega|$ distinct β-configurations of s realized in models of T which have power $\aleph_\beta$ we say s is *normal.* If not, s is *abnormal.*

3.11 Definition. If T has no ENI-depth 2 types, the *full configuration* of a model M is

$$\{k_r : k_r = \dim(r, M), r \in R^e\} \cup$$
$$\{l_{\bar{T},s} : l_{\bar{T},s} = |\{b : b \in s(M) \text{ and } \overline{v}_{[b],M} = \bar{T}\}|, s \in S^e\}.$$

Recall that $\bar{T}$ ranges over configurations.

The key point of the following proof is the selection of an infinite set of trite 0-configurations for a type s such that any subset of them can be realized.

3.12 Theorem (Vaught's Conjecture for ω-stable T). *A countable ω stable theory has either at most $\aleph_0$ or $2^{\aleph_0}$ countable models.*

Proof. The successive reductions in this chapter have shown that if T has fewer than $2^{\aleph_0}$ countable models then each model is uniquely determined by its full configuration. Let $\langle R^e, S^e, Q^e\rangle$ be the ENI-frame for T. Suppose some $s \in S^e$ is normal. Let $\langle \bar{T}_i : i < \omega\rangle$ be a set of distinct trite configurations for s. Since $D - A$ is finite for each $s \in S^e$, given any infinite independent set of realizations of s, all but finitely many are independent from D over A. For any $W \subseteq \omega$, let M_W be prime over an extended basis such that if b realizes s and $b \downarrow_A D$, $|J_b| \in \bar{T}_i$ for some $i \in W$. Then $M_U \approx M_W$ implies $W = U$ so T has $2^{\aleph_0}$ countable models.

If every $s \in S^e$ is abnormal, every countable model of T is prime over one of the countably many ENI-full configurations and we finish. There are infinitely many full configurations since the full configuration counts the number of times each s-configuration is realized.

3.13 Historical Notes. The proof of Vaught's conjecture first appears in [Shelah, Harrington, & Makkai 1984]. The proof here depends greatly on the analysis of [Bouscaren 1984]. Numerous conversations with Steve Buechler further aided the preparation of this section. Bouscaren [Bouscaren 1984] has shown that the analysis expounded here yields Martin's conjecture for ω-stable countable T. To state Martin's conjecture we must define a certain extension of first order logic. The language L^* is the fragment of $L_{\omega_1,\omega}$ generated by the consistent types of T. Now, Martin's conjecture asserts that if T has fewer than $2^{\aleph_0}$ countable models then for every countable model M of T, the L^* theory of M is $\aleph_0$-categorical. The next problem is to show that an ω-stable countable theory T satisfies the strong Martin conjecture: either T has fewer than $2^{\aleph_0}$ countable models and for every countable model M of T, the L^* theory of M is $\aleph_0$-categorical or T has $2^{\aleph_0}$ distinct complete extensions in $L^*(T)$.

4. The Existence of Abnormal Types

In this section we describe Shelah's example of a countable theory which has ENI-depth two but only $\aleph_0$ countable models. This phenomenon arises from the existence of abnormal types. Thus, this example justifies the attention paid to abnormal types in the proof of the Vaught and Morley conjectures for ω-stable theories. This completes the survey in Section XVII.3 of prototypic depth 2 theories.

Locally the pathology we are investigating is manifested by the existence of a type which is abnormal (Definition 3.10). From a more global and less formal perspective the situation is this. We will find a countable ω-stable theory which has only countably many countable models but which has a non-principal type which is orthogonal to the empty set. At first blush, we hope to be able to pick the dimensions of infinitely many copies of such a type independently. In the example at hand we can not.

There are three parts to our description of the example. First, we will informally describe the axioms of the theory and show they induce certain definable relations in each model of the theory. Then, we will show that any theory whose models admit these definable relations provides the required example. Finally, we describe the prime model of the required theory.

4.1 Definition. Let the language L contain unary relation symbols P_i for $i < \omega$, binary relations symbols E, R, and a 5-ary relation symbol G.

The first part of the description (through Paragraph 4.5) contains an informal partial axiomatization of the theory. On each model M of T, the relation E will define an equivalence relation and R will be an asymmetric relation on the structure M/E. We will denote the E-equivalence classes in M by α, β etc. and write $R(\alpha, \beta)$ if for one (and hence any) $a \in \alpha$, $b \in \beta$, $R(a, b)$ holds. By a cycle in M/E we mean a finite sequence of points $\langle \alpha_0, \dots \alpha_{n-1} \rangle$ with $n \geq 4$ which are distinct except that $\alpha_0 = \alpha_{n-1}$ and for each $i < n$ either $R(\alpha_i, \alpha_{i+1})$ or $R(\alpha_{i+1}, \alpha_i)$. If $M \models T$ then M/E shall contain no cycles. For each $i < \omega$, there is a unary predicate P_i which intersects each E-equivalence class in an infinite set. The P_i shall be disjoint; if $P_i(a)$ holds, we say a has order i. We say a has finite order to mean $P_i(a)$ holds for some $i < \omega$ without specifying the i. An element which satisfies none of the P_i is said to have infinite order.

The relation between two R-adjacent classes α and β in a model of T is determined by a map $\gamma_{\alpha,\beta}$ which depends on a point a from α and a subset $\{b, c\}$ from β. The graph of the function $\gamma_{\alpha,\beta}$ is defined by the formula $g(a, b, c, x, y)$. The relation G is symmetric in the second and third coordinates; that is, $G(a, b, c, x, y) \leftrightarrow G(a, c, b, x, y)$.

4.2 Definition. (Fig. 2). The E equivalence classes α and β are *properly linked* via $\langle a, b, c \rangle$ with $a \in \alpha$ and $b, c \in \beta$ if there exists a unique map $\gamma_{\alpha,\beta}$ such that

i) $\gamma_{\alpha,\beta}$ is a $1-1$ onto map from $\alpha - \{a\}$ to $\beta - \{b, c\}$.

ii) For every $i < \omega$, $\bigwedge_{j<i} \neg P_j(a) \rightarrow \bigwedge_{j<i} (\neg P_j(b) \wedge \neg P_j(c))$.

iii) For every $i < \omega$, $P_i(a) \rightarrow (P_i(b) \vee P_i(c))$.

iv) For every $j < \omega$,

$$(\forall x)[(\neg P_j(x) \wedge P_j(\gamma_{\alpha,\beta}(x)) \rightarrow (P_j(a) \wedge [(P_j(b) \wedge \neg P_j(c)] \vee [\neg P_j(b) \wedge P_j(c)])].$$

v) For every $j < \omega$,

$$(P_j(a) \wedge P_j(b) \wedge \neg P_j(c)) \rightarrow ((\exists! x)[\neg P_j(x) \wedge P_j(\gamma_{\alpha,\beta}(x))]).$$

v′) For every $j < \omega$,

$$(P_j(a) \wedge P_j(c) \wedge \neg P_j(b)) \rightarrow ((\exists! x)[\neg P_j(x) \wedge P_j(\gamma_{\alpha,\beta}(x))]).$$

vi) For every $i, j < \omega$,

$$(\forall x)((\neg P_i(x) \wedge P_i(\gamma_{\alpha,\beta}(x)) \wedge P_i(b)) \rightarrow [P_j(c) \leftrightarrow P_j(x)]).$$

vi′) For every $i, j < \omega$,

$$(\forall x)((\neg P_i(x) \wedge P_i(\gamma_{\alpha,\beta}(x)) \wedge P_i(c)) \rightarrow [P_j(b) \leftrightarrow P_j(x)]).$$

It easily follows from ii) that if a has infinite order so do both b and c. It is possible for b or c to have infinite order while a does not. If both b and c have

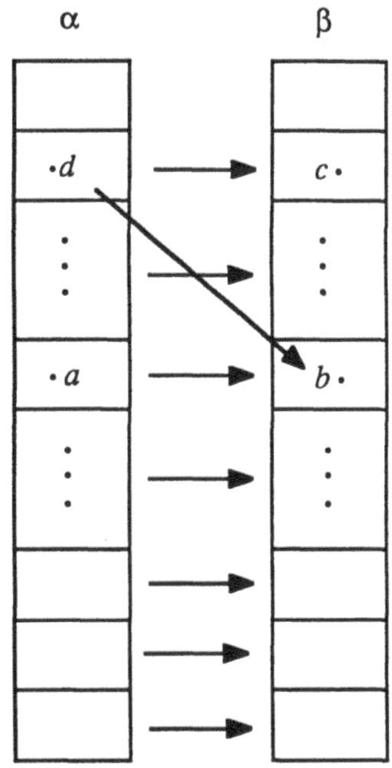

Fig. 2. α and β are properly linked; a, b, and c have finite order.

infinite order so does a. These axioms imply for all j, $P_j(x) \leftrightarrow P_j(\gamma_{\alpha,\beta}(x))$ holds with at most one exception. Moreover, this exception can occur only if a and thus one of b or c has finite order. Note that the conditions in Definition 4.2 are symmetric in b and c.

4.3 Uniqueness of Proper Linkings. i) There is a unique proper link between α and β if they are R-adjacent.

ii) If $G(a,b,c,x,y)$ defines a nonempty relation on $\alpha \times \beta$ then α and β are properly linked via $\langle a,b,c\rangle$.

By condition 4.3 ii), the fact that α and β are properly linked by $\langle a,b,c\rangle$ can be expressed by a single first order sentence. Note however that the properties of a proper linking described in Definition 4.2 can only be expressed by an infinite conjunction.

Each model of T is partitioned by the connected components of M/E. In order to calculate the number of models we will analyze the possible structure of a component. For this we require the following notation.

4.4 Notation. i) For any equivalence class α, q_α denotes the type of an element of α which is in no P_i (has infinite order).

ii) We write $\dim(q_\alpha, M)$ for the cardinality of the set of elements of infinite order in α.

When the choice of M is clear we write $\dim(q_\alpha)$ for $\dim(q_\alpha, M)$.

After the fact we will justify the notation in Definition 4.4 ii) by showing that for any model M of T and any α, the realizations of q_α in M, denoted $q_\alpha(M)$, form a set of indiscernibles. This indiscernibility will be a special case of the proof that the isomorphism type of any component of a model of T is determined by the dimensions of the equivalence classes in that component. In fact, we will show that if $\dim(q_\alpha) = \dim(q_\beta)$ then any isomorphism between α and β extends to an isomorphism of the components containing them. Such an isomorphism can be easily constructed by going back and forth between the components once the existence of enough proper linkings is stipulated. We so stipulate in the next family of axioms.

4.5 Existence of Proper Linkings. 1) If α is R-adjacent to β then α and β are properly linked.

2) We must specify that each α has many R-successors.

   i) For each equivalence class α, and each $a \in \alpha$ there exist infinitely many β which are R-successors of α and for each β unique elements $b, c \in \beta$ such that α and β are properly linked by $\langle a, b, c\rangle$.

   ii) When a has finite order i, we must make this requirement more specific.

      a) There are infinitely many β and for each β unique b, c each having order i so that α and β are properly linked by $\langle a, b, c\rangle$.

      b) For each $d \in \alpha$ with order $j > i$, there exist infinitely many β which are R-successors of α and for each β unique elements $b, c \in \beta$ such that α and β are properly linked by $\langle a, b, c\rangle$ and $\gamma_{\alpha,\beta}(d)$ has order i.

3) We must specify that each β has many R-predecessors.

   i) For each equivalence class β and each $b, c \in \beta$ there exist infinitely many α which are R-predecessors of β and for each α a unique element $a \in \alpha$ such that α and β are properly linked by $\langle a, b, c\rangle$.

   ii) If b has finite order i, we must make this requirement more specific. For every $c \neq b$ with order greater than or equal to i, and for every $e \neq b \in \beta$ with order i, there are infinitely many α and for each α a unique a with order i so that α and β are properly linked by $\langle a, b, c\rangle$ and $\gamma^{-1}_{\alpha,\beta}(e)$ has the same order as c.
   To achieve this effect with a single sentence even in the case that the order of c differs from that of b note that in view of Definition 4.2 it suffices to assert $\neg P_i(\gamma^{-1}_{\alpha,\beta}(e)))$. The symmetry between b and c makes it unneccessary to repeat condition 3 substituting c for b.

The calculation of the number of countable models follows easily from the following result.

4.6 Claim 1. *Each countable component, C, of a model, M, of T falls into exactly one of the following three classes.*

i) *For each* $\alpha \subset C$, $\dim(q_\alpha) = 0$.
ii) *For each* $\alpha \subset C$, $\dim(q_\alpha) = \aleph_0$.
iii) *For each* $\alpha \subset C$, $1 \leq \dim(q_\alpha) < \aleph_0$.

In the terminology of Definition 3.4, Claim 1 asserts that the unique type over the empty set sets admits only three 0-configurations: $\{0\}, \{\aleph_0\}, \omega$. The claim follows from the following more specific assertion.

4.7 Claim 2. *Suppose* $R(\alpha, \beta)$ *and* α *and* β *are properly linked by* $\langle a, b, c\rangle$.

i) $\dim(q_\alpha) \leq \dim(q_\beta) \leq \dim(q_\alpha) + 1$.
ii) *If* $\dim(q_\alpha) = 0$ *or* $\dim(q_\beta) = 0$ *then* $\dim(q_\alpha) = \dim(q_\beta)$.
iii) *If* $\dim(q_\alpha) \geq 1$ *then for some* β *properly linked to* α, $R(\alpha, \beta)$ *holds and* $\dim(q_\beta) = \dim(q_\alpha) + 1$.
iv) *If* $\dim(q_\beta) > 1$ *then for some* α *properly linked to* β, $R(\alpha, \beta)$ *holds and* $\dim(q_\beta) = \dim(q_\alpha) + 1$.

Proof. We must examine the cases depending on which of a, b, c have infinite order.

a) All three have infinite order. Note that this case occurs if a has infinite order. In this situation $\gamma_{\alpha,\beta}$ is a $1-1$ onto map from $q_\alpha(M) - \{a\}$ onto $q_\beta(M) - \{b, c\}$. Thus, $\dim(q_\beta) = \dim(q_\alpha) + 1$.

b) Suppose a and b have finite order, i, while c has infinite order. By Definition 4.2 v) we can choose a d with $\neg P_i(d) \wedge P_i(\gamma_{\alpha,\beta}(d))$. By vi) d has infinite order. In this case $\gamma_{\alpha,\beta}$ is a $1-1$ onto map from $q_\alpha(M) - \{d\}$ onto $q_\beta(M) - \{c\}$. Thus, $\dim(q_\beta) = \dim(q_\alpha)$.

b′) Suppose a and c have finite order, i, while b has infinite order. Now, the analysis is just as in case b) interchanging the roles of b and c with appeal to v′) and vi′) instead of v) and vi). We find $\dim(q_\beta) = \dim(q_\alpha)$.

c) All three have finite order. If they all have the same order condition 4.2 iv) implies that $\gamma_{\alpha,\beta}$ preserves the P_i and $\dim(q_\beta) = \dim(q_\alpha)$. If b and c have different orders we proceed as in case b) or b′) depending on which order is larger. In either case $\gamma_{\alpha,\beta}$ is a $1-1$ onto map from $q_\alpha(M)$ onto $q_\beta(M)$. Thus, $\dim(q_\beta) = \dim(q_\alpha)$.

In all three cases conclusion i) holds. If $\dim(q_\alpha) = 0$ or $\dim(q_\beta) = 0$ it is impossible (invoking Definition 4.2 ii) in the second case) to choose a to realize q_α. Thus, conclusion ii) holds. If $\dim(q_\alpha) \geq 1$ and $\dim(q_\beta) > 1$, we can choose a, b and c to have infinite order. Applying the appropriate one of the existence conditions (4.5) we conclude iii) and iv).

4.8 Exercise. Show that if $\dim(q_\beta) = 1$ then for any R-predecessor α of β, $\dim(q_\alpha) = 1$.

4.9 Theorem. *Let* T *satisfy the axioms given in this section. Then* T *is complete and has countably many countable models.*

Proof. It is easy to conclude Claim 1 from Claim 2. The isomorphism class of a component is determined by the category from Claim 1 in which it fits. In fact, if α and α' are two equivalence classes with the same dimension any $1-1$ map between α and β which preserves the P_i can be extended to an isomorphism between the components which contain α and α'. This isomorphism is constructed equivalence class by equivalence class using the existence conditions 4.5 and inducting on the distance in M/R from α. But then a countable model of T is determined by the number of components of each of the three kinds it has. Thus T has only countably many countable models.

It remains to show that there is a consistent theory satisfying the conditions described above. For those with sufficient intuition, the prescription in those conditions will lead directly to the model. This path is smoothed if one thinks of M/R as a tree where any class is chosen as a base point and whether an R-adjacent point is a successor or a predecessor is indicated by coloring the edge connecting them.

For a more concrete representation, we will explicitly construct the prime model of the theory. The first difficulty in such a construction is to explicitly describe a partial order (A, R) which is isomorphic to the required structure $(M/E, R)$. The following notation will enable us to overcome this problem. The structure we define here is the solution to the proportion

$$\omega : Z = \omega^{<\omega} : x.$$

4.10 Notation. Let V_ω denote the set of all partial functions with domain an initial segment of the integers and range contained in ω. Now $(V_\omega, \subseteq)$ is the required partial order.

In the following definition we require several auxilliary functions. Let δ be a map from $\omega \times \omega$ onto the direct product of ω^5 with the set A of subsets of ω with one or two elements. We require the following properties of δ.

$$\text{For each } n, (\lambda x)\delta(n, x) \text{ maps } \omega \text{ bijectively onto } A \times \omega^5$$

$$\text{For each } m, (\lambda x)\delta(x, m) \text{ maps } \omega \text{ bijectively onto } A \times \omega^5$$

If $\delta(m, n)$ picks out a two element subset of ω we write $\delta_l(m, n)$ (respectively $\delta_h(m, n)$) for the lesser (greater) of the two element subset. If $\delta(m, n)$ picks out a one element subset of ω we just let $\delta_l(m, n) = \delta_h(m, n)$. The 5-tuple is denoted $\langle \delta_0(m, n), \delta_1(m, n), \delta_2(m, n), \delta_3(m, n), \delta_4(m, n) \rangle$. The function δ can easily be constructed by analogy with the enumeration of the rational numbers.

We will define a structure M so that E and R are the equivalence relation and graph described after Definition 4.1 and the P_i are the unary relations. Suppose α and β are properly linked by $\langle a, b, c \rangle$, the order of a is r, and the order of c is s. Let g be the partial function whose graph is obtained by fixing three elements and then projecting G on the last two coordinates. Then $(\lambda x)g(a, b, c, x)$ is $\gamma_{\alpha,\beta}$.

Let $W = \omega \times \omega$. The universe of M is $V_\omega \times V_\omega \times W$. In the following we let s, t range over elements of V_ω and $w = \langle m, n \rangle$ range over W.

- $E(\langle s, t, w \rangle, \langle s', t', w' \rangle)$ iff $s = s' \wedge t = t'$.
- $R(\langle s, t, w \rangle, \langle s', t', w' \rangle)$ iff for some $m, n \in \omega$, $s' = s^\frown m$ and $t = t'^\frown n$.

- $P_i(\langle s,t,\langle m,n\rangle\rangle)$ iff $m=i$.

We say that the E class of $\langle s,t,w\rangle$ is indexed by $\langle s,t\rangle$.

Finally we define the partial function g which connects those equivalence classes which are properly linked by $\langle a,b,c\rangle$. Suppose that the two classes are indexed by $\langle s,t\rangle$ and $\langle s',t'\rangle$. Let $s'=s^\frown m$ and $t=t'^\frown n$. The elements a,b,c,d, and $e=\gamma_{\alpha,\beta}(d)$ have the following explicit representations.

$$a=\langle s,t,\delta_l(m,n),\delta_0(m,n)\rangle$$
$$b=\langle s',t',\delta_l(m,n),\delta_1(m,n)\rangle$$
$$c=\langle s',t',\delta_h(m,n),\delta_2(m,n)\rangle$$
$$d=\langle s,t,\delta_h(m,n),\delta_3(m,n)\rangle$$
$$e=\langle s',t',\delta_l(m,n),\delta_4(m,n)\rangle$$

The relation G will only hold when the first three coordinates are the a,b,c which properly link two successive E-classes. Then, $G(a,b,c,x,y)$ holds exactly when $x=\langle s,t,u,v\rangle$ and $v=\langle s',t',u,v\rangle$ with the following exceptions.

i) There is no y so that $G(a,b,c,a,y)$ holds.
ii) There is no x so that $G(a,b,c,x,b)$ or $G(a,b,c,x,c)$ holds.
iii) For any $i,j,k\in\omega$ let $\sigma_{i,j,k}$ be a bijection between $\omega-\{i\}$ and $\omega-\{j,k\}$. Then if x has the form $\langle s,t,\delta_l(m,n),v\rangle$, $G(a,b,c,x,y)$ holds if $y=\langle s',t',\delta_l(m,n),\sigma_{\delta_0(m,n),\delta_1(m,n),\delta_4(m,n)}(v)\rangle$.
iv) $G(a,b,c,d,e)$ holds.

Now checking that $[a]E$ and $[b]E$ are properly linked by a,b,c and then that M has the properties described in Paragraphs 4.1 through 4.5 completes the formal proof that T is consistent.

To give a little more insight into the example consider the following proposition.

4.11 Proposition. *For any model M of T and any α, the realizations of q_α in M, denoted $q_\alpha(M)$, form a set of indiscernibles.*

This proposition can be deduced as a special case of the general argument for extending maps which preserve the P_i from one equivalence class to the entire component. The following alternative argument also displays some of the structure of the automorphism group of M. For this we require some further notation.

4.12 Notation. Let L_n denote the language containing E, R, G, and the P_i for $i<n$.

4.13 Lemma. *Fix $n<\omega$. Let β be an E-class of M. If π is a permutation of β which fixes $P_i(M)$ as a set for $i<n$ then π extends to an automorphism $\hat{\pi}$ of $M|L_n$.*

Proof. We will define the extension of π to the R-predecessors of β. Suppose α precedes β with α indexed by $\langle s,t\rangle$ and β indexed by $\langle s',t'\rangle$. Suppose $s' = s^\frown m$ and $t = t'^\frown n$. Let

$$\pi(b) = \langle s', t', l_0, l_1\rangle$$

and

$$\pi(c) = \langle s', t', k_0, k_1\rangle.$$

Choose n' so that

$$\delta_l(m,n') = \min(l_0, k_0)$$
$$\delta_h(m,n') = \max(l_0, k_0)$$
$$\delta_1(m,n') = l_1$$
$$\delta_2(m,n') = k_1.$$

Now our strategy is to map α to the class α' indexed by $\langle s, t'^\frown n'\rangle$. We extend the domain of the partial automorphism π to α by defining

$$\hat{\pi}(a) = \langle s, t^\frown n', \delta_l(m,n'), \delta_0(m,n')\rangle$$

and

$$\hat{\pi}(\langle s,t,u,v\rangle) = \langle s, t^\frown n', {\gamma_{\alpha',\beta}}^{-1} \circ \pi \circ \gamma_{\alpha,\beta}(\langle s,t,u,v\rangle)\rangle.$$

There is no conflict between the two clauses of the definition since a is not in the domain of $\gamma_{\alpha,\beta}$. There are in fact infinitely many possible choices for n'.

We have seen how to extend π to the R-predecessors of β. The situation for the R-successors is similar. An easy induction using the same procedure extends π to an automorphism of M.

Lemma 4.13 shows that if we fix pointwise a finite number of the sets $P_i(M) \cap \beta$, then the automorphism group of M acts transitively on the remainder of β. This ensures that there are no definable relations on $q_\alpha(N)$ for any $N \models T$ and any α and demonstrates Proposition 4.11.

To show that T is ω-stable we require another observation about the automorphism group of M.

If two elements, d and e, of M are in the same P_i, say P_n, then there is an automorphism taking one to the other. For, the structure of M/E (ignoring G and the P_i) is the structure of a directed graph without cycles such that each element has infinitely many predecessors and infinitely many successors. It is easy to construct inductively automorphisms which guarantee that the automorphism group of such a structure acts transitively on it. To find an automorphism of M taking d to e, choose an automorphism π_0 of M/E taking the E-class of d to the E-class of e. Then π_0 induces an automorphism of M which we also denote by π_0. Let π_1 stabilize $P_i(M) \cap \beta$ for $i \leq n$ and map $\pi_0(d)$ to e. By Lemma 4.13 we extend π_1 to an automorphism of M. Then $\pi_1 \circ \pi_0$ is the required automorphism of M mapping

d to e. This yields that there are only countably many 1-types over the empty set.

4.14 Theorem. *The theory* $T = \mathrm{Th}(M)$ *is* ω*-stable.*

Proof. As we have just argued that there are only countably many one types over the empty set, it suffices to show that each type over the empty set has only countably many extensions over the countably saturated model of T. There are, in fact, just three basic kinds of non-algebraic type. For each of them, there are only countably many completions to an element of $S(M)$. To verify this in detail one must consider each of the a, b, c, d, e associated with a particular realization.

i) $p \supseteq q = \{E(x, m_0) : \text{some } m_0 \in M\} \cup \{x \neq m_0 : m_0 \in M\}$.
ii) $p \supseteq q = \{\neg E(x, m) : m \in M\} \cup \{R(x, m_0) : \text{some } m_0 \in M\}$.
iii) $p \supseteq q = \{\neg E(x, m) : m \in M\} \cup \{\neg R(x, m) : m \in M\}$.

This completes the description of the basic example. One other necessary feature of the example follows from the proof by Bouscaren (Proposition 4 of [Bouscaren 1983]) that if $U(t(\overline{a}; \emptyset))$ is finite then each $q \in S(\overline{a})$ is normal.

4.15 Exercise. Complete the proof of Theorem 4.14.

4.16 Exercise. Verify directly that in this example any $p \in S(\emptyset)$ has infinite U rank.

4.17 Exercise. Show that T has ENI-depth 2.

The example given does not show that the abnormal types affect the computation of the spectrum in uncountable powers as T has $2^{\aleph_1}$ models of power $\aleph_1$. Harrington suggested the following variation to solve this difficulty. Add to the language unary functions which guarantee that all the P_i have the same cardinality. Observe the following.

4.18 Theorem. *If a theory T' is an expansion of theory satisfying conditions 4.2, 4.3 and 4.5 and all the P_i have the same cardinality then T has only countably many models of power $\aleph_1$.*

Proof. There are now two categories of component: those in which each P_i is countable and those in which each P_i is uncountable. The first category divides into the four cases discussed in Claim 1. In the second category, there is the additional case that all the q_α have dimension $\aleph_1$.

4.19 Historical Notes. This example was discovered by Shelah in 1981. Lascar and Bouscaren had solved the Vaught conjecture for countable ω-stable T modulo the assumption that T admits no abnormal types. Moreover, they had given sufficient conditions, e.g., α_T is finite, for there to be no abnormal types. In the summer of 1984 Harrington and Cherlin presented the example at the Forking Festival. This writeup differs from that presentation primarily in the choice of description. The example was described as the model completion of a theory satisfying certain conditions;

we have specified the prime model. In addition to discussions with Buechler, Hrushovsky, and Harrington, we owe a great debt to Alan Mekler for correcting errors in the penultimate version.

5. *The Morley Conjecture for ω-Stable Theories*

We turn now to the proof of Morley's conjecture for countable ω-stable T. In Chapter XVII we reduced to the case that T has NDOP and has finite depth, say m. Thus, if $M \models T$, M can be decomposed by a normal tree of countable models $\langle \bar{A}, < \rangle$ with depth m. We can assume that T has fewer than 2^λ models of power λ for each λ. Thus, by Theorem XV.2.12 we can assume every type with positive depth is trivial. The analysis in Chapter XVII showed that M was completely determined by the associated ω-labeled tree where $t(a; A^-)$ labels $A \in \bar{A}$ when $A = A^-[a]$. The resulting upper bound, $\beth_{m-1}(|\beta + \omega|^{|\omega|})$, on the number of models in power $\aleph_\beta$ is not exact for some theories; it is computed on the worst case assumption that there are infinitely many pairwise orthogonal ENI types over each relevant $A \in \bar{A}$ and that all significant types are normal. A finer analysis must consider the situation where there are only finitely many types over some A and where some of the types may be isolated or abnormal. In Section 2 we computed lower bounds on the number of models if some type p of height r and depth two has R_p, S_p, or some $Q_{p,s}$ infinite. In this section we compute the upper bounds in each case. Moreover, we extend the analysis to deal with abnormal types and with one further problem. Recall that the strategy in Section 2 was to construct models from trees that were $\aleph_1$-ample up to height r and to label the nodes of height r. Ideally, we would just apply this strategy with $r = m - 1$. However, the nodes of height $m - 1$ might create few models while the nodes of height $m - 2$ create many. This problem is illustrated by the following two examples.

5.1 Example. i) The difficulty arises already at depth 2. Let the models of T be a disjoint union of models of the theory T' which has infinitely many disjoint unary predicates and a model of $\mathrm{Th}(Z, S)$ in each predicate and the theory T'' of a single equivalence relation with infinitely many infinite classes. Then in power $\aleph_1$, T has $2^{\aleph_0}$ models given by the depth one theory T', not the $\aleph_0$ models given by the depth two theory T''.

ii) The following, somewhat more complicated example, illustrates the problem at depth 3. Let T be the theory of the disjoint union of the theory T_2 from Examples XVII.3.8 and REF_2, the theory of two refining equivalence relations. Then the depth of T is determined by REF_2 as 3. But T has $|\beta + \omega|^{|\beta+1|^{|\omega|}}$ models of power $\aleph_\beta$ as determined by the depth 2 theory T_2, not the $|\beta + \omega|^{|\beta+1|}$ models of REF_2. In particular, there are $2^{2^{|\omega|}}$ models of power $\aleph_1$, not $2^{|\omega|}$.

Note, however, that while the fragment of the theory that has lower depth may dominate the spectrum function for small values of β, for large β the spectrum function is determined by the types of greater depth. This partially explains why we were able to complete the analysis for large values of $\aleph_\beta$ with less effort in Chapter XVII.

In the light of the example in Section 4 and the analysis in Section 2 we extend the list from Definition 2.9 of kinds of depth two types.

5.2 Definition. i) The depth two type $p \in S(X)$ has kind IV if R_p, S_p, and Q_p, are all finite but some $s \in S_p^e$ is normal.

ii) The depth two type $p \in S(X)$ has kind V if R_p, S_p, and Q_p are all finite and every $s \in S_p^e$ is abnormal.

The next definition and the following two lemmas deal with the problem of abnormal types. Suppose $\bar{A}$ and $\bar{B}$ represent M. By Section XVII.4 we establish a quasiisomorphism between $\bar{A}^1$ and $\bar{B}^1$. In fact, since the depth is finite we can guarantee that this map is an isomorphism of trees. We will label the top nodes of $\bar{A}$ and $\bar{B}$ by full configurations (see the next definition). To see these labels are preserved by the isomorphism of trees, recall that it is induced by nonorthogonality. Thus we would like to show configurations are preserved by nonorthogonality. Lemma 5.5 almost accomplishes that aim and comes close enough to permit us to make the desired calculations. After establishing this preservation result, we finish the computation of the spectrum paying attention to the problem described in Example 5.1.

5.3 Definition. Let $p \in S(X)$ have depth two. Let a realize p. The *full configuration* of a in a model M is

$$\{k_r : k_r = \dim(r, M), r \in R_a\} \cup$$
$$\{l_{\bar{T},s} : l_{\bar{T},s} = |\{b : b \in s(M) \text{ and } \overline{v}_{[b],M} = \bar{T}\}|, s \in S_a\}.$$

Here we need the β-configurations from Definition 3.4 in order to describe models of power $\aleph_\beta$.

5.4 Definition. Let p_1, p_2 be depth 2 types. Then p_1 and p_2 are *similar*, written $p_1 \sim p_2$ if for arbitrary a_1 realizing p_1 and b_1 realizing $s_{a_1} \in S_{a_1}$, there exist a_2 realizing p_2 and b_2 realizing $s_{a_2} \in S_{a_2}$, such that nonorthogonality induces $1-1$ correspondences between R_{a_1} and R_{a_2}, S_{a_1} and S_{a_2}, and between Q_{a_1,b_1} and Q_{a_2,b_2}. For simplicity we denote all these maps as α.

We show that nonorthogonal depth two types are similar and have almost the same configuration.

5.5 Lemma. *Let $\bar{A}$ be an* **AT**-*decomposition of M and suppose $X \in \bar{A}$. Let $p_1, p_2 \in S(X)$ have depth two and Kind IV or V. If $p_1 \not\perp p_2$ then*

i) $p_1 \sim p_2$

ii) *For each a_1 realizing p_1, there is an a_2 realizing p_2 such that $a_1 \not\downarrow_X a_2$ and*

a) *For each* $r \in R_{a_1}$, $\dim(r, M) = \dim(\alpha(r), M) \mod \aleph_0$.
b) *For each* $s \in S_{a_1}$, $\dim(s, M) = \dim(\alpha(s), M) \mod \aleph_0$.
c) *For each* $s \in S_{a_1}$, *each realization* b_1 *of* s, *and each* $q \in Q_{a_1,b_1}$, *there is an integer* k *and a* b_2 *realizing* $\alpha(s)$ *such that*

$$\dim(q, M) = \dim(\alpha(q), M) + k.$$

Proof. Just as we refined our picture of the frame of a model at the beginning of Section 3, we can invoke the tools of Section 1 to clarify the description of a frame at a realization of p_1. Since p_1 has kind IV or V, S_{p_1} is finite. Thus, we can find a type $\tilde{p}_1$ such that if a_1 realizes p_1 there is a finite sequence $\tilde{a}_1$ containing a_1 and realizing $\tilde{p}_1$ such that $\tilde{a}_1$ is atomic over a_1, $a_1 \rhd_X \tilde{a}_1$, and every member of S_{a_1} is strongly based on $\tilde{a}_1 \cup X$.

Since p_1 and p_2 are trivial we can find by Lemma XVI.2.12 a realization $a_2 \in X[a_1]$ of p_2 with $a_1 \not\downarrow_X a_2$. Just as in the preceding paragraph we can choose an $\tilde{a}_2$ on which all members of S_{a_2} are based. By Lemma 1.14 each member of S_{a_1} needs $\tilde{a}_2/X$. Thus, since nonorthogonality is trivial we establish a $1-1$ correspondence between S_{a_1} and S_{a_2}. This correspondence sends s to $\hat{s}$ with $s \not\perp \hat{s}$. Applying Theorem XIV.2.8 we have condition b). To establish the similarity of p_1 and p_2 we note now that since each s_a has depth one, Lemma 1.14 allows us to establish the required $1-1$ correspondence between Q_{a_1,b_1} and Q_{a_2,b_2}. Finally, we establish the correspondence between R_{a_1} and R_{a_2}. For if r is in R_{a_1} and not in R_{a_2} then r is nonorthogonal to some member $\hat{q}$ of Q_{a_2}. But this is impossible. As, $\hat{q}$ is nonorthogonal to some $q \in Q_{a_1}$ and $q \perp r$. This proves i) and a) and b) of ii).

Condition c) follows from the next lemma. We already established the correspondence between a_1 and a_2 and between s and $\alpha(s) = \hat{s}$. Since s and $\hat{s}$ are both strongly based on subsets of $X[a_1]$, we can take them both as types over $X[a_1]$. To simplify our notation and since we don't use any special properties of the construction of $X[a_1]$, we write N for $X[a_1]$. It is important for the following argument to remember that $\overline{v}_{[b]}$ is the collection of $\overline{v}_{b'}$, the vector of dimensions of types based on b', for $b' \in [b]$.

5.6 Lemma. *Let* $s, \hat{s} \in S(N)$ *be nonorthogonal trivial depth one types. There is a* $\overline{k} = \overline{k}_{s,\hat{s}}$ *such that for any* $c \in s(M)$ *there is a* $\hat{c}$ *realizing* $\hat{s}$ *with* $c \not\downarrow_N \hat{c}$ *and* $\overline{v}_{[\hat{c}],M} = \overline{v}_{[c],M} + \overline{k}$.

Proof. For ease of reading we find a k_q for a particular $q \in Q_s$. The required $\overline{k}$ is then the finite sequence of the k_q's. For isolated $q \in Q_b$ there is nothing to prove. For each $b \in s(M)$ and each $q_b \in Q_b^e$, we can choose by triviality and Lemma XVI.2.12 a $\hat{b} \in N[b]$ which realizes $\hat{s}(M)$ and as in Proposition 3.5 a $q_{\hat{b}}$ such that $b \not\downarrow_N \hat{b}$ and $q_b \not\perp q_{\hat{b}}$. Let k_q denote $\dim(q_b, M) - \dim(q_{\hat{b}}, M)$. We show the result first for c which depend on b over N and then show that the argument trivially extends to an arbitrary c realizing s. Since $t(\hat{b}; N \cup b)$ is principal, for any $c \in s(M)$ with $c \not\downarrow_N b$, we choose $\hat{c}$ so that $t(b^\frown\hat{b}; N) = t(c^\frown\hat{c}; N)$. Thus, choosing $N[c]$ to contain $\hat{c}$, we have

$$\dim(q_c, N[c]) - \dim(q_{\hat{c}}, N[c]) = \dim(q_b, N[b]) - \dim(q_{\hat{b}}, N[b]).$$

But,

$$\dim(q_c, M) = \dim(q_c, N[c]) + \dim(q_c^{N[c]}, M)$$

and

$$\dim(q_{\hat{c}}, M) = \dim(q_{\hat{c}}, N[c]) + \dim(q_{\hat{c}}^{N[c]}, M).$$

Moreover, by Theorem XIV.2.7, $\dim(q_c^{N[c]}, M) = \dim(q_{\hat{c}}^{N[c]}, M)$. So

$$\dim(q_c, N[c]) - \dim(q_{\hat{c}}, N[c]) = \dim(q_c, M) - \dim(q_{\hat{c}}, M).$$

Similarly,

$$\dim(q_b, N[b]) - \dim(q_{\hat{b}}, N[b]) = \dim(q_b, M) - \dim(q_{\hat{b}}, M) = k_q.$$

This implies $\dim(q_c, M) - \dim(q_{\hat{c}}, M) = k_q$ for any $c^\frown c'$ realizing $t(b^\frown b'; N)$ with $b \not\downarrow_N c$. Thus, $\overline{v}_{\hat{c}} = \overline{v}_c + \overline{k}$. But since c was an arbitrary member of $[b]$, we have $\overline{v}_{[\hat{b}]} = \overline{v}_{[b]} + \overline{k}$.

Now suppose c realizes s and $c \downarrow_N b$. To finish the lemma we must find a $\hat{c}$ with $\overline{v}_{[\hat{c}]} = \overline{v}_{[c]} + \overline{k}$. But this is immediate. For, $N[b]$ and $N[c]$ are isomorphic by an isomorphism which fixes N; taking $\hat{c}$ as the image of $\hat{b}$ under that isomorphism and repeating the previous proof completes the argument.

Let p_1 and p_2 be nonorthogonal and suppose a_1 and a_2 are assigned as in Lemma 5.5. Lemma 5.6 implies that the set of $\bar{T}$ which appear in the full configuration of a_1 (i.e. $l_{\bar{T},s} \neq 0$ (cf. Definition 5.3)) is the translation by a constant sequence $\overline{k}$ of the $\bar{S}$ which appear in the full configuration of a_2. In particular, if $s_1, s_2 \in S(N)$, $s_1 \not\perp s_2$ then s_1 is abnormal iff s_2 is abnormal.

In Section XVII.4 we computed the spectrum of a countable ω-stable T if $\mathrm{dp}(T)$ is infinite and for cardinals $\aleph_\beta$ where β is infinite. We now settle the case for arbitrary uncountable β of T with finite depth m.

To construct models of T, we construct partially labeled trees of height $m-2$ with the top nodes labeled by the full configuration of the type at that node. To demonstrate the lower bounds, we insist when defining the tree that $\mathrm{ht}(A) = k$ if and only if $\mathrm{dp}(t(A; A^-)) = m - k$. Thus, nonorthogonality preserves height. By constructing $\aleph_1$-ample trees, we guarantee as in Section XVII.2 that if $\bar{A}, \bar{B}$ both represent a model M, then for each $A \in \bar{A}$ of height $m-2$, there is an $\hat{A} \in \bar{B}$ also of height $m-2$ such that $A \not\downarrow_{A^-} \hat{A}$. The lower bounds follow since by Lemmas 5.5 and 5.6 the full configuration at A differs from that at $\hat{A}$ by a constant function. We fill out this sketch with

5.7 Theorem. *Let T be an ω-stable countable theory with NDOP and suppose $\mathrm{dp}(T) = m < \omega$. Consider those $p \in S(N)$, for some model N of T, which have height $m-2$.*

The spectrum function of T is determined by the following cases.

i) If some such p has kind I then

$$I^*(\aleph_\beta, AT) = \min(2^{\aleph_\beta}, \beth_{m-1}(|\beta + 1|^{|\omega|})).$$

ii) If no p of height $m-2$ has kind I but there is one with kind II, kind III, or kind IV then

$$I^*(\aleph_\beta, AT) = \min(2^{\aleph_\beta}, \beth_{m-1}(|\beta+\omega|)).$$

iii) If every type of height $m-2$ has kind V then

$$I^*(\aleph_\beta, AT) = \min(2^{\aleph_\beta}, \beth_{m-2}(|\beta+\omega|^{|\beta+1|})).$$

Proof. Let M be a model of T with power $\aleph_\beta$. Let $(\bar{A}, <)$ be a representation of M and let $\bar{A}^l$ be the $\kappa(\beta)$-labeled tree obtained by restricting $(\bar{A}, <)$ to the nodes of height at most $m-2$ and labeling each node $A^-[a]$ of height $m-2$ by the full configuration of a in M. Thus, $\kappa(\beta)$ denotes the number of possible labels for the representing tree of a model with power $\aleph_\beta$; later, we calculate the value of $\kappa(\beta)$ in each case. Ideally, we would now construct a $1-1$ correspondence between the set of such labeled trees and the models of power $\aleph_\beta$. We can not quite do this. But, we have a map σ (for surjective) from the class, $T_{\beta,\kappa}$, of such trees onto the class of models of power $\aleph_\beta$. To obtain $\sigma(\bar{A}^l)$, choose points to witness the configurations described by the labels and take the model prime over the resulting tree. Thus, $T_{\beta,\kappa}$ is an upper bound.

We will show, by a separate argument in each case, that $T'_{\beta,\kappa}$, the number of κ-labeled $\aleph_1$-ample trees of height $m-2$ provides a lower bound on the number of models of power at most $\aleph_\beta$. For, if the $\aleph_1$-ample trees $\bar{A}^l$ and $\bar{B}^l$ both represent a model M, Theorem XVII.4.7 implies the unlabeled trees are isomorphic. By an ad hoc argument in each case, we will show that this isomorphism preserves labels.

Writing κ for $\kappa(\beta)$, Lemma XVII.3.17, yields the upper bound:

$$|T_{\beta,\kappa}| = P_{\beta,m-3,\kappa} = \beth_{m-3}(|\beta+\omega|^{\kappa}).$$

By Theorem 2.8, we have the lower bound:

$$|T'_{\beta,\kappa}| = \beth_{m-2}(\kappa).$$

In every case

$$\kappa(\beta) \geq |\beta+\omega|^{|\beta+1|}. \tag{$*$}$$

For, there is at least one type q of height m which needs a type s. We obtain $|\beta+\omega|^{|\beta+1|}$ labels for realizations of p which are preserved by the isomorphism of $\bar{A}^l$ and $\bar{B}^l$. Namely for each f mapping $|\beta+1|$ into the set of cardinals less than or equal $\aleph_\beta$ and each $\gamma < \beta+1$ let

$$|\{b \in M_f : \dim(q_b) = \aleph_\gamma\}| = f(\gamma).$$

Now $(*)$ implies that the upper and lower bounds yield the same function of $\kappa(\beta)$, namely, $\beth_{m-2}(\kappa(\beta))$. To complete the proof we just need to compute $\kappa(\beta)$ in each case and verify that $T'_{\beta,\kappa}$ is indeed the lower bound.

Case i). As noted in Exercise 2.12, the lower bound from Theorem 2.11 equals the upper bound from Theorem XVII.3.19 so we have nothing to prove.

Case ii). If p has kind II or kind III we obtain the following upper bound on $\kappa(\beta)$. Since R_p is countable the number of labels contributed by R_p is at most $|\beta+\omega|^{\omega}$ and the number contributed by each s is at most $|\beta+\omega|^{(|\beta+\omega|^k)}$ (which is raised to the ω again in case III since there are infinitely many s). In either case $I^*(\aleph_\beta, AT) \leq \beth_{m-3}(|\beta+\omega|^{\kappa(\beta)}) \leq \beth_{m-1}(|\beta+\omega|)$. For kind II this is the lower bound computed in Theorem 2.11 ii) and we finish. However, if p has kind III the computation in Section 2 provides only a lower bound of

$$\beth_{m-2}(|\beta+1|^{|\omega|}).$$

But, we saw just before beginning the case analysis that there is always a lower bound of

$$\beth_{m-2}(|\beta+\omega|^{|\beta+1|}).$$

So there are $\beth_{m-2}(\sup(|\beta+1|^{|\omega|}, |\beta+\omega|^{|\beta+1|})) = \beth_{m-1}(|\beta+\omega|)$ models of power $\aleph_\beta$ as required.

If R_p, S_p and each $Q_{p,s}$ are finite $\kappa(\beta)$ is bounded by the product $|\beta+\omega|^k$, contributed by R_p, times $|\beta+\omega|^{(|\beta+\omega|^k)}$, contributed by each s, which again yields: $I^*(\aleph_\beta, AT) \leq \beth_{m-3}(|\beta+\omega|^{\kappa(\beta)}) \leq \beth_{m-1}(|\beta+\omega|)$. We note now that in the case of kind IV this upper bound is attained while we show below that for kind V it is not. If some p with depth $m-2$ has kind IV, then some $s \in S_p^e$ is a normal type. Thus there are $|\beta+\omega|$ distinct configurations for realizations of s. We can distinguish a realization a of p by the set of s-configurations which occur β times in the full configuration of a. From Lemma 5.6 we see that if $\{\bar{T}_i : i < |\beta+\omega|\}$ appears in a label for M then the set of partial configurations associated with $\alpha(s)$ in another representation of M differs from the set of $\bar{T}_i$ by some constant $\bar{k}$. Thus, there are $2^{|\beta+\omega|}$ possible labels for the depth $m-2$ types and clearly, $2^{|\beta+\omega|}$ of these differ by nonconstant functions so there are $\beth_{m-2}(2^{|\beta+\omega|}) = \beth_{m-1}(|\beta+\omega|)$ models as required.

iii) Since for each p, S_p is finite, we label each realization of p by the number of realizations b of each $s \in S_p$ that have a given configuration for Q_b. Since each member of S_p^e is abnormal, there are only $|\beta+1|$ possible configurations so for some finite k , there are $|\beta+\omega|^{|\beta+k|}$ labels. This upper bound on $\kappa(\beta)$ is the same as our general lower bound so in this case:

$$I^*(\aleph_\beta, AT) = \beth_{m-2}(|\beta+\omega|^{|\beta+1|}).$$

There are more kinds than there are distinct spectrum functions. This illustrates the disadvantage of coding our structure results as simple enumeration problems. The following exercise shows that this kind of coding can reflect some further structure.

5.8 Exercise. Let T be an ω-stable theory of depth m. Show that if every type of height $m-2$ is of kind IV then $I(\aleph_\beta, S) = \beth_{m-2}(|\beta+\omega|^{|\beta+1|})$ but if T has a type of depth $m-2$ with kind II, then $I(\aleph_\beta, S) = \beth_{m-1}(|\beta+\omega|)$.

5.9 Historical Notes. The conjectures considered in this chapter were solved by Shelah in [Shelah, Harrington, & Makkai 1984] and [Shelah 198?] (cf. [Harrington & Makkai 1985]). Neither of the 'main gap' papers deal explicitly with the finite depth case. The first detailed discussion of this case appears in Saffe [Saffe 1983]. Our proof here combines the techniques of [Harrington & Makkai 1985] with the analysis by Bouscaren [Bouscaren 1984] of depth 2 theories. This proof and exposition rely on many conversations with Harrington, Buechler, and Saffe. We have answered a question of Saffe [Saffe 1983] by providing a single classification which explains the behavior of the spectrum function both at $\aleph_0$ and in uncountable cardinals.

The proof here still falls somewhat short of the assignment of invariants discussed in Section I.1. That is, rather than assigning to each model a tree of ordinals, we have only assigned a set of trees and then computed the spectrum by a fluke of cardinal arithmetic. The only complete solution of that problem so far requires the additional assumption that T is $\aleph_0$-categorical. Buechler accomplishes this in [Buechler 1985b] by interweaving the geometrical ideas of Zilber and [Cherlin, Harrington, & Lachlan 1985] with Shelah's analysis.

Bibliography

Abakumov A.I., Palyutin E.A., Taitslin M.A. and Shishmarev Yu.E.

1972 *Categorical Quasivarieties.* **Algebra and Logic**, vol. 11 (1972), pp. 1–21. Russian original: pp.3–38.

Adamson I.T.

1982 **Introduction to field theory. (2nd edition)**. Cambridge Press, 1982. 181 pp. At: *166*

Ahlbrandt G.

1984 **Totally categorical theories of modular type**. University of Illinois at Chicago, 1984. 108 pp. Dissertation. At: *144*

198? *Almost strongly minimal totally categorical theories.* In: **Proceedings of Logic Colloquium 85, Paris. North Holland**. Preprint.

Ahlbrandt G. and Baldwin J.T.

198? *Categoricity and generalized model completeness.* Preprint. 11 pp.

Ahlbrandt G. and Ziegler M.

1986 *Quasi-finitely axiomatizable totally categorical theories.* **Annals of Pure and Applied Logic**, vol. 30 (1986), pp. 63–83. At: *184*

Andler D.

1975 *Semi-minimal theories and categoricity.* **The Journal of Symbolic Logic**, vol. 40 (1975), pp. 419–438.

Ash C.J. and Nerode A.

1981 *Functorial properties of algebraic closure and Skolemization.* **Journal of the Australian Mathematical Society**, vol. 31 (1981), pp. 136–141.

Ash C.J. and Rosenthal J.

1986 *Intersections of algebraically closed fields.* **Annals of Pure and Applied Logic**, vol. 30 (1986), pp. 103–121.

Baizhanov B.S.

1980 *Questions on the spectra of totally transcendental theories of finite rank (Russian).* In: **Theory of Models and its Applications**, ed. by Baizanov B. **Kazakhskii Gosudarstvennyi Universitet**, 1980. pp. 25–44.

1980a *Some properties of totally transcendental theories (Russian).* In: **Theory of Models and its Applications**, ed. by Baizanov B. **Kazakhskii Gosudarstvennyi Universitet**, 1980. pp. 14–24.

Baizhanov B.S. and Omarov B.

1979 *On finite diagrams.* In: **Teorija Nereguljarnyh Krivyh v Razlichnyh Geometri**, ed. by Taiclin M.A. **Kazakhskii Gosudarstvennyi Universitet**, 1979. pp. 11–15.

1979 *Restriction of a theory to a formula.* In: **Teorija Nereguljarnyh Krivyh v Razlichnyh Geometri**, ed. by Taiclin M.A. **Kazakhskii Gosudarstvennyi Universitet**, 1979. pp. 9–11.

Baldwin J.T.

1971 **Countable theories categorical in uncountable power.** Simon Fraser University, 1971. 81 pp. Dissertation. At: *62*

1972 *Almost strongly minimal theories.* **The Journal of Symbolic Logic**, vol. 37 (1972), pp. 487–493. At: *182*

1972a *Almost strongly minimal theories II.* **The Journal of Symbolic Logic**, vol. 37 (1972), pp. 657–660.

1973 *α_T is finite for $\aleph_1$-categorical T.* **Transactions of the American Mathematical Society**, vol. 181 (1973), pp. 37–51. At: *163*

1973a *The number of automorphisms of models of $\aleph_1$-categorical theories.* **Fundamenta Mathematicae**, vol. 83 (1973), pp. 1–6.

1974 *Atomic compactness in $\aleph_1$-categorical Horn theories.* **Fundamenta Mathematicae**, vol. 83 (1974), pp. 263–268.

1975 *Conservative extensions and the two cardinal theorems for stable theories.* **Fundamenta Mathematicae**, vol. 88 (1975), pp. 7–9. At: *62, 212*

1979 *Stability theory and algebra.* **The Journal of Symbolic Logic**, vol. 44 (1979), pp. 599–608.

1981 *Definability and the hierarchy of stable theories.* In: **Logic Year 1979-80 The University of Connecticut**, ed. by Lerman M., Schmerl J.H. and Soare R.I. **Springer-Verlag**, 1981. pp. 1–15. At: *112*

1984 *First order theories of abstract dependence relations.* **Annals of Pure and Applied Logic**, vol. 26 (1984), pp. 215–243. At: *256, 267*

1984a *Strong saturation and the foundations of stability theory.* In: **Logic Colloquim '82**, ed. by Lolli J. and Longo G. & Marcja A. **North Holland**, 1984. pp. 71–85. At: *71, 79, 240*

1985 *Definable second order quantifiers.* In: **Model-Theoretic Logics**, ed. by Barwise J. and Feferman S. **Springer-Verlag**, 1985. pp. 445–477.

Baldwin J.

1986 *Classification Theory: 1986.* In: **Classification Theory: Chicago, 1985**, ed. by Baldwin J.T. **Springer-Verlag**, 1987. Preprint. At: *2*

Baldwin J.T. and Berman J.

1981 *Elementary classes of varieties.* **Houston Journal of Mathematics**, vol. 7/4 (1981), pp. 473–492. At: *172*

1986 *Concrete representations of lattices and the fundamental order.* In: **Classification Theory: Chicago, 1985**, ed. by Baldwin J. **Springer-Verlag**, 1987. Preprint.

Baldwin J.T. and Blass A.

1974 *An axiomatic approach to rank in model theory.* **Annals of Mathematical Logic**, vol. 7 (1974), pp. 295–324. At: *45, 158*

Baldwin J.T., Blass A., Glass A.M.W. and Kueker D.W.

1973 *A 'natural' theory without prime model.* **Algebra Universalis**, vol. 3 (1973), pp. 152–155.

Baldwin J.T. and Harrington L.

198? *Trivial pursuit: Remarks on the main gap.* **Annals of Pure and Applied Logic** Preprint. At: *330, 350, 364, 377*

Baldwin J.T. and Kueker D.W.

1980 *Ramsey quantifiers and the finite cover property.* **Pacific Journal of Mathematics**, vol. 90 (1980), pp. 11–19. At: *25*

1981 *Algebraically prime models.* **Annals of Mathematical Logic**, vol. 20 (1981), pp. 289–330. At: *192*

Baldwin J.T. and Lachlan A.H.

1971 *On strongly minimal sets.* **The Journal of Symbolic Logic**, vol. 36 (1971), pp. 79–96. At: *6, 15, 17, 24, 95, 271, 311*

1973 *On universal Horn classes categorical in some infinite power.* **Algebra Universalis**, vol. 3 (1973), pp. 98–111. At: *94*

Baldwin J.T. and McKenzie R.

1982 *Counting models in universal Horn classes.* **Algebra Universalis**, vol. 15 (1982), pp. 359–384. At: *94*

Baldwin J.T. and Rose B.I.

1977 *$\aleph_0$-categoricity and stability of rings.* **Journal of Algebra**, vol. 45 (1977), pp. 1–16. At: *98*

Baldwin J.T. and Saxl J.

1976 *Logical stability in group theory.* **Journal of the Australian Mathematical Society**, vol. 21 (1976), pp. 267–276. At: *95, 97*

Baldwin J.T. and Shelah S.

1983 *The structure of saturated free algebras.* **Algebra Universalis**, vol. 17 (1983), pp. 191–197.

1985 *Second order quantifiers and the complexity of theories.* **Notre Dame Journal of Formal Logic**, vol. 26 (1985), pp. 229–302. At: 7, *130, 344, 350, 356, 364*

Baudisch A.

1980 *Application of the theory of Shirshov-Witt in the theory of groups and Lie algebras. Addendum.* **Prepr Akad Wiss DDR ZI Math Mech**, vol. P-30/80 (1980), pp. 9.

1982 *Decidability and stability of free nilpotent Lie algebras and free nilpotent p-groups of finite exponent.* **Annals of Mathematical Logic** (1982), pp. 1–25.

1983 *On Lascar rank in non-multidimensional theories ω-stable theories.* In: **Proceedings of the Third Easter Conference on Model Theory**. 1983.

1984 *Magidor-Malitz quantifiers in modules.* **The Journal of Symbolic Logic**, vol. 40 (1984), pp. 1–8.

1986 *On elementary properties of free Lie algebras.* **Annals of Pure and Applied Logic**, vol. 30 (1986), pp. 121–136.

198? *On preservation of dimensional order property and depth.* In: **Proceedings of Logic Colloquium 85, Paris**. Preprint.

198?a *Lascar rank in non-multidimensional theories.* Preprint.

198?b *Two hierarchies of dimension.* Preprint.

Baudisch A. and Rothmaler P.

1984 *The stratified order in modules.* In: **Proceedings of the 2nd. Frege Conference, Schwerin (GDR), 1984. Akademie-Verlag, Berlin**, 1984.

Baur W.

1976 *Elimination of quantifiers for modules.* **Israel Journal of Mathematics**, vol. 25 (1976), pp. 64–70. At: *30*

Baur W., Cherlin G.L. and Macintyre A.

1979 *Totally categorical groups and rings.* **Journal of Algebra**, vol. 57 (1979), pp. 407–440. At: *97, 98, 168*

Bekenov M.I.

1980 *Theories with a basis*. In: **Theory of Models and its Applications**, ed. by Baizanov B. **Kazakhskii Gosudarstvennyi Universitet**, 1980. pp. 45–47.

Bekenov M.I. and Herre H.

1979 *A remark concerning a generalized Morley-rank*. Preprint, 1979. 13 pp.

Bekenov M.I. and Mustafin T.G.

1979 *On rank functions, definable and indecomposable types in stable theories*. **Soviet Mathematics. Doklady**. vol. 20 (1979), pp. 333–336.

1981 *Properties of nonsplit types in stable theories*. **Sibirskij Matematicheskij Zhurnal Akademiya Nauk SSSR. Sibirskoe Otdelenie**, vol. 22 (1981), pp. 27–34,228.

Belegradek O.V.

1972 *Categoricity in nondenumerable powers and* $\aleph_1$*-homogeneous models*. **Algebra and Logic**, vol. 11 (1972), pp. 71–73. Russian original: pp.125–129, 237.

1973 *Almost categorical theories*. **Siberian Mathematical Journal**, vol. 14 (1973), pp. 191–198.

1974 *Algebraically closed groups (Russian)*. **Algebra i Logika**, vol. 13 (1974), pp. 239–255,363.

1978 *On nonstable group theories (Russian)*. **Izvestiya Vysshikh Uchebnykh Zavedenij. Matematika**, vol. 8(195) (1978), pp. 41–44.

Beljaev V.Ja.

1975 *On theories with Ehrenfeucht's property*. **Siberian Mathematical Journal**, vol. 16 (1975), pp. 141–143. Russian original: pp.175–177,198.

Benda M.

1974 *Remarks on countable models*. **Fundamenta Mathematicae**, vol. 81 (1974), pp. 107–119.

Berline C.

1981 *Stabilité et algèbre I, II, III, IV*. In: **Théories Stables: 1978-9**, ed. by Poizat B. **Université Pierre et Marie Curie, Institut Henri Poincaré**, 1981.

1982 **Groupes superstables**. Université de Paris 7, 1982. 68 pp. Thèse de Doctorat d'Etat.

1983 *Déviation des types dans les corps algébriquement clos*. In: **Théories Stables**, ed. by Poizat B. **Université Pierre et Marie Curie, Institut Henri Poincaré**, 1983. pp. 10. At: *39, 71, 169*

1986 *Superstable groups; a partial answer to conjectures of Cherlin and Zil'ber*. **Annals of Pure and Applied Logic**, vol. 30 (1986), pp. 45–63.

Berline C. and Lascar D.

1986 *Groupes superstables*. **Annals of Pure and Applied Logic**, vol. 30 (1986), pp. 1–45. At: *98, 162, 166, 168, 169*

Berthier D.

1975 *Stability of non-model-complete theories; products, groups*. **The Journal of the London Mathematical Society**, vol. 11 (1975), pp. 453–464.

Birkhoff G.

1949 **Lattice Theory**. American Mathematical Society, 1949. 283 pp. At: *275*

Blum L.

1968 **Ph.D. Thesis**. Massachusetts Institute of Technology, 1968. 45 pp. At: *229*

Bouscaren E.

1983 *Countable models of multidimensional $\aleph_0$-stable theories*. **Journal of Symbolic Logic**, vol. 48 (1983), pp. 377–383. At: *290, 366, 391*

1984 *Martin's Conjecture for ω-stable theories*. **Israel Journal of Mathematics**, vol. 49 (1984), pp. 15–26. At: *363, 377, 380, 383, 397*

Bouscaren E. and Chatzidakis Z.

1978 *Modèles premiers et atomiques: théorème des deux cardinaux dans les théories totalement transcendantes*. In: **Groupe d'Etude de Théories Stables. 2ème Année: 1978/79**, ed. by Poizat B. **Université Pierre et Marie Curie, Institut Henri Poincaré**. 1978. pp. 11.

Bouscaren E. and Lascar D.

1983 *Countable models of non-multidimensional theories*. **Journal of Symbolic Logic**, vol. 48 (1983), pp. 197–205. At: *224, 225, 290, 291, 302, 311, 315, 366*

Bridge J.

1977 **Beginning Model Theory. The completeness theorem and some consequences**. Oxford: The Clarendon Press, 1977. VII+143 pp. At: *15*

Buechler S.

1982 *Expansions of models of ω-stable theories.* **Journal of Symbolic Logic**, vol. 49/2 (1982), pp. 470—-477.

1984 *Geometry of weakly minimal types.* **Journal of Symbolic Logic**, vol. 50 (1984), pp. 1044–1054. At: *183*

1984a *Kueker's conjecture for superstable theories.* **Journal of Symbolic Logic**, vol. 49/3 (1984), pp. 930–934. At: *252, 311*

1984b *Recursive definability and resplendency in ω-stable theories.* **Israel Journal of Mathematics**, vol. 49 (1984), pp. 26–34. At: *226*

1984c *A note on normalization.* 1984. Unpublished note. At: *180*

1985 *Coordinatization in superstable theories I: stationary types.* **Transactions of the American Mathematical Society**, vol. 288 (1985), pp. 101–114.

1985a *One theorem of Zil'ber's on strongly minimal sets.* **Journal of Symbolic Logic**, vol. 50 (1985), pp. 1054–1062.

1985b *Invariants for ω-categorical, ω-stable theories.* **Israel Journal of Mathematics**, vol. 52 (1985), pp. 65–81. At: *183, 184, 398*

1986 *Locally modular theories of finite rank.* **Annals of Pure and Applied Logic**, vol. 30 (1986), pp. 83–95. At: *183, 327, 369*

1986a *Maximal chains in the fundamental order.* **Journal of Symbolic Logic**, vol. 51 (1986), pp. 323–326.

198? *The fundamental order and relativized continuity of rank.* Preprint.

198?a *A theorem on stable unidimensional theories.* **Journal of Symbolic Logic** Preprint. At: *293*

198?b *The classification of small weakly minimal sets I.* In: **Classification Theory: Chicago, 1985**, ed. by Baldwin J. **Springer-Verlag**, 1987. Preprint.

198?c *The classification of small weakly minimal sets II.* Preprint.

Chang C.C. and Keisler H.J.

1973 **Model Theory**. North-Holland. 1973. XII+550pp. At: *13, 15, 21, 22, 24, 55, 212*

Chatzidakis Z.

1981 *Forking et rangs locaux selon Shelah.* In: **Groupe d'Etude de Théories Stables. 2ème Année: 1978/79**, ed. by Poizat B. **Université Pierre et Marie Curie, Institut Henri Poincaré**. 1981. pp. 12.

1984 **Model theory of profinite groups**. Yale University, 1984. Dissertation.

Cherlin G.L.

1983 *Totally categorical structures*. In: **Proceedings of the International Congress of Mathematicians, Warsaw 1983**. 1983. pp. 301–306.

1976 **Model Theoretic Algebra. Selected topics**. Springer-Verlag, 1976. IV+234 pp. LNM 521.

1978 *Superstable division rings*. In: **Logic Colloquium 77**, ed. by Macintyre A., Pacholski L. and Paris J. **North-Holland**. 1978. pp. 99–111. At: *168*

1979 *Groups of small Morley rank*. **Annals of Mathematical Logic**, vol. 17 (1979), pp. 1–28. At: *98*

1979a *Stable algebraic theories*. In: **Logic Colloquium '78**, ed. by Boffa M., Dalen D. van and Mc Aloon K. **North-Holland**. 1979. pp. 53–74.

Cherlin G.L., Harrington L. and Lachlan A.H.

1985 *$\aleph_0$-categorical, $\aleph_0$-stable structures*. **Annals of Pure and Applied Logic**, vol. 2 (1985), pp. 103–135. At: *45, 156, 162, 163, 170, 181, 184, 295, 309, 327, 398*

Cherlin G. and Lachlan A.H.

1986 *Stable finitely homogeneous structures*. **Transactions of American Mathematical Society**, vol. 296 (1986), pp. 815–850. At: *184*

Cherlin G.L. and Reineke J.

1976 *Categoricity and stability of commutative rings*. **Annals of Mathematical Logic**, vol. 9 (1976), pp. 367–399. At: *97, 98*

Cherlin G.L. and Rosenstein J.G.

1978 *On $\aleph_0$-categorical abelian by finite groups*. **Journal of Algebra**, vol. 53 (1978), pp. 188–226.

Cherlin G.L. and Shelah S.

1980 *Superstable fields and groups*. **Annals of Mathematical Logic**, vol. 18 (1980), pp. 227–270. At: *169*

Chikhachev S.A.

1977 *$\aleph_1$-categorical commutative rings*. **Siberian Journal of Mathematics**, vol. 18 (1977), pp. 644–649. Russian original: pp. 908–914.

Chudnovskii G.V.

1970 *Problems of the theory of models, related to categoricity*. **Algebra and Logic**, vol. 9 (1970), pp. 50–74. German original: 80–120 (Algebra i Logika).

Cooper G.R.

1980 **On complexity of complete first order theories**. Simon Fraser University, 1980. 117 pp. Dissertation.

1982 *On complexity of complete first order theories*. **Zeitschrift für Mathematicsche Logik und Grundlagen der Mathematik**, vol. 28 (1982), pp. 93–136.

Delon F.

1978 *Types localement isolés et théorème des deux cardinaux dans les théories stables dénombrables*. In: **Groupe d'Etude de Théories Stables. 2ème Année: 1978/79**, ed. by Poizat B. **Université Pierre et Marie Curie, Institut Henri Poincaré**. 1978. pp. 10.

1981 *Types sur $C((X))$*. In: **Groupe d'Etude de Théories Stables. 2ème Année: 1978/79**, ed. by Poizat B. **Université Pierre et Marie Curie, Institut Henri Poincaré**. 1981. pp. 29.

1984 *Espaces Ultramétriques*. **Journal of Symbolic Logic**, vol. 49 (1984), pp. 405–424.

198? *Idéaux et sur les corps séparablement clos*. Preprint.

Dickmann M.A.

1973 *The problem of non-finite axiomatizability of $\aleph_1$-categorical theories*. In: **Proceedings of the Bertrand Russell Memorial Logic Conference**, ed. by Bell J., Cole J., Priest G. and Slomson A. 1973. pp. 141–216.

Duret J.-L.

1977 *Instabilité des corps formellement réels*. **Canadian Mathematical Bulletin**, vol. 20 (1977), pp. 385–387.

1980 *Les corps faiblement algébriquement clos non séparablement clos ont la propriété d'indépendance*. In: **Model Theory of Algebra and Arithmetic.Proceedings of the Conference on Applications of Logic to Algebra and Arithmetic**, ed. by Pacholski L., Wierzejewski J. and Wilkie A.J. **Springer-Verlag**, 1980. pp. 136–162. LNM 834.

1980 *Les corps pseudo-finis ont la propriété d'indépendance*. **Acad. des Sciences de Paris. Comptes Rendus**, vol. 290 (1980), pp. 981–983.

Ehrenfeucht A.

1957 *On theories categorical in power*. **Fundamenta Mathematicae**, vol. 44 (1957), pp. 241–248.

Ehrenfeucht A. and Mostowski A.W.

1956 *Models of axiomatic theories admitting automorphisms*. **Fundamenta Mathematicae**, vol. 43 (1956), pp. 50–68. At: *24*, *126*

Ehrsam S.

1978 *Exposé élémentaire de la théorie de la stabilité*. In: **Groupe d'Etude de Théories Stables. 2ème Année: 1978/79**, ed. by Poizat B. **Université Pierre et Marie Curie, Institut Henri Poincaré**. 1978. pp. 30.

1978 *L'ordre fondamental, le théorème de la borne, le théorème de la relation d'équivalence finie*. In: **Groupe d'Etude de Théories Stables. 2ème Année: 1978/79**, ed. by Poizat B. **Université Pierre et Marie Curie, Institut Henri Poincaré**. 1978. pp. 16.

1978 *Théories ω_1-catégoriques*. In: **Groupe d'Etude de Théories Stables. 2ème Année: 1978/79**, ed. by Poizat B. **Université Pierre et Marie Curie, Institut Henri Poincaré**. 1978. pp. 6.

Eklof P.C. and Fisher E.R.

1972 *The elementary theory of abelian groups*. **Annals of Mathematical Logic**, vol. 4 (1972), pp. 115–171. At: *30*

Eklof P.C. and Sabbagh G.

1970/71 *Model completions and modules*. **Annals of Mathematical Logic**, vol. 2 (1970/71), pp. 251–295. At: *30*

Enderton H.B.

1972 **A mathematical introduction to logic**. Academic Press, 1972. XIII+295 pp. At: *171*

Engeler E.

1959 *A characterization of theories with isomorphic denumerable models*. **Notices of the American Mathematical Society**, vol. 6 (1959), pp. 161. At: *24*

Erimbetov M.M.

1974 *Categoricity in power and nonbicardinality of a formula of finite rank*. **Algebra and Logic**, vol. 13 (1974), pp. 283–286. Russian original: pp. 493–500.

1975 *Complete theories with 1-cardinal formulas* (*Russian*). **Algebra i Logika**, vol. 14 (1975), pp. 245–257,368.

1981 *A property of countable stable theories*. **Sibirskij Matematicheskij ZhurnalAkademiya Nauk SSSR. Sibirskoe Otdelenie**, vol. 22 (1981), pp. 81–86,229.

Eršov Ju.L.

1974 *Theories of nonabelian varieties of groups*. In: **Proceedings of the Tarski Symposium**, ed. by Henkin L. & et.al. **American Mathematical Society**, 1974. At: *176*

Feferman S. and Vaught R.L.

1959 *The first order properties of products of algebraic systems.* **Fundamenta Mathematicae**, vol. 47 (1959), pp. 57–103.

Felgner U.

1975 *$\aleph_1$-kategorische Theorien Nicht-kommutativer Ringe.* **Fundamenta Mathematicae**, vol. 82 (1975), pp. 331–346. At: *98*

1975a *On $\aleph_0$-categorical extra-special p-groups.* In: **Six Days of Model Theory**, ed. by Henrard P. **Paul Castella, Editions**, 1975. pp. 175–196.

1977 *Stability and $\aleph_0$-categoricity of nonabelian groups.* In: **Logic Colloquium 76**, ed. by Gandy R.O. and Hyland J.M.E. **North-Holland**. 1977. pp. 301–324.

1978 *$\aleph_0$-categorical stable groups.* **Mathematische Zeitschrift**, vol. 160 (1978), pp. 27–49.

1980 *Kategorizität.* **Jahresbericht der Deutschen Mathematiker-Vereinigung**, vol. 82 (1980), pp. 12–32.

1980a *The model theory of FC-groups.* In: **Mathematical Logic in Latin America, Proceedings of the IV Latin American Symposium on Mathematical Logic**, ed. by Arruda A.I., Chuaqui R. and Costa da N.C.A. **North-Holland**. 1980. pp. 163–190.

Forrest W.K.

1979 *Some basic results in the theory of ω-stable theories.* **Zeitschrift für Mathematische Logik und Grundlagen der Mathematik**, vol. 25 (1979), pp. 513–520.

1983 *The theory of affine constructible sets.* **Zeitschrift für Mathematicsche Logik und Grundlagen der Mathematik**, vol. 29 (1983), pp. 97–135.

Fuhrken G.

1966 *Minimal- und Primmodelle.* **Archiv für Mathematische Logik und Grundlagenforschung**, vol. 9 (1966), pp. 3–11.

Garavaglia S.

1979 *Direct product decomposition of theories of modules.* **The Journal of Symbolic Logic**, vol. 44 (1979), pp. 77–88. At: *30:*

1980 *Decomposition of totally transcendental modules.* **The Journal of Symbolic Logic**, vol. 45 (1980), pp. 155–164. At: *30:*

1981 *Forking in modules.* **Notre Dame Journal of Formal Logic**, vol. 22 (1981), pp. 155–162.

Givant S.

1978 *Universal Horn classes categorical or free in power*. **Annals of Mathematical Logic**, vol. 15 (1978), pp. 1–53. At: *45, 184*

1979 *A representation theorem for universal Horn classes categorical in power*. **Annals of Mathematical Logic**, vol. 17 (1979), pp. 91–116. At: *45, 184*

Grätzer G.

1979 **Universal Algebra, Second Edition**. Springer-Verlag, 1979. xviii+582 pp. At: *45*

Grossberg R.

198? *A categoricity theorem for theories in* $L_{\omega_1,\omega}$ *which have the amalgamation property*. Preprint.

Grossberg R. and Shelah S.

1983 *On universal locally finite groups*. **Israel Journal of Mathematics**, vol. 44 (1983), pp. 289–302.

1986 *On the number of nonisomorphic models of an infinitary theory which has the infinitary order property*. **Journal of Symbolic Logic**, vol. 51 (1986).

198? *A nonstructure theorem for an infinitary theory which has the unsuperstability property* Preprint. At: *31*

Gross W.

1977 *Dimension and finite closure*. **Journal of the Australian Mathematical Society. Series A**, vol. 23 (1977), pp. 421–430.

Gurevich Y. and Schmitt P.H.

1984 *The theory of ordered abelian groups does not have the independence property*. **Transactions of the American Mathematical Society**, vol. 284 (1984), pp. 171–182.

Harnik V.

1971 **Stable theories and related concepts (Hebrew)**. Hebrew University of Jerusalem, 1971. iv+61 pp. Dissertation.

1975 *A two cardinal theorem for sets of formulas in a stable theory*. **Israel Journal of Mathematics**, vol. 21 (1975), pp. 7–23. At: *212*

1975 *On the existence of saturated models of stable theories*. **Proceedings of the American Mathematical Society**, vol. 52 (1975), pp. 361–367. At: *91*

1985 *Stability theory and set existence axioms*. **Journal of Symbolic Logic**, vol. 50 (1985), pp. 123–137. At: *79*

Harnik V. and Harrington L.

1984 *Fundamentals of forking*. **Annals of Pure and Applied Logic**, vol. 26 (1984), pp. 245–286. At: *15, 62, 71, 79, 130, 158, 180*

Harnik V. and Ressayre J.-P.

1971 *Prime extensions and categoricity in power*. **Israel Journal of Mathematics**, vol. 10 (1971), pp. 172–185. At: *126*

Harrington L. and Makkai M.

1985 *An exposition of Shelah's 'Main Gap' — counting uncountable models of ω-stable and superstable theories*. **Notre Dame Journal of Formal Logic**, vol. 26 (1985), pp. 139–177. At: *30, 52, 233, 302, 317, 325, 331, 339, 346, 350, 351, 352, 356, 358, 364, 367, 397*

Hart B.

1986 *NOTOP revisited*. In: **Classification Theory: Chicago, 1985**, ed. by Baldwin J. **Springer-Verlag**, 1987. Preprint. At: *347, 350*

Herre H.

1979 *Transzendente Theorien*. Preprint, 1979. 23 pp.

1980 *Modelltheoretische Eigenschaften endlichvalenter Graphen*. **Z. für Mathematische Logik und Grundlagen der Mathematik**, vol. 26 (1980), pp. 51–58.

Herre H., Mekler A.H. and Smith K.

1983 *Superstable Graphs*. **Fundamenta Mathematicae**, vol. 118 (1983), pp. 75–79.

Herstein I.N.

1968 **Noncommutative rings**. The Mathematical Association of America, 1968. 199 pp. At: *169*

Hodges W.

1981 *Encoding orders and trees in binary relations*. **Mathematika. A Journal of Pure and Applied Mathematics**, vol. 28 (1981), pp. 67–71. At: *79*

1983 *On constructing many non-isomorphic algebras*. In: **Proceeding "25. Arbeitstagung über Allgemeine Algebra Darmstadt 1983**. 1983. pp. 67–77. At: *31*

1985 **Building Models by Games**. London Mathematical Society, 1985. 311 pp.

198? **Model Theory**. Cambridge University Press. At: *15, 31, 216, 339*

Hodges W., Lachlan A.H and Shelah S.

1977 *Possible orderings of an indiscernible sequence*. **Bulletin of London Mathematical Society**, vol. 9 (1977), pp. 212–215.

Hrushovski E.

1986 **Contributions to stable model theory**. University of California, Berkeley, 1986. ii+108pp. Dissertation. At: *25, 70, 98, 141, 153, 166, 261, 293, 311*

198? *Kueker's conjecture for stable theories*. Preprint. At: *141, 311*

198?a *Countable unidimensional theories*. Preprint. At: *311*

198?b *Finitely based stable theories*. Preprint.

198?c *Almost orthogonal types and definable groups*. Preprint.

198?d *Locally modular regular types*. In: **Classification Theory: Chicago, 1985**, ed. by Baldwin J. **Springer-Verlag**, 1987. At: *98, 153*

198?e *Unidimensional theories are superstable*. In preparation.

Hrushovski E. and Pillay A.

1986 *Weakly normal groups*. In: **Proceedings of Logic Colloquium 85, Paris. North Holland**, 1986. Preprint. At: *184*

Humphreys J.

1975 **Linear Algebraic Groups**. Springer-Verlag, 1975. 247pp.

Jacobson N.

1985 **Basic Algebra I (2nd edition)**. W.H. Freeman, 1985. 499pp.

Jonsson B.

1956 *Universal relational systems*. **Mathematica Scandinavica**, vol.4 (1956), pp.193–208.

1960 *Homogeneous universal relational systems*. **Mathematica Scandinavica**, vol.8 (1960), pp.137–142.

Keisler H.J.

1966 *Some model theoretic results for ω-logic*. **Israel Journal of Mathematics**, vol.4 (1966), pp.249–261.

1967 *Ultraproducts which are not saturated*. **The Journal of Symbolic Logic**, vol.32 (1967), pp.23–46. At: *25*

1971 *On theories categorical in their own power*. **The Journal of Symbolic Logic**, vol.36 (1971), pp.240–243.

1976 *Six classes of theories*. **Journal of the Australian Mathematical Society**, vol.21 (1976), pp.257–266. At: *89*

1978 *The stability function of a theory*. **The Journal of Symbolic Logic**, vol.43 (1978), pp.481–486. At: *89*

1986 *Choosing elements in a saturated model*. In: **Classification Theory: Chicago, 1985**, ed. by Baldwin J. **Springer-Verlag**, 1987.

198? *Measures and forking.* **Annals of Pure and Applied Logic**

Keisler H.J. and Morley M.

1967 *On the number of homogeneous models of a given power.* **Israel Journal of Mathematics**, vol. 5 (1967), pp. 73–78.

Kierstead H.A.

1980 *Countable models of ω_1-categorical theories in admissible languages.* **Annals of Mathematical Logic**, vol. 19 (1980), pp. 127–175.

Kirby L.A.S. and Pillay A.

1986 *Weakly definable types.* **Transactions of the American Mathematical Society**, vol. 295 (1986), pp. 542–564. At: *62*

Knight J.F.

1977 *Skolem functions and elementary embeddings.* **The Journal of Symbolic Logic**, vol. 42 (1977), pp. 94–98.

1978 *An inelastic model with indiscernibles.* **The Journal of Symbolic Logic**, vol. 43 (1978), pp. 331–334.

1978a *Prime and atomic models.* **The Journal of Symbolic Logic**, vol. 43 (1978), pp. 385–393.

1981 *Algebraic independence.* **Journal of Symbolic Logic**, vol. 46 (1981), pp. 377–384.

1986 *Saturation of homogeneous respendent models.* **Journal of Symbolic Logic**, vol. 51 (1986), pp. 222–224.

Knight J.F., Pillay A. and Steinhorn C.

1986 *Definable sets in ordered structures II.* **Transactions of the American Mathematical Society**, vol. 295 (1986), pp. 593–606.

Kolchin E.R.

1973 **Differential Algebra and Algebraic Groups**. Academic Press, 1973. xvii +446 pp. At: *229*

Korec I. and Rautenberg W.

1976 *Model-interpretability into trees and applications.* **Archiv für Mathematische Logik und Grundlagenforschung**, vol. 17 (1976), pp. 97–104.

Kucera T.

1984 **Applications of Stability-theoretical methods to theories of modules**. Mcgill University, 1984. 218 pp. Dissertation.

Kueker D.W.

1970 *Back-and-forth arguments and infinitary model theory.* In: **Infinitary Logic: In Memoriam Carol Karp**, ed. by Kueker D.W. **Springer-Verlag**, 1970. pp. 17–72. LNM 492. At: *205*

1987 *Small extensions of countable models.* Preprint. At: *311*

Kunen K.

1980 **Set Theory, An Introduction to Independence Proofs.** North Holland, 1980. xvi + 317 pp. At: *214*

Lachlan A.H.

1971 *The transcendental rank of a theory.* **Pacific Journal of Mathematics**, vol. 37 (1971), pp. 119–122.

1972 *A property of stable theories.* **Fundamenta Mathematicae**, vol. 77 (1972), pp. 9–20. At: *212*

1973 *On the number of countable models of a countable superstable theory.* In: **Logic, Methodology and Philosophy of Science IV**, ed. by Suppes P., Henkin L., Moisil G.C. and Joja A. **North-Holland**. 1973. pp. 45–56. At: *6, 271*

1974 *Two conjectures regarding the stability of ω-categorical theories.* **Fundamenta Mathematicae**, vol. 81 (1974), pp. 133–145. At: *6, 105, 107, 180, 183*

1975 *Theories with a finite number of models in an uncountable power are categorical.* **Pacific Journal of Mathematics**, vol. 61 (1975), pp. 465–481. At: *99, 311*

1975a *A remark on the strict order property.* **Zeitschrift für Mathematische Logik und Grundlagen der Mathematik**, vol. 21 (1975), pp. 69–70.

1976 *Dimension and totally transcendental theories of rank 2.* In: **Set Theory and Hierarchy Theory**, ed. by Marek W., Srebrny M. and Zarach A. **Springer-Verlag**, 1976. pp. 153–183.

1978 *Spectra of ω-stable theories.* **Zeitschrift für Mathematische Logik und Grundlagen der Mathematik**, vol. 24 (1978), pp. 129–139. At: *99, 311*

1978 *Skolem functions and elementary extensions.* **Proceedings of London Mathematical Society**, vol. 18 (1978), pp. 1–6.

1980 *Singular properties of Morley rank.* **Fundamenta Mathematicae**, vol. 108 (1980), pp. 145–157. At: *162, 163, 176*

1984 *On countable stable structures which are homogeneous for a finite relational language.* **Israel Journal of Mathematics**, vol. 49 (1984), pp. 69–153. At: *182*

1986 *Binary homogeneous structures I.* **Proceedings of London Mathematical Society**, vol. 52 (1986), pp. 385–411. At: *182*

1986a *Binary homogeneous structures II.* **Proceedings of London Mathematical Society**, vol. 52 (1986), pp. 412–426.

198? *Structures coordinatized by indiscernible sets.* **Annals Pure and Applied Logic** Preprint.

Lachlan A.H. and Shelah S.

1984 *Stable structures homogeneous for a finite binary language.* **Israel Journal of Mathematics**, vol. 49 (1984), pp. 155–180. At: *184*

Lang S.

1971 **Algebra**. Addison Wesley Publishing Co. 1971. 526pp. At: *166, 169*

Lascar D.

1973 *Types définissables et produits de types.* **Académie des Sciences de Paris. Comptes Rendus**, vol. 276 (1973), pp. A1253–A1256. At: *62, 212*

1975 *Définissabilité dans les théories stables.* **Logique et Analyse. Nouvelle Série**, vol. 18 (1975), pp. 489–507.

1975 *Quelques remarques sur la notion de bifurcation (forking) (English summary).* **Académie des Sciences de Paris. Comptes Rendus**, vol. 280 (1975), pp. A607–A610.

1976 *Ranks and definability in superstable theories.* **Israel Journal of Mathematics**, vol. 23 (1976), pp. 53–87. At: *6, 45, 63, 71, 158, 162, 163, 237, 271*

1976 **Définissabilité de types en théorie des modèles**. Université Paris 7, 1976. Thèse de Doctorat d'Etat. At: *163*

1977 *Définissabilité dans les théories stables.* In: **Six Days of Model Theory**, ed. by Henrard P. **Paul Castella, Editions**, 1977. pp. 257–275.

1977a *Généralisation de l'ordre de Rudin-Keisler aux types d'une théorie.* In: **Colloque International de Logique. Editions du Centre National de la Recherche Scientifique (C.N.R.S.)**, 1977. pp. 73–81.

1979 *Les modèles dénombrables d'une théorie superstable ayant des fonctions de Skolem.* **Académie des Sciences de Paris. Comptes Rendus**, vol. 289 (1979), pp. 655–658.

1981 *Les modèles dénombrables d'une théorie ayant des fonctions de Skolem.* **Transactions of the American Mathematical Society**, vol. 268/2 (1981), pp. 345–366. At: *7, 365, 366*

1981a *Le problème de la classification des modèles dénombrables.* In: **Journées de théorie des modèles à Opheylissem (Belgique), mars 1981**.

1982 *On the category of models of a complete theory*. **Journal of Symbolic Logic**, vol. 47/2 (1982), pp. 249–266.

1982a *Ordre de Rudin-Keisler et poids dans les théories ω-stables*. **Zeitschrift für Mathematische Logik und Grundlagen der Mathematik**, vol. 28 (1982), pp. 413–430. At: *145, 147, 155, 261*

1983 *Les corps différentiellement clos dénombrables*. In: **Théories Stables**, ed. by Poizat B. **I.H.P.** 1983. pp. 9.

1984 *Relation entre le rang U et le poids*. **Fundamenta Mathematicae**, vol. 121 (1984), pp. 117–123. At: *155, 240, 249, 271, 279*

1984a *Sousgroupes d'automorphismes d'une structure saturée*. In: **Logic Colloquim '82**, ed. by Lolli J. and Longo G. & Marcja A. **North Holland**, 1984. pp. 123–134.

1985 *Quelques précisions sur la D.O.P. et la profondeur d'une théorie*. **The Journal of Symbolic Logic** (1985), pp. 316–330. At: *322, 325, 351*

1986 **Introduction à la stabilité**. Longman, 1986. 220 pp.

198?a *Les groupes ω-stable de rang fini*. Preprint.

Lascar D. and Andler D.
198? *Classification des modèles et complexité d'une théorie*. In: **Encyclopedia Universalis**.

Lascar D. and Poizat B.
1979 *An introduction to forking*. **The Journal of Symbolic Logic**, vol. 44 (1979), pp. 330–350. At: *55, 64, 65, 71, 78, 79, 225*

Laskowski M.C.
198? *Uncountable theories which are categorical in a higher power*. Preprint. 31 pp.

Łos J.
1954 *On the categoricity in power of elementary deductive systems and related problems*. **Colloquium Mathematicum**, vol. 3 (1954), pp. 58–62. At: *4*

Loveys J.
1984 *The uniqueness of envelopes in $\aleph_0$-categorical, $\aleph_0$-stable structures*. **Journal of Symbolic Logic**, vol. 49 (1984), pp. 1171–1184.

1985 **Certain weakly minimal theories**. Simon Fraser University, 1985. 104 pp. Dissertation.

Macintyre A.

1971 *On ω_1-categorical theories of abelian groups*. **Fundamenta Mathematicae**, vol. 70 (1971), pp. 253–270. At: *94, 97*

1971 *On ω_1-categorical theories of fields*. **Fundamenta Mathematicae**, vol. 71 (1971), pp. 1–25. At: *169*

Macintyre A. and Shelah S.

1976 *Uncountable universal locally finite groups*. **Journal of Algebra**, vol. 43 (1976), pp. 168–175.

Makkai M.

1984 *A survey of basic stability theory with emphasis on orthogonality and regularity*. **Israel Journal of Mathematics**, vol. 49 (1984), pp. 181–238. At: *52, 148, 155, 176, 225, 238, 242, 249, 252, 271, 279, 285, 290, 291, 302, 346*

Makowsky J.A.

1971 **Kategorizität und endliche Axiomatisierbarkeit**. ETH Zürich, 1971. Diplomarbeit. At: *182*

1972 *Note on almost strongly minimal theories (Russian summary)*. **Bulletin de l'Académie Polonaise des Sciences**, vol. 20 (1972), pp. 529–534.

1974 *On some conjectures connected with complete sentences*. **Fundamenta Mathematicae**, vol. 81 (1974), pp. 193–202.

Mal'cev A.I.

1971 **The Metamathematics of Algebraic Systems, Collected Papers: 1936-1967**. North Holland, 1971. Benjamin F. Wells (translator).

Manaster A.B.

1975 **Completeness, Compactness, and Undecidability: an introduction to mathematical logic**. Prentice Hall: Eaglewood Cliffs;Prentice Hall, 1975. VI+154 pp. At: *284*

Mangani P. and Marcja A.

1980 *Shelah rank for boolean algebras and some application to elementary theories I*. **Algebra Universalis**, vol. 10 (1980), pp. 247–257.

Marcus L.

1972 *A minimal prime model with an infinite set of indiscernibles*. **Israel Journal of Mathematics**, vol. 11 (1972), pp. 180–183.

1975 *A type-open minimal model*. **Archiv für Mathematisches Logic und Grundlagenforschung**, vol. 15 (1975), pp. 17–24.

1980 *The number of countable models of a theory of one unary function*. **Fundamenta Mathematicae**, vol. 108 (1980), pp. 171–181. At: *366*

Marczewski E.

1966 *Independence in abstract algebras. Results and problems.* **Colloquium Mathematica**, vol. 14 (1966), pp. 169–188. At: *45*

Marjca A. and Toffalori C.

1984 *On Cantor-Bendixson spectra containing (1,1) I.* In: **Models and Sets (Proceedings, Logic Colloqium Aachen, 1983). Springer Verlag**, 1984. pp. 331–350. LNM 1103.

1985 *On Cantor-Bendixson spectra containing (1,1) II.* **Journal of Symbolic Logic**, vol. 50 (1985), pp. 611–618.

Marsh W.E.

1966 **On ω_1-categorical and not ω-categorical theories**. Dartmouth College, 1966. Dissertation. At: *24*

Martin G.A.

1985 **Two classification problems: Rank one structures which coordinatize $\aleph_0$-categorical $\aleph_0$-stable structures; reducts of algebraically closed fields**. Rutgers University, 1985. vii+123 pp. Dissertation.

Mazoyer J.

1975 *Sur les théories catégoriques finiment axiomatisables (English summary).* **Académie des Sciences de Paris. Comptes Rendus**, vol. 281 (1975), pp. A403–A406.

Meirembekov K.A.

1980 *Remarks on theories with a bounded spectrum.* In: **Theory of Models and its Applications**, ed. by Baizanov B. **Kazakhskii Gosudarstvennyi Universitet**, 1980. pp. 65–72.

Mekler A.H.

1981 *Stability of nilpotent groups of class 2 and prime exponent.* **Journal of Symbolic Logic**, vol. 46 (1981), pp. 781–788.

1983 *Model complete theories with a distinguished substructure.* **Proceedings American Mathematical Society**, vol. 75 (1983), pp. 294–299.

198? *Classification theory and stationary logic.* **Canadian Journal of Mathematics** Preprint.

Millar T.

1981 *Stability, complete extensions, and the number of countable models.* In: **Aspects of Effective Algebra**, ed. by Crossley J.N. **Upside Down A Book Company**, 1981. pp. 196–205.

Miller A.

1977 **Ph.D. Thesis**. University of California, Berkeley, 1977. Dissertation.

1981 *Vaught's conjecture for theories of 1 unary operation.* **Fundamenta Mathematicae**, vol. 111 (1981), pp. 135–141. At: *366*

Morley M.

1963 *On theories categorical in uncountable powers.* **Proceedings of the National Academy of Sciences of the United States of America**, vol. 49 (1963), pp. 213–216.

1965 *Categoricity in power.* **Transactions of the American Mathematical Society**, vol. 114 (1965), pp. 514–538. At: *4, 15, 24, 52, 62, 91, 126, 163, 188, 192, 311*

1965 *Omitting classes of elements.* In: **The Theory of Models.** ed. by Addison J.W., Henkin L. and Tarski A. **North-Holland.** 1965. pp. 265–273. At: *212*

1967 *Countable models of $\aleph_1$-categorical theories.* **Israel Journal of Mathematics**, vol. 5 (1967), pp. 65–72.

1968 *Partitions and models.* In: **Proceedings of the Summer School in Logic**, ed. by Loeb M.H. **Springer-Verlag**, 1968. pp. 109–158. At: *22*

1970 *The number of countable models.* **The Journal of Symbolic Logic**, vol. 35 (1970), pp. 14–18. At: *6*

Morley M. and Vaught R.L.

1962 *Homogeneous universal models.* **Mathematica Scandinavica**, vol. 11 (1962), pp. 37–57.

Mortimer M..

1974 **Some topics in model theory**. Bedford College, 1974. 120pp. Dissertation.

Murzin F.A.

1974 *Totally transcendental and $\aleph_1$-categorical theories (Russian).* 1974. pp. 145–147.

Mustafin T.G.

1970 *Models of totally transcendental theories (Russian. Kazakh summary).* **Izvestiya Akademii Nauk Kazakhskoj SSR.** vol. 1971/1 (1970), pp. 37–43.

1977 *A strong base of elementary types of theories.* **Siberian Mathematical Journal**, vol. 18 (1978), pp. 961–969. Russian original:1977, pp.1356–1366.

1980 *A non-two-cardinal set of stable types.* **Mathematical Notes of the Academy of Sciences of the USSR**, vol. 27 (1980), pp. 253–259. Russian original: pp. 515–525.

1981 *On the number of countable models of a countable complete theory (Russian).* **Algebra i Logika**, vol. 20 (1981), pp. 69–91.

1981a *Principles of normalization of formulas.* **Sibirskij Matematicheskij ZhurnalAkademiya Nauk SSSR. Sibirskoe Otdelenie**, vol. 22 (1981), pp. 158–169,237.

1981b *Rank functions in stable theories.* **Siberian Mathematical Journal**, vol. 21 (1981), pp. 815–824.

1981c *Theories with a two-cardinal formula.* **Algebra and Logic**, vol. 19 (1981), pp. 438–442. Russian original: pp. 676–682,746.

1985 *Classification of superstable theories by rank functions.* **Algebra and Logic**, vol. 24 (1985), pp. 171–209. Russian original: pp. 42–65.

Mustafin T.G. and Taimanov A.D.

1970 *Countable models of theories which are categorical in power $\aleph_1$ but not in power $\aleph_0$ (Russian).* **Algebra i Logika**, vol. 9 (1970), pp. 559–565.

Narkiewicz W.

1964 *On a cerain class of abstract algebras.* **Fundamenta Mathematicae**, vol. 54 (1964), pp. 115–124. At: *45*

Nesin A.

1985 **Groups of finite Morley rank**. Yale University, 1985. 119 pp. Dissertation. At: *98*

1985 *A survey of results of groups of finite Morley rank.* Preprint, 1985.

198? *Bad Groups.* Preprint.

198?a *Non-solvable groups of Morley rank 3.* Preprint. 26 pp.

198?b *Solvable groups of finite Morley rank.* Preprint. 20 pp.

198?c *Zilber's results.* Preprint. 17 pp.

Neumann B.H.

1954 *Groups covered by permutable subsets.* **Journal of the London Mathematical Society**, vol. 29 (1954), pp. 236–248. At: *27*

Omarov A.I.

1979 *Δ_t-minimality and the finite cover property.* In: **Teorija Nereguljarnyh Krivyh v Razlichnyh Geometri**, ed. by Taiclin M.A. **Kazakhskii Gosudarstvennyi Universitet**, 1979. pp. 74–76.

Omarov B.

1980 *Ranks and Deg of countable ω_0-categorical theories*. In: **Theory of Models and its Applications**, ed. by Baizanov B. **Kazakhskii Gosudarstvennyi Universitet**, 1980. pp. 61–65.

Pabion J.F.

1971 *Une notion de transcendance dans les structures relationnelles*. **Académie des Sciénces de Paris. Comptes Rendus**, vol. 272 (1971), pp. A1541–A1544.

Paillet J.L.

1978 *Théories complètes $\aleph_0$-catégoriques et non finiment model-complétables*. **Académie des Sciences de Paris. Comptes Rendus**, vol. 286 (1978), pp. 1167–1170.

Paljutin E.A.

1971 *Models with countable-categorical universal theories*. **Algebra and Logic**, vol. 10 (1971), pp. 15–21.

1971a *N^*-homogeneous models (Russian)*. **Sibirskij Matematicheskij ZhurnalAkademiya Nauk SSSR. Sibirskoe Otdelenie**, vol. 12 (1971), pp. 920–921.

1980 *Categorical Horn classes I (Russian)*. **Algebra i Logika**, vol. 19 (1980), pp. 582–614,617.

1973 *Categorical quasivarieties of arbitrary signature*. **Siberian Journal of Mathematics**, vol. 14 (1973), pp. 904–916. Russian original: pp.1285–1303. At: *45, 184*

1976 *The description of categorical quasivarieties*. **Algebra and Logic**, vol. 14 (1976), pp. 86–111. Russian original: pp.145–185. At: *45, 184*

Peretjat'kin M.G.

1973 *On complete theories with a finite number of denumerable models*. **Algebra and Logic**, vol. 12 (1973), pp. 310–326. Russian original:pp.550–576,618.

1980 *A theory with three countable models*. **Algebra i Logika**, vol. 19 (1980), pp. 224–235,251. Russian original: pp.224–235.

1980 *An example of an ω_1-categorical complete finitely axiomatizable theory (English translation)*. **Algebra and Logic**. vol. 19 (1980), pp. 202–229.

Pillay A.

1978 *Number of countable models*. **The Journal of Symbolic Logic**, vol. 43 (1978), pp. 492–496.

1980 *Instability and theories with few models*. **Proceedings of the American Mathematical Society**, vol. 80 (1980), pp. 461–468.

1980a *Theories with exactly three countable models and theories with algebraic prime models.* **The Journal of Symbolic Logic**, vol. 45 (1980), pp. 302–310.

1981 *Number of models of theories with many types.* In: **Groupe d'Etude de Théories Stables. 2ème Année: 1978/79**, ed. by Poizat B. **Université Pierre et Marie Curie, Institut Henri Poincaré**. 1981. pp. 10. At: *31, 216*

1981a *A class of $\aleph_0$-categorical thoeries.* **Zeitschrift für Logik und Grundlagen der Mathematik**, vol. 27 (1981), pp. 411–418.

1981b *Finitely generated models of ω-stable theories.* **Bulletin de la Société Mathématique de Belgique**, vol. 33 (1981), pp. 83–91.

1981c *Prime models over subsets.* In: **Groupe d'Etude de Théories Stables. 2ème Année: 1978/79**, ed. by Poizat B. **Université Pierre et Marie Curie, Institut Henri Poincaré**. 1981. pp. 8.

1982 *Dimension theory and homogeneity for elementary extensions of a model.* **Journal of Symbolic Logic**, vol. 47/1 (1982), pp. 147–160. At: *249*

1982a *Weakly homogeneous models.* **Proceedings of the American Mathematical Society**, vol. 86 (1982), pp. 126–132. At: *315*

1983 *Countable models of stable theories.* **Proceedings of the American Mathematical Society**, vol. 89 (1983), pp. 660–672. At: *6, 180, 184, 271*

1983a **An introduction to stability theory**. Clarendon Press, Oxford, 1983. xi + 146 pp. At: *65, 89, 110, 112, 163, 196, 204*

1983b *A note on finitely generated models.* **Journal of Symbolic Logic**, vol. 48 (1983), pp. 163–166.

1983c *A note on tight stable theories.* **Archiv für Mathematische Logik und Grundlagenforschung**, vol. 23 (1983), pp. 147–152.

1983d *$\aleph_0$-categoricity over a predicate.* **Notre Dame Journal of Formal Logic**, vol. 24 (1983), pp. 527–536.

1983e *Well-definable types over subsets.* In: **Théories Stables**, ed. by Poizat B. **I.H.P.** 1983. pp. 4. At: *71*

1983f *The models of a non-multidimensional ω-stable theory.* In: **Théories Stables**, ed. by Poizat B. **I.H.P.** 1983. pp. 22. At: *295, 311*

1984 *Regular types in non-multidimensional ω-stable theories.* **Journal of Symbolic Logic**, vol. 49 (1984), pp. 880–891. At: *249*

1984a *Countable modules*. **Fundamenta Mathematicae**, vol. 121 (1984), pp. 125–132.

1986 *Superstable groups of finite rank without pseudoplanes*. **Annals of Pure and Applied Logic**, vol. 30 (1986), pp. 95–101. At: *184*

1986a *Simple Superstable theories*. In: **Classification Theory: Chicago, 1985**, ed. by Baldwin J. **Springer-Verlag**, 1987.

198? *Forking, normalisation and canonical bases*. **Annals of Pure and Applied Logic** Preprint. At: *71, 130, 176*

198?a *Stable theories, pseudoplanes and the number of countable models*. **Annals of Pure and Applied Logic** Preprint. At: *184*

198?b *On groups and fields definable in O-minimal structures*. Preprint.

Pillay A. and Poizat B.

198? *Pas d'imaginaires dans l'infini*. Preprint. 7 pp.

Pillay A. and Prest M.

1983 *Forking and pushouts in modules*. **Journal of the London Mathematical Society**, vol. 46 (1983), pp. 365–384.

198? *Modules and stability theory*. **Transactions of the American Mathematical Society** Preprint.

Pillay A. and Shelah S.

1985 *Classification theory over a predicate I*. **Notre Dame Journal of Formal Logic**, vol. 26 (1985), pp. 361–376.

Pillay A. and Srour G.

1984 *Closed sets and chain conditions in stable theories*. **Journal of Symbolic Logic**, vol. 49 (1984), pp. 1350–1362. At: *184*

Pillay A. and Steinhorn C.

1984 *Definable sets in ordered structures*. **Bulletin of the American Mathematical Society**, vol. 11 (1984), pp. 159–161.

1985 *A note on non-multidimensional superstable theories*. **Journal of Symbolic Logic**, vol. 50 (1985), pp. 1020–1024. At: *112, 117, 225*

1986 *Definable sets in ordered structures I*. **Transactions of the American Mathematical Society**, vol. 295 (1986), pp. 565–593. At: *2*.

Podewski K.-P. and Reineke J.

1974 *An ω_1-categorical ring which is not almost strongly minimal*. **The Journal of Symbolic Logic**, vol. 39 (1974), pp. 665–668.

Podewski K.-P. and Ziegler M.

1978 *Stable graphs*. **Fundamenta Mathematicae**, vol. 100 (1978), pp. 101–107.

Poizat B.

1977 **Déviation des types**. Université de Paris 6, 1977. 83 pp. Thèse de Doctorat d'Etat.

1978 *Une preuve par la théorie de la déviation d'un théorème de J. Baldwin*. **Académie des Sciences de Paris. Comptes Rendus. Série A**, vol. 287 (1978), pp. 589–591. At: *163*

1978a **Groupe d'Etude de Théories Stables (Bruno Poizat). 1ère Année: 1977/78. Exposés 1 a 9**. Université Pierre et Marie Curie, Institut Henri Poincaré. Paris: Secretariat Mathématique. 1978. 104 pp.

1978b *Rangs des types dans les corps différentiels*. In: **Groupe d'Etude de Théories Stables. 2ème Année: 1978/79**, ed. by Poizat B. **Université Pierre et Marie Curie, Institut Henri Poincaré. Paris: Secretariat Mathématique**. 1978. pp. 13.

1978c *Suites d'indiscernables dans les théories stables*. In: **Groupe d'Etude de Théories Stables. 2ème Année: 1978/79**, ed. by Poizat B. **Université Pierre et Marie Curie, Institut Henri Poincaré. Paris: Secretariat Mathématique**. 1978. pp. 7.

1978d *Une théorie ω_1-catégorique à α_T fini*. In: **Groupe d'Etude de Théories Stables. 2ème Année: 1978/79**, ed. by Poizat B. **Université Pierre et Marie Curie, Institut Henri Poincaré. Paris: Secretariat Mathématique**. 1978. pp. 3.

1980 *Le rang U selon Lascar*. In: **Théories Stables**, ed. by Poizat B. **I.H.P**. 1980.

1980a *Les équations différentielles vues par un logicien*. In: **Sem Equat Deriv Part Hyperb & Holom(1979/1980)**. 1980.

1981 *Sous-groupes définissables d'un groupe stable*. **The Journal of Symbolic Logic**, vol. 46 (1981), pp. 137–146. At: *93, 97, 168, 169*

1981a *Exercices en stabilité*. In: **Groupe d'Etude de Théories Stables. 2ème Année: 1978/79**, ed. by Poizat B. **Université Pierre et Marie Curie, Institut Henri Poincaré. Paris: Secretariat Mathématique**. 1981. pp. 5.

1981b *Modèles premiers d'une théorie totalement transcendante*. In: **Groupe d'Etude de Théories Stables. 2ème Année: 1978/79**, ed. by Poizat B. **Université Pierre et Marie Curie, Institut Henri Poincaré. Paris: Secretariat Mathématique**. 1981. pp. 7.

1981c *Théories instables.* **The Journal of Symbolic Logic**, vol. 46 (1981), pp. 513–522.

1981d **Groupe d'Etude de Théories Stables (Bruno Poizat). 2ème Année: 1978/79. Exposés 1 à 11.** Université Pierre et Marie Curie, Institut Henri Poincaré. Paris: Secretariat Mathématique. 1981. ii+115 pp.

1983 *Pairs of structures stable.* **Journal of Symbolic Logic**, vol. 48 (1983), pp. 239–249. At: *25*

1983a *Une théorie de Galois imaginaire.* **Journal of Symbolic Logic**, vol. 48 (1983), pp. 1151–1171. At: *176*

1983b *Post-scriptum à "Théories instables".* **Journal of Symbolic Logic**, vol. 48/1 (1983), pp. 60–62.

1983c *Groupes stables avec types génériques réguliers.* **Journal of Symbolic Logic** (1983), pp. 339–355.

1983d **Groupe d'étude de théories stables (Bruno Poizat). 3è Année: 1980-82.**. Université Pierre et Marie Curie, Institut Henri Poincaré. Paris: Secretariat Mathématique. 1983. 115 pp.

1984 *Deux remarques à propos de la propriété de recouvrement fini.* **Journal of Symbolic Logic**, vol. 49 (1984), pp. 803–807. At: *25*

1985 **Cours de théorie des modèles**. Nur al-Mantiq wal-Ma'rifah 82 Rue Racine, 69100 Villeurbanne, France, 1985. 584 pp. At: *15, 91, 145, 176*

198? *La structure géométrique des groupes stables.* In: **Humboldt Universität Seminarberichte Nr.60**. pp. 205–217.

198?a *A propos de groupes stables.* In: **Proceedings of Logic Colloquium 85, Paris**. Preprint.

Prest M.

1985 *The generalized RK-order orthogonality and regular types for modules.* **Journal of Symbolic Logic**, vol. 50 (1985), pp. 202–219. At: *226*

198? **Model theory of modules**. Preprint. At: *25, 29, 30*

198?a *Pure injectives and T-injective hulls of modules.* Preprint.

Reineke J.

1975 *Minimale Gruppen.* **Zeitschrift für Mathematische Logik und Grundlagen der Mathematik**, vol. 21 (1975), pp. 357–359.

Ressayre J.-P.

1969 *Sur les théories du premier ordre catégorique en un cardinal.* **Transactions of the American Mathematical Society**, vol. 142 (1969), pp. 481–505. At: *62*

Reyes G.E.

1970 *Local definability theory.* **Annals of Mathematical Logic**, vol. 1 (1970), pp. 95–137.

Robinson A.

1956 **Complete Theories**. North Holland, 1956. At: *192*

Robinson A.

1959 *On the concept of a differentially closed field.* **Research Council of Israel. Bulletin. Section F. Mathematics and Physics**, vol. 8 (1959), pp. 113–128.

Rogers P.

1981 *Preservation of saturation and stability in a variety of nilpotent groups.* **The Journal of Symbolic Logic**, vol. 46 (1981), pp. 499–512.

Rose B.I.

1980 *On the model theory of finite dimensional algebras.* **Proceedings of London Mathematical Society**, vol. 40 (1980), pp. 21–39.

1978 *The $\aleph_1$-categoricity of strictly upper triangular matrix rings over algebraically closed fields.* **The Journal of Symbolic Logic**, vol. 43 (1978), pp. 250–259.

Rosenstein J.G.

1971 *A note on a theorem of Vaught.* **The Journal of Symbolic Logic**, vol. 36 (1971), pp. 439–440.

Rosenthal J.W.

1972 *A new proof of a theorem of Shelah.* **The Journal of Symbolic Logic**, vol. 37 (1972), pp. 133–134.

1975 *Partial $\aleph_1$-homogeniety of the countable saturated model of an $\aleph_1$-categorical theory.* **Zeitschrift für Mathematische Logik und Grundlagen der Mathematik**, vol. 21 (1975), pp. 307–308.

1979 *On the dimension theory of $\aleph_1$-categorical theories with the non-trivial strong elementary intersection property.* **Zeitschrift für Mathematische Logik und Grundlagen der Mathematik**, vol. 25 (1979), pp. 359–362.

Rothmaler P.

1978 *Total transzendente abelsche Gruppen und Morley-Rang.* Preprint, 1978. 45 pp.

1982 *A note on the finite cover property in modules*. Preprint. Akad Wiss DDR ZI Math Mech: Berlin, 1982.

1983 *Another treatment of the foundations of forking theory*. In: **Proceedings of the Third Easter Conference on Model Theory**. 1983. pp. 25. At: *79*

1983 *Stationary types in modules*. **Zeitschrift für Mathematische Logik und Grundlagen der Mathematik** (1983), pp. 445–464.

1983 *Some model theory of modules I. On total transcendence of modules*. **Journal of Symbolic Logic**, vol. 48 (1983), pp. 570–574.

1983 *Some model theory of modules II. On stability and categoricity of flat modules*. **Journal of Symbolic Logic**, vol. 48 (1983), pp. 970–985.

1984 *Some model theory of modules III. On infiniteness of sets definable in modules*. **Journal of Symbolic Logic**, vol. 49 (1984), pp. 32–46.

Rothmaler P. and Tuschik H.P.

1980 *A two cardinal theorem for homogeneous sets and the elimination of Malitz quantifiers*. In: **Prepr Akad Wiss DDR ZI Math Mech**. 1980. pp. 26.

Rowbottom F.

1964 *The Łos conjecture for uncountable theories*. **Notices of the American Mathematical Society**, vol. 11 (1964), pp. 248. At: *62*

Rubin M.

1974 *Theories of linear order*. **Israel Journal of Mathematics**, vol. 17 (1974), pp. 392–443. At: *7, 9, 366*

Ryll-Nardzewski C.

1959 *On the categoricity in power* $\leq \aleph_0$. **Bulletin de l'Académie Polonaise des Sciences**, vol. 7 (1959), pp. 545–548. At: *24*

Sabbagh G.

1975 *Catégoricité en* $\aleph_0$ *et stabilité: constructions les preservant et conditions de chaine*. **Académie des Sciences de Paris. Comptes Rendus**, vol. 280 (1975), pp. A531–A535. At: *98*

1975 *Catégoricité et stabilité: Quelques exemples parmi les groupes et anneaux (English summary)*. **Académie des Sciences de Paris. Comptes Rendus**, vol. 280 (1975), pp. A603–A606.

1975 *Sur les groupes qui ne sont pas réunion d'une suite croissante de sous-groupes propres (English summary)*. **Académie des Sciences de Paris. Comptes Rendus**, vol. 280 (1975), pp. A763–A766.

Sacks G.E.

1972 **Saturated model theory**. W.A. Benjamin, Addison-Wesley. 1972. XII+335 pp. At: *15*

1972 *The differential closure of a differential field*. **Bulletin of the American Mathematical Society**, vol. 78 (1972), pp. 629–634.

1979 *Effective bounds on Morley rank*. **Fundamenta Mathematicae**, vol. 103 (1979), pp. 111–121.

Saffe J.

1981 **Einige Ergebnisse über die Anzahl abzählbarer Modelle superstabiler Theorien**. Fakultät für Mathematik und Naturwissenschaftender Universität Hannover, 1981. 55 pp. Dissertation. At: *117, 356, 364*

1982 *A superstable theory with the dimensional order property has many models*. In: **Proceedings of the Herbrand Symposium: Logic Colloquium '81, Held in Marseilles, France, July 1981**, ed. by Stern J. **North-Holland Publishing Co**. 1982. pp. 281–286. At: *339*

1983 *One theorem on the number of uncountable models*. Preprint 31. USSR Academy of Sciences,Siberian Division, Institute of Mathematics, Novosibirsk, 1983.

1983 *The number of uncountable models of ω-stable theories*. **Annals of Pure and Applied Logic**, vol. 24 (1983), pp. 231–261. At: *295, 311, 330, 351, 364, 377, 397*

1984 *Categoricity and ranks*. **Journal of Symbolic Logic**, vol. 49 (1984), pp. 1379–1382. At: *163*

198? *An introduction to regular types*. Preprint.

198?a *Normalization and forking*. Preprint. At: *180*

Saffe J., Paljutin E.A and Starchenko S.S.

1985 *Models of superstable Horn Theories*. **Algebra and Logic**, vol. 24 (1985), pp. 171–209. Russian original: pp.278–327.

Shelah S.

1969 **Categoricity and classes of models (Hebrew)**. Hebrew University of Jerusalem, 1969. ii+69 pp. Dissertation. At: *212*

1969a *Stable theories.* **Israel Journal of Mathematics**, vol. 7 (1969), pp. 187–202. At: *62*

1970 *Finite diagrams stable in power.* **Annals of Mathematical Logic**, vol. 2 (1970), pp. 69–118. At: *201*

1970a *On theories T categorical in | T |.* **The Journal of Symbolic Logic**, vol. 35 (1970), pp. 73–82.

1971 *Stability, the f.c.p. and superstability; model theoretic properties of formulas in first-order theories.* **Annals of Mathematical Logic**, vol. 3 (1971), pp. 271–362. At: *54, 55, 62, 71, 91, 161, 212*

1971a *The number of non-isomorphic models of an unstable first-order theory.* **Israel Journal of Mathematics**, vol. 9 (1971), pp. 473–487. At: *31*

1971b *On the number of non almost-isomorphic models of a theory in a power.* **Pacific Journal of Mathematics**, vol. 36 (1971), pp. 811–818.

1972 *Uniqueness and characterization of prime models over sets for totally transcendental first-order-theories.* **The Journal of Symbolic Logic**, vol. 37 (1972), pp. 107–113. At: *126, 229*

1972a *A combinatorial problem; stability and order for models and theories in infinitary languages.* **Pacific Journal of Mathematics**, vol. 41 (1972), pp. 247–261.

1973 *Differentially closed fields.* **Israel Journal of Mathematics**, vol. 16 (1973), pp. 314–328.

1973a *There are just four second-order quantifiers.* **Israel Journal of Mathematics**, vol. 15 (1973), pp. 282–300.

1974 *Remark to — "local definability theory" of Reyes.* **Annals of Mathematical Logic**, vol. 2 (1974), pp. 441–448. At: *62*

1974a *Categoricity of uncountable theories.* In: **Proceedings of the Tarski Symposium. An International SymposiumHeld to Honor Alfred Tarski on the Occasion of his 70th Birthday**, ed. by Henkin L. et al. **American Mathematical Society**, 1974. pp. 187–203.

1975 *The lazy model theorist's guide to stability.* **Logique et Analyse. Nouvelle Série**, vol. 18 (1975), pp. 241–308. At: *30, 97, 226, 229*

1975a *The monadic theory of order.* **Annals of Mathematics**, vol. 102 (1975), pp. 379–419. At: *344*

1975b *Why there are many nonisomorphic models for unsuperstable theories*. In: **Proceedings of the International Congress of Mathematicians**. ed. by James R.D. **Canadian Mathematical Congress**, 1975. pp. 259–263. At: *31*

1975c *Categoricity in* $\aleph_1$ *of sentences in* $L_{\omega_1\omega}(Q)$. **Israel Journal of Mathematics**, vol. 20 (1975), pp. 127–148.

1978 **Classification Theory and the Number of Nonisomorphic Models**. North-Holland. 1978. XVI+544 pp. At: *3, 4, 5, 6, 12, 25, 30, 45, 52, 55, 68, 79, 80, 88, 89, 90, 91, 101, 107, 108, 112, 117, 118, 120, 126, 132, 148, 153, 158, 159, 161, 163, 176, 188, 192, 193, 195, 196, 201, 204, 208, 209, 212, 214, 216, 225, 231, 232, 233, 237, 240, 242, 244, 249, 251, 252, 256, 261, 265, 271, 281, 282, 290, 291, 293, 295, 302, 311, 339, 365*

1978a *End extensions and numbers of countable models*. **The Journal of Symbolic Logic**, vol. 43 (1978), pp. 550–562. At: *7, 365, 366*

1978b *On the number of minimal models*. **The Journal of Symbolic Logic**, vol. 43 (1978), pp. 475–480.

1979 *On uniqueness of prime models*. **The Journal of Symbolic Logic**, vol. 44 (1979), pp. 215–220. At: *229*

1980 *Independence results*. **The Journal of Symbolic Logic**, vol. 45 (1980), pp. 563–573. At: *31*

1980a *Simple unstable theories*. **Annals of Mathematical Logic**, vol. 19 (1980), pp. 177–203. At: *33, 132, 136*

1982 *The spectrum problem I,* $\aleph_\epsilon$*-saturated models the main gap*. **Israel Journal of Mathematics**, vol. 43 (1982), pp. 324–356. At: *30, 52, 153, 233, 252, 317, 325, 330, 331, 339, 342, 346, 350, 351, 356, 364*

1982a *The spectrum problem II, Totally transcendental theories and the infinite depth case*. **Israel Journal of Mathematics**, vol. 43 (1982), pp. 357–364. At: *277*

1982b *Better quasi-orders for uncountable cardinals*. **Israel Journal of Mathematics**, vol. 42 (1982), pp. 177–226. At: *31*

1983 *Constructions of many complicated uncountable structures and Boolean algebras*. **Israel Journal of Mathematics**, vol. 45 (1983), pp. 100–146. At: *31*

1983a *Classification theory for non-elementary classes I, the number of uncountable models* $\psi \in L_{\omega_1\omega}$. *Part A*. **Israel Journal of Mathematics**, vol. 46 (1983), pp. 212–240. At: *33, 317*

1983b *Classification theory for non-elementary classes I, The number of uncountable models $\psi \in L_{\omega_1\omega}$. Part B.* **Israel Journal of Mathematics**, vol. 46 (1983), pp. 241–173.

1984 *On universal graphs without instances of CH.* **Annals of Pure and Applied Logic**, vol. 26 (1984), pp. 75–87.

1985 *Classification of first order theories which have a structure theory.* **Bulletin of the American Mathematical Society**, vol. 12 (1985), pp. 227–232. At: *2*

1985a *Monadic logic: Löwenheim numbers.* **Annals of Pure and Applied Logic**, vol. 28 (1985), pp. 203–216.

1985b *On the possible number of* no(M) = *the number of non-isomorphic models $L_{\infty,\lambda}$-equivalent to M of power λ, for singular λ.* **Notre Dame Journal of Formal Logic**, vol. 26 (1985), pp. 36–50.

1986 *Monadic logic: Hanf numbers.* In: **Around Classification Theories of Models**. **Springer Verlag**, 1986. pp. 203–222. LNM 1182. At: *33*

1986a *Classification over a predicate II.* In: **Around Classification Theories of Models**. **Springer Verlag**, 1986. pp. 1–47. LNM 1182.

1986b *On countable theories with model-homogeneous models only.* In: **Around Classification Theory**. **Springer-Verlag**, 1986. pp. 269–272. LNM 1182.

1986c *On almost categorical theories.* In: **Classification Theory, Chicago 1985.**, ed. by Baldwin J. **Springer-Verlag**, 1987. Preprint. At: *311, 330*

198? **Classification theory and the number of nonisomorphic models (2nd edition)**. North-Holland. Preprint. At: *6, 31, 110, 112, 219, 263, 282, 339, 345, 346, 350, 397*

198?a *The spectrum problem III, Universal theories.* **Israel Journal of Mathematics** Preprint.

198?b *Existence of many $L_{\infty,\lambda}$-equivalent non-isomorphic models of T of power λ.* **Annals of Pure and Applied Logic** Preprint.

198?c *On the numbers of strongly $\aleph_\epsilon$-saturated models of power λ.* Preprint. 20 pp. At: *71*

198?d *The number of pairwise non-elementary embeddable models.* **The Journal of Symbolic Logic** Preprint. At: *31*

Shelah S, Harrington L. and Makkai M.

1984 *A proof of Vaught's conjecture for totally transcendental theories.* **Israel Journal of Mathematics**, vol. 49 (1984), pp. 259–278. At: *7, 306, 311, 366, 371, 377, 383, 397*

Shoenfield J.R.

1967 **Mathematical Logic**. Addison-Wesley Publishing Co. 1967. VII+344 pp. At: *15*

Siegel C.L.

1949 **Transcendental Numbers**. Princeton University Press, 1949. viii + 162 pp. At: *62*

Sierpinski W.

1965 **Cardinal and Ordinal Numbers. (2nd revised edition)**. Panstwowe Wydawnictwo Naukowe (PWN), 1965. 491 pp.

Sikorski R.

1964 **Boolean Algebras. (2nd edition)**. Springer-Verlag, 1964. X+237 pp. At: *11, 106*

Smith K.W.

1981 *Stability and categoricity of lattices.* **Canadian Journal of Mathematics**, vol. 33 (1981), pp. 1380–1419.

Steel J.R.

1978 *On Vaught's conjecture.* In: **Cabal Seminar 76-77: Proceedings of the Caltech-UCLA Logic Seminar 1976-77**, ed. by Kechris A.S. and Moschovakis Y.N. **Springer-Verlag**, 1978. pp. 193–208. At: *7, 366*

Svenonius L.

1959 *$\aleph_0$-categoricity in first-order predicate calculus.* **Theoria (Lund)**, vol. 25 (1959), pp. 82–94. At: *24*

1963 *On the denumerable models of theories with extra predicates.* In: **The Theory of Models**. ed. by Addison J.W., Henkin L. and Tarski A. **North-Holland**. 1963. pp. 376–389.

Szmielew W.

1955 *Elementary properties of abelian groups.* **Fundamenta Mathematicae**, vol. 41 (1955), pp. 203–271. At: *30*

Thomas S.

1983 **Classification theory of simple locally finite groups**. Bedford College, University of London, 1983. 89 pp. Dissertation. At: *98*

198? *Model theory of locally finite groups, I.* Preprint. 18 pp.

198?a *Model theory of locally finite groups, II.* Preprint.

Tsuboi A.

1985 *On the number of independent partitions.* **The Journal of Symbolic Logic**, vol. 50 (1985), pp. 809–815.

1985 *On theories having a finite number of non-isomorphic countable models.* **The Journal of Symbolic Logic**, vol. 50 (1985), pp. 806–808.

Urbanik K.

1959/60 *A representation theorem for Marczewski's algebras.* **Fundamenta Mathematicae**, vol. 57 (1959/60), pp. 147–167. At: *45*

1965 *On algebraic operations in idempotent algebras.* **Colloquium Mathematicum**, vol. 13 (1965), pp. 129–157. At: *45*

1966 *Linear independence in abstract algebras.* **Colloquium Mathematica**, vol. 14 (1966), pp. 233–255. At: *45*

Van der Waerden B.L.

1949 **Modern algebra**. Frederick Ungar Publishing Co., New York, 1949. 264 pp. At: *35, 45*

Vaughn J.

1985 **$\aleph_1$-categorical locally modular theories**. University of Illinois at Chicago, 1985. 67 pp. Dissertation. At: *106, 180*

Vaught R.L.

1954 *Applications of the Löwenheim-Skolem-Tarski theorem to problems of completeness and decidability.* **Indagationes Mathematicae**, vol. 16 (1954), pp. 467–472. At: *4*

1961 *Denumerable models of complete theories.* In: **Infinitistic Methods.Proceedings of the Symposium on Foundations of Mathematics. Panstwowe Wydawnictwo Naukowe (PWN)**, 1961. pp. 303–321. At: *20, 188, 192, 202, 212*

1961a *A Loewenheim-Skolem theorem for two cardinals.* **Notices of the American Mathematical Society**, vol. 8 (1961), pp. 239.

1965 *A Löwenheim-Skolem theorem for cardinals far apart.* In: **Theory of Models (Proceedings of 1963 International Symposium at Berkeley)**, ed. by Addison J.A. and Henkin L. & Tarski A. **North Holland**, 1965. pp. 390–401. At: *211, 212*

Wagner C.W.

1982 *On Martin's Conjecture.* **Annals of Mathematical Logic**, vol. 22 (1982), pp. 47–68.

Weil A.

1955 *On algebraic groups of transformations.* **American Journal of Mathematics** (1955), pp. 355–391.

Wheeler W.

1980 *Model theory of strictly upper triangular matrix rings.* **Journal of Symbolic Logic**, vol. 45 (1980), pp. 205–206.

Whitney H.

1935 *On the abstract properties of linear dependence.* **American Journal of Mathematics**, vol. 57 (1935), pp. 509–533. At: *35, 45*

Wierzejewski J.

1974 *A note on stability and products.* **Bulletin de l'Académie Polonaise des Sciences**, vol. 22 (1974), pp. 875–876.

1976 *On stability and products (Add & corr ibid 97*51-52).* **Fundamenta Mathematicae**, vol. 93 (1976), pp. 81–95.

1976a *Remarks on stability and saturated models.* **Colloquium Mathematicum**, vol. 34 (1976), pp. 165–169.

1977 *A note on products and degree of types.* **Zeitschrift für Mathematische Logik und Grundlagen der Mathematik**, vol. 23 (1977), pp. 431–434.

Wood C.

1973 *Prime model extensions of differential fields, characteristic.* **Notices of American Mathematical Society**, vol. 20 (1973), pp. 445. At: *229*

1976 *The model theory of differential fields revisited.* **Israel Journal of Mathematics**, vol. 25 (1976), pp. 331–352. At: *229*

1979 *Notes on the stability of separably closed fields.* **The Journal of Symbolic Logic**, vol. 44 (1979), pp. 412–416. At: *229*

Woodrow R.E.

1978 *Theories with a finite number of countable models.* **The Journal of Symbolic Logic**, vol. 43 (1978), pp. 442–455.

Zariski O. and Samuel P.

1958 **Commutative Algebra: Volume I**. Van Nostrand, 1958. viii+329 pp. At: *67*

Ziegler M.

1984 *Model theory of modules.* **Annals of Pure and Applied Logic**, vol. 26/2 (1984), pp. 149–213. At: *25, 30*

Zil'ber B.I.

1974 *The transcendental rank of the formulas of an $\aleph_1$-categorical theory.* **Mathematical Notes of the Academy of Sciences of the USSR**, vol. 15 (1974), pp. 182–186. Russian original: pp. 321–329. At: *163*

1974a *The theory of a free abelian group is stable (Russian).* **Algebra i Logika**, vol. 13 (1974), pp. 713.

1977 *The structure of models of categorical theories and the problem of axiomatizability (in Russian).* VINITI, Dep No. 2800-77, 1977. 80 pp.

1977a *Groups and rings, the theory of which is categorical (Russian. English summary).* **Fundamenta Mathematicae**, vol. 95 (1977), pp. 173–188.

1980 *Totally categorical theories, structural properties and the non-finite axiomatisability.* In: **Proceedings of the Conference on Model Theory of Algebra and Arithmetic, Karpacz, Poland, 1979**, ed. by Pacholski L., Wierzejewski J. and Wilkie A.J. **Springer-Verlag,Berlin**, 1980. pp. 381–410. LNM 834.

1980a *Strongly minimal countably categorical theories.* **Siberian Journal of Mathematics**, vol. 21 (1980), pp. 219–230. Russian original: pp.98–112. At: *184*

1980b *On the problem of finite axiomatizability for theories categorical in all infinite powers.* In: **Theory of Models and its Applications**, ed. by Baizanov B. **Kazakhskii Gosudarstvennyi Universitet**, 1980. pp. 47–60. At: *45, 126, 184*

1981 *On the problem of finite axiomatizability for theories categorical in all infinite powers.* In: **Researches in theoretical programming**, ed. by Baizanov B. **Kazakhskii Gosudarstvennyi Universitet**, 1981. pp. 69–75.

1981a *Totally categorical structures and combinatorial geometries. (English translation).* **Doklady Akademii Nauk SSSR** (1981), pp. 149–151. Russian original: pp.1039–41. At: *45, 181, 184*

1983 *The structure of models of uncountably categorical theories.* In: **ICM 82 Proceedings, Warsaw.** 1983.

1983a *Structural problems of models of $\aleph_1$-categorical theories.* In: **Proceedings of the ICLMPS, Salzburg, 1983**. 1983.

1984 *Strongly minimal countably categorical theories II.* **Siberian Journal of Mathematics**, vol. 25 (1984), pp. 396–412. Russian original: pp.71–88. At: *184*

1984a *Strongly minimal countably categorical theories III.* **Siberian Journal of Mathematics**, vol. 25 (1984), pp. 559–572. Russian original: pp.63–77. At: *184*

Subject Index

Symbol Index

The symbols are organised according to subject matter as follows:

1. Set theoretic notations
2. Basic model theoretic notations
3. Distinguished sets of formulas, theories or types
4. Relations among types, elements and sets.
5. Functions and operators defined in stability theory
6. Classes of models
7. Ranks
8. Specific theories
9. Notations about trees (of models)
10. Notations from algebra
11. Many sorts and L^{eq}
12. Special notations from Chapter XVIII

1. Set theoretic notations

2. Basic model theoretic notations

3. Distinguished sets of formulas, theories or types.

4. Relations among types, elements and sets.

5. Functions and operators defined in stability theory.

6. Classes of models

7. Ranks

8. Specific theories

9. Notations about trees (of models)

10. Notations from algebra

11. Many sorts and L^{eq}

12. Chapter XVIII

For EU product safety concerns, contact us at Calle de José Abascal, 56–1°,
28003 Madrid, Spain or eugpsr@cambridge.org.

www.ingramcontent.com/pod-product-compliance
Lightning Source LLC
LaVergne TN
LVHW102009210926
848483LV00010B/240

* 9 7 8 1 1 0 7 1 6 8 0 9 1 *